eBook Bonus.

With the purchase of this book, you also receive access to its electronic version. Use the following personal code to download the electronic pages.

2018j-pugl2-
6r82n-u00ne

Take advantage of the eBook Bonus on your computer, smartphone, or tablet. Go to **www.hanserpublications.com/ebookbonus** and register with your personal access code.

Gebhardt, Hötter

Additive Manufacturing

Andreas Gebhardt
Jan-Steffen Hötter

Additive Manufacturing

3D Printing for Prototyping and Manufacturing

HANSER

Hanser Publishers, Munich
Hanser Publications, Cincinnati

The Authors:

Prof. Dr.-Ing. Andreas Gebhardt, Managing Director of CP – Centrum für Prototypenbau GmbH, Erkelenz/Düsseldorf

Professor, FH Aachen University of Applied Sciences
Subject area: high-performance methods in production engineering and rapid prototyping

Jan-Steffen Hötter, M.Eng., Project Engineer, FH Aachen University of Applied Sciences

Distributed in the Americas by:
Hanser Publications
6915 Valley Avenue, Cincinnati, Ohio 45244-3029, USA
Fax: (513) 527-8801
Phone: (513) 527-8977
www.hanserpublications.com

Distributed in all other countries by:
Carl Hanser Verlag
Postfach 86 04 20, 81631 München, Germany
Fax: +49 (89) 98 48 09
www.hanser-fachbuch.de

Cataloging-in-Publication Data is on file with the Library of Congress

Editor: Cheryl Hamilton
Production Management: Jörg Strohbach
Coverconcept: Marc Müller-Bremer, www.rebranding.de, München
Coverdesign: Stephan Rönigk
Typesetting: Kösel Media GmbH, Krugzell
Printed and bound by Kösel, Krugzell
Printed in Germany

ISBN: 978-1-56990-582-1
E-Book ISBN: 978-1-56990-583-8

Foreword

Since the late 1980s, in fact, for more than 25 years, Additive Manufacturing (AM) has been penetrating the world of manufacturing. When the layer-based technology emerged, it was called Rapid Prototyping (RP). This was the best name for a technology that could not fabricate anything but sticky and brittle parts, which could only be used as prototypes. The process was not even "rapid," although it allowed the making of time- and money-consuming tools to be avoided. With the creation of the first prototype by RP, a significant amount of time and money could be saved.

The initial process was called *stereolithography* and it was based on photo-polymerization, which first processed acrylates and then epoxies later on. In the following years, new layer-based processes were developed and an extended range of materials became qualified for AM applications, and all of them were plastics.

Around the turn of the millennium, processes for making metal parts were introduced to the market. With this development, the focus of manufacturers as well as of the users changed from just prototyping to manufacturing because of improved processes, materials, software, and control. The challenge was then to make final parts.

Today all classes of engineering materials, such as plastics, metals, ceramics, and even nontraditional materials, such as food, drugs, human tissue, and bones, can be processed using 3D printers.

There is still a long way to go, but due to vibrant activities concerning all aspects of 3D printing worldwide, this high-speed development is incomparable to the expansion of any fabrication technology in the past.

There are two main reasons for intense interesting in this technology for somebody active in the field of product development and production:

First, to stay competitive, one should be able to judge the capabilities of existing, new, and emerging AM processes in comparison to traditional manufacturing processes and process chains. The task is not just a matter of speeding up the process but to improve the way we do engineering design towards "designing for AM." This makes completely new products possible and shifts the competition of traditional manufacturing towards a new level of lightweight design, as well as resource-saving and environmentally friendly mass production of individual parts.

Second, people begin to understand that AM is not just capable of revolutionizing our way of designing and producing parts, but able to affect many aspects of our daily lives.

AM touches upon legal aspects, such as product reliability and intellectual property rights, as compared to the digital entertainment market. AM also brings even more challenges as parts can cause significant problems like physical injuries or even death, which music and videos do not do.

Digital data, including not only technical data such as a blue print, but the exact information for creating the product, can easily be sent all over the world and encounter every imaginable hurdle, such as frontiers, embargos, custom fees, export regulations, and many more. This requires us to rethink the well-functioning world of today.

Many of the questions raised, if not the majority, need to be decided by people who are not technicians. The better that those involved understand the technical part and the more thorough their information, the better decisions they will be qualified to make.

Consequently, this book was written to support the product developers and people who are responsible for the production, as well as others who are involved in the process of realizing the enormous challenges of this technology.

Aachen in March 2016 *Andreas Gebhardt*

About the Authors

Andreas Gebhardt, born in 1953, studied mechanical engineering at the Technical University Aachen, Germany (RWTH), where he received his Engineering Diploma (Dipl-Ing). In 1986 he passed his doctoral exam (Dr-Ing) at the same university with a thesis on the "Simulation of the transient behavior of conventional power plants."

In 1986, Mr. Gebhardt was appointed general manager of a company that specialized in engine refurbishment. In 1991, he moved to general manager at the LBBZ GmbH, a service bureau on laser material processing, where in 1992, he started working on rapid prototyping. When in 1997, the CP Center of Prototyping GmbH, an Additive Manufacturing Service Bureau was founded, he transferred there as a general manager.

With the beginning of the spring term in 2002, Mr. Gebhardt was appointed Professor for Advanced Fabrication Technology and Rapid Prototyping at the Aachen University of Applied Sciences (FH Aachen) where he established an AM Team and Lab called the GoetheLab for Additive Manufacturing. Since 2002, Mr. Gebhardt has also been a guest professor at the City College of the City University of New York (CCNY).

In 2012, Mr. Gebhardt was elected Dean of the Department of Mechanical Engineering and Mechatronics, FH Aachen. In November 2014, he was appointed extraordinary Professor at the Tshwane University of Technology (TUT), Pretoria, RSA.

Mr. Gebhardt is Chairman of the AM Research Committee (FA13) of the German Welding Association (DVS) and he heads the team of the "Aachen Center of 3D Printing," a joint research group of FH Aachen and Fraunhofer ILT AM specialists.

Since 2004 Mr. Gebhardt has been the editor of the peer-reviewed, open access online journal on AM called the RTeJournal.

Jan-Steffen Hötter, born in 1987, received his Bachelor's Degree (B.Eng.) and Master's Degree in Mechanical Engineering (M.Eng.) from the Aachen University of Applied Sciences, Aachen, Germany. He established the Metal Laser Sintering Lab and Team under the umbrella of the GoetheLab, which he now is heading. He is engaged in the Aachen Center of 3D Printing and coordinates the AM work of the Institute for Tool-less Production (IWF GmbH). Mr. Hötter is a member of the VDI Committee "Additive Manufacturing," and gives guest lectures at several German universities.

Acknowledgements

The interdisciplinary character and the enormous developmental speed of AM in general, and of the layer-based fabrication processes and machines in particular, make it almost impossible for an individual to display this discipline correctly, completely, and entirely up-to-date.

We are therefore very thankful for the enormous assistance from many people.

The practical orientation of this book mainly is backed up by the contribution of the management and the staff of the AM Service-Bureau CP-GmbH, mainly from Besima Sümer, Christoph Schwarz, and Michael Wolf.

Major help came from the GoetheLab team of the Aachen University of Applied Sciences.

First of all, a special thanks goes to the whole metal group of the Goethelab team that supported the entire process in every chapter. We want to thank Philipp Ginkel, Prasanna Rajaratnam, Simon Scheuer, Patrycja Wiezik, Alina Richter, and Niklas Kessler. A special thanks goes to Alexander Schwarz, who focused on the correct formatting of this book and supported the authors in organizational questions.

Additionally, we thank our colleagues

- Miranda Fateri, who personally worked on some chapters and contributed together with her Glass- and Ceramic Team,
- Laura Thurn and her Plastic and Fabber Team, and
- Julia Kessler, who brought in not only professional knowledge concerning her medical research but her management skills to shape the teamwork.

As this book is based on four editions in German, our appreciation goes to all who, since the late 1990s, have helped to make and optimize the topic and who are listed in the preceding German editions.

Thanks to all members of the professional committees of the VDI, Association of German Engineers and DVS, German Welding Association (FA 13), which we are members of.

We thank countless colleagues (who must stay nameless in this content) whom we met on conferences, exhibitions, and meetings for countless discussions and suggestions. In case we forgot anyone, we sincerely apologize.

Thanks to the Hanser team and to Mrs. Monika Stüve for her support over the years.

Andreas Gebhardt

Jan-Steffen Hötter

Contents

1 Basics, Definitions, and Application Levels

To understand the characteristics and the capabilities of additive manufacturing (AM), it is very helpful to take a look at the systematics of manufacturing technologies in general first.

1.1 Systematics of Manufacturing Technologies

Orientated on the geometry only, manufacturing technology in general is divided into three fundamental clusters [Burns, 93, AMT, 14]:[1]

1. subtractive manufacturing technology,
2. formative manufacturing technology, and
3. additive manufacturing technology.

With subtractive manufacturing technology, the desired geometry is obtained by the defined removal of material, for example, by milling or turning.

Formative manufacturing means to alter the geometry in a defined way by applying external forces or heat, for example, by bending, forging, or casting. Formative manufacturing does not change the volume of the part.

Additive manufacturing creates the desired shape by adding material, preferably by staggering contoured layers on top of each other. Therefore it is also called layer (or layered) technology.

The principle of layer technology is based on the fact that any object, at least theoretically, can be sliced into layers and rebuilt using these layers, regardless of the complexity of its geometry.

[1] In Germany, manufacturing technology is divided into six main categories, and each of them is subdivided into various subcategories [DIN 8580], [Witt, 06].

Figure 1.1 underlines this principle. It shows the so-called sculpture puzzle, in which a three-dimensional (3D) object has to be assembled from more than 100 slices. Therefore the layers have to be arranged vertically in the right sequence using a supporting stick.

Figure 1.1 Principle of layer technology, example: sculpture puzzle (Source: HASBRO/MB Puzzle)

Additive manufacturing (AM) is an automated fabrication process based on layer technology. AM integrates two main subprocesses: the physical making of each single layer and the joining of subsequent layers in sequence to form the part. Both processes are done simultaneously. The AM build process just requires the 3D data of the part, commonly called the virtual product model.

It is a characteristic of AM that not only the geometry but the material properties of the part as well are generated during the build process.

■ 1.2 Systematics of Layer Technology

In this section the commonly used terms in AM are addressed. The related characteristics as well as their interdependency and the hierarchical structure are discussed.

In this book the generally accepted so-called generic terms are used, and alternatively used names are mentioned.

Generic terms and brand names have to be distinguished from each other. If they are mixed, which happens quite often, this frequently leads to confusion. As brand names are important in practice, they are addressed, explained, and linked to the generic terms in Chapter 3, where the AM machines are presented.

1.2.1 Application of Layer Technology: Additive Manufacturing and 3D Printing

Additive manufacturing is the generic term for all manufacturing technologies that automatically produce parts by physically making and joining volume elements, commonly called voxels. The volume elements are generally layers of even thickness.

Additive manufacturing is standardized in the US (ASTM F2792) and in Germany (VDI 3405), and is commonly used worldwide.

As alternative terms, additive manufacturing (technology) and additive layer manufacturing (ALM) have minor acceptance.

3D printing is about to replace all other names, including additive manufacturing, and to become the generally accepted generic term for layer technology in the near future. This is mainly because it is very easy to understand. Everyone who can operate a text editor (a word processor) and a 2D office printer easily understands that he or she will be able to print a 3D object using a 3D design program (a part processor) and a 3D printing machine, regardless of how it works.

NOTE: Additive manufacturing and 3D printing are used as equal generic terms in this book. While in Chapter 1 this is expressed by always writing additive manufacturing/3D printing (or AM/3DP). In the following chapters only additive manufacturing or AM is used in order to shorten the text volume.

Beginners should realize that 3D Printing is also the brand name of a family of powder binder processes (see Section 3.6), originally developed by MIT and licensed to Z-Corporation (now 3D Systems), Voxeljet, and others.

1.2.2 Characteristics of Additive Manufacturing

Layer technologies in general and additive manufacturing in particular show special characteristics:

- The geometry of each layer is obtained solely and directly from the 3D computer-aided design (CAD) data of the part (commonly called a virtual product model).
- There are no product-related tools necessary and consequently no tool change.
- The material properties of the part are generated during the build process.

- The parts can be built in any imaginable orientation. There is no need for clamping, thus eliminating the clamping problem of subtractive manufacturing technologies. Nevertheless, some processes need support structures, and the orientation of the part influences the parts' properties.
- Today, all AM processes can be run using the same so-called STL (or AMF) data structure, thus eliminating data exchange problems with preprocessors as used in subtractive manufacturing.

Additive manufacturing/3D printing therefore ensures the direct conversion of the 3D CAD data (the virtual product model) into a physical or real part.

As scaling can be done simply in the CAD file, parts of different sizes and made from different materials can be obtained from the same data set. As an example, the towers of a chess set shown in Fig. 1.2 are based on the same data set but made with different AM machines and from different materials. The range of materials includes foundry sand, acrylic resin, starch powder, metals, and epoxy resin.

Figure 1.2 Additive manufacturing. Scaled towers of a chess set, based on the same data set but made with different AM machines and from different materials.
Small towers, from left to right: PMMA (powder-binder process, Voxeljet), metal (laser sintering, EOS), acrylate, transparent (stereolithography, Envisiontec; height approx. 3 cm).
Big towers, from left to right: foundry sand (powder-binder process, Voxeljet), starch powder (powder-binder process, 3D Systems; height approx. 20 cm) (Source: machine manufacturers)

One of the biggest AM parts of all is the tower shown in Fig. 1.3 with a height of approximately 2.5 m, which is higher than the general manager of the Voxeljet Company, Mr. I. Ederer.

Figure 1.3 Chess tower made from foundry sand, height approx. 2.5 m, powder-binder process (Source: Voxeljet)

By contrast, Fig. 1.4 shows a tower made by micro laser sintering. It is approximately 5 mm high.

Figure 1.4 Tower made from metal, height approx. 5 mm, micro laser sintering (Source: EOS/3D Micromac)

AM/3DP allows manufacturing of geometric details that cannot be made using subtractive or formative technologies. As an example, the towers on Fig. 1.2 contain spiral staircases and centered double-helix hand rails. The details can be seen on a cutaway model displayed in Fig. 1.5.

Figure 1.5
Internal details of the rear right tower on Fig. 1.2 (Source: 3D Systems)

Another example of geometries that cannot be manufactured using subtractive or formative technologies is shown in Fig. 2.5.

All AM/3DP processes mentioned here will be explained in detail in Chapter 3.

■ 1.3 Hierarchical Structure of Additive Manufacturing Processes

For a proper definition of the terms used, it is very helpful to distinguish the technology and its application from each other. Subtractive manufacturing, for example, marks the technology level, and drilling, grinding, milling, and so on are the names for its application (or the application level).

The technology of additive manufacturing/3D printing is divided in two main application levels: *rapid prototyping* and *rapid manufacturing*. Rapid prototyping is the application of AM/3DP to make prototypes and models or mock-ups, and rapid manufacturing is the application to make final parts and products.

The manufacturing of tools, tool inserts, gauges, and so on usually is called *rapid tooling*. The term often is regarded as an independent hierarchical element or application level, but effectively it is not. Depending on how a tool is made, it represents a prototype or a product (Fig. 1.6).

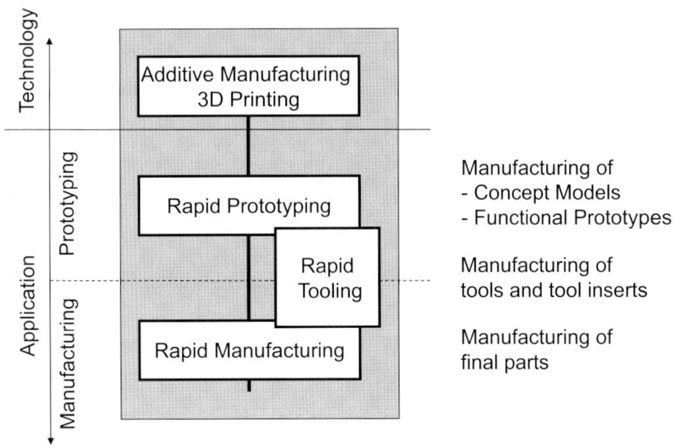

Figure 1.6 Basic structure of additive manufacturing/3D printing technology and its subcategories rapid prototyping, rapid manufacturing, and rapid tooling

1.3.1 Rapid Prototyping

Rapid prototyping (RP) is the application of AM/3DP technology to make prototypes, models, and mock-ups, all of them being physical parts but not products. They only mimic isolated properties of the latter product in order to verify the engineering design and to allow the testing of selected product capabilities and thus to improve and speed up the product development process. The goal is to preplan a part to make it as simple as possible in order to get it quickly and cheaply. Therefore, rapid prototyping parts generally cannot be used as final products.

As prototypes differ from products, serial identical prototypes (which are not products but prototypes) do not exist, although the term is used to underline a strategy.

Rapid prototyping again is subdivided into *solid imaging* or *concept modeling*, and *functional prototyping*.

Solid imaging or *concept modeling*: If a rapid prototyping part is made mainly for 3D visualization, it is called a solid image, a concept model, a mock-up, or even a rapid mock-up. The idea behind it is to generate a 3D picture or a statue (Fig. 1.7). To highlight this aspect, the parts are also called *show-and-tell models*.

If a part has a single or some of the functionalities of the latter product, it can be used to verify this aspect of the engineering design. Consequently it is called a functional prototype (and the process *functional prototyping* accordingly).

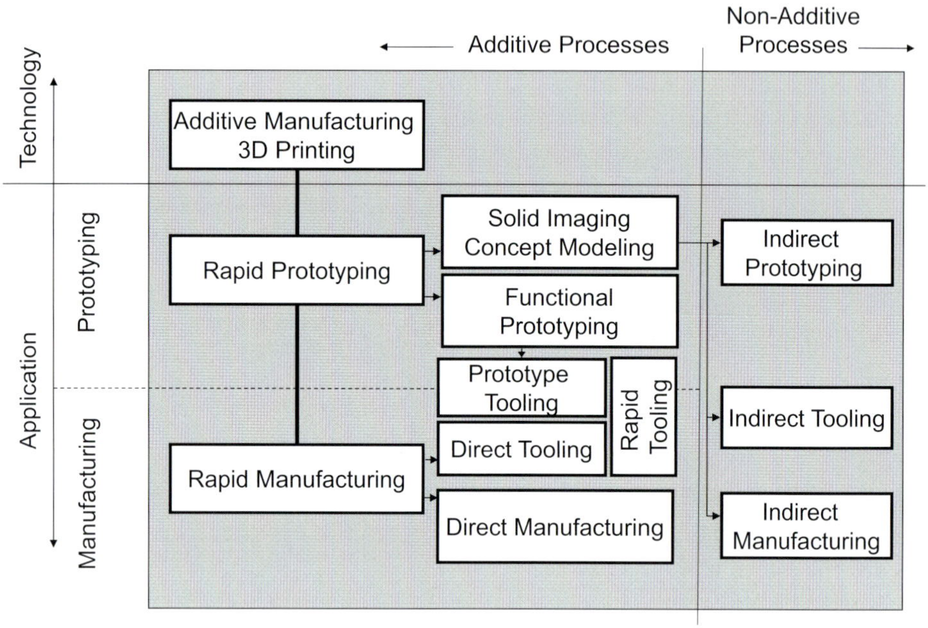

Figure 1.7 Basic structure of the AM/3D printing technology: application levels rapid prototyping, rapid manufacturing, and rapid tooling and its subcategories

A sample of each category is displayed in Fig. 1.8. The scaled data control model of a convertible roof system (made from polyamide by laser sintering) can be regarded as a typical concept model. The air-outlet nozzle of a passenger car (made by laser stereolithography from epoxy resin) is a functional prototype that supports the testing of the car's climate control.

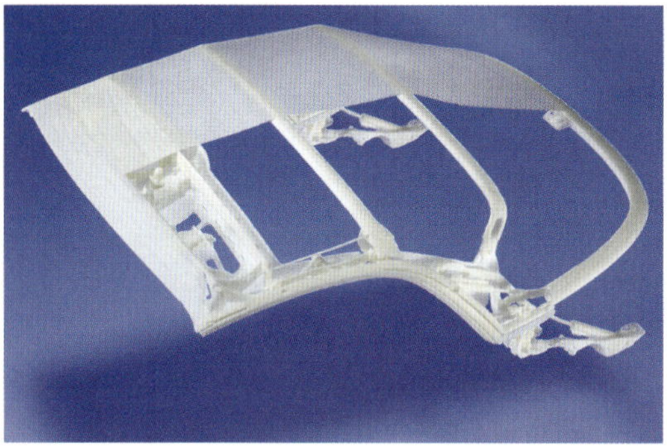

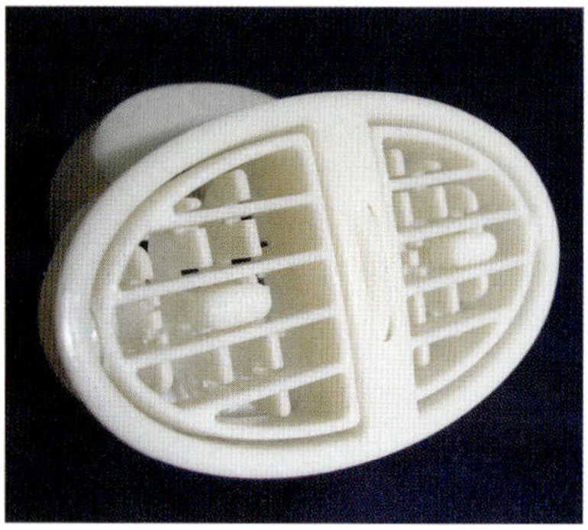

Figure 1.8 Rapid prototyping: concept model or solid image (left), laser sintering (Source: CP GmbH); functional prototype (right), laser stereolithography (Source: 3D Systems)

The corresponding AM processes are presented in detail in Chapter 3.

1.3.2 Rapid Manufacturing

Rapid manufacturing (RM) names the application of the AM/3D printing technology to make final parts or products, often called series products, even if they are one-offs. (A deeper discussion can be found in Chapter 6.) The parts can be positives like connectors as well as negatives like cavities. Making positives or parts is called *direct manufacturing*, and the additive manufacturing of negatives or cavities, such as tools and tool inserts, is called *direct tooling*; see Fig. 1.7.

1.3.2.1 Rapid Manufacturing—Direct Manufacturing

Additive manufacturing or 3D printing of final parts or products is called *direct manufacturing (DM)*. Frequently and for historical reasons it is also called *rapid manufacturing (RM)* and complies directly with the main term. Often the terms *e-manufacturing, digital manufacturing, tool-less fabrication,* and others are used.

Direct manufacturing is based on the same technology as rapid prototyping and, at least until today, uses the same machines. The goal is to make final products. Whether the goal can be reached or not depends on the degree of accomplishment of the required mechanical and technological properties. This again depends on the machines, processes, and materials available. Further, whether the needed accuracy can be reached and if a competitive price can be achieved are essential.

As an example of direct manufacturing, Fig. 1.9 shows a three-element dental bridge (left). The associated process chain will be shown in Chapter 3, and applications will be discussed in Chapter 6.

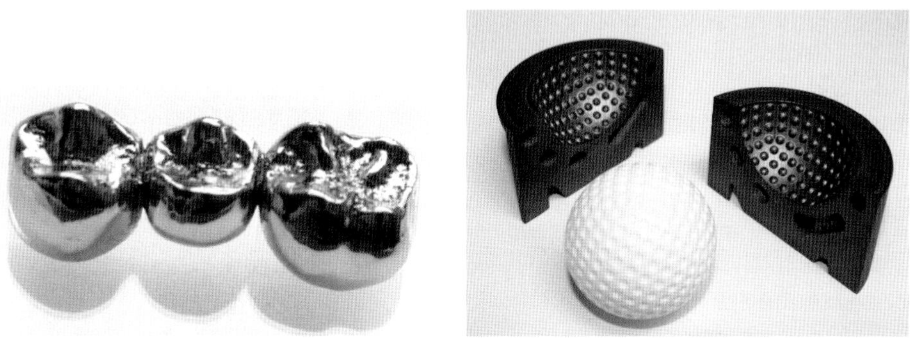

Figure 1.9 Rapid manufacturing: direct manufacturing of a three-element dental bridge (left) (Source: GoetheLab FH Aachen/Sokalla); direct tooling for making golf balls (right) (Source: EOS GmbH)

1.3.2.2 Rapid Manufacturing—Rapid Tooling (Direct Tooling—Prototype Tooling)

Additive manufacturing/3D printing of a high number tools is a rapid manufacturing process. It is called *direct tooling* or, to underline the character of additive manufacturing, *direct rapid tooling* (Fig. 1.7). Figure 1.9 (right) shows a tool for making golf balls.

Direct tooling has to be distinguished from prototype tooling. Prototype tooling is the name for processes to make tools and tool inserts from model or prototype materials, for example from stereolithography resin (see AIM, Section 5.4.1.1). Therefore, prototype tooling is basically a part of functional prototyping. Prototype tooling often is called *bridge tooling* as well because it bridges the gap between prototype tools and series tools, mainly for small-batch production. Prototype tooling is characterized by quickly and easily made low-volume tools.

As can be seen in Fig. 1.7, rapid tooling does not indicate a separate application level of the additive manufacturing/3D printing technology but integrates different tooling applications in the sense of a vertical structuring element.

Historically, the making of tools and tool inserts first were based on rapid prototyping and thus were realized before modern direct manufacturing was available. For marketing reasons, the new application was published as an advancement of rapid prototyping and thus was given a new name: *rapid tooling*. Because of this, there are still publications that structure additive manufacturing/3D printing with a separate subcategory called rapid tooling.

1.3.3 Related Nonadditive Processes: Indirect or Secondary Rapid Prototyping Processes

The terms *indirect prototyping*, indirect manufacturing, and *indirect tooling* do not indicate additive manufacturing/3D printing processes, although it is suggested by the names. A process is called indirect if it uses additive-manufactured masters without being an additive process itself. The best-known examples are copying processes like vacuum casting (also called room-temperature vulcanization (RTV) or reaction injection molding (RIM).

If indirect processes lead to parts (positives), the process is called indirect prototyping, if final products are the result, for example cast parts obtained from (lost) AM/3DP master models, it is called indirect manufacturing, and if tools or tool inserts (negatives) are made, it is called indirect tooling (for examples see Fig. 1.10(a)(b)).

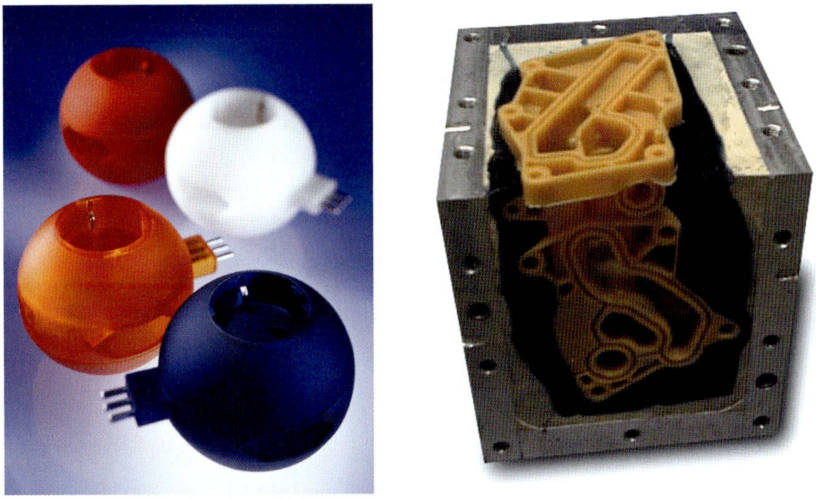

Figure 1.10 Indirect Processes: Indirect Prototyping (left), Indirect Tooling (right)

The above-mentioned copying processes are frequently called rapid prototyping, rapid manufacturing, or rapid tooling as well; sometimes they are named a little bit more precisely as secondary rapid prototyping processes (or applications). In practice, often the adjective "rapid" is used although the processes are nonadditive. Mostly this is done to make it more attractive to customers.

The application of additive manufacturing/3DP for making rapid prototyping parts and final products is discussed in detail in Sections 4.3.2 to 4.3.6; for making tools and tool inserts it is discussed in detail in Section 5.3.

1.3.4 Rapid Prototyping or Rapid Manufacturing?

Whether we talk about *rapid prototyping* or *rapid manufacturing*, it often causes intensive discussions. In fact, the same part can be a prototype or a product and consequently can be made using rapid prototyping (functional prototyping) or rapid manufacturing (direct manufacturing) processes.

A part that was designed for additive manufacturing from polyamide, is made by AM/3DP from polyamide, and that finally shows all relevant properties as designed is a product. In contrast, a part that was designed for plastic injection molding of polyamide and then is made using additive manufacturing and polyamide at best is a prototype if the geometry is exact.

In this context, the material used in AM/3DP does not play a role, as long as it is identical in engineering design and production. A part made from paper, wax, starch powder, or gypsum definitely can be a product.

1.3.5 Diversity of Terms

Today, additive manufacturing and 3D printing are regarded as generic terms for layer technologies. Besides this, many other terms are in use. They change with time, and even their meaning sometimes changes with time.

With the first introduction of additive manufactured parts and processes in the late 1980s, the generic name was rapid prototyping.

At that time the term rapid prototyping was correct. First, using rapid prototyping processes, parts could be made much faster. This was mainly because for the first time parts could be made directly because tools could be avoided, thus saving time and money. Second, the parts could not be used as products but as prototypes at best. This was mainly because of the materials available and because of poor processes.

As a kind of reminiscence of rapid prototyping, even today many processes are named with the adjective "rapid," mostly to underline their speed, even if they are nonadditive.

Besides additive manufacturing/3D printing, the most prominent term is *rapid technology* or rapid technologies, but additive manufacturing or, very rarely, formative manufacturing is used as well.

Another family of terms used is linked to layer-based manufacturing. Names like layer manufacturing or more often *additive layer manufacturing* (ALM) are used. As an equivalent to manufacturing, the terms fabrication or production are used as well.

In addition, in the literature some terms are often used that refer to a specific ability of a process. For example, *solid free-form manufacturing* (SFM) underlines that solids are made that are traced by free-form surfaces. *Desktop manufacturing* marks the manufacturing in an office environment (on the desk).

Besides the full names, three- and four-letter abbreviations are used. Often they lead to confusion rather than to precision. The most-often used ones are explained in the text or listed in the glossary.

In practice, the terms often are not properly used. Frequently, not even generic names and brand names are distinguished from each other. For example, many people call it stereolithography if they speak about AM/3DP in general.

1.3.6 How Fast Is Rapid?

Many terms and definitions contain the term rapid or fast. But rapid is a relative term. It only becomes a quality term if "how fast or fast in comparison to what?" is added.

The term itself is kind of dangerous because it suggests that rapid processes are basically faster than any other manufacturing processes. This is not true and cannot be generalized. The speed depends on the geometry. If a board of 250 × 250 × 10 mm is needed, one cuts it from a semifinished bulk. No additive manufacturing/3D printing process will be faster.

Under special conditions, additive manufacturing processes are faster than nonadditive ones, for example if tools can be avoided, if a volume-independent flexible output is required, if complex geometries are needed, or if individualized parts are wanted.

But the term *rapid* does bear a practical advantage. It is accepted as a synonym of today's computerized and therefore automated processes for mainly making prototypes. It is self-explanatory, which is one of the most important attributes of a term. That is why people will continue calling AM/3DP rapid prototyping, rapid tooling, and rapid manufacturing, as it is done sometimes in this book as well.

1.4 Integration of Additive Manufacturing in the Product Development Process

Additive manufacturing can be used not only to make parts, but due to their special abilities (see Section 1.2.2, "Characteristics of Layer Technologies"), they are suited to improve established processes, realizing new product features, and supporting new product development strategies.

1.4.1 Additive Manufacturing and Product Development

Industrial product formation involves the time period from the first product idea to the introduction of the product to the market. It includes the development of the product, the development and the fabrication of the production facility, and the production of the product itself.

The goal of all manufacturers is to keep this time span as short as possible and therefore to optimize all subprocesses. Additive manufactured parts are particularly suitable to shortening the product development process and to improving it at the same time. The biggest influence comes from the fact that expensive and time-consuming tools can be avoided with the use of additive manufacturing/3D printing.

This effect is enforced if in any phase of product development the optimal AM/3DP process is used. To identify it, it is favorable to have a correlation between the application levels of the additive manufacturing technology (as displayed on Fig. 1.7) and the phases of the product development process on Fig. 1.11. This shows an internationally accepted structure, although this particular process chain was taken from the guidelines of the German Association of Mechanical Engineers, VDI [VDI 2221].

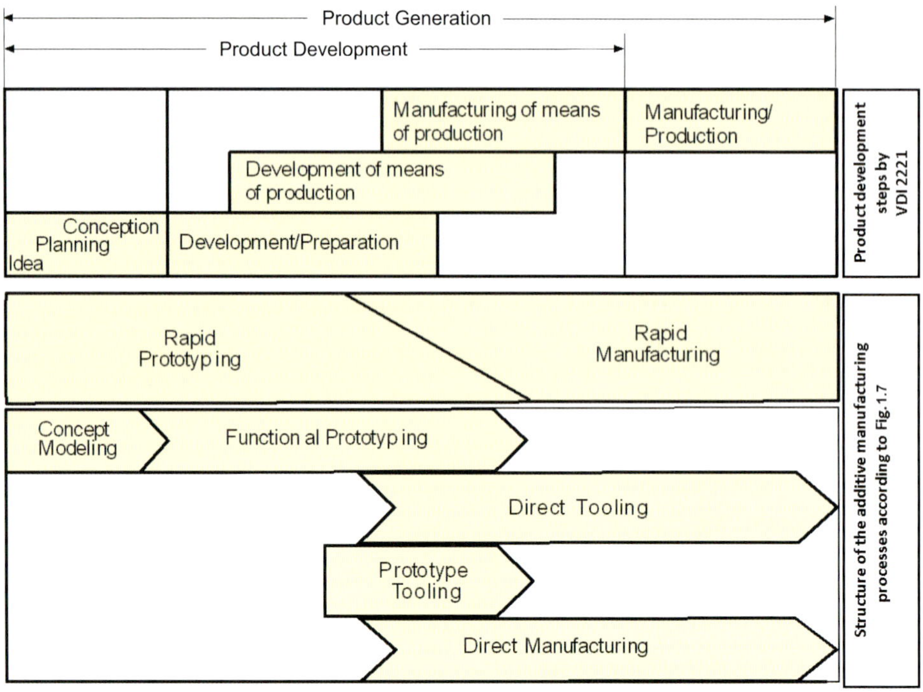

Figure 1.11 Phases of product development in correlation to the application levels of AM/3DP (Fig. 1.7).
The internationally accepted process chain following the guidelines of the German Association of Mechanical Engineers, VDI [VDI 2221] (top) and its correlation to the application levels rapid prototyping and rapid manufacturing (middle) and its subcategories concept modeling, functional prototyping, prototype tooling, direct tooling, and direct manufacturing (bottom)

Product development covers rapid prototyping and rapid manufacturing (of the production facility as well as of the product; see Fig. 1.11, middle). In detail, ideation and conception are improved by concept modeling, while the engineering design and the technical preparation are supported by functional prototyping. The fabrication of the products is done by direct manufacturing.

Rapid tooling supports the making of tools and tool inserts. In the prototyping phase, prototype tooling is applied, and in the production phase direct tooling is used (see Fig. 1.11, bottom).

A more detailed structure is discussed later in section 4.3.1. The capabilities of the models (model classes or categories, to be defined later) are linked to the AM families defined in Chapter 3.

In practice, the definitions are not as sharp as displayed here, and their transitions are smooth. In addition, depending on the product, not every phase is addressed in each product development process.

1.4.2 Additive Manufacturing for Low-Volume and One-of-a-Kind Production

AM/3DP is done layer by layer and without product-dependent tools. It does not matter how many parts are manufactured at once in one build space (as long as they fit in) and whether they are identical or not. Therefore, AM/3DP enables a volume-independent production, even with a mixed production containing different volumes and one-offs.

1.4.3 Additive Manufacturing for Individualized Production

AM/3DP shows not only the volume effect mentioned in Section 1.4.2, but also supports the manufacturing of individually modified products. This is the realization of the strategy of customizing, which basically cannot be done using tools.

The individualization of a product can be done by using CAD systems or by the integration of 3D scans, in so-called reverse engineering. As an example, hearing aid shells made by AM/3DP are displayed in Fig. 1.12 that show the production and the finished parts.

Besides this more professional approach, there are part libraries available, even on the Internet. Along with easy to operate (low-cost or shareware) 3D CAD systems, this provides easy access to the needed data, even for private and semiprofessional use.

Using AM, virtually everybody can be a manufacturer.

Figure 1.12 Individualized production or customizing: manufacturing of individualized hearing aid shells. Upside-down production, left; finished parts, right (Source: 3D Systems)

1.5 Machines for Additive Manufacturing

Today, machines for AM/3DP are called *3D printers.*

Rarely, the term "fabricator" is used to indicate that the final products are made, and in contrast, the machines for making prototypes are called "prototypers."

As indicated on Fig. 1.7, parts and machines are classified according to the application levels. They can be roughly divided into three machine classes. Machines that make

- concept models are called *personal 3D printers* or *fabbers,*
- functional prototypes are called *professional 3D printers* or *office printers,* and
- final products are called *production 3D printers.*

This interrelation is displayed on Table 1.1.

Table 1.1 Classification of machines for Additive Manufacturing/3D Printing

	Machine class			
Designation *	**Prototyper**		**Fabricator**	
	Personal 3D printer		Professional 3D Printer	Production machine Production 3D printer
	Fabber	Personal printer	Office printer	Production printer (Shop floor machine)
Use	Private, domestic	Semiprofessional office	Professional office or workshop	Professional, production or job shop
Usage levels (Fig. 1.7)				
▪ Prototypes Display models	X			
▪ Concept models		X		
▪ Functional prototypes			X	
▪ Final products				X
Construction material	Plastic	Plastic	Plastic	Plastic, metal, ceramic
Price level	500 to 4000 € [1 € ≈ 1.07 USD]	1000 to 10,000 €	20,000 to 70,000 €	130,000 to > 800,000 €
Example	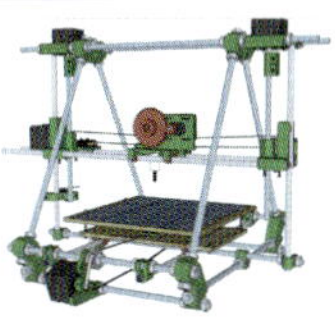 Prusa Mendel	 Dimension	 Objet	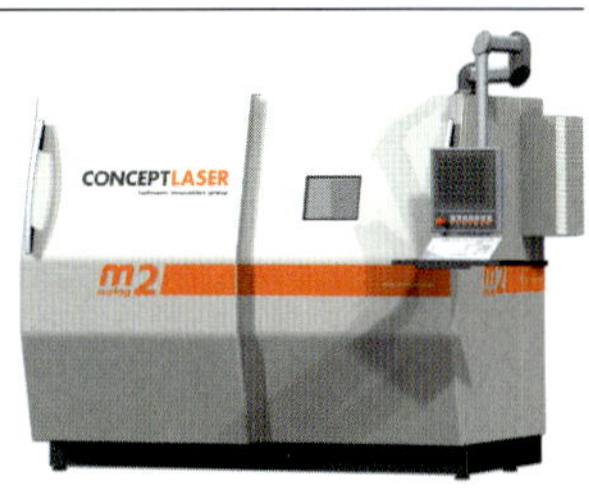 Concept SLM

* The classifications are not standardized and are handled differently by different manufacturers and users. Those used here are based on the recommendation from 3D Systems and follow a system often used in practice.

The machine classes again are linked to characteristic specifications.

Personal 3D Printer

In analogy to personal computers, small, simply engineered, shareware-based, easy to operate (even on a desk), and inexpensive 3D printers are called personal 3D printers. Today they mostly use plastics, but other materials will be available in the near future.

Fabber

Personal 3D printers in private use are commonly called fabbers. Fabbers are mostly assembled by the owner and operator himself (DIY printer). The term "fabber" is an abbreviation of fabricator. The process is called fabbing. The philosophy behind it is to enable everybody to fabricate almost everything by himself. Corresponding web-based blogs and Internet communities are established (the fabbing community). They support the exchange of information regarding the construction and the operation of fabbers and propagate new kinds of cooperation, such as cloud-fabbing. Members mutually train themselves. To start with, it is sufficient to be able to operate a personal computer and to own a kitchen table as infrastructure.

Professional 3D Printer

Professional 3D printers or office printers are compact, easy to operate, and service friendly. Even semiskilled people can operate it in an office environment.

Professional 3D printers have cartridge-supported material changing systems. The parts can be taken out without contamination of the operators or the environment. Removal of the support structures and cleaning for the most part can be done automatically or semiautomatically using special washing media and devices.

Operators are trained by tutorials or by a short training session, mostly in-house. There are no special requirements concerning the infrastructure.

Production 3D Printer

Production 3D printers (or factory work floor printers) enable a continuous high-quality and high-volume output. Production 3D printers focus on productivity. Generally, they have big build chambers, material handling systems, and automated devices for postprocessing.

Production printers are heavy, have big footprints, and cause emissions that are typical for production machines. Installation, maintenance, and service require professional support. Operators need to be trained intensively, mainly supported by the machine manufacturer.

Machine Classes and Part Properties

In general, there is a correlation between the machine classes and the part properties that can be obtained. A little bit more detailed than in Table 1.1, the characteristics of parts and the capabilities of printers (machine classes) are linked as follows:

- Personal 3D printers are preferred for making concept models (show-and-tell models) and less loadable parts with limited details and reduced geometric freedom. At least today, they are limited to manufacturing parts from plastics.
- Professional 3D printers are preferred for making concept models and functional prototypes with good details and reasonable loading capacity. Often they are the basis for making small series of end products with the help of secondary rapid

prototyping processes. Like personal 3D printers (at least today), they are limited to manufacturing parts from plastics.

- Production 3D printers are used to make final products or series parts. They can be one-offs as well as series of any volume, including a mixed production of different parts and volumes. Materials are plastics, ceramics, or metals. Usually the parts need postprocessing to achieve their final properties and complexion.

But, the correlation by trend between the machine classes and the capability of the parts cannot be regarded as principles. On one hand, there are fabbers that produce plastic parts that can be used as products, for example, individually shaped clamps (Fig. 1.13). On the other hand, there are production printers that make, for example, hollow turbine blades that only can be used for cold-air testing (Fig. 1.14).

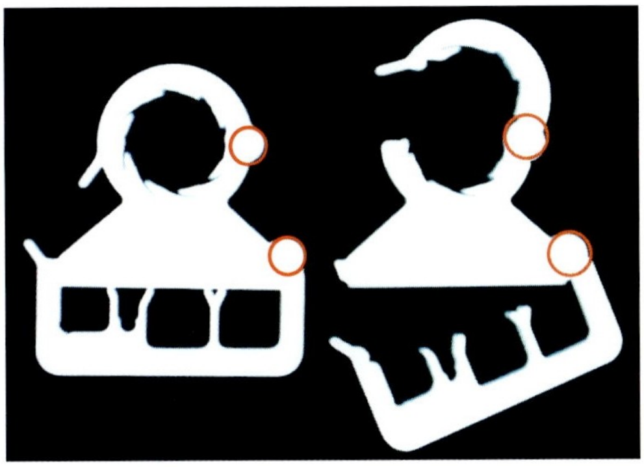

Figure 1.13 Articulated cable clamp, a final product made by laser sintering from polyamide in one piece. Closed on the left and open on the right (Source: EOS)

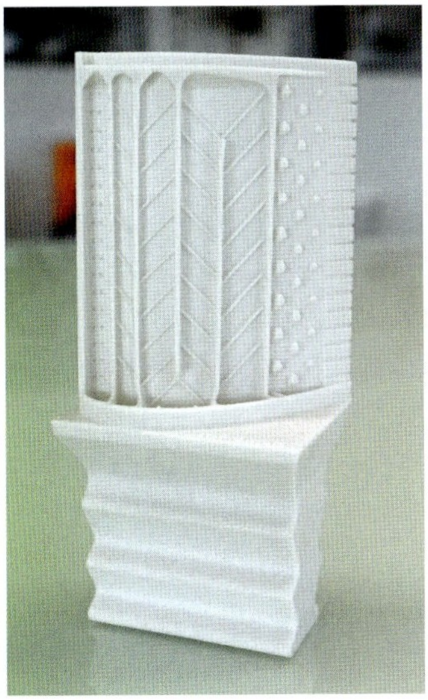

Figure 1.14
Cutaway model of a gas turbine blade, gypsum ceramics, made with a powder-binder process (Source: IwF-Aachen)

Characteristics of the Additive Manufacturing Process

Additive manufacturing machines on the market today are developed at a high rate of speed. New processes that are currently in the laboratory stage or under development will break into the market while simultaneously, tried and tested systems will be upgraded within a relatively short time.

Because the equipment presently on the market will be obsolete or approaching obsolescence over a relatively short period, the physical and technological bases of the various processes are portrayed and discussed in detail in this chapter. Chapter 3, "Machines for Rapid Prototyping, Direct Tooling, and Direct Manufacturing," then shows which industrially offered installations derive from which fundamental processes. This approach not only facilitates the assessment of the current processes, but it also supplies the basis for the assessment of future industrial processes.

This division is also intended to separate the representation of basic principles valid for a longer term from machine concepts that change more quickly. In reality, however, overlaps and repetitions are unavoidable.

■ 2.1 Basic Principles of the Additive Manufacturing Process

All additive manufacturing (AM) models are built by joining single layers of equal thickness. The layer is shaped (contoured) in an *x-y* plane two-dimensionally. The third dimension results from single layers being stacked up on top of each other, but not as a continuous *z* coordinate. In the strictest sense, additive manufacturing processes are therefore 2½D processes.

The models are therefore three-dimensional forms that are very exact on the building plane (*x-y* direction) and owing to the described procedure are then stepped in the *z* direction, whereby the smaller the *z* step is, the more the model looks like the original. Figure 2.1 shows an example of a three-dimensional model of a plastic and the resulting shift model, which is marked by the stair-step effect.

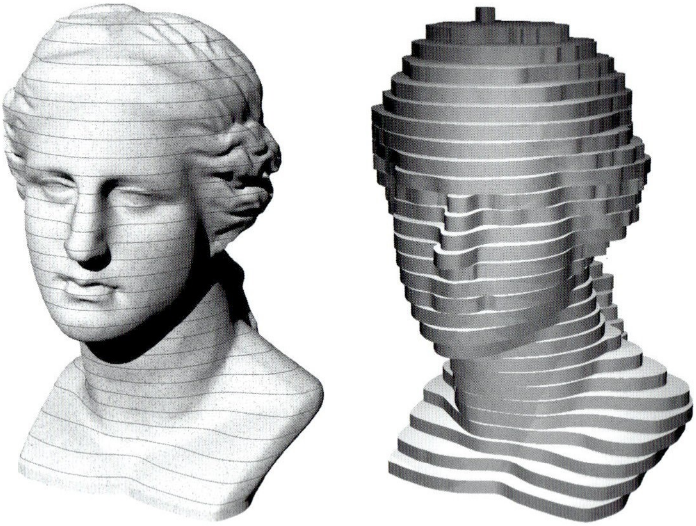

Figure 2.1 Stepped surface as a result of the layering process. Three-dimensional solid model (left) with marked equidistant layers and the created layer model (right) (Source: FH-Aachen)

The stair-step effect is a typical characteristic of the additive manufacturing process that can never be entirely eliminated but can be reduced by decreasing the layer thickness. Figure 2.2 shows the proportions of a real model with a layer thickness of 0.125 mm.

Figure 2.2 Stair-step effect in a stereolithography component (layer thickness 0.125 mm). (Source: CP-GmbH)

Today, the layer thickness is between 0.1 and 0.05 mm. Machines that are used for macroscopic components (characteristic dimensions of several to 100 mm) have a minimum thickness of 0.016 mm, and microcomponents have layer thicknesses up into the 5 nm range. A larger layer thickness, up to around 0.2 mm, is most often used for fabbers or for reducing the building time on other additive manufacturing machines. The consequence of a large layer thickness is low precision. Layer milling processes use plates with a thickness up to 40 mm (See Section 3.4.6 "Layer Milling Process (LMP), Zimmermann").

Depending on the type of contouring (scanning, plotting, and so on) and the chosen AM process, the object is contoured continuously in the building plane (*x-y* plane). If it is not, the secant effect (see also Fig. 2.12) or the stair-step effect occurs on the boundary, which develops lower in *x-y* direction than in the *z* direction. Therefore, models built by an AM process have different precisions in the *x-y* direction than in the *z* direction. As can be seen theoretically in Fig. 2.3, a drilling hole is built parallel to the *x-y* plane and another one perpendicular to the *x-y* plane. It is assumed that the circular contours in the layers are generated continuously.

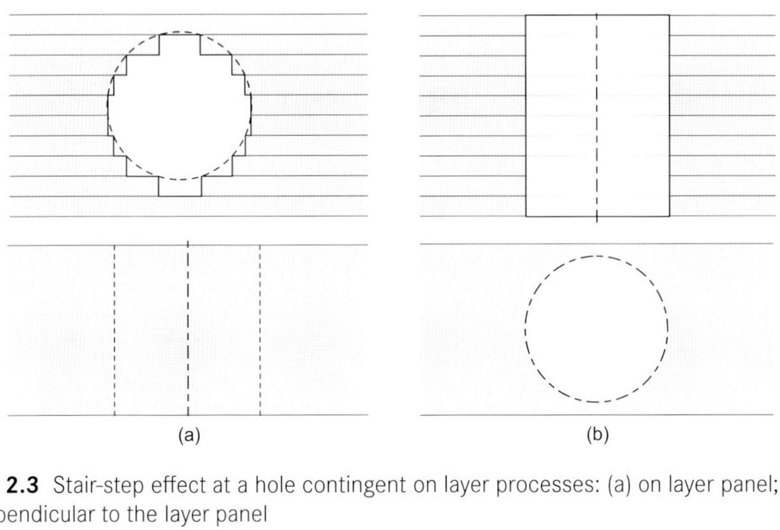

Figure 2.3 Stair-step effect at a hole contingent on layer processes: (a) on layer panel; (b) perpendicular to the layer panel

Although all AM processes known today work in this way as 2½D processes, some processes (e.g., extrusion processes) are in principle 3D processes, which means they can add incremental volume elements at any chosen point of the model. This method has not been implemented to date.

The layer milling process is the only one that can be contoured continuously in the *z* direction (compare to Section 3.4.6 "Layer Milling Process (LMP), Zimmermann").

The characteristic feature of additive manufacturing processes is that the physical models are produced directly from computer data. In principle, it is therefore unimportant from where the data are provided as long as they describe a 3D volume completely. Data from CAD construction, from the process of measuring and reverse engineering, or from other measurements (such as computer tomography (CT), magnetic resonance tomography (MRT), and 3D tracking systems) may be used equally as well if the relevant evaluation programs enable the preparation of the measured values in the 3D data models.

In this way, model making has become an integral part of the computer-integrated product development. From the product development aspect, additive manufacturing models can, therefore, be regarded as three-dimensional plots or facsimiles of the corresponding CAD data. The decisive advantage, in contrast to the classic manual or semiautomatic model-making processes, lies in the fact that the data remain unaltered (with the exception of the generated supports) by the model making. As a result, no data need to be taken from the model. Because the making of AM models does not alter the common database, additive manufacturing processes have become the most important elements of modern product development strategies such as simultaneous engineering. The 3D data model is used as a product model and as a basis for production at the same time.

The advantage of AM processes over nonadditive manufacturing computer-controlled processes is that the AM machines are using the same data type, the so-called STL data (STL means standard transformation language, see Section 2.2.2.1 "STL Data"). Nonadditive manufacturing numerically controlled processes, such as computer numeric control (CNC) milling, usually use a system-specific data set.

The generation of layer information is based on a purely computer-oriented CAD model (Fig. 2.4). The CAD model is cut into layers with equal thickness by mathematical methods. This layer information is used to generate the physical single layers in an AM machine; the total sum of the single layers forms the physical model. The merging of the physical layers happens during the generation of the next layer or after completing the layer. This depends on the AM process. The finished physical model can be a prototype or the final product.

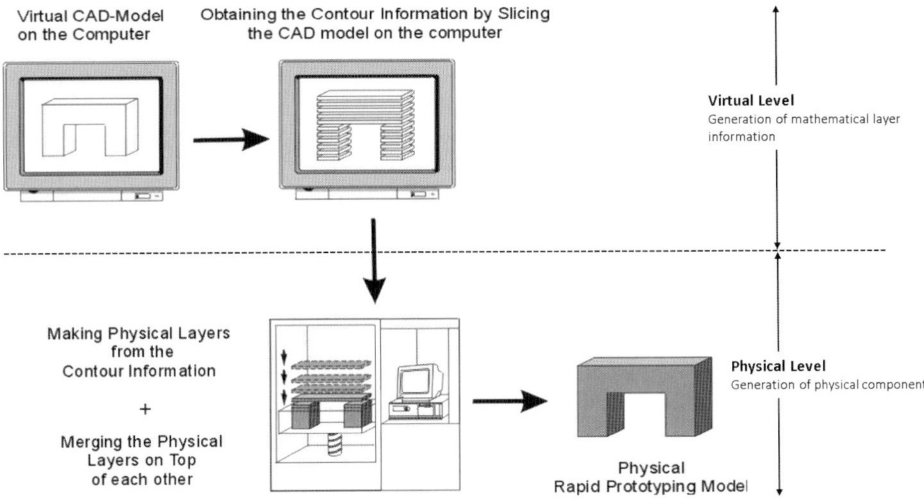

Figure 2.4 Concept of additive manufacturing

With the method of layer generating it is possible to realize geometrically complex structures that are impossible or expensive to produce by conventional methods. As a result, new designs can be realized, like the mathematical compositions by George Hart (www.georgehart.com). In Fig. 2.5 a (mathematical) 4D object can be seen that consists of 120 flattened dodecahedra and 600 tetrahedra.

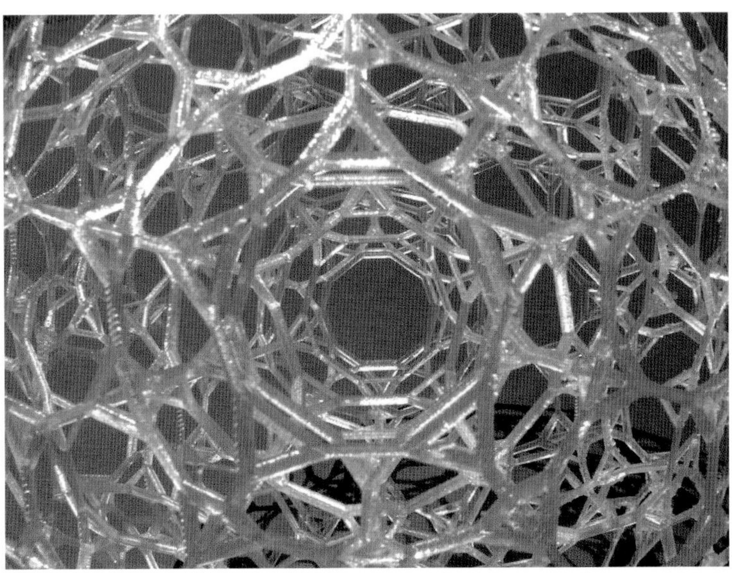

Figure 2.5 Complex structure of polymer printing, only additive manufacturing geometry. (Source: George Hart)

"Rapid prototyping" or "additive manufacturing processes" are classified into two fundamental process steps:

1. Generation of the mathematical layer information (Section 2.2)
2. Generation of the physical layer model (Section 2.3)

2.2 Generation of Layer Information

The additive manufacturing process copies the virtual CAD data into a physical model, so the generation of the mathematical slice information for the production of the single layers is most important in order to produce an accurate part. It is divided into three steps:

1. Description of the geometry by a 3D data record (Section 2.2.1)
2. Generation of the geometrical information of each layer (Section 2.2.2)
3. Illustration of the geometrical layer information on each layer (Section 2.4)

2.2.1 Description of the Geometry by a 3D Data Record

2.2.1.1 Data Flow and Interfaces

The production of models and prototypes by means of additive manufacturing processes requires that the geometry of the component is available as a 3D data record. This is achieved in most industrial applications by construction on a 3D CAD system or by other measurements. The data are produced independently of the production and need to be prepared and transferred via an interface to the machines. The data are called the digital or virtual product model.

To build the components further, process- and installation-specific calculations are necessary in addition to geometrical data and material parameters. The data are set with the help of programs called front-end software or additive manufacturing software. Front-end software is a part of AM machines and is supplied by the manufacturer. AM software is offered by a third party. The software types interface with each other (Fig. 2.6).

The STL format has been established as an industry standard (STL interface, Section 2.2.2), but other formats are also used (Sections 2.2.2.2 and 2.2.2.3).

Before starting production of an AM model, the 3D model first has to be placed on the building platform in the optimal direction by the front-end or additive manufacturing software. It is also necessary to add the process-specific data like the supports. To achieve the best use of the machine, several components are built simultaneously on the platform. In some AM processes the components are placed into and one above each other.

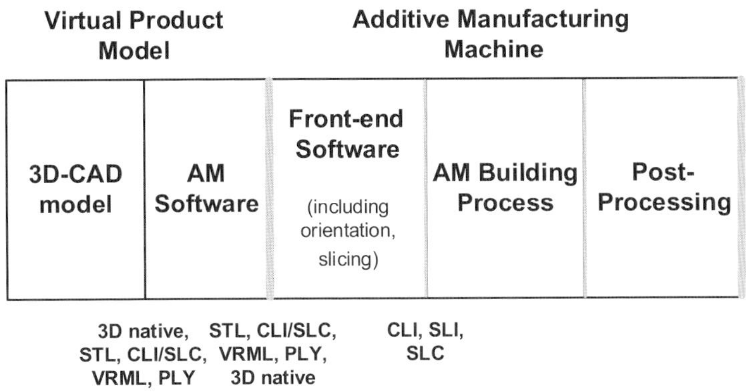

Figure 2.6 Data flow during additive manufacturing

Third-party software contains optional features that allow, for example, the derivation of 3D data from 2D sketches. Some of this software is just used to display the 3D model. This kind of software is trimmed of the needed features of CAD software and can transform the model into the STL format.

When the model is included with the geometrical and directional information, it is necessary to establish the machine and material parameters required for the control of the rapid prototyping process in order to proceed with the building process.

In any case, the geometrical information of the entire body, like the layer thickness and the contour data for every layer, must exist to produce the 3D model. The generation of equal layer thicknesses with mathematical methods is called slicing. Most machines slice the whole 3D model and use the layer information in the batch mode. Depending on the capacity (speed and memory) of the computers the machines generate each layer shortly before the physical generation of each previous layer. Today, the so-called adaptive slice method ("slicing on the fly") is installed on the machines. This method allows measurement of the current geometrical height and generates the next layer depending on that value. In that way, geometrical errors in the *z* direction that depend on propagation errors will be reduced or avoided completely.

The data is the basis for manufacturing the AM model, and after postprocessing the model is ready for use.

In addition to this direct path, there are also alternatives in the design and production. It is possible to deliver the geometrical data in SLI/SLC (Section 2.2.2.2), PLY (Section 2.2.2.3), or AMF (Section 2.2.2.4) format instead of STL format.

Neutral interfaces are also used to transfer data. The reason is in the advantages when the construction is finished and the model has to be modified during the production process. Modification of an STL model is possible but not as straightforward as in a full-fledged CAD system.

The most important neutral interfaces that are integrated into 3D CAD systems are listed below:

- IGES Initial graphic exchange specification
- VDAIS Association of Automotive Manufacturers – IGES interface
- VDAFS Association of Automotive Manufacturers – Surface interface
- DXF Drawing exchange format
- HPGL Hewlett-Packard graphics language
- SET Standard déchange et de transfer [Standard Exchange and Transfer]
- STEP Standard for the Exchange of Product Model Data.

IGES is one of the worldwide standards for geometrical interfaces. This format exhibits a lot of varieties that should be described in detail.

VDAIS is an IGES interface from the Association of Automotive Manufacturers. This interface also includes a lot of varieties, so in each particular case one must examine carefully which interface formulation the data exchange is based on.

VDAFS is specialized for freeform surfaces and therefore has importance in the automotive industry.

HPGL is a contour-oriented plotter format that is especially used in the context of direct contouring in CAD for AM processes

STEP is an interface that is becoming increasingly established after a long test phase. In addition to the geometrical information, other information can be exchanged with the STEP format. The STEP format represents an approach to exchanging the original CAD model and not only the geometry information.

2.2.1.2 Modeling 3D Bodies in a Computer by Means of 3D CAD

The creation of a 3D body is the indispensable prerequisite for the production of an additive manufacturing model. Therefore, the application of AM processes is linked especially close with CAD processes. For this reason, 3D CAD processes will be looked into only as far as is absolutely necessary to understand the fundamental relationships in the production of rapid prototyping models.

Every CAD system uses certain data elements and data structures to describe a component in detail. The data record includes not only the component geometry but also the materials, the quality of the surface, the production process, and much more. The component geometry therefore comprises only one part of the information. The complete information registered in the database of a CAD system for a component is called a CAD model (the product to be made). If the geometric description of a component is 3D, then it is called a 3D CAD model or digital product model.

By choosing a certain CAD system the user commits himself to its database. The structure and the data elements decide to a high degree the quality of a CAD system and its compatibility with other systems via an interface. The CAD system also defines the type, extent, and quality of the AM process.

2.2.1.2.1 CAD Model Types

CAD models are defined by model types regardless of the kind of CAD system. As shown in Fig. 2.7, not all types that display a 3D model are suitable for an AM process; some types do not have sufficient information.

Dimension of CAD Elements	Element	Type of CAD Model
0D	Point 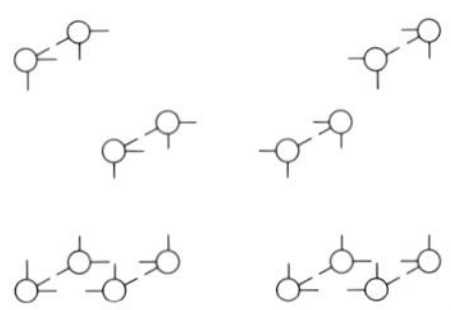	Corner Model
1D	Line 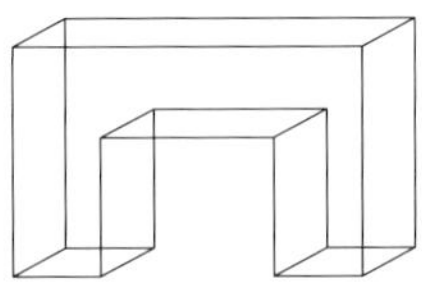	Edge Model
2D	Surface 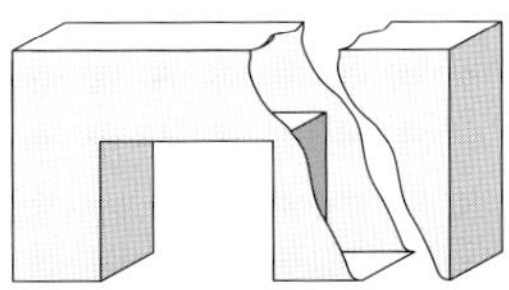	Surface Model
3D	Solid / Volume 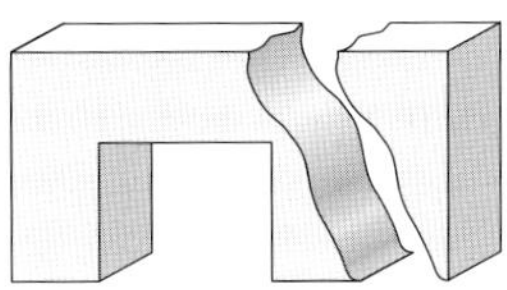	Solid or Volume Model

Figure 2.7 CAD elements and model types

The corner model defined by points is of less practical importance and not usable for AM processes. It is used, for example, as an intermediate model for the semiautomatic transformation of grid data or 2D CAD models into 3D CAD models.

Owing to its small amount of data, the edge model enables a fast graphic representation of 3D elements even with low-performance computers. Its importance is therefore growing again in connection with virtual reality (VR) applications and digital mock-ups (DMU). The most important disadvantage of the edge model is the missing information about the exact position of the surfaces and the volumes. For this reason it cannot be recommended as a basis for the production of AM models.

All CAD systems that process components as surface models in their geometrical databases are in principle suitable for issuing data via an additive manufacturing interface. When a component is defined by its external surface, the user is usually able to calculate the exact component volume as well. This is usually achieved by assigning and storing an additional normal vector for each surface pointing away from the inside of the component. For the complete description of a component, therefore, it is absolutely necessary that the orientation of the component volume is known.

Solids are optimal for the modeling of CAD models that (among other things) are also used for additive manufacturing. The orientation of the volume is preset exactly and need not be explicitly defined by the user.

Solids are differentiated into

- basic solids,
- surface determining models, and
- hybrid models.

For basic solids, the component is reproduced in the CAD system by combining basic bodies (so-called geometric primitives) such as cuboids, globes, and cylinders by means of Boolean operations. The CSG (constructive solid geometry) tree is part of the database and reflects the component's history of origin. The basic solid includes the CSG tree, but it does not contain any information about the single surfaces. If a basic solid is issued as an STL file, the contours of the basic body are calculated as a first step. Mistakes caused by inexactitudes in fitting neighboring surfaces are therefore impossible.

With surface determination models in the extreme case, only the details concerning single surfaces and the position of the volume are stored. The position of the volume, also defined by a normal vector standing on each surface and pointing to the outside, does not need to be defined by the user. This model type also enables bodies with extremely complex outlines to be described, which is otherwise possible only with the aid of surface models.

CAD systems do not usually work on the sole basis of one of the two described solids but rather in a combination of both model types. The advantages of both model types are combined in so-called hybrid models, which include elements of the basic solid as well as those of the surface model. CAD systems that work with hybrid models usually generate faultless STL data, as the edges of the surfaces used for the issuing are exactly fitted by the system itself when the resulting hybrid model is generated.

2.2.1.2.2 Demands on CAD Systems

When judging CAD systems as the basis for additive manufacturing processes, many of the properties discussed in the preceding as possible should be taken into consideration. The basic requirements for the systems are to develop a 3D model and the possibility to transfer this model for additive manufacturing via the interface (compare to Section 2.2.2). The transformation via neutral interfaces into another CAD system and from there via an STL interface to the AM machine should always be regarded as a detour.

The following features have been established as advantageous for the selection of CAD systems:

- *Parametric 3D designs*
  Instead of using fixed measurements, parameters are agreed on that can be correlated with each other by mathematical functions.
- *Hybrid models*
  Hybrid models combine the advantages of both solids and surface models and are therefore very well suited for rapid prototyping applications.
- *Continuous database*
  The CAD system and all associated modules must always refer to a common, compulsory database.
- *Redundancy avoidance*
  A continuous database will avoid redundancies; that the database will be free of data unnecessarily stored more than once. Such multiple storages are to be avoided in view of the storage capacity, speed, and clarity of the program.
- *Open system*
  It needs to be guaranteed that the systems can be linked with specialized modules from independent producers (reverse engineering, CT modeling, and additive manufacturing software).
- *Associativity*
  The internal architecture of the CAD systems must ensure that any alterations cause all dependencies to be checked and modified where necessary.

Further, when choosing a CAD system, one should, for example, assess how easy it is to learn to work with it and its level of support in performing certain tasks, which may be specific for special branches.

The market offers a vast number of CAD systems that often differ only in specialties typical for specific fields of activity. It is increasingly probable that all important CAD systems either possess an STL interface already or will obtain one in the near future. Usually other complementary rapid prototyping interfaces are available, for example, SLC, PLY, VRML, or HPGL.

Appendix A3.1 includes a list of CAD systems that possess a rapid prototyping interface. The special properties of the CAD systems listed are deliberately omitted. The Internet addresses provided will ensure that the interested reader can inform himself of the current situation speedily and reliably.

2.2.1.3 Generating 3D Models from Measurements

Normally it doesn't matter how the 3D data will be generated. Figure 2.8 shows several other alternative possibilities for obtaining a 3D model.

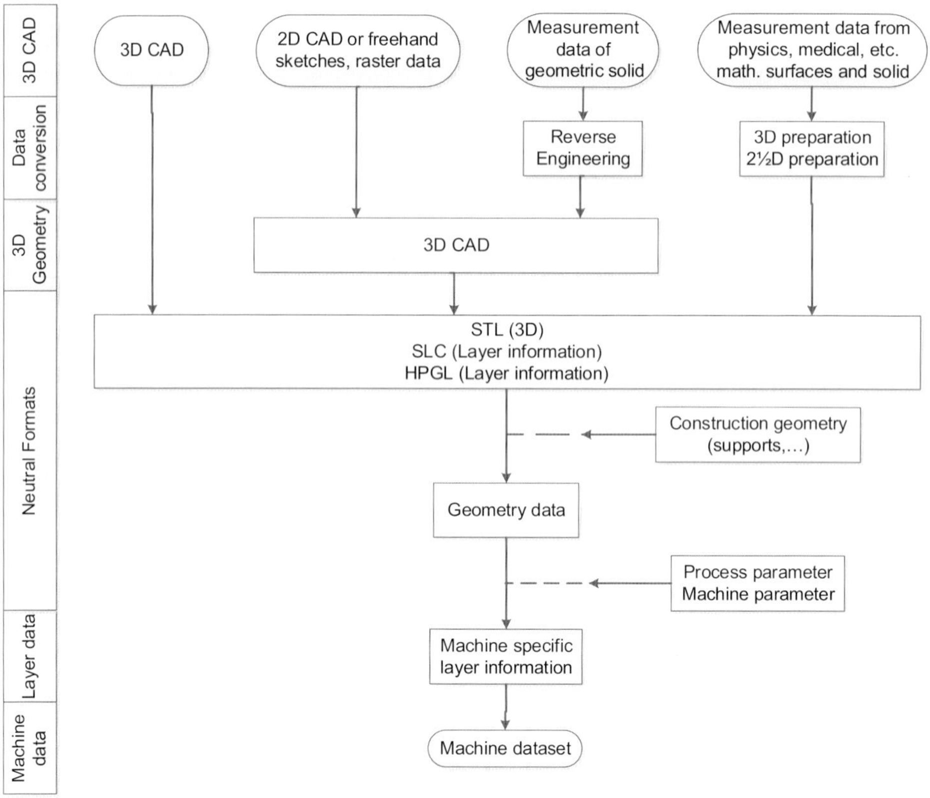

Figure 2.8 Generalized representation of the data paths for additive manufacturing

In addition to 3D CAD designs, 2D CAD sketches, manual sketches, and anything similar can in general be used as input data. To transform them into AM models, they will require a final conversion into 3D volume information via various intermediate operations. If a body is described only by means of 2D elements, then they require conversion into a 3D format by preparing them with a 3D CAD. In most cases this is done using a manual 3D CAD drawing that is no less time-consuming than a new design. When the layer thickness is known, some systems can process the required contour data directly. For example, the 2D data in the HPGL format are used for transmitting.

If, on the other hand, 3D measurement data exist, for example from a coordinate-measuring device, then the measured data can be converted into a 3D CAD model with the aid of special program systems.

In the field of mechanical engineering, this conversion of point data (considered geometrically inferior) to surface data (geometrically higher quality) is called "reverse engineering." The subject of reverse engineering is far too complex to be discussed exhaustively here. An overview is given in [Back, 05].

Even point clouds can be used for additive manufacturing when the format is exchanged in a neutral data format for the model description. Use of them is made in particular when only the physical model is required and no CAD drawing, such as for body scanning for producing sculpture (Fig. 6.12).

Furthermore, 3D measurement data are obtained from computer tomography (CT) scanners, which are more and more established in the technical industry; for example, they are used for nondestructive testing and also to generate 3D data. The CT data can be converted into 3D volume data that can be used for the AM process (Section 4.3.4, "Rapid Prototyping in medicine"). Application fields include the individual repair of defective parts, for which the defective geometries have to be gathered. Another application is the production of spare parts that have no existing production documents, for example spare parts for a vintage car.

In particular for private use of fabber also libraries are used that allows to download the data directly from the Internet. Depending on the provider the data can be modified.

2.2.2 Generation of Geometrical Layer Information on Single Layers

To produce three-dimensional models by layer-oriented additive manufacturing processes, the 3D CAD solid must be mathematically split into the same layers as those produced physically by the AM machine. This process is known as "slicing."

There are two basic methods of doing this: triangulation, which leads to the STL format, and direct cutting in the CAD system, which leads to the CL (SLI) format.

Either the surface is covered with very small triangles, thereby enabling the real 3D geometry to be converged to any degree desired and to be cut into layers at will (STL formulation), or at defined points cuts are made directly into the CAD model (contour-oriented or SLC formulation).

2.2.2.1 STL Format

In order to obtain an STL data set of the part, the surfaces of the part are approximated by triangles. Volume elements exhibit at least two surfaces, the inner and outer surfaces. Both of them differ only by the normal vectors. The definition of the surface by triangles is called triangulation or tessellation. This leads to the so-called STL data. It is regarded as a de facto industry standard for AM processes, but it is actually nowhere standardized.

This contributed to the fact that this process, long before it was discovered for additive manufacturing, was used for shading and thus for the visualization of three-dimensional CAD lattice models. Decisive for the establishment of the STL format as an interface for additive manufacturing was the early publication of the interface formulation. The STL interface, which has been known since then as the stereolithography interface, could be used by both machine manufacturers and free software businesses. This was especially beneficial to the development of special software that is offered by independent developers and made a lasting contribution to the user-friendliness of additive manufacturing systems.

The STL data contains the normal vector (positive direction outward, away from the volume) and the coordinates of the three vertices of each triangle (Fig. 2.9). An ASCII or binary file can be created. The amount of data is much lower for binary files, but ASCII files are comparatively easy to read and control in the source code. Figure 2.9 shows such a triangle patch and the corresponding ASCII data set per triangle enclosed by the commands FACET and ENDFACET.

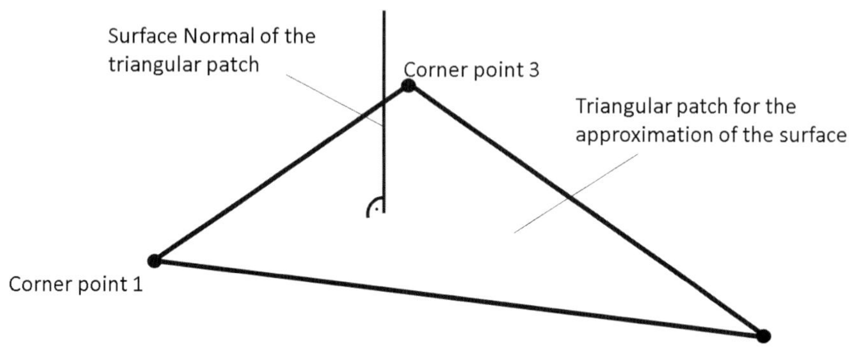

FACET NORMAL 0.000000e+00 0.000000e+00 -	**Surface Normal of the triangular patch 1**
OUTER LOOP	Start of the point data
VERTEX 9.999990e+00 9.999990e+00	Triangular facet 1; corner point 1
VERTEX 9.999990e+00 1.100001e+01	Triangular facet 1; corner point 1
VERTEX 1.100001e+01 9.999990e+00	Triangular facet 1; corner point 1
ENDLOOP	End of the point data
ENDFACET	End of facet description for facet 1

Figure 2.9 Definition of triangle patches in STL format and the associated ASCII data set [Hoffmann, 95] [BRITE/EuRAM, 94]

Figure 2.10 shows the triangular patches coated on a real component, called the triangulated surface.

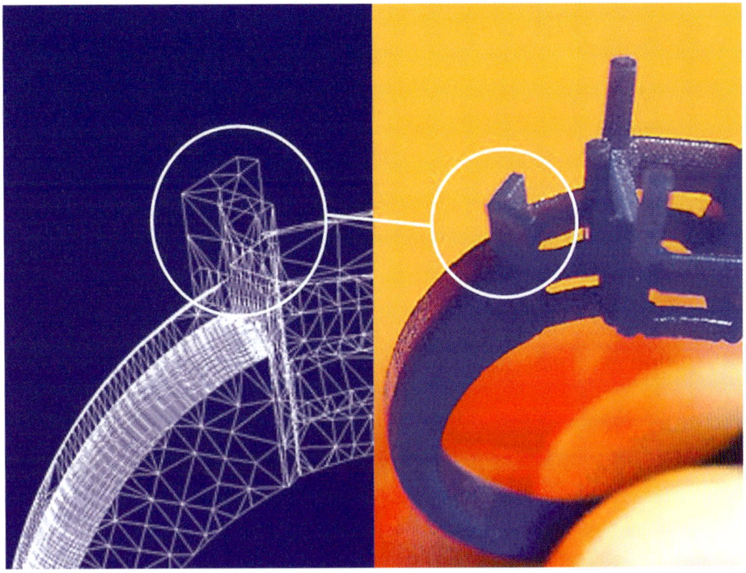

Figure 2.10 Triangulated surface and associated manufactured component (Source: 3D Systems)

The STL formulation, however, possesses practical advantages:

Given that the surface is based on triangles, it is possible to cut the model at any desired *z* coordinate. Also, when the CAD model is not available, STL models permit any desired scale at random without reversing into the CAD.

Because the intersection contains only data elements of a type that can be described by relatively simple means, syntax errors of the ASCII version in the programming of the interface are very easy to recognize and eliminate, and therefore they pose practically no problem.

In contrast to contour-oriented intersections, smaller errors may be repaired relatively easily. It is also an advantage that a triangle provides a higher quality of geometric information than does the contour vector.

The STL formulation also has disadvantages:

- It generates a large volume of data, especially when the surface quality is improved by refining the net of triangles.
- STL files contain only geometrical information. Information about color, texture, material, or other characteristics of the physical model are missing.

2.2.2.1.1 Errors in STL Files

During the transformation of the CAD internal geometry data into STL files, different errors can occur that affect the quality of the physical component. The errors are categorized by Hoffmann [Hoffmann, 95] as

- construction errors,
- transforming errors, and
- description errors.

Construction errors are based on unnecessary data inside the component that are the result of combining the single elements incorrectly in the CAD system (Fig. 2.11). These errors are problematic for the AM process. For example, LLM processes include unnecessary cuts because of these errors. The consequences range from additional expenses during the building process to the total loss of the part. Construction errors do not affect components that are produced by polymerization and sintering processes.

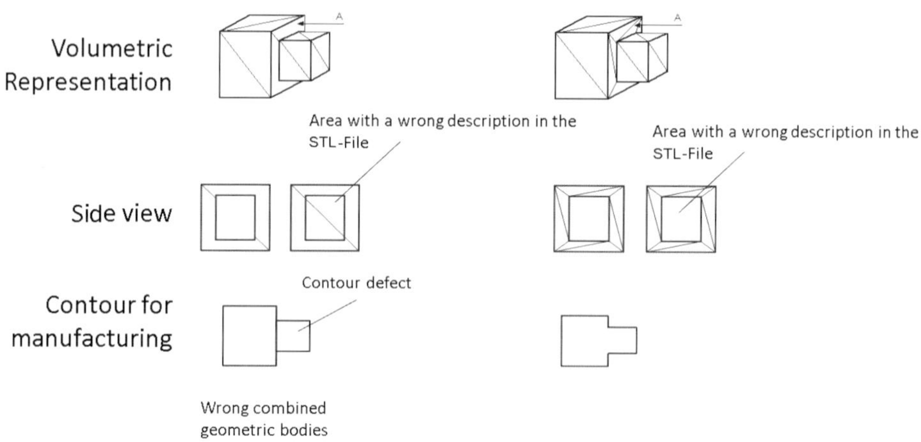

Figure 2.11 Effects of merged faulty geometric bodies [Hoffmann, 95]

Transforming errors exist when the convergence of the mathematically exact contour (as provided by the CAD) by triangles is inaccurate and the number of transforming errors is larger the lower the number of triangles chosen. In Fig. 2.12(a), this fundamental secant error is demonstrated in the example of errors appearing in the convergence of a circle by (f/4), eight (f/8), and twelve (f/12) secants. Figure 2.12(b) shows the consequences on the modeling of a globe surface.

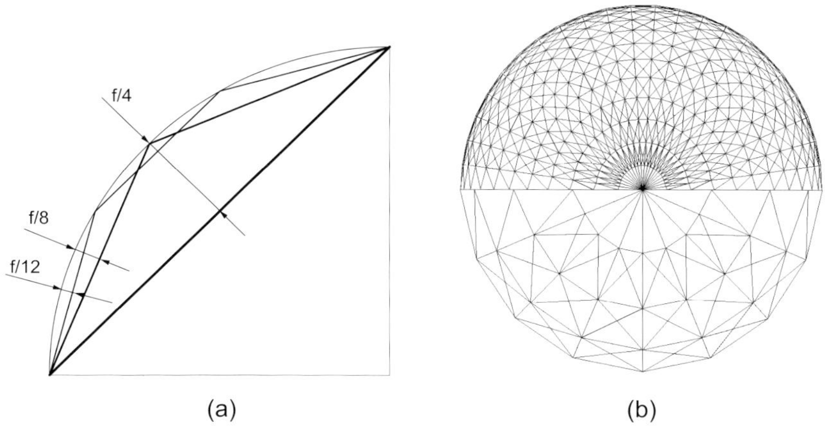

Figure 2.12 (a) Secant error at the approach of a circle by 4 (f/4), 8 (f/8), or 12 (f/12) line segments; (b) Influence of the number of triangles in the modeling of the surface of a sphere (STL)

With the increased accuracy in defining the surface made possible by increasing the number of triangles, the amount of data increases enormously. In one published example the growth factor was 22 [Donahue, 91]. Critics continually cite this as one of the disadvantages of the STL formulation. Although this is correct in principle, it should be remembered that if alternative processes are used, for example the contour-oriented formulation, the closed curves must also be displayed as polygonal drawings and the amount of data resulting from this kind of representation also grows enormously with the growing demand for accuracy.

In practical terms, the fineness of the triangulation is not problematic if approved settings are used.

Description errors are primarily attributable to three causes:

1. gaps between triangle patches (boundary error),
2. double triangle patches (overlap), and
3. incorrect orientation of individual patches (disorientation).

Figure 2.13 shows this error schematically.

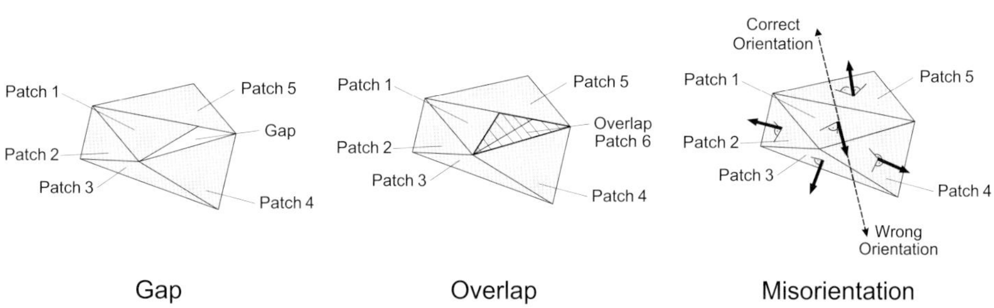

Figure 2.13 Description error: gap, overlap, and incorrect orientation

Gaps (and double triangle patches as a special form of gaps) are the result of inaccurate boundaries that border on each other. The existence of differing resolution densities of geometries can cause boundary errors on the edge on the opposite side. These are called "naked edges."

Such defects are irrelevant for visualization and also for processing with cutter diameters in the range of millimeters. In applications with lasers that exhibit a beam diameter of 0.1 mm, such defects have a negative effect. Description errors make the production process difficult or in some cases impossible.

When the surface is oriented incorrectly, the normal vector points to the inside of the model. In general, the human eye can assign such surfaces correctly, but for generating machine data these results are problematic. The result is that the inner and outer sides cannot be separated (Fig. 2.13).

When the machine-specific layer information is generated, all gaps have to be closed. This process is called the repair of the data set. Normally, special modules of front-end software do this automatically. While repairing semiautomatically, manual intervention leads to faster and better results.

Repairing data are limited. The sample shown in Fig. 2.14(a) still allows for an easy repair. On the other hand, Fig. 2.14(b) shows a sample that is likely not fixable. The best solution for such errors is to avoid mistakes during the construction phase of the CAD model.

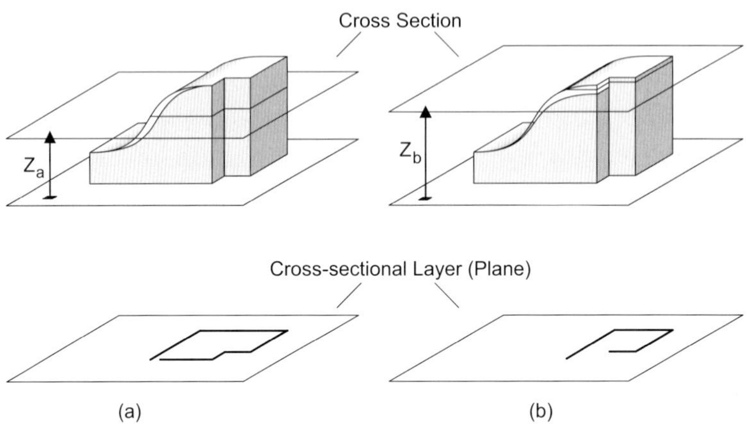

Figure 2.14 Influence of imperfectly bordered deterministic surface models on the production of layer models

2.2.2.2 CLI/SLC Format

The CLI (SLI, SLC) interface, also called the contour-oriented interface, assumes the geometry data for each individual layer of the component that is to be produced. In this case the CLI (common layer interface) interface is a cross-system,

system-neutral form of data transfer. In contrast, a slice file contains additional machine-specific information on the geometric data and is not exchangeable between the various additive manufacturing systems.

All contour-oriented formulations define the contour cuts directly into the respective *z* coordinate of the 3D CAD model and cut at the mathematically precise object [Fockele, 94]. These constitute the 2D information. The established plotter software formats are based on HPGL files (Hewlett Packard graphic language) or similar standard formulations.

The slicing is generated directly in CAD, respectively in the layer. The information from the 3D CAD is sufficient to specify the hatches, in addition to the contour, and thus can fully describe the contour. A distinction is made between the outer and internal boundaries and the hatches of the cross-sectional areas.

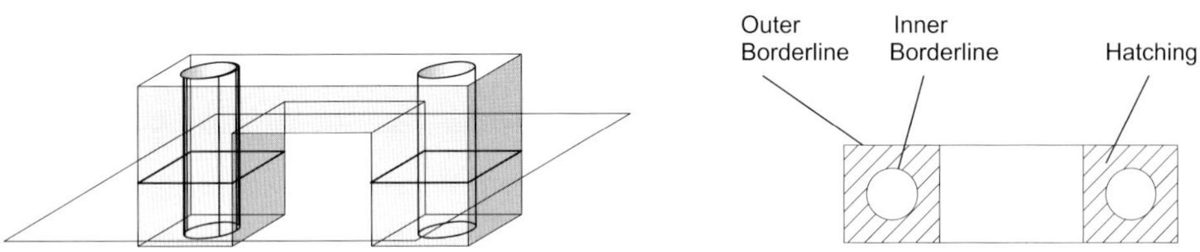

Figure 2.15 Direct contour generation from the 3D CAD system

Along with the advantages of the generated layer information, there are also disadvantages.

An essential disadvantage in practice for the SLC formulation is that the contour-oriented data sets, in contrast to STL files, cannot later be scaled without re-editing in the (original) CAD because they are isolated layer data unrelated to the height coordinate.

For optimizing the component, the layer data cannot be edited without the CAD.

The alternative transfer of the complete CAD data assumes that in the field of additive manufacturing the respective CAD system is operated and controlled. Mostly more than two CAD-Systems are excluded for reasons of cost. There is then just the process of transfer, which also runs the risk of errors.

Especially in the development phase and for secrecy reasons, the entire CAD data set for contract manufacturing should not be rendered because much more information would be shared than is necessary for additive manufacturing.

As a solution, newer front-end software offers functionally reduced, surface-modeled CAD programs, and has a limited but sufficient functionality.

The contour-oriented formulation follows the polyline instructions. The data set contains, for each layer, the polylines of the outer contour (mathematically positive)

and the inner contour (mathematically negative) in 1/100 mm. In the structure of the data set these parameters are to be noted (Fig. 2.16):

- The numerical data are integers and the dimension is agreed in the Unit command.
- Before the polyline data the *z* height of the respective layer is given.
- The polyline instructions contain only the *x* and *y* coordinates of the individual points.
- The contours should be closed (the start and end points are identical and should be stated).
- The order of points determines the type of contour (inner or outer contour) and therefore the implementation of the model construction machine.
- In a plane, any number of contours can exist (the basic requirement is that an outer contour must exist too).
- The distances between the layers should be constant.

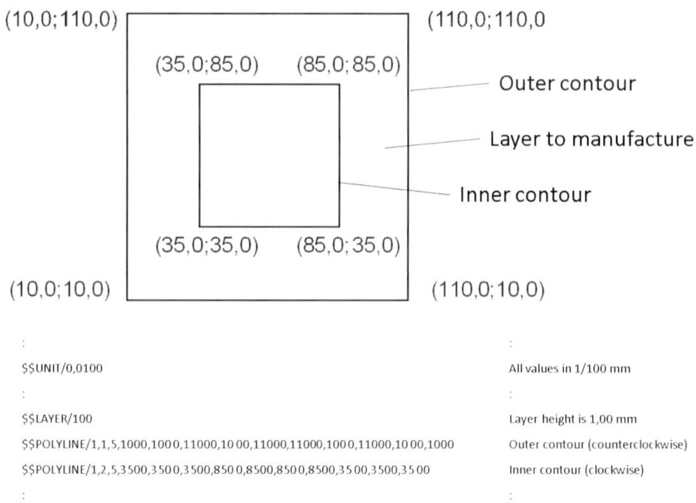

Figure 2.16 Representation of contours in the CLI data format

The cross section of all building levels are contained in the CLI file. It is important to define unambiguously the inner and outer contours. If this is not the case, the work piece will not be produced correctly in stereolithography, for example, because the laser beam did not melt the material and therefore does not harden (Fig. 2.17).

In the case of incorrectly defined contours, a faulty component is produced. An incorrect outer contour with an incorrect inner contour means that the entire volume cannot be built. An incorrect inner contour results in the internal volume being built even though the component should actually be hollow (Fig. 2.17).

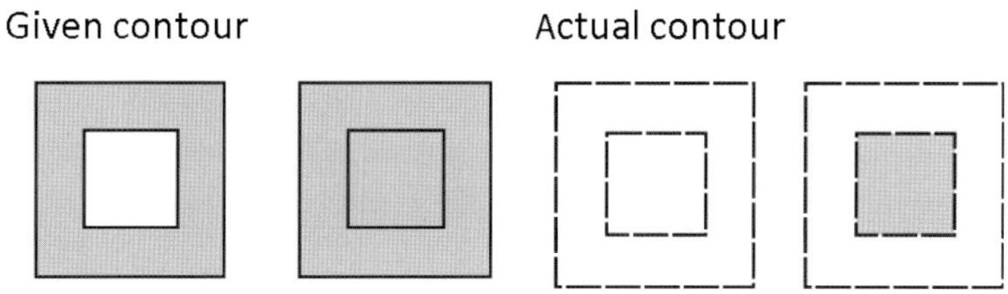

Contour definition	ok	incorrect	incorrect	incorrect
Outer contour	ok	ok	incorrect	incorrect
Inner contour	ok	incorrect	ok	incorrect

Figure 2.17 Errors in the CLI File and its impact on the component [Hofmann, 95]

2.2.2.3 PLY and VRML Formats

Additive manufacturing machines that produce colored components or handle different materials cannot be controlled with traditional geometry-oriented STL data. As well, machines that will enable to produce components with properties that are continuously variable over the cross section in the future. Although the STL data structure provides sufficient space for additional information, in any case, adding more information would always affect the whole triangle. Color or material information or continuously changing properties stored in bitmaps cannot be presented in this way. Approaches for extending the STL format [Swaelens, 99] have therefore not been used.

For the production of colored components today, the "polygon file" format (PLY) and the "virtual reality modeling language" data format (VRML) are used instead of the STL format.

These formats are based on the development of interactive Internet avatars, web-oriented PC games with 3D virtual worlds and the related character animations for moving the figures. In addition to the triangulated mesh envelope and the geometric information, color assignments are also edited.

The VRML data format was developed in 1995 as version 1.0 and in 1997 as version 2.0. VRML 2.0 is the same as VRML 97. VRML 1.0 represented a suitable STL data compressor for the Internet. Unlike its predecessor, VRML 2.0 binds the color information (Fig. 2.18) to the object geometry, in addition to a wide range of multimedia information. VRML 97 has been standardized since December 1997 by ISO (ISO/IEC DIS 14 772 1, 1997) [VRML, 97]. It therefore has the potential to replace the STL formulation, which is only a de facto standard.

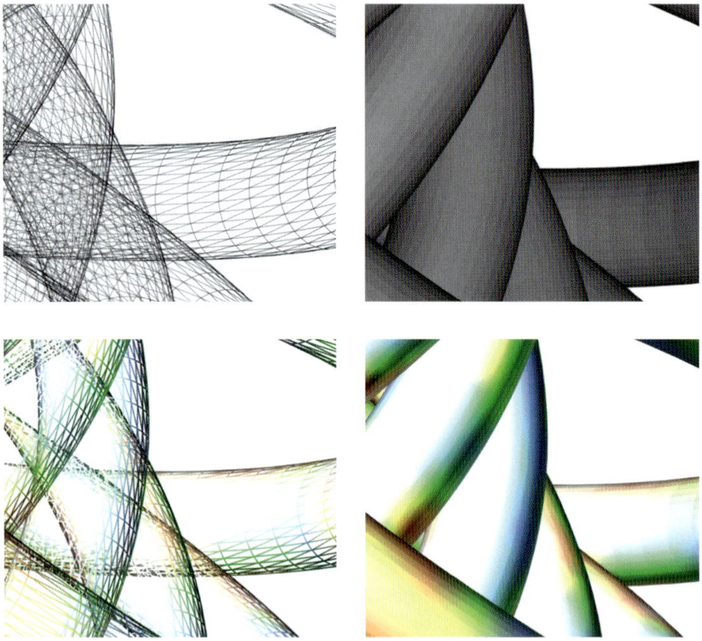

Figure 2.18 Object geometry as triangulated mesh monochrome (top left), light assignment (top right), calculation of color texture and light reflection on the triangular mesh (bottom left), and body with lighting and texturing (bottom right). (Source: FH-Aachen, RP-Labor/F.-M. Schmidt)

Color information has different origins as calculated algorithms. It is based on object color, object material assignments (luminescence, transparency, reflection, roughness), object illumination (light or shade species), and the associated interactions of optical laws within a virtual scene (so-called radiosity ray tracing). These factors substantially influence the so-called texturing at the object geometry by choosing the material (Fig. 2.19).

Figure 2.19 Object geometry with different material in the same lighting conditions; (Source: FH-Aachen, RP-Labor/F.-M. Schmidt)

A detailed discussion of the VRML format and its interaction with additive manufacturing machines can be found in [Ming, 02].

2.2.2.4 AMF Format

The additive manufacturing file format (AMF) is a standard for data sets that contains information about different materials, colors, textures, and other physical and technological properties in addition to the geometry. With the AMF, machines can be controlled that generate material mix, texture, and color directly in the process. The AMF format supports the production of multicolored components made with gradient materials or made with textures or microstructures. The ASTM (American Society for Testing and Materials) published this as F2915-12 “Standard Specification for Additive Manufacturing File Format (AMF) Version 1.1” on May 2, 2011.

The format is XML-based and is therefore independent from the platform. The AMF format is compatible with classic STL files. The transmitted function is then reduced to the geometry. The individual functions are activated by a header in the data set. The AMF format is forward compatible, which means it will allow future extensions.

There are five main commands that define the geometry and the properties of the model. It is possible to use individual commands for both the total component as well as the individual structure elements. The commands are described in the following:

1. `<object>` This command defines one or more material volumes. These are clearly marked with identification numbers (ID). In an AMF file there must be at least one object; more than one is optional for an assembly module.
2. `<material>` This is an optional command that defines the material that can be assigned to a specific volume. Different materials are clearly defined by an ID.
3. `<texture>` The texture command is optional and allows the assignment of colors and textures.
   a) Texture maps: Texture maps assign colors and materials to specific surface areas or volumes. This is similar to the texturing of graphics.
   b) Colors: Colors are defined by red, green, blue, and the alpha channel. Each channel is assigned a value between 0 and 1, which indicates the intensity of the color. These values may exist as constants or made as a formula depending on the geometrical location. The alpha channel controls the transparency and allows subordinate colors to shine through. The `<color>` element can be used on materials, objects, volumes, vertexes, or triangles. The rank order of colors is reversed such that the color of a triangle has the highest priority.
4. `<constellation>` If several components are included in the AMF file, the constellation command specifies the order in which the components should be printed.

5. `<metadata>` Within the metadata area is additional information such as author, project name, copyright, CAD system used, and so on.

In addition to the simple rectilinear edges of the STL triangles, the AMF format has the possibility of curved triangle edges, which allow a better approximation of the geometry. This allows for a reduction in the amount of data while maintaining the same geometry resolution.

AMF is available for free and helps to minimize the number of proprietary solutions and to keep one of the important advantages of the STL format: the machine-independent use.

Other benefits are in added functions, such as the definition of units of length, compatibility with FE applications, limiting the number of possible prints, information about the space orientation, and password encryption.

With the development of the AMF format, the major disadvantages of the STL format were eliminated, but it is still based on triangles. The amount of data has been reduced through new compression algorithms, but the fundamental disadvantages of the STL formulation are still there. The AMF format is struggling to gain broad acceptance, though has not yet among the major manufacturers.

2.3 Physical Principles for Layer Generation

All AM processes work in two steps sequentially or simultaneously when generating the physical layer model:

1. generation of a cross section (*x*-*y* plane), and
2. joining this layer with the preceding one (*z* direction).

Joining the layers in the *z* direction with one another is achieved in the same way as joining them in the *x*-*y* direction, with the exception of layer lamination processes. The energy or the amount of binder necessary for the joining is proportioned in such a way that not only the layer itself but also part of the preceding layer is affected and thus joined to the new layer. In layer lamination processes, the layers are cut out of foils with predetermined thicknesses (*z* increment) and glued on top of each other.

To implement the additive manufacturing principle, several fundamentally different physical processes are suitable. Based on systematic observation, the difference has proven to be the physical state of the starting material as it is introduced into the machine.

1. Solidification of liquid materials (Section 2.3.1)
   - Solidification, preferably by polymerization of liquid or viscous materials

2. Generation from the solid phase (Section 2.3.2)
   - Incipiently or completely melted and solidified powders, powder mixtures, or granules (sintering and melting processes)
   - Cutting from foils or ribbons, milling from slabs or other materials (layer process)
   - Incipiently or completely melted and solidified solid materials (extrusion processes)
   - Conglutination of granules or powders by additional binders (3D printing processes)
3. Precipitation from the gaseous phase (Section 2.3.3)
   - Physical deposition from aerosols
   - Chemical deposition from the gas phase

2.3.1 Solidification of Liquid Materials

In the field of additive manufacturing technology, currently non- or low-cross-linked monomers of the type acrylate, epoxy, or vinyl are used as liquid original materials in ambient conditions. They are locally cross-linked by ultraviolet radiation to form solid layers and components. The processes are called photopolymerization, stereolithography, or stereography.

2.3.1.1 Photopolymerization—Stereolithography (SL)

All processes in which the underlying mechanism is the solidification of liquids are based on the concept of (photo)-polymerization. They use a viscous monomer with few or no cross-links that is interspersed with suitable photoinhibitors. Exposure to ultraviolet radiation sets off a spontaneous polymerization, in the course of which the liquid monomer becomes a solid polymer. This process, which in principle also works with all UV light sources and under sunlight, is adjusted to the special requirements of additive manufacturing processes regarding exposure strategy.

The laser scanning process is the oldest and still the most accurate process. A fine laser beam forms the contour of the respective cross section on the surface of a resin bath and generates locally the critical energy density that is required for the polymerization and thus the desired solidification. In current industrial methods, a single laser beam provides the required energy in the focus. Especially in microtechnology, the two-photon process is applied.

There are two methods in the mask process. The entire cross section is imaged onto a transparent mask in the lamp-mask process and is projected by means of strong UV lamps through this mask onto the surface of the resin bath. In the projections process, a powerful projector (beamer) performs both functions and projects the information layer directly on the surface to be exposed.

In the nozzle-lamp process, the component is produced by means of a nozzle and then polymerized by means of a UV lamp.

2.3.1.2 Basic Principles of Polymerization

Polymerization is a chain reaction in which unsaturated molecules are linked into macromolecules (polymers) [Michaeli, 06]. The liquid mixture of single molecules (monomers) is converted into cross-linked, cured plastics.

The monomers consist mainly of hydrocarbon compounds and possess double linkages that can be broken by heat or catalysts. A chain reaction follows, in the course of which the monomers are linked into polymer chains. Depending on the type of reactive particles bringing the polymerization about, so-called radical, cation, and anion polymerizations are distinguishable. For use in AM processes, the radical and cation polymerizations are relevant. There are three different phases in polymerization:

1. initial reaction,
2. propagation reaction, and
3. termination reaction.

During the initial reaction, an external energy input causes the initiators contained in the resin to disintegrate into their radicals, which are then left with a single unpaired electron on their outer shell [Christen, 49]. Because the particle seeks to find a partner for its electron, radicals are extremely reactive.

During the propagation or growth reaction, the radicals react with the double-linked hydrocarbon in the monomer. Since the double linkage is broken, the monomer is also acquiring a single electron on one side of its outer shell. The new group reacts further with other monomers to form long polymer chains. The chain reaction is brought to an end by the termination reaction. The termination is caused by

- interlocking of two polymer chains into one,
- reaction of the polymer chain with the radical of an initiator,
- transfer of a hydrogen atom to a macro radical, or
- elimination of a bondable hydrogen atom.

The solidification of acrylates is based on radical polymerization. In the early days of additive manufacturing (stereolithography), acrylates were used almost exclusively because of their higher reactivity toward epoxy resins. With further development of the systems, acrylates were increasingly replaced by the epoxy resins, which provide greater accuracy due to less shrinkage. With the spread of polymer printing systems and mask processes that use lamps with much lower intensities instead of lasers, the acrylate-based materials are significant again.

The linkage of monomers to polymers is explained in the example of the polymerization of ethylene to polyethylene (Fig. 2.20).

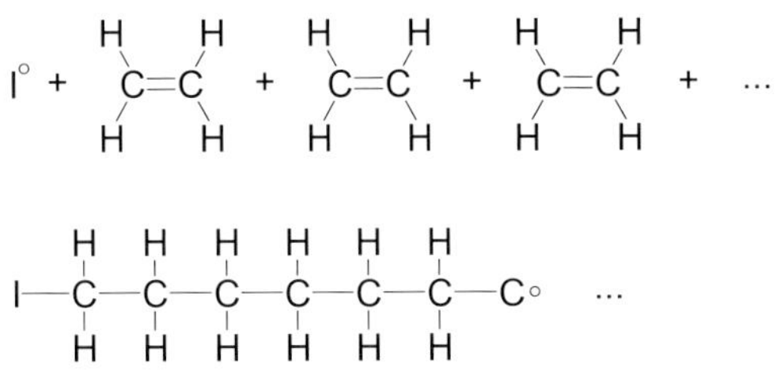

Figure 2.20
Polymerization of ethylene to polyethylene

In cationic polymerization, positively charged ions (cations) are always found at the end of a chain. During the initial reaction, cations are generated from a catalyst. In most cases the catalysts are acids.

During the propagation reaction, ionic linkages are formed between the cation and the monomer. Further monomers can then be added to the end of the chain, which is now cationic.

The termination of the chain is brought about by the addition of an anion from the catalyst. The termination of a chain by the linkage of two polymer chains is impossible because all reactive chain ends are positively charged.

An example of a cationic polymerization is the polymerization of isobutylene to polyisobutylene (see Fig. 2.21).

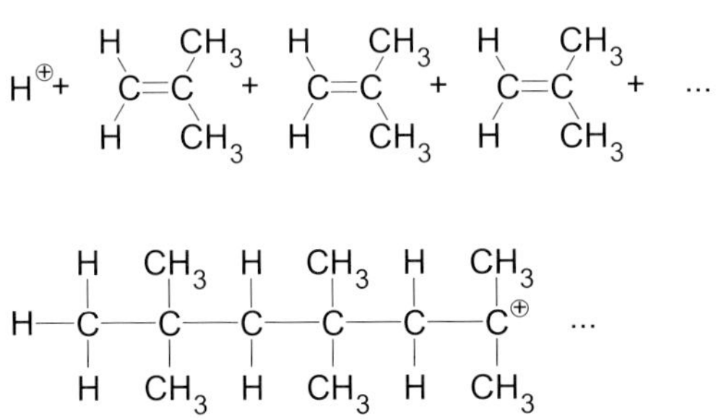

Figure 2.21
Polymerization of isobutylene to polyisobutylene

The solidification of epoxy resins and vinyl ether resins is based on cationic polymerization.

In stereolithography it is necessary to spatially limit the polymerization to allow specific single areas to be cured and thus produce detailed components. The monomer is mixed with an initiator that disintegrates into two radicals or ions via the action of a photon of a certain wavelength. Because the polymerization can be started only with the aid of radicals or ions, it is limited to areas affected by photon radiation.

The presence of a degradative reaction as opposed to the growth reaction causes the monomers to separate from the polymer chains. The speed of this degradative reaction increases as the temperature increases, so at a certain ceiling temperature, growth and decomposition are in equilibrium. Because polymerization reactions are highly exothermic (50 to 100 kJ mol^{-1}), large polymerization installations have to be cooled continuously to keep them under the ceiling temperature. It is not necessary to cool the resin in additive manufacturing installations because the polymerized amounts are too small. To achieve a temperature of approximately 25°C to 30°C, which is beneficial for the flowability of the resin, the resin container may even need to be heated, depending on the temperature of the environment.

In addition to the degree of polymerization, it is necessary to take into account the so-called polymerization rate, which defines the ratio of monomers not participating in the polymerization to the total mass. In contrast to the polymerization degree, the polymerization rate grows with increased viscosity.

When monomers are linked to polymer chains, the material is densified and the resin shrinks. This shrinkage causes tensions in the consolidated resin that can lead to warping and cracking. The extent of volume shrinkage for epoxy resins is around 0.06%; for acrylic resins it is 0.6%.

2.3.1.2.1 Laser-Induced Polymerization

The partial polymerization by laser radiation occupies a special position.

In the following passage some specialties are discussed that result from the use of laser-radiation sources for photopolymerization. It is the basis of laser stereolithography and historically is the root of all additive manufacturing processes and thus also of all additive manufacturing processes.

Depth of the Cure Track

The local degree of polymerization and the rate of polymerization depend on the number of photons that pass through a certain activation cross section of the resin, thereby potentially reacting with the initiators.

From a critical surface energy (critical energy) onward, so many photons react with the resin that it is transformed from a liquid to a solid state. This transformation point is called the gel point. At first the resin does not have any mechanical stability; only after the surface energy is increased is the resin sufficiently polymerized to carry mechanical strain. The boundary surface inside the resin between the solid and the liquid state is formed by that surface in which the surface energy exactly corresponds to the critical energy.

For the surface energy irradiated on average onto the resin surface ($z = 0$), which is also the maximum energy affecting the resin surface, the following applies:

$$E_{max} = \frac{P_L}{v_s + h_s} \tag{2.1}$$

with

E_{max} = surface energy on the surface at $z = 0$

P_L = average laser performance

v_s = speed of the laser beam

h_s = hatch width

The absorption within the resin follows the Beer-Lambert equation:

$$E(z) = E_{max} + \exp\left(-\frac{z}{D_p}\right) \tag{2.2}$$

with

$E(z)$ = surface energy in the depth z

E_{max} = surface energy on the surface at $z = 0$

D_p = optical penetration depth of the resin

The optical penetration depth D_p of a material is defined as the path length after which the intensity of a transmitted beam has dropped to the $1/e$-fold part, or its energy to the $1/e^2$-fold part.

With the aid of the Beer-Lambert equation and the definition of critical energy E_c, the cure depth C_d, down to which the resin is cured, is given as

$$C_d = z(E_c) = D_p + \ln\left(\frac{E_{max}}{E_c}\right) \tag{2.3}$$

With the first relationship between the laser performance and energy on the surface it follows:

$$C_d = D_p \cdot \ln\left(\frac{P_L}{v_s \cdot h_s \cdot E_c}\right) \tag{2.3a}$$

As C_d is proportional to the logarithm of E_{max}, a straight line results if a logarithmic representation is employed. The gradient of the straight line is defined by the value of D_p. The critical energy density, at which the cure depth is zero, is defined by the intersection of the straight line with the abscissa. Because the curve is dependent only on the resin constants E_c and D_p, and on the maximum energy density, it is characteristic for a certain kind of resin and is also called a working curve (Fig. 2.22) [Heller, 95].

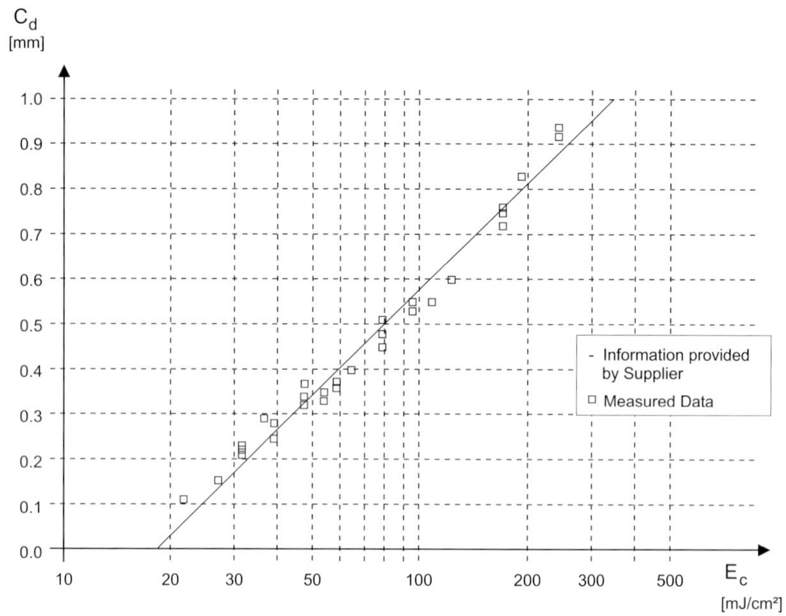

Figure 2.22 Cure depth (C_d) depending on density of surface energy (E_c) and of resin parameter; working curve is of resin HS 660; although today the resin HS 660 has no technical meaning, the figure clearly shows the principles of the relationships that are also valid for current resins.

Given the resin parameters E_c, D_p, and a constant laser performance P_L, the cure depth of the resin can be established on the basis of the speed of the laser.

Contour of the Cure Track

In addition to the cure depth, the width and the form of the cured track are significant. This calculation cannot be based on an average surface energy; the energy distribution of the Gaussian beam must be taken into account. For this purpose, a system of coordinates is fixed with its *x* axis in the direction of the laser speed vector and its *z* axis in the direction of the beam. The surface fixed by the *x* and the *y* axis coincides with the resin surface. A cutting plane is laid through the zero point onto which the contour of the cure track is to be calculated. The distance between the midpoint of the laser beam and a point Q on the cutting plane is called *r* (Fig. 2.23).

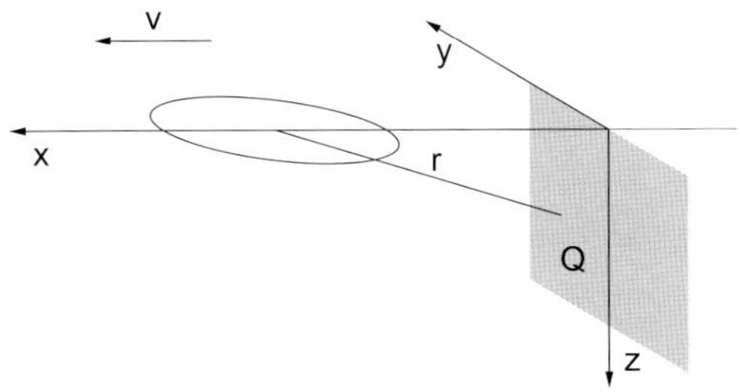

Figure 2.23 Coordinate system for the calculation of the track geometry

For the intensity distribution in the Gaussian beam and with the relation between the intensity in the beam midpoint I_0 and the laser performance P_L at a Gaussian beam, then

$$I_0 = \frac{2 \cdot P_L}{\pi \cdot \omega_0^2} \tag{2.4}$$

with

ω_0 = beam radius on the resin surface

It follows:

$$I(r,0) = \frac{2 \cdot P_L}{\pi \cdot \omega_0} \cdot \exp\left(\frac{-2 \cdot r^2}{\omega_0^2}\right) \tag{2.5}$$

The surface energy at a specific point on the section plane (y, z = 0), over which a laser beam runs with the intensity $I(r, 0)$, equals the temporal integration over intensity. By replacing I_0 according to Eq. 2.4, the surface energy on the resin surface results.

The periphery is to be found exactly where $E(y, z)$ equals the critical energy E_C. By equating $E(y^*, z^*) = E_C$ and transforming, we arrive at the following equation:

$$\left(\frac{2}{\omega_0^2}\right) \cdot y^{*2} + \left(\frac{1}{D_p}\right) \cdot z^* = \ln\left[\sqrt{\frac{2}{\pi}} \cdot \left(\frac{P_L}{\omega_0 \cdot v_s \cdot E_c}\right)\right] \tag{2.6}$$

Equation 2.6 is the definition of a parabola. The cure track therefore has the geometry of a parabolic cylinder, as shown in Fig. 2.24. Pulsed lasers produce paraboloids of rotation, called voxels.

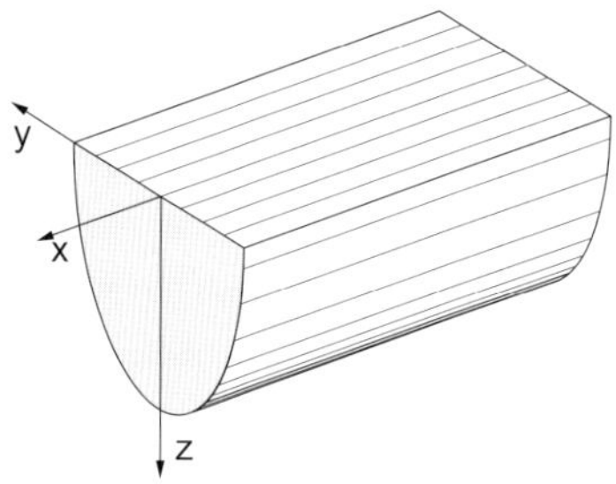

Figure 2.24
Parabolic shape of cure track for a Gaussian beam exposure on a photopolymer

From Eq. 2.6, the relations for the track width L_W and the cure depth C_d can also be derived [Jacobs, 92], [Faßbender, 94].

Optimization of the Layer Thickness

Knowing the geometry of the cure track enables a layer thickness to be determined in which the time needed for curing a certain volume is minimal. For a reasonable coverage of the volume with parabolic cure tracks, it is assumed that the cure track has a distance of 0.02 mm to both the lower layer and the adjoining layer.

With d as layer thickness for the curing time of a certain volume, the following proportionality results:

$$T \sim \frac{1}{v_s \cdot d \cdot \left(L_w + 0.02\text{mm}\right)} \tag{2.7}$$

With the cure depth $C_d = d - 0.02$ mm and insertion of the relation for the track width L_w it follows:

$$T \sim \left\{ d \cdot \sqrt{\frac{2}{\pi}} \cdot \frac{P_L}{\omega_0 \cdot E_c} \cdot \exp\left(-\frac{d - 0.02\text{mm}}{D_p} \right) \cdot \left[\frac{\omega_0}{\sqrt{2}} \cdot \sqrt{\frac{d - 0.02\text{mm}}{D_p}} + 0.02\text{mm} \right] \right\}^{-1} \tag{2.8}$$

Equation 2.8 is shown graphically in Fig. 2.20. The minimum of the function is $d = 0.3706$ mm for an assumed optical penetration depth $D_p = 0.25$ mm. As only proportionalities were investigated, there are no resulting absolute values for the ordinate.

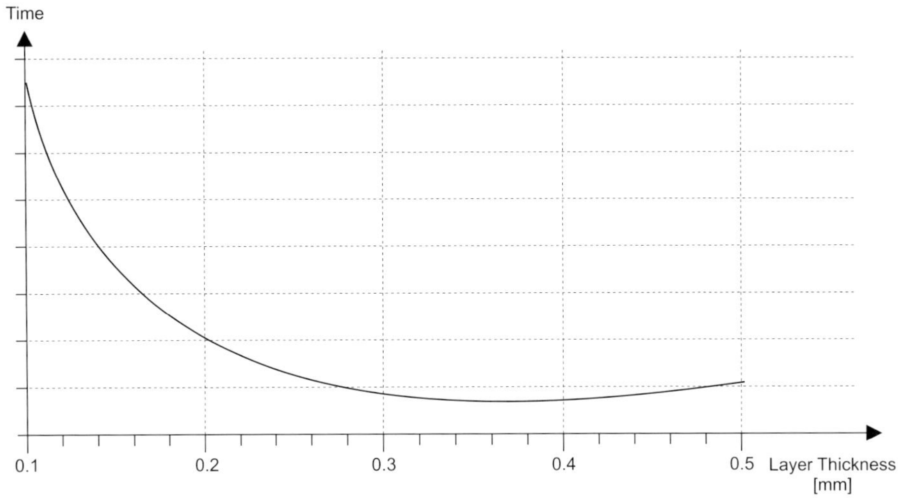

Figure 2.25 Required time for curing depending on the layer thickness

The penetration depth D_p of 0.2 to 0.3 mm for different resins lies within the technically sensible dimension of 0.1 to 0.5 mm. If deeper layers are required, the surface energy needs to be increased, which simultaneously lowers the scan speed. The aim, however, is to expose layers as thin as possible with the highest possible scan speed. To increase this, attempts have been made to make resins with greater optical penetration depth available. As described hereinafter, the variation of the depth of penetration also has an influence on the resistance component.

If resins with a greater optical penetration depth are used, it should be taken into account—especially in combination with higher powered lasers—that leakage radiation can result far more easily in undesired polymerization than previously, having a negative influence on the accuracy of the components and above all on the aging of the resin.

Effect of the Penetration Depth on the Stability of the Component

By increasing the penetration depth D_p of the resin with a constant cure depth C_d, the required surface energy on the surface of the resin is lowered. Accordingly, fewer photons are absorbed and the rate of polymerization decreases, which results in less stability of the component. The relevant value, excess energy E_x is introduced here. This is a measure for the amount of energy available for the polymerization in addition to E_c. A useful definition for E_x is given in Eq. 2.9 [Jacobs, 92]:

$$E_x = \left(\frac{1}{C_d}\right) \cdot \int_0^{C_d} \left(E(z) - E_c\right) \cdot dz \qquad (2.9)$$

After implementing Eq. 2.2 (the Beer-Lambert equation) it follows that:

$$\frac{E_x}{E_c} = \left(\frac{D_p}{C_d}\right) \cdot \left[\exp\left(\frac{C_d}{D_p}\right) - 1\right] - 1 \tag{2.10}$$

This function is illustrated graphically in Fig. 2.26. The following fundamental relationships derive from Eq. 2.10 and Fig. 2.26:

- The excess energy is directly proportional to the critical energy E_c. A reduction of E_c to achieve shorter construction times will result in a direct decrease of stability during the green phase.
- By reducing the penetration depth or raising the layer thickness, the stability in the green state is increased. The effect is especially prominent with values of $C_d/D_p > 3$.

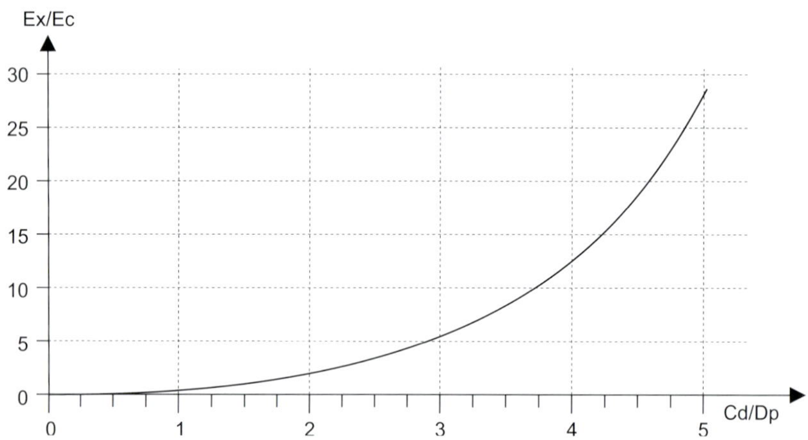

Figure 2.26 Excess energy depending on the cure depth and the optical penetration depth

The fundamental relationships discussed here are valid for all photopolymerization processes, especially for laser-supported stereolithography processes. These relationships explain also the fundamental advantages and disadvantages of stereolithography processes that—with process-specific restrictions—are in principle valid for all industrial machines discussed later.

2.3.1.2.2 Advantages of Stereolithography

Stereolithography, also known as stereography, is presently (2016) the most accurate of all additive manufacturing processes. Its accuracy is limited by the machine, but not by physical limits. For example, the minimal depictable land widths are in principle a function of the laser beam diameter. The fineness of the z stepping is not limited by the process. It is limited by the wettability of a solid layer by the (following) liquid monomer layer, expressed as the relationship of the volume power (proportional to the layer thickness) and the surface tension. Thin layers consequently tend to "rip." There are more cost-related reasons that are decisive in practice; for example, thin layers extend the construction period and therefore increase cost.

It is in principle possible to contour the boundary of the *x-y* planes in the *z* direction by appropriate control (five-axis) and exposure strategies (variation of pulse-pause relationship and laser performance) and thus to achieve a quasi-continual *z* modeling.

Stereolithography not only allows the production of internal hollow spaces, as do nearly all of the other AM processes, but it also permits their complete evacuation as a result of the process technology. For this, a drainage opening is necessary that should obviously be much smaller than the diameter of the hollow space.

Stereolithography materials were formerly all transparent or translucent, and therefore they allowed the visual assessment of internal hollow spaces. This is useful, for example, in many flow studies and medical applications. Today most of the materials are opaque and nearly white like Somos 14120 (DSM), Accura SI 25 (Systems), and RenShape SL7580.

In processes with a resin bath, only individual regions of the component are provided with supports. In general, the volume of the support is significantly less than that of the component and may be affected by the operator. The entire volume of the component will always be solidifying in polymer printing processes and film transfer imaging processes (FTI) (Section 3.1.12, "Film Transfer Imaging-3D Systems") and consist of either the construction material or the supporting material. The volume of the support structures is therefore quite high. The support material is waste.

Noncross-linked monomers can be reused, and completely polymerized resin can be treated as household garbage. But liquid monomer is hazardous waste.

Complex models, or those of larger dimensions than the construction chamber, can be assembled from single partial models into arbitrary, complex, complete models. If the same photosensitive resin is used as binder and UV radiation sources are used for local curing, the section points are unnoticeable in respect to their mechanical-technological properties, and they are also invisible to the eye.

The models can be finished by sand blasting and polishing and, to a certain extent, by machining and coating.

2.3.1.2.3 Disadvantages of Stereolithography

Owing to its process technology, stereolithography is restricted to photosensitive material. In resin development the focus is on cross-linkability with UV radiation. The usual primary properties such as resistance to extension, elasticity, temperature stability, and so forth are of secondary importance. Furthermore, material development is limited to stereolithographic usage because the apportionable costs of the product offset this.

Stereolithography is a two-step process in which the models are first solidified to a high percentage (>95%) in the actual stereolithography machine; afterwards the green product must be cleaned with solvents (TPM (tripropylene glycol monomethyl ether), isopropanol) and it is placed into a postcuring oven to build up further cross-linkages with the help of UV light until it is cured completely.

Printer and mask processes are single-stage processes. They polymerize the component completely and therefore work without cross-linking after the process.

When making stereolithography models, unsupported structures and certain critical angles of overlapping model parts cannot be realized without support. Therefore supports are needed. These support structures shall be placed in the context of model preparation using appropriate programs. The component and the supports of laser-based processes consist of the same material. The supports have to be removed manually from the green product or from the cured model.

In the printer or mask processes, the supports are usually made of a thermoplastic hard wax that is washed out after the building process. It can also be removed automatically or semimanually with the help of solvents. Because all areas of the building volume that do not belong to the component are filled with supporting material, a separate supporting structure is not necessary.

In addition to the construction process, the application of stereolithography limits the removal of the supports, and depending on the process, the storage, handling, and disposal of solvents are necessary.

To a small extent, photosensitive acrylates absorb oxygen, whereas epoxy resins are hygroscopic; this has to be taken into account when storing and processing the material. The models tend to creep even after being completely cured. After a few days or weeks, unsupported walls show sagging that disappears if the model is turned over or supported. The newest epoxy resins show these characteristics less prominently.

2.3.2 Generation from the Solid Phase

2.3.2.1 Melting and Solidification of Powders and Granules: Laser Sintering (LS)

Powders or granules in a powder bed are the materials used for the formation of a defined solid layer. They are partially melted or melted in the respective layer level by an energy source and solidified after cooling into a solid body. Energy sources can be energetic single beams such as laser beams, electron beams, or an infrared panel heater.

The processes are called *sintering processes* in reference to the nonadditive manufacturing diffusion-controlled sintering process. For many years, laser-based processes were exclusively commercialized and were called laser sintering (LS) or selective laser sintering (SLS). It is called even-beam melting or simply melting as a generic term because today electron and infrared rays are also used.

In the classical nonadditive manufacturing sintering, two neighboring particles are joined by a substance exchange. For this, high temperature and high pressure over a relatively long period of time are needed. The sintering process is dominated by the mechanism of surface diffusion. It begins in the form of a neck at the contact point of the particles. With progressive sintering, a material transport takes place—preferably along the grain boundaries—and continues in the inside of the particles (grain boundaries, volume, and lattice diffusion). The sintering that is used in additive manufacturing does not require the two essential components of the classical sintering process: high pressure and long time. It must therefore be assumed that the additive manufacturing sintering does not occur or is not diffusion-controlled. Only a short thermal activation of the nonadjacent particles in the powder bed takes place. When these are melted on the surface after cooling, it results in a more or less porous component. When the particles are completely melted, a dense component is formed.

The result is usually slightly porous components. For plastics, this is often for process reasons intended to avoid warpage and internal stresses (Section 2.3.1.1). But metal parts are dense. Therefore, most of the processes with metal powder melt the powder completely. It is also called laser or, more generally, beam melting.

The sintering process will be described essentially by the interaction between the viscosity of the melted particle areas and their surface tension. Both of these (opposite) effects are dependent on the temperature and the material. Some details are therefore considered in the following sections in the context of the materials.

Fundamental studies on the mechanisms of selective laser sintering, which also address the interaction with the practical application, originate from Alscher [Alscher, 95] and Nöken [Nöken, 97]. Ader [Ader, 06] has revised the theory with a focus on ceramics. The results of the work of Hansjosten (see Chapter 9.2, “Contour Crafting”) underscore the model concept of the complete melting in the production of metal parts.

2.3.2.1.1 Materials for Sintering and Melting

In principle, all materials that can be melted and, after cooling, solidify again (as far as possible with a constant volume) can be used for the sintering process. Because, in contrast to the classic sintering process, the parameter pressure and temperature have to be kept very low. Only plastic material and wax were sintered at first, mainly owing to their low melting points.

Plastic Powder

The low temperature range of up to approximately 200°C favors the sintering of plastic materials. The low heat conductibility is also advantageous. It helps to limit the melting bath locally and to prevent the "growth" of models when neighboring particles being sintered on. Although the surface tension of the molten mass–low compared to that of metals–impedes the sintering process, it fosters the wetting of the model, thereby minimizing the danger that macroscopic globules are formed and separated from the layer.

Crystalline and amorphous plastics behave completely different. The vast majority of materials (metals included) solidify in crystalline form. Their elementary particles, atoms, or molecules, are set in defined, spatially regular arrangements.

Glass, resin, or pitch solidify amorphously (i. e., with no fixed form). Although their elementary particles are also practically stationary, they possess no recognizable defined structure. Physically they should be considered as an undercooled molten mass. The effect of devitrification is well known: an unwanted crystallization causes amorphous transparent glass to crystallize and become opaque.

Amorphous plastics are characterized by a broad temperature range within which they soften or convert into another state of aggregation without sudden alterations to their mechanical-technological properties (Fig. 2.27(a)). Crystalline plastics change their state of aggregation and thereby all of their important mechanical-technological properties in such a narrow temperature range (often only 0.1°C) that it is called a "melting point" (Fig. 2.27(b)).

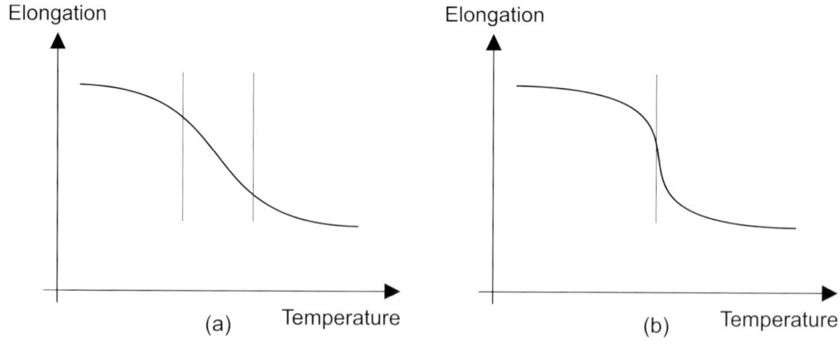

Figure 2.27 Physical properties of (a) amorphous and (b) crystalline plastic

For the behavior of components during the additive manufacturing process, whether and by how much they shrink and the achievable properties such as strength, which is the fundamental material behavior, is crucial.

Solidifying amorphous polymers form from cooling molecular chains, leading to a denser packing, but not to crystallization. The density therefore varies only slightly (shrinkage is about 0.3–0.8 %) [Alscher, 95]. Polycarbonate (PC) and polystyrene (PS) are polymer sintering materials. Sintered components from solidifying amorphous polymers have therefore a comparatively high porosity and a low density. They do not have high strength but are very suitable for the manufacture of lost forms that are melted out during a casting process (see also Section 6.5 and Fig. 1.28).

Crystalline solidifying polymers for sintering are only semicrystalline. When they cool over a wide temperature range they start to crystallize, and this is the reason for the densest possible packing. There remain, however, amorphous areas that still allow a slow flow (creep) [Alscher, 95]. During the cooling of semicrystalline polymers, unequal temperature fields lead to uncontrolled crystallization. They affect different degrees of changes in the volume, which again lead to different degrees of shrinkage and warpage. In this context, the characteristic of "rolling" corners, the so-called "curl," is caused by these effects.

Uncontrolled crystallization can be avoided if the processing takes place within a temperature range in which the viscous flow of the plastic is locally induced by the laser beam and the solidification process does not take place. Therefore, the solidification begins by crystallization. This is possible if the entire chamber is preheated to a few degrees under the melting temperature and is also kept over the crystallization temperature. With the differential scanning calorimetry (DSC) method, the proceedings can be backtracked easily. During heating and cooling, the melting heat and the heat of crystallization are easy to detect and locate through the process window (Fig. 2.28 [Drummer, 11]).

This process is referred to as quasi-isothermal laser sintering. Only when the entire component is completed and enclosed in the surrounding powder bed it is cooled slowly so that local stresses are avoided.

Polyamide is one of the most often used sintering materials. Types PA 12 or PA 11 are less common. PA 6 is in the developmental stage.

To increase the component strength, sintered materials are mixed with glass beads. They are then called glass-filled, although the analogous term in injection-molding material means fiber-reinforced materials and also the type PA 6. There are recent attempts to develop a suppository form to approach the mechanical properties of fiber-reinforced materials to the filling while maintaining the safety of the process stability (Section 3.2.3).

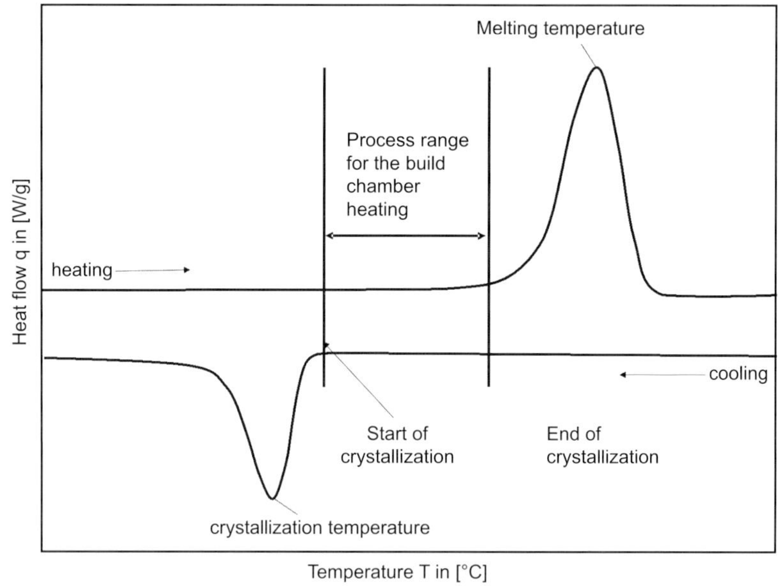

Figure 2.28 Schematic DSC curve of semicrystalline thermoplastics

Aluminum or carbon-filled sintered materials are also available. Sintered materials with aluminum or carbon have the same material designation as injection molded materials, but can have other mechanical properties than filled injection components/materials.

Multicomponent Metal-Polymer Powder

The first step to sintering metal powders is made via a polymeric-linked metal powder. The process is therefore fundamentally not very different from the fusion of plastic powders. The polymeric shells are sintered on and "glue" the integrated metal particles together to form a so-called green product. The component exhibits just the strength that is needed for the process.

In a subsequent oven process, the polymer contents are simultaneously driven off by heating and the metal parts are connected by diffusion processes and microwelding into a solid body shape.

In another furnace process with a reducing atmosphere, a low-melting metal is infiltrated. Nowadays these two subprocesses are combined into one. The process is multilevel, and the process and machines are very complex.

From the viewpoint of sintering theory, this process belongs to the category of plastic sintering processes.

Basically, all materials can be processed in the same manner in plastic sintering machines. Polymeric-linked sands, ceramics, and the above-described metals are technically realized.

Multicomponent Metal-Metal Powder

When low-melting metal powder—instead of polymeric binder—is mixed mechanically with the high-melting component, the process becomes safer and faster. The binder function is then taken over by the low-melting component. The mechanical-technical properties of the model, however, are very different from those of currently used series alloys. These models can therefore be used only to a limited extent as functional models. Figure 2.23 shows the basic interrelationships of polymeric linked powders and of multicomponent powders in the sintering process. Today this type of material is not required.

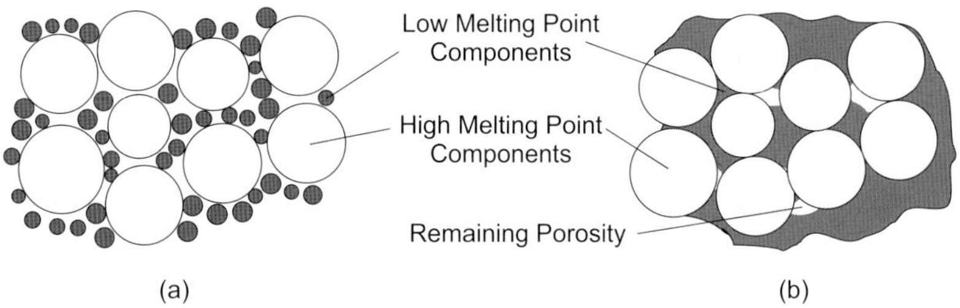

Figure 2.29 The principle of liquid phase sintering with multicomponent metal powder: (a) original material, (b) sintered state. (Source: FHG-ILT)

Single-Component Metal Powders and Ceramics

Processes with single-component metal powders are established worldwide for the additive manufacturing production of components with the product character (final products). In addition, metals and ceramics are also suitable for laser sintering. The powder can be completely melted to achieve dense components. Therefore, the proceeding is also called the melting process, SLM, or selective laser melting.

In their melted state, metals have a higher surface tension and a lower viscosity than do plastic materials. The driving force here is the surface tension and the opposing force is the viscosity, both of which are more favorable than with plastic materials. However, the melting points and therefore the process temperature are much higher, although the process works without preheating. In total, the build chamber remains generally cold. Higher thermal conductivity favors local hardening, stress, warpage, and cracks. In order to counteract this, the melting process operates with supports.

Poor wettability and the high oxidation tendency of metals are also sometimes problems. In order to obtain good results, process or material databases should be developed for each material-machine-combination.

All machinery manufacturers offer powder materials. In addition to this, a large market is developing for metal powder (commercial metal powder) with independent manufacturers.

Such developments are the subject of worldwide intensive research, especially at the Fraunhofer-Institut für Lasertechnik (ILT) (Institute of Laser Technology, Aachen, Germany) and Produktionstechnologie (IPT) (Institute of Production Technology, Aachen, Germany) and at the Fraunhofer-Institut für Werkstoff- und Strahltechnik (IWS) (Institute of Material and Beam Processing Technology, Dresden, Germany). The machines of Concept Laser and SLM Solutions are based on this research, as are the factories of Renishaw.

In principle two paths are followed: the direct sintering of metallic materials according to the process of selective laser sintering described previously, and the laser-supported generation derived from the classic coating process. The aim is the treatment of commercial metal powders.

Laser sources, such as a fiber laser with excellent beam quality, good controllability, and optimized inert gas and scanning strategies, have led to efficient and relatively cold processes (<100°C in the powder bed). Some concepts were implemented industrially that operate with a high-temperature building chamber (>900°C) (Phenix) and preferably with ceramics, but they also handle metals.

2.3.2.1.2 Advantages and Disadvantages of Laser Sintering

Advantages

Sintering and melting can be applied to all materials that behave thermoplastically. This allows for the possibility of an unlimited choice of materials in comparison to stereolithography. The resulting components are mechanically and thermally loadable. Depending on the material and application, they can be used as a functional model or even as final products. Sintering processes are therefore the key to direct tooling, direct manufacture of tools and tool inserts, and direct manufacturing, which is the additive manufacturing production of end products.

Powder that is not adapted thermally can be partially recovered and recycled into the process, usually with the addition of unused powder. The recycling rate depends on the degree of damage and for polyamide powders it is between 30 and 50% waste powder in the new building process. Metal powders can be worked up by sieving and depending on the process also completely reused.

Sintering and melting are basically a one-step process. A post-cross-linking is not necessary. For plastics, no support structures are needed because the nonsolidified powder remains in the powder bed and supports the component. In contrast, metal processes use supports to absorb the thermally induced mechanical stresses.

Supports are therefore used intentionally for temperature control (including for plastics).

Sometimes also bases are used, corresponding to the cooling iron during casting.

To clean the plastic components, only the removal of loose particles with a brush and a light sandblasting are necessary. The supports of metal components must be removed mechanically. In addition to this, heat treatments are performed. Depending on the component and mode of application, a machining rework is necessary. Solvents are not required. The components are ready after the postprocessing.

Disadvantages

The achievable accuracy of the model is basically limited by the size of the powder particles and the beam diameter. When it is an infrared process, the accuracy depends on the illustration quality of the mask. The material and its absorption properties and heat conductibility define the possible construction speed and the necessary laser power.

The models tend to "grow" depending on the relationship of "applied power" and "conductibility," that is, neighboring particles in the powder bed not belonging to the model are glued on by heat conduction and the model grows a "fur" so that the component is geometrically inaccurate.

Internal hollow spaces are more difficult to clean than with stereolithography. It is unavoidable that loose or only slightly glued particles remain on the model. These particles impair accuracy and could possibly detach themselves at inconvenient moments when the model is later used. These components therefore cannot be sterilized for medical use in operating theaters.

To avoid oxidation, the sintering process takes place within an inert gas (nitrogen) atmosphere. This means that the installation technology becomes more complex and the operating costs are higher.

Because the sintering process takes place at near the melting temperature, the entire powder bed needs to be preheated evenly to near this temperature to achieve an efficient sintering process. For plastics, the entire powder bed is equally preheated to a few degrees below the melting temperature. The temperature must be kept within restricted limits (a few degrees centigrade). The heating and cooling processes are very time-consuming. When melting, the same applies to the melting temperature.

2.3.2.1.3 Proprietary or Commercial Powders?

Among all users of additive manufacturing processes, the consensus seems to be that so-called commercial powders are preferable for sintering or melting. The refusal to use a proprietary powder that was specially developed and exclusively marketed by the manufacturer is usually due to both economic and technical arguments; the economic arguments are based on the fear of buying a material for which there is no alternative.

On the technical argument side, it is argued that sometimes proprietary material systems have emerged that are not used anywhere else in other production technologies or applications. Therefore it would have invested many years relying on such materials. The designer would have a problem when the valid documents for these materials are unavailable in the required extent.

Both arguments are correct but also valid for commercial powders. It must be clear that the properties of the additive manufacturing component are generated not only by the powder, but equally through the process and the combination of the powder and the additive manufacturing process (Section 1.2). Changed production parameters but using same material and the same machine will result in different component properties.

The fact that the user may use a well-known commercially available powder in the process does not guarantee which properties of the additive manufacturing component the user will get at the end.

Finally, the user has only two alternatives, regardless of which material is processed in the additive manufacturing machine. The user can request that the manufacturer hand over extensive documentation of the mechanical properties of the produced components with the respective machine and guarantee the values. Alternatively, the user determines these properties as part of a process optimization himself. However, an optimization in-house requires not only the relevant expertise, but also a machine that allows for the variation of the necessary parameters. During this time it may not be used for production.

In practice, the user does not keep changing the material. Typically he commits himself to one or perhaps a few material systems, which are then used as quality-assured in their daily routine. The independence of a very special material is therefore of an advantage in the phase in which a user finds a suitable material system. At this stage, he is generally also still looking for a suitable machine or an efficient service provider.

After all, the materials for additive manufacturing sometimes differ significantly from the nominal (chemical) composition, same as the ones used for other manufacturing processes. Criteria are the flowability guaranteed by geometry and aggregates, moisture absorption, and possible outgassing of additives (reusability). Under these aspects, the commercially available material may not necessarily be the more suitable material. The demand for universally useful, commercially available powders in an additive manufacturing machine is therefore technically unfounded. When users derive the product specifications on their own, without further verification of the type of powder, it is even dangerous in terms of product safety.

These considerations are valid also for other additive manufacturing processed materials. All successful applications like dental (Section 6.5.2.2.1) and the hearing aid industry (Section 6.5.2.3.2) use special material-machine systems.

2.3.2.2 Cutting from Foils: Layer Laminate Manufacturing (LLM)

The most simple method of producing 3D models is to split them into 2D contoured layers, cut these layers out, and then to assemble them into 3D models. This is no longer a purely additive process. Bernard's precise term for all additive manufacturing processes that, in addition to the purely additive steps, also include denuding (milling) or separating (cutting) steps is subtractive/additive processes [Bernard, 98]. We will not follow this refined terminology, as the discussion here is focused primarily on the generation of models from layers; in practice, scientifically exact, refined terminology does not enlighten further.

To obtain sufficiently detailed models, the layers must be very thin and exactly positioned onto one another. Therefore, in most processes, the new layer is first glued to the preceding one, that is, onto the already finished part of the model; only then is it contoured. Layer model-making processes are known as layer laminate manufacturing (LLM). More common is the term laminated object manufacturing (LOM). This is not a generic term but a registered trade name that in this way became a synonym for the entire family of models (also see Chapter 3).

Lasers are very commonly used for contouring. Hot-wire cutters, knives, and milling machines are also used. Especially when contouring by milling machine, the layers–in addition to being glued together–are also pinned together over center holes contained in the contour. Metal layers are joined by laser welding, deposition welding, and ultrasonic welding. Plastic foils are glued or polymerized layer by layer together.

2.3.2.2.1 Advantages and Disadvantages of Layer Processes (LLM)

Advantages

LLM processes are especially suited for the generation for large models and for compact model parts without elaborate shading techniques. Layer processes work fast in relation to the model volume, and they are not very complicated from the machine-technical aspect. They work with practically any material. If paper is used, material costs are low and disposal is not a problem.

A continuous contouring of the *x-y* boundaries in the *z* direction is, in principle, possible by means of a more complicated (five-axis) control of the laser or by milling.

The use of a laser for contouring the layer geometry has the advantage that–theoretically–an almost unlimited range of material is available. With the aid of the laser beam, all materials known today can be cut. By using correspondingly optimized laser cutting processes, cutting speeds well in excess of those presently realized are in principle possible (see Fig. 2.30) [Gebhardt, 95].

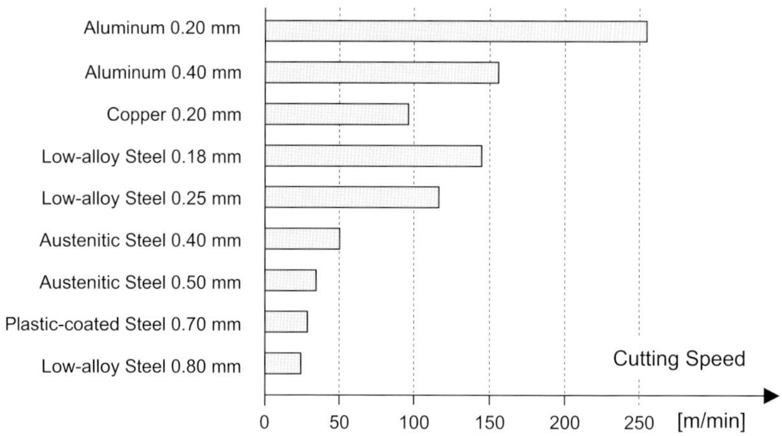

Figure 2.30 Reachable cutting speed of high-speed cutting with laser beam

Disadvantages

All LLM process models have the disadvantage of having very different mechanical-technological properties in the layer direction and at right angles to the layer direction.

It is also a disadvantage that parts not required in the models, such as internal cutouts, either have to be removed manually from each layer or remain in the model as solid bodies that subsequently need removal (depending on the geometry) by means of complicated cleaning measures.

Slightly inclined slopes are realizable only with difficulty, as the superfluous material not belonging to the model cannot easily be removed. Paper models have to be impregnated after being unpacked. All in all, final finishing costs are very considerable.

Most machines allow the use of only one material thickness, resulting in the production of models with only constant *z* distances. However, there is no compulsory reason why the material that is available in several thicknesses should not be changed during the running construction process.

The unused material is waste. For "shell character" models, the ratio of model to waste can be as high as 1:9. Under these circumstances, a favorable material price is put in perspective.

2.3.2.3 Melting and Solidification out of the Solid Phase: Fused Layer Modeling (FLM)

The processes in which the layer of the solid phase is melted are the industrially more important extrusion processes and ballistic processes.

2.3.2.3.1 Extrusion Processes and Ballistic Processes

Extrusion Processes

The extruded processes have become known as fused deposition modeling, but this is a trademark of the manufacturer Stratasys. The generic name is extrusion process or in English fused layer modeling (FLM).

The extrusion processes belong to the well-known (described next) thermal plastic-extrusion processes and include variants that operate with pastes or foams that are physically or chemically processed from solid basic materials.

Of particular importance are the thermal extrusion processes. The applied volume solidifies through thermal conduction into the component. The method is well suited for materials with low thermal conductivity such as waxes and plastics, including higher melting material like polysulfide. Theoretically, there is no limitation on the usable materials. Practically it requires high melting temperatures, both in terms of melting and of temperature gradients on the model. The required preheating and isolation of the work space also take correspondingly higher mechanical engineering effort. It is a fully 3D-capable process, depending only on the control of the nozzle because the material supply and material application are done by one nozzle. The relatively large cross section of material limits the achievable degree of detail of the models. With a suitable machine design, these disadvantages can be reduced as far as possible.

A satisfactory joining between the extruded hot material and the already finished part of the model takes place only when the material is "crimped." The circular cross section of the material is applied to an ellipse. Then the balance of volume and surface forces can be set up so that the smoothest possible connection is reachable. This requires that the nozzle axis must always be in the z plane and also includes a specific procedural angle with the x-y plane.

The application of an extruded viscous volume includes the problem that the extruded strand has a beginning and an end. When the contour is closed, a joint line results (Fig. 2.31). The lines are particularly noticeable when they are at each layer at the same location.

For the extrusion process a number of interesting plastics were developed, which are like most common construction materials, including ABS and polyphenylsulfone (PPSF, 2003), the first high-performance plastic for additive manufacturing. The materials often available in colors so that components can be produced in different colors.

The processes require supports that have to be constructed onto the component and removed after the building process. There are different user-friendly solutions for removing the supports, such as breaking off or detaching in warm water. In this case, the support material is usually a water-soluble wax consisting of polyethylene glycol.

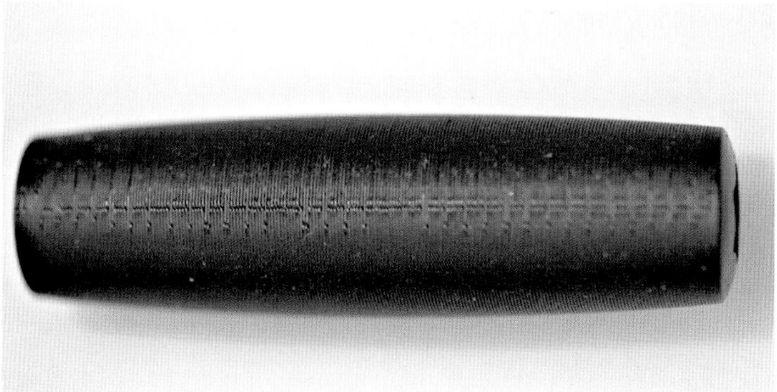

Figure 2.31 Joint line on extrusion processing. (Source: CP-GmbH/FH-Aachen)

The disadvantage of some machines is the complicated threading of the prefabricated wires. Cassettes for the wire feed and tablets make this procedure easier today.

For fabbers, there is a very large range of cost-effective materials from different manufacturers, which also include nylon and WPC (wood plastic components). Fabbers mostly use the extrusion process and millwork wires made of ABS or PLA, with diameters of usually 0.175 mm or 3 mm. Most of them work with a nozzle and without supports. This limits the manufacturable geometry, particularly in relation to approach angle and undercuts.

Ballistic Processes

A variant of melting from the solid phase followed by solidification by cooling of the component is when the molten material is applied discontinuously in the form of single droplets on the model. In contrast to extruded processes, ballistic methods operate without contact, but they are not free from force effects on the component, model, or part. The difference from the printing process is that the material is partially melted or melted fully. While it is usually cold during printing and initially remains in the liquid phase, it is then partially vaporized or react and leaves solids.

For a homogeneous component, the temperature of the material globules must be such that they can melt a sufficiently large location after landing and thus bond in this way with the component. The advantage compared to the extrusion process is that on small globules the approaches are not visible, and a much higher degree of detail can be realized.

The globules can applied at any angle on the model, and this allows one to theoretically apply materials with other colors or with other mechanical properties in an additional step. The surfaces can be locally smoothed or otherwise improved by applying other materials or other sphere diameters.

Basically, the material range covers all materials that behave thermoplastically and includes, at least theoretically, the metals.

2.3.2.3.2 Advantages and Disadvantages of Extrusion Processes

Advantages

The use of single-nozzle extrusion processes enables relatively large amounts to be applied within a relatively short time. The resulting structures are solid. Model materials are used that are very similar or even identical to those series of materials used in later production. The technical realization is comparatively simple. The material is completely used, with no waste. Solvents and similar agents are unnecessary. In contrast to most other rapid prototyping processes, the application of various materials within one construction process, or even one layer, is possible. The number of simultaneously applicable materials is limited only by the fact that the corresponding number of extrusion heads must be fitted geometrically into the machine and controlled by process techniques. The process just uses the material that is necessary for the component. The unprocessed material is within the component, so there is no damage and it remains fully usable.

The process is achievable with machines that can be set up and operated in a variety of environments.

Disadvantages

The main disadvantage of single-nozzle extrusion is that structures finer than the extrusion width cannot be produced. The same goes for details that, in the extreme case, may not be smaller than double the track width. This means that it is impossible to produce very fine grooves and especially fine ribs. The start of the extrusion always incurs a scab that, depending on the material, remains externally visible even after the contour is closed. Some materials tend to form filaments or condensate. The nozzles tend to clog up, requiring the installation of a suitable cleaning mechanism. The components are very rough and have pronounced anisotropies, especially for the low-price machines and fabbers, due to the large nozzle cross section.

2.3.2.4 Conglutination of Granules and Binders: 3D Printing

Powder-binder processes generate solid layers by gluing granules with a liquid binder. The so-called binder processes is also known as 3D printing processes because they closely follow the 2D printing process of inkjet printers. This process is patented by MIT. The starting point is again a bed of granules or powder. The powder particles are, however, glued together by an external binder that is injected by a nozzle directly on and in the powder layer following the contours. The result is that layer by layer a powder bed arises with the component located therein. The loose powder supports the component and is easily blown off after completion of the building process. For some time there has been a trend to call all such processes that are user-friendly 3D printing and that the machines for this process should also be called 3D printers. From a marketing perspective and according to the current standards (VDI, ASTM), this may be true, but as a systematic generic name this is wrong.

2.3.2.4.1 Advantages and Disadvantages of Powder-Binder Processes

Advantages

A major advantage of this process is that the range of materials is theoretically unlimited because it depends essentially on the properties of the binder. It can hence not only be used for processed powder that behaves thermoplastically, but also, for example, for foods such as nuts or temperature-sensitive pharmaceutically active substances. Via the print head, different binders can be integrated into a building process, and components with locally different mechanical and technological properties can be realized. The range of the binder is very large. In addition to water-based systems, which represent the majority of applications today, all kinds of resins and even living cells can be introduced via carrier liquids.

It is quite conceivable to modify the composition of the powder layer from layer to layer or within a layer, but currently this is not industrially implemented.

The 3D printing offers the possibility to "compose" locally the materials with requirements. In the future, manufacturing materials with locally varying properties will take the place of the isotropic materials used nowadays. Such materials are also called graded materials. The first technical implementations represented by the 3D printing process are to those that manufacture components with color and textured (pattern) surfaces.

The current disadvantage is that the additive manufacturing building process provides only a fixed-shape component that is not a final fixed-shape component. It must be subsequently infiltrated to be ready for use. The processes therefore have two stages.

In a downstream process, the binder is driven off from the metal component, presintered in the oven, and also infiltrated additionally. This not only limits the range of usable materials, but it also causes problems of warpage and scaling (shrinking). However, this can be controlled with enough experience. Another negative aspect is that the available powder-binder combinations that are used as construction materials for end products usually are not available and therefore there is no record of similar construction materials. Because the material properties are largely determined by the infiltration, information for designers about the materials' properties is practically unavailable. The machines are robust, which is not necessarily true for the print heads.

The failure rates for the individual nozzles are very low. Print heads with many nozzles (>100) still have a large theoretical default risk for the total print head. The print heads are relatively expensive.

2.3.3 Solidification from the Gas Phase

When generated from the gas phase, particles (solid material) are deposited on a substrate from gaseous materials or materials react similarly. Physical and chemical deposition differ. During physical deposition, the particles are resolved in the liquid phase of a two-phase flow. The particles are transported with the flow and deposited by the locally evaporating liquid. In chemical deposition, reactive gases are excited by external energy sources to local reactions, and consequently defined solids are deposited.

Both processes have the goal of providing the finest structures.

2.3.3.1 Aerosol Printing Process

The aerosol printing process uses aerosols that are atomized into fine droplets as carrier liquids. The carrier liquids are filled with solid particles with maximum dimensions of 200 nm. This is also known as the ink liquid, which is transported in a carrier gas and applied in a geometrically defined manner via a nozzle system on a substrate. It can be used on a wide variety of substrates. In contrast to other conventional coating processes, in microtechnology no masks and no etching processes are necessary. However, in most cases the carrier liquid has to be vaporized or the solid components should be melted or otherwise fixed.

This process allows components to be made of different materials such as metals, ceramics, or plastics in different areas and layers.

Aerosol printing processes are characterized by the deposition of the bound materials in the carrier fluid and are therefore rightly called printing processes. They differ from 3D printing by the fact that the inks have no binding function and that they print on substrates and not in a powder bed. They differ from ballistic processes by the fact that they work with a carrier liquid that evaporates during the process. Aerosol printing is so close to both processes that in the future hybrid forms can be expected.

2.3.3.1.1 Advantages and Disadvantages of the Aerosol Printing Process

Advantages

The main advantage of the aerosol process is the wide variety of materials that can be used, including even living cells. In addition, different materials can be joined together that would not be possible with traditional methods. The realization of microstructures, such as of silver on Kapton film[1], is possible.

[1] Kapton is a brand name of a fabricator

Disadvantages

The materials must be prepared to be used in a steady atomization. The buildup rates are low. Many materials need a thermal treatment after the application; metals for example are sintered by a laser. The processes are oriented more 2D than 3D, which means they are optimized for complex structures on 2D substrates and are less suitable for the construction of complex 3D structures.

2.3.3.2 Laser Chemical Vapor Deposition (LCVD)

In this chemical deposition process, a laser locally triggers a chemical reaction. The basics are given in the book by Bauerle [Bauerle, 86]. Lehmann and staff [Lehmann, 94] have applied the principle to the production of a microstructure. They used an aluminum- ($AlH_3N(CH_3)_3$) and oxygen-containing (N_2O) gas.

Two subcritical laser beams are crossed so that in the common exposed volume the required critical value for the reaction is exceeded and a chemical reaction is triggered. Solid aluminum oxide is formed as a result. With this so-called LCVD method, alumina rods of 5–20 µm in diameter and with a growth rate of up to 80 µm/s can be generated. An argon-ion laser is used. By modifying the processes with a spatially fixed reaction zone and moving the construction platform against the direction of growth with a growth rate, the finest structures can be built without a support structure.

Figure 2.32 shows such a microstructure of aluminum oxide. In a further process step, the structure can be metallized in whole or partially, and it is then electrically conductive.

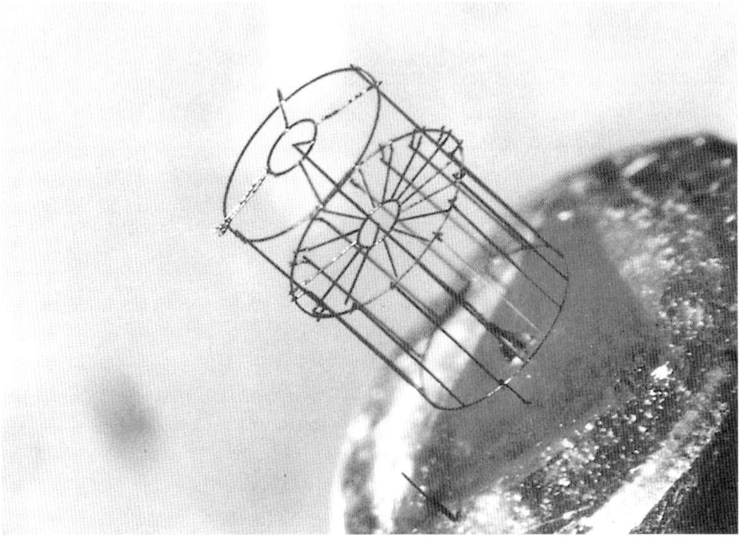

Figure 2.32 Microstructure produced by LCVD process. (Source: MPI-Göttingen)

The process allows the production of extremely filigreed structures, such as those required in microsystems engineering and biomedical engineering. It can also become the basis for an extreme improvement of the surface. It is theoretically conceivable to metallize an additively manufactured component and then use the LCVD technique to improve surfaces locally in the appropriate processing chamber. This train of thought is supported by the fact that the process is basically 3D-capable. The condition for building the structure is simply the accessibility for the laser beam.

The LCVD process, which is still in development for additive manufacturing applications, is characterized by two features:

- It is a fully 3D-capable process.
- It is not just a prototyping process, but an actual production process of microstructures for which no conventional alternatives exist.

2.3.4 Other Processes

In addition to the processes mentioned, variants exist that until now could not be developed into a machine that was usable by industry. Polymerization can be achieved by two laser beams intersecting at the point of polymerization (beam interference solidification, BIS) or—as in industrial polymerization processes—by heat (thermal polymerization, TP). Another idea worth considering is triggering polymerization by the projection of holographic images onto photosensitive materials, that is, no longer in layers but along predetermined 3D curves (holographic interference solidification, HIS).

Among the methods of gluing different foils together, in principle, it should be possible to combine the foil and the polymerization process in such a way that different foils are glued together by polymerization (solid foil polymerization, SFP). A detailed representation of such processes can be found in an older but still excellent book [Kruth, 91].

In most processes that use lasers, variants with electron radiation or other energy sources such as infrared emitters are also conceivable.

Two other interesting ideas are based on lesser known physical effects: sonoluminescence and electroviscosity.

2.3.4.1 Sonoluminescence

Sonoluminescence is the concentration of sound energy to such a degree that hollow spaces in liquids, especially air bubbles in water, radiate ultrashort light flashes in the UV range. It is at present not yet known for certain whether the light energy is strong enough to trigger a photopolymerization. Should this be the case, this process would bring the great advantage that the UV radiation necessary for polymerization would not need to penetrate a material layer of a certain thickness and thereby reducing its strength; it could instead be generated directly at the required point of polymerization. Spatial structures could be constructed by such sound fields and their interferences.

2.3.4.2 Electroviscosity

Electroviscosity is the ability of certain materials to alter their viscosity within broad limits while under the influence of powerful electromagnetic fields. This can go so far that liquids completely solidify. The process is not yet in use in commercial industry, primarily because the increase in viscosity cannot be produced technically and economically over a longer period of time. However, this method opens new perspectives for additive manufacturing processes. Electroviscosity enables liquids to be solidified along a defined 3D curve that can be calculated using potential equations. They form thereby a continuous contour that can be filled in with photosensitive or other rapidly solidifying materials. This can be done within a relatively short time. After solidification, the contour is removed by liquefaction and, on the basis of the new contour calculated from theoretical potential equations, another part of the model can be generated. Such a process opens the possibility of working entirely in 3D, but it depends, at least according to current knowledge, on theoretically calculated three-dimensional potential curves.

2.4 Elements for Generating the Physical Layer

2.4.1 Moving Elements

The moving element defines the geometry of the component. Therefore it has to generate the inner and outer contours of each layer and if necessary, the intervening space. It demands high movement speeds and also a high path accuracy and repeatability. A (galvo) scanner or (x-y) plotter and sometimes combinations of both are used.

2.4.1.1 Plotter

Plotters are essentially x-y-positioning systems with separate axes. Timing belts and racks are used as drivers. Plotters are available in different sizes having characteristic dimensions in a range of a few millimeters to several meters. The variety is too large to be treated here in detail. The building platforms of most additive manufacturing machines have characteristic dimensions of several hundreds of millimeters. Plotters used in the cutting operation nowadays have the following technical data: cutting speeds between 10 and 100 mm/s (tangential plotters in the diagonal area to 1500 mm/s) and more; repeat accuracies of ± 0.1 mm; and pressing forces from about 0 to 600 grams.

2.4.1.2 Scanner

Scanners or galvos are freely programmable deflection systems that move the laser beam on the building plane. The term galvo merely refers to a motor-driven mirror unit rotating around its longitudinal axis according to the galvanometer principle, and is only one element of a scanner unit. Two galvos are aligned orthogonal to one another (one each for the *x* and *y* directions) to reach every point on the build platform. In this arrangement, the focus describes in fact a spherical shell. The projection is therefore in the center of the building plane, and with increasing distance it becomes increasingly blurred, especially so in the corners of the building space.

The answer to the blurriness is a so-called F-theta or flat-field lens that provides the focus outside the optical axis. The focus therefore remains for any deflection in one building plane.

Figure 2.33 shows the scanning optics, consisting of the beam expander, two motorized galvo mirrors, and the F-theta lens. The arrangement of the galvo in front of the flat-field lens is also called preobjective scanning.

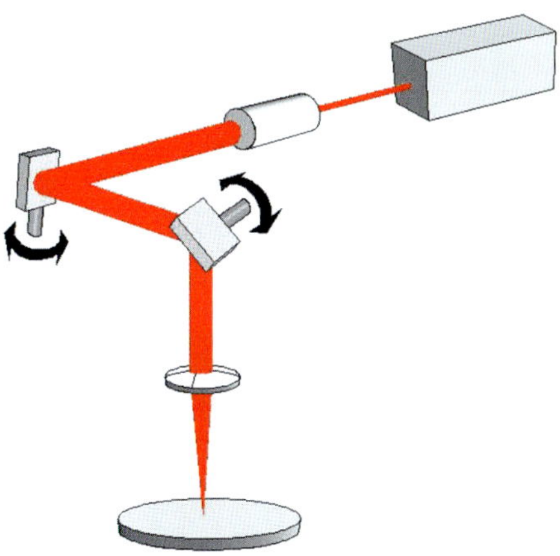

Figure 2.33
Scanning optics for imaging of the layer information on the building plane, consisting of the beam, two galvo mirrors, and an F-theta lens

The mirrors shall be lightweight to allow fast scanning speeds. To achieve fast scanning speeds, their mass inertia can be reduced by cutting off the corners, which gives them their characteristic shape. Mirrors are advantageously made of single-crystal silicon as a substrate material and a coating with the lowest possible absorption losses. The coating must be matched to the laser wavelengths used.

The trend is to use laser scanners with digital servo electronics. This is possible due to an improved encoder with a significantly improved dynamic response and increased accuracy. The electronics also allows communication between the system controller and the scanner and thereby enables online diagnostic procedures.

2.4.1.3 Simultaneous Robots (Delta Robots)

Delta robots are a variant of the parallel robots. There are three universal joints fixed to the frame-mounted rotary axes. The result is a spatial parallelogram guiding the moving element and the contouring element (the effector). This new design has only been used in the Delta Rostock fabber (Fig. 2.34). There the contouring element is an extrusion nozzle. Delta robots are relatively simple in construction, light, and fast. Their accuracy is sufficient for fabber applications.

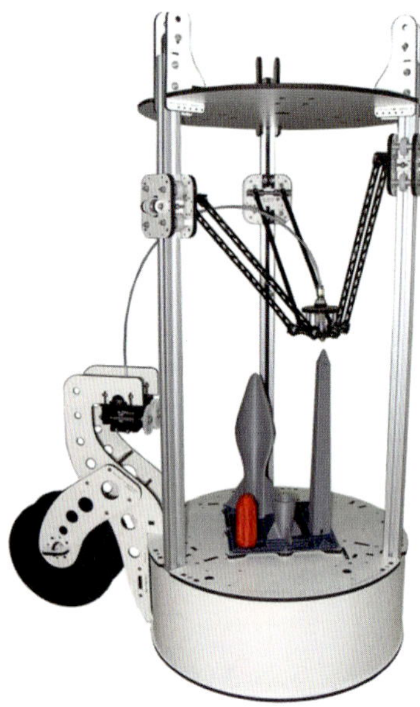

Figure 2.34 Delta Rostock fabber with parallel kinematics

2.4.2 Generating and Contouring Elements

Generating elements generate the layer. Depending on the process, the elements also do the contouring.

The elements differ in that there may or may not be a mechanical interaction with the layer. Laser and nozzles are among the noncontact processes (nonimpact). The cutting blades, milling cutter, and to a lesser extent, the extruder, place mechanical forces on the component. These are called impact processes.

2.4.2.1 Laser

The laser is of special importance in prototyping because of its ability to focus a high energy density onto an extremely small operating diameter. Especially in connection with the scanners for the polymerization, melting, or vaporizing (cutting) of materials, it can be necessary to introduce energy with high positioning and tracking accuracy and repeatability that can produce pinpoint accuracy.

Depending on the absorption behavior of the materials, lasers with different wavelengths are used.

For the cutting of paper or plastics or the sintering or melting of plastics, a CO_2 laser (wavelength 10,600 nm) is preferred.

Polymerization requires a laser at wavelengths below 500 nm. For the processing of metals, even shorter wavelengths are advantageous.

Ar^{++} ion and HeCd lasers were the first lasers used in machines because of their favorable wavelengths despite their poor efficiency, and were used until the late 1990s. Today only solid-state lasers are installed in machines. Thus, the efficiency and the service life are significantly improved and the age-related gradual decline in performance largely eliminated.

The latest designs, such as diode-pumped solid-state lasers or disk or fiber lasers, have significantly contributed to improving the beam quality and dynamics. Today with fiber lasers, beam diameters in the range of 20 to 50 microns are realized. In addition, the beam cross section can be varied quickly with adjustable optical elements and thus reduced for boundaries and enlarged for filling areas. An elaboration of the laser principle, which due to the few types of lasers that were preferred for use in additive manufacturing production in 2000, was located in the appendix to the German second edition of this book. Given the numerous types that are used today, this has been omitted. A comprehensive current representation with a focus on production with lasers can be found for example in Poprawe [Poprawe, 04].

We differentiate between three processes when projecting the geometrical layer information onto the layer: vector, raster, and mask processes. The basic principles of these methods are also found in other contouring processes; for example, the vector process is used in cutting plotters when knives are in action. Masks can also be used in the process of using lamps or projectors for polymerization. The vector process uses either a (mirror) scanner or *x-y* plotters. In the vector process, the single contour elements are generated continuously from basic geometric elements such as straight lines, circular arcs, and so forth on the basis of the standardized plotter software. It can be assumed, therefore, that a circular contour parallel to the layer will appear as a polygonal curve in the STL formulation and as a continuous circle in the SLC formulation. The laser beam can be pulsed, as in most polymerization and sintering processes (the processes are discussed in Section 2.3), or work continuously (continuous wave, CW), as in layer laminate processes. The vector process is the most accurate process, but also the slowest. The duration of the generation of a layer depends in particular on the complexity of the layer.

In the raster process, the contour is still generated using the same geometric information, but—as with an older television picture—line by line. As a result, a stair-stepping effect appears from line to line similar to that in the *z* direction. The height of the steps is determined by the width of the effective track, and in the case of a laser, by the width of the laser beam. As the effective width of the laser beam may be up to 0.3 mm. Such raster processes will, for nonrectangular geometries, have a much higher tendency to generate a stair-stepping effect in the layer surface than that of the design-dependent layers in the *z* surface. The vector process enables the contour to be moved by half a beam diameter (beam width compensation), resulting in very accurate contourings; the raster process cannot do this. The duration of the generation of a layer in the raster process does not depend on the complexity of the layer. The process is faster than the vector process, but slower than the mask process.

The third possibility is the mask process. Here a mask is produced to be geometrically similar but on a smaller scale and is then screened by an energy source. As with a diapositive, an authentic-scale contour is reproduced on the layer surface. The accuracy is limited by the precision with which the mask can be made. Mask processes operate with transparent, electrostatically or sequentially coated masks for LCD (liquid crystal display) technology monitors or directly with DLP (digital light processing) projectors, but also with mechanical masks. The exposure of an entire layer occurs all at once. The exposure time is therefore not dependent on the complexity of the layer, and the process operates quickly.

Figure 2.35 shows the three principal processes for the reproduction of the geometric layer information onto the layer.

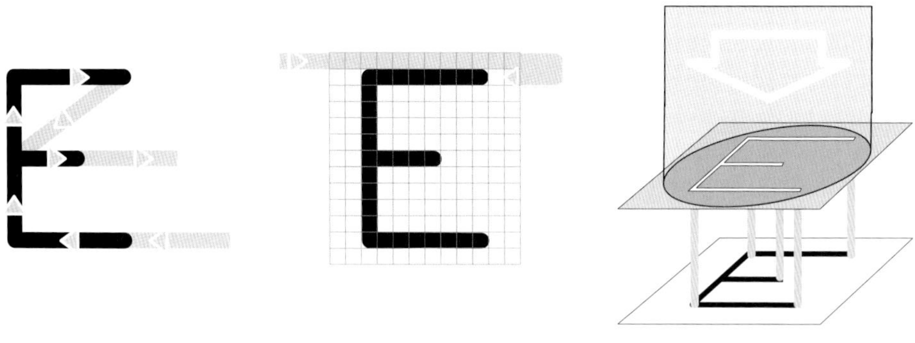

(a) (b) (c)

Figure 2.35 (a) Vector, (b) raster, and (c) mask processes to image the geometric information onto the layer

2.4.2.2 Nozzles

Bubble-Jet or Thermal Nozzles

A bubble-jet print head is composed of an ink reservoir that is connected through a capillary with the actual nozzle (Fig. 2.36). The ink channel is electrically heated shortly before the nozzle. As a result, the ink vaporizes locally. The resulting vapor bubble abruptly displaces a defined volume of ink that is fed due to the pressure conditions through the nozzle onto the print destination. In the case of additive manufacturing it is sprayed on the powder bed.

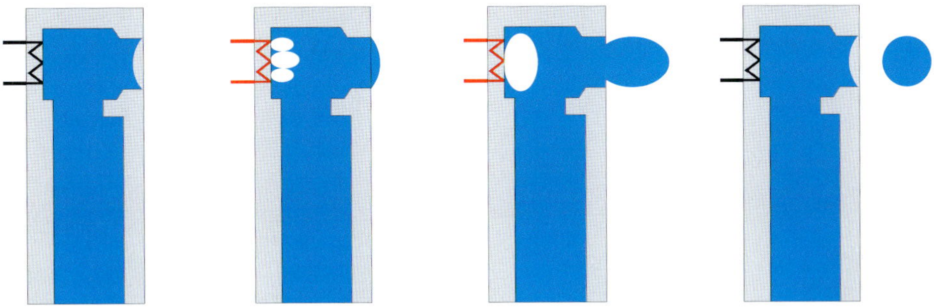

Figure 2.36 Principle of bubble-jet nozzles. (Source: Tecchannel (www.tecchannel.de))

In a print head, up to several hundred nozzles can be accommodated per print head, with the nozzle having diameters of about 20 to 50 microns. The drop volume is approximately 4 to 30 picoliters, and the temperature at the heating element is 300 °C. The thermal system is comparatively slow. A frequency of about 4 kHz is standard, with about 10 kHz as the upper limit. The Photo-Ret-III process of Hewlett-Packard reaches a frequency of approximately 18 kHz. From 408 nozzles, approximately 7.3 million droplets per second are applied with a volume of 5 picoliters. However, the process prints 29 points, one above the other, and has been optimized for 2D printing on paper or foil.

It is often considered a disadvantage of thermal print heads that the drop size cannot be varied. That is not true. By sequential heating, preferably by two downstream heaters connected in series, two different-sized vaporization zones can be created with appropriate tuning, which joins into one and thus carries away a significant amount of liquid. However, it is very difficult to form very different large drops. In addition, the drop rate is quite low. Thermal print heads for 2D printing are usually replaced together with the ink reservoir so that the wear of the nozzle does not affect the print result in practice. Each color (cyan, magenta, yellow, and black) may have its own reservoir and head, or an integrated head may be used for all colors, depending on the manufacturer and the use of the device. There are also systems with separate nozzles and reservoirs.

In the integration of printing systems in additive manufacturing machines, larger amounts of fluid have to be used, and the controllers are modified accordingly. Therefore, the standard ink reservoirs are replaced with binder tanks that are connected via tubing systems. The commercial print heads of 2D printers are usually used as nozzles.

Piezo Nozzle

A pure (electro)mechanical process can be more precise and especially faster than a thermal controlled one. The deformation occurs in approximately 5 microseconds, the reverse deformation as well, so frequencies up to 30 kHz can be achieved. The droplet size can be adjusted by the applied voltage and quickly varied. The ink channels and hence the drop volume are smaller (2 pl (pL)) than in thermal systems, so a point diameter of 0.035 microns can be achieved on the paper, which corresponds to a resolution of 800 dpi. Currently, the mechanical positioning systems set the limits more than the print heads.

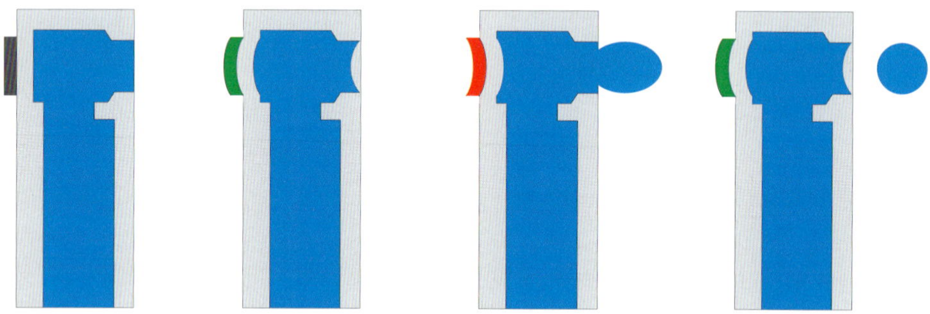

Figure 2.37 Principle of a piezo nozzle. (Source: Tecchannel (www.tecchannel.de))

With a piezo nozzle, highly viscous liquids and binders that are thermally sensitive can also be processed. Moreover, an independent preheating of the fluid is possible. Thus is created the optimal conditions for additive manufacturing machines that can print photopolymers or other high-viscosity fluids. Coinciding with the "firing" of a droplet, a negative pressure can be generated by applying a negative voltage to the piezo quartz in the ink channel. The result is that the ink is pulled back slightly at the nozzle. This "meniscus" effect (drastic pull ejection meniscus control) also causes a bias of the droplet surface, which favors the "firing" of the next droplet and at the same time suppresses the formation of secondary droplets.

Piezo heads are especially sensitive to air or gas inclusions, because due to its compressibility, the impulse needed for ejection can be resorbed (or absorbed). The problem corresponds to the case of the venting of water pipes. Piezo elements and their control usually require more space than thermal jets. Therefore a typical piezo print head has fewer nozzles (50 to 100, maximum 200) than does a thermal print head.

Piezo heads are robust but also more expensive than thermal print heads. They are not usually exchanged in printers, but are static and therefore also optimally adjusted. When refilling the ink there is no risk of maladjustment. Today's systems usually still have replaceable print heads because of the higher wear in additive manufacturing machines. Particularly maintenance-friendly designs offer replaceable single nozzles or nozzle groups.

Accuracy

The minimum droplet size can define the accuracy of a printing process in combination with the positioning in the *y* direction (row direction) and in the *x* direction (line spacing). High-quality printing requires a correspondingly high repeatability. The system also includes very high mechanical requirements.

Powder-binder systems are also limited in their accuracy by the size of the powder particles. Common are particle sizes of 50 microns (up to 100 µm). The lower limit is approximately 20 microns. When colored components are produced by such systems and their texture is defined by the binder, the surface shows after a possible grinding operation for finishing large areas of colored powder. At preferably white powder the component accordingly appears in pastel colors.

Materials

The different inks should not be discussed at this point because this happens later in the description of the individual machines. However, the development of a 2D printing technique that has primarily UV-stable colors as a goal should be pointed out. In this context, RCP colors (radiation curable pigmented inks) are developed whose coloring constituents polymerize by UV radiation to stable pigments. This allows one to print almost all weather-resistant materials. In view of today's additive manufacturing processes that print photopolymers, there arise interesting perspectives for the production of colored components.

2.4.2.3 Extruder

Extruders used in additive manufacturing operate according to the system of constant volume. A geometrically defined material, mostly as a wire but also in the form of blocks, is placed in an electrically heated chamber. There it is converted by the heat into the viscous state. The chamber is also called a "liquifier," although liquefaction must be avoided in order to achieve a defined material application. A supplied volume of the same amount is extruded through a nozzle and coated as a layer onto the component. Nowadays, machines have only one nozzle for the building material. Because the material is available colored, differently colored components can be manufactured. However, the color cannot be changed within one component, so textures cannot be achieved. Because all commercialized extrusion processes work with supports, a second nozzle for the support material is usually installed in the machine. The use of multiple, parallel operating extrusion nozzles is generally possible but currently only applied by fabbers. There are also constructions that apply differently colored material. For industrial machines this has not been realized yet. Unlike printing systems, extruders operate quasi-continuously; in other words, they apply the material in one go, but interrupt the extruding during the positioning phase. The distance between the outlet of the nozzle and the surface of the component must always be lower than the extruded material cross section so that the viscous material can be "crimped" on the partially completed structure. The cross section will become oval. Extruder systems work with little force, but not entirely without contact. Extrusion systems are very sensitive to air and gas pockets. Another problem is the water consumption of the building material, which is expressed as vapor bubbles with splashes.

2.4.2.4 Cutting Blade

Contouring with blades draws on the technique of 2D cutting plotters for paper and film. For this, *x-y* plotters are used. Cutting speeds of up to 300 mm/s can be reached. Processes with cutting blades are not nonimpact processes. The blades reach a pressing force of about 30 to 300 g.

Films made of paper, plastic, and ceramic are preferably contoured with blades. A cutting resolution of 0.025 mm per step and an accuracy of 0.1 mm (Roland SP 300) can be achieved.

All systems for contouring have in common that not only the characteristics of the contouring element, for example, the nozzle diameter, the laser beam diameter, or the width of the blade, but equally also the precision of the *x-y* (*z*) handling system used determine the quality of the component.

2.4.2.5 Milling Cutter

If cutters are used in additive manufacturing, the optionally modified milling machines and processing are not any different from conventional milling. Therefore, and in view of the large number of possible machine concepts, machining with milling cutters will not be discussed further. A very extensive and also hands-on book was written by Degner [Degner, 02]. Schulz [Schulz, 96] describes the aspects of high-speed machining in his book. Wirth [Wirth, 02] describes the model milling process and practical implementation from the CAD program.

2.4.3 Layer-Generating Element

An additive manufacturing machine is equipped with moving elements and generating elements that are geared to each other. Together they form the layer-generating element and thus constitute the additive manufacturing heart of the machine. Just a few combinations have been proven in practice. Table 2.1 shows the most common combinations. Any other special solutions are discussed with the corresponding machine in Chapter 3.

Table 2.1 Most Common Combinations of Layer-Generating Elements

Physical principle		Additive manufacturing family	Moving element	Generating/contouring element
Photopolymerization	2.3.1.1	Stereolithography	Scanner	Laser
		Polymer printing	Plotter	Printing head
Melting and solidification of powders and pellets	2.3.2.1	Sintering Melting	Scanner	Laser
Cutting from foils/films and joining	2.3.2.2	LLM	Plotter	Laser, knife, milling tool
Melting and solidification out of the solid phase	2.3.2.3	FLM	Plotter	Extruder, nozzle
Conglutination of granules with binders	2.3.2.4	3D Printing	Plotter	Printing head
Generation from the gas phase	2.3.3.1	Aerosol printing	Plotter	Nozzle

The layer-generating element is actually the tool of the additive manufacturing machine. It firmly belongs to the machine and would not change depending on the component, which is why it is said that additive manufacturing works without tools. Strictly speaking, they do not need a tool that is individually adapted to the production task. It is not necessary to change the tool when a skull model and a radio panel are produced one after the other, just as with a 2D printer: if a notice is to be printed instead of a commendation, it is not necessary to change the machine settings.

Because the layer-generating element has the function of the tool, it is important to determine and control the parameters that define the operational capability of the tool and also develop strategies to operate in the allowed parameter fields.

For all elements, the designers make use of additive manufacturing systems that have preferably been proven on large-series technology. Around 26 million installed inkjet and laser printers form an economically solid basis for the development of applications adapted accordingly. Only through the use of components, whose development had been financed through sales from other markets, was the development of additive manufacturing equipment possible. It is also the only way to finance it in the future. In about 15 years (1988–2003), additive manufacturing machines increased to 10,000 systems worldwide.

2.5 Classification of Additive Manufacturing Processes

The professional literature contains a number of different representations concerning the systematics of additive manufacturing processes. Many are oriented only to the currently available methods or to isolated properties of such processes. They agree, for the moment, but that quickly reaches the limits when new processes need to be integrated. All additive manufacturing processes (including all additive manufacturing processes) are in the end manufacturing processes. Therefore their nomenclature should be based on the standardization of the manufacturing processes (DIN 8580).

Additive manufacturing processes are not always unique to a main group due to their variety. Basically they are in the main group 1, creating cohesion or archetypes (DIN 8581).

As constituent manufacturing processes, they would fall under the aspect "cohesion increasing" assigned to the main group 5 "Coating," and because of many variants of the process, assigned to other groups and subgroups. After weighing the pros and cons, a classification is possible based on the standardization of archetypes (DIN 8581). The additive manufacturing methods are classified in the first outline level according to the state of aggregation of their original material, and in the following level, by the appearance in the sense of a semifinished product. In the third level, the mechanism of layer formation is shown, and the fourth level contains the generic description of the process.

In Fig. 2.38, the current processes can be clearly classified to date while maintaining the proven structure.

The industrial processes and their manufacturers or brand names allocated to the generic designations are discussed in Chapter 3. There the classification found in Fig. 2.38 continues, including the industrial processes and their product names.

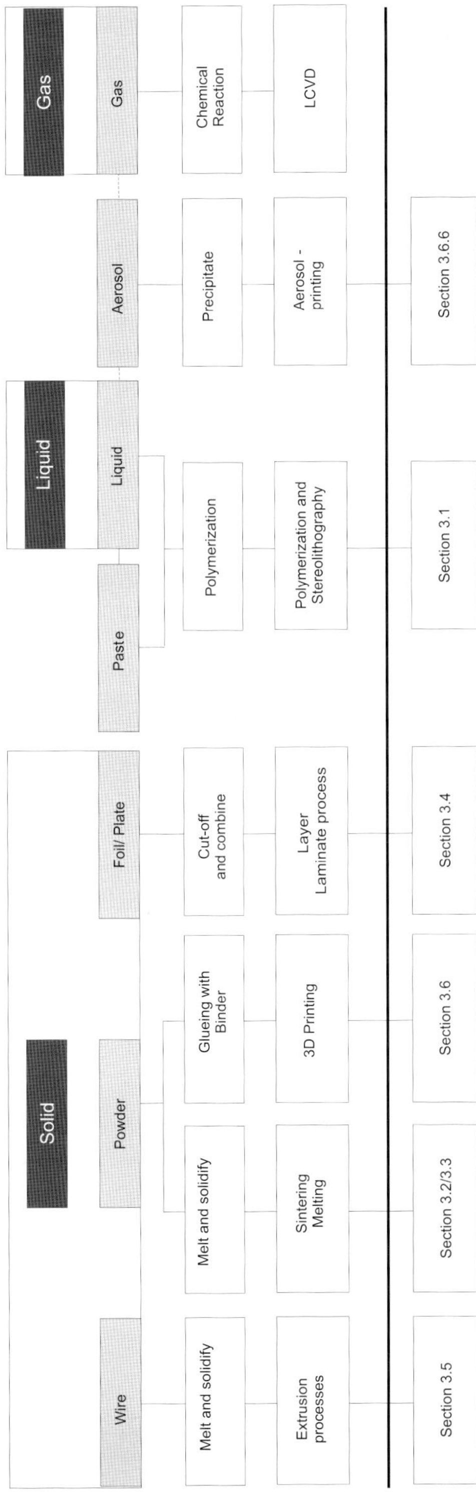

Figure 2.38 Classification of additive manufacturing processes depending on aggregate state of the basic material referring to DIN 8580

The symbol illustrated by Fig. 2.39 shows a graphical representation of the additive manufacturing process.

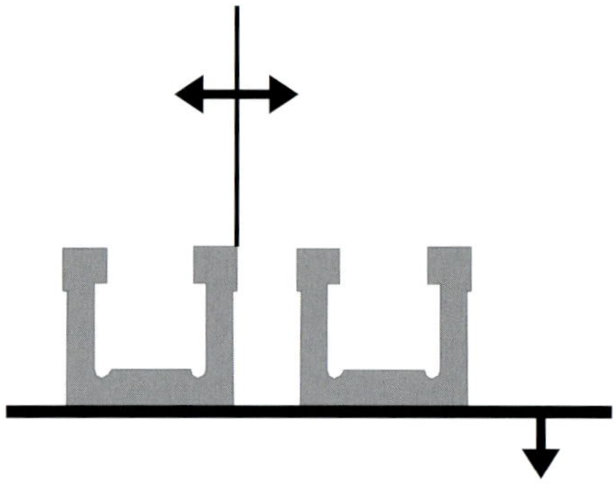

Figure 2.39
Icon for the graphical representation of the additive manufacturing process

2.6 Summary Evaluation of the Theoretical Potentials of Rapid Prototyping Processes

This evaluation of basic physical processes and their deducible advantages and disadvantages provide an insight into foreseeable development trends. The question of theoretical development limitations for individual processes can at least be answered in the sense of a general trend. It is necessary to consider the limiting factors concerning the main points:

- Materials
- Model properties
- Details
- Accuracies
- Surface quality
- Development potential

The most important AM processes, photopolymerization (stereolithography), melting of powders (laser sintering), cutting of foils (layer process), melting and solidification of wire materials (extrusion processes), ballistic processes, and injection of binders into powders (3D printer), are included.

Every manufacturing process has its own advantages and disadvantages. It would therefore be advisable to use the advantages of an additive manufacturing process when a conventional process does not achieve the best results. Relating to surfaces and precision, this policy should be followed: If it is possible to mill, it should be milled.

2.6.1 Materials

A prerequisite for stereolithography is the presence of a photopolymer. Therefore, usable materials will always be limited to photopolymers, that is, plastic materials. This does not exclude all kinds of filling materials being used to modify these plastics. Filled resins improve the material behavior of the plastic components. This leads to higher strengths and higher temperature resistance. With an increasing proportion of filling material, the monomer acts less and less like the function of a material and becomes like the carrier medium, having the properties of a binder. If it is expelled in the downstream furnace processes, and the ceramic or metallic fillers form a monolithic component. Therefore, in the future, different materials are expected for photopolymerization that are "composed" by fillers.

The range of materials that is usable for sintering processes is theoretically limited only because these materials need thermoplastic properties in the widest sense. This means they must be meltable, and after solidification they must be able to regain, at least mostly, their former volumes and material properties. Of course, high-melting materials such as metals require a much higher complexity for their machine-technical realization. Coated particles, as they are used, for example, during the sintering of foundry sands, showing the path to more complex material developments. The use of granules compiled from various component powder grains can begin, in addition to a thermal also a chemical process, so that new material properties can be created.

The range of materials that can be used for layer processes is basically unlimited because all currently known materials can be cut by a laser and all known technical materials are machinable. The problems are usually the joint between the layers and the anisotropic properties of the component that often result. Better results are shown by diffusion welding processes, although these are technically complex processes, the ultrasonic welding method and the perspective powder-based, diffusion-controlled surface welding process. The layer laminate process for spatially curved components shows the way for the production of composite parts.

Regarding the upper process temperature, the range of materials that can be used for extrusion and ballistic processes is theoretically limited only because the materials cannot decompose and the nozzle material itself should not melt in the process. The materials shall have, in the pasty state, a sufficiently high intrinsic strength and result in a homogeneous component without "weld line effects." A too-high moisture absorption has the risk of local evaporations as a result. Thread formation must be avoided. Multiple simultaneous or parallel operating extrusion heads are already implemented (for fabbers) and allow the building of multicolor and multimaterial components. Here, too, the limitation is not so much theoretical but lies more in the technical realization.

The extrusion of pastes or foams enables new applications. The deposition of rapid-hardening (polymer) cement as formwork for the walls of buildings is promoted as an example for additive manufacturing building construction (Contour Crafting, B. Khoshnevis, www.contourcrafting.org).

The 3D printer can use the widest range of material because a large number of binder-powder mixtures is conceivable. The processes are limited in their accuracy mainly by the particle size. In addition to the current, dominant single- or multi-component powders with homogeneous particles, there is an enormous potential for the application of filled materials in the context of the previously discussed granules. Because 3D printing processes operate at room temperatures, they are also available for temperature-sensitive components such as active cell cultures, pharmaceuticals, food, and a range of relevant sensors.

Deposition processes create new opportunities, especially for microstructures. In aerosols, various materials can be transported and deposited in certain areas and in layers. The safe storage of the required materials and the production of the respective necessary aerosols still requires significant development effort. The same applies to the postprocessing of the functional layers with lasers, for example.

All processes operating with a liquid plastic bath or a powder bed need to contain more material than is necessary for the model, depending on the size of the building chamber. This has economic consequences. The filling material for a large stereolithography installation can easily cost about 10,000 € (just under $11,000), but there are also logistical consequences regarding long-term stability. Processes that operate with small amounts of material, preferably in a cartridge, avoid these disadvantages, but the materials are usually more expensive.

2.6.2 Model Properties

Basically there are two major groups of materials for the additive manufacturing process, leading to different component properties: model materials and engineering or construction materials. Model materials are closely linked to a machine or a process and allow combinations with the production of components with defined geometries. Usually model materials do not occur in end products. They are preferably used for the production of prototypes, so-called design models, for which the subsequent simulation of the mechanical properties is not important. Examples are 3D printing materials consisting of starch-powder and water-based binders and subsequently infiltrated epoxy resin.

Engineering materials are equivalent to most materials of the same name used in nonadditive manufacturing processes for producing end products. Examples are the polyamide resins or polystyrene for the (laser) sintering of plastics or CoCr powder for manufacturing metal components. Due to the layer construction process and associated anisotropies and other effects, additive manufactured components must be expected to have different mechanical properties from components of nonadditive manufacturing processes, even when the materials are nominally equal.

The mechanical properties of the component differ from the foundation of the material designation's expected properties, as much they refer with model materials instead of referring to engineering materials.

2.6.3 Details

When assessing the ability to depict details, a differentiation is made between the capability of describing a thin wall or hollow spaces. With regard to wall thickness, the polymerization process is limited by the realizable laser beam diameter. One must bear in mind, however, that polymerization processes are considered to be among the most accurate of processes and that laser beam diameters below 10 µm will be achieved in the near future (fiber laser).

Polymerization processes need support, which must be removed after the building, therefore limit the number of internal hollow spaces. Where the supports are attached to the part surface, often the associated trademark remains. That's not the case if the supports can be washed out or if no supports are required due to geometry. Internal hollow spaces are limited in the extrusion process due to the necessary supports being limited with regard to cleaning.

Regarding the representation of fine details, the extrusion process shows less potential. The nozzle diameter and the resulting diameter of the extruded material define the minimum wall thickness.

Due to the nanoscale dimensions of the transported materials, aerosol printing processes have special potential for the imaging of fine structures, but they are not now suitable for real 3D structures.

In sintering processes, the minimum wall thickness and the internal hollow spaces are limited by the diameter of the powder particles and the laser beam diameter. Getting the powder completely out of the hollow spaces depends on the geometry of the model. In the metal sintering process, the supports and their removability limit the process.

In layer laminate processes, in which the laser beam diameter is also important, the problem arises that very thin walls and internal hollow spaces are difficult to remove from the mold. In some process variants (especially in the metal layer process) the superfluous materials are removed after each layer. In some variants of the layer milling, this limitation is no longer required because here the chips of each individual layer are discharged directly in the course of contouring.

2.6.4 Accuracy

In the polymerization and the layer laminate processes, accuracy in the construction plane (*x-y* plane) is limited by the scanner or plotter unit. In the sintering process, on the other hand, the diameter of the powder particles, and in the extrusion and ballistic process, the nozzle diameters and the resulting diameter of the applied material, are the limiting factors. Limiting factors for processes that work with nozzles are the nozzle itself or the plotter unit.

In polymerization processes with free surfaces, there is the additional problem that for very thin layers, the wetting can no longer be guaranteed.

In layer processes, the ability to contour in the *z* direction is limited by the fact that the ratio between the cut glue layer and the effective model-building material layer becomes too large. In addition there is a minimum foil thickness below which the transport of the foils can no longer be guaranteed. Milling layer processes can also have constant accuracies on very tall models because the milling machines permit this. Also in the layer process, the tolerances in the thickness of the used plates and films are detected as are the costs of the tolerable materials.

In the extrusion process, the limiting factors are the small nozzle diameter and wire diameter, the amount of heat, which is not enough for melting, and the poorly controlled process. In contrast to this, the ballistic process can work with microdrops. In the aerosol printing process, the drop sizes of the materials used limit the accuracies.

2.6.5 Surface Quality

The qualities of surfaces are basically limited by the same factors as those affecting continual *z* contouring. How they differ are, on one hand, the surfaces that are generated by layers, meaning, is the stepping in the *z* direction, and on the other hand, surfaces that are dependent on the material. Surfaces of layers that are cut out match the delivered foils. The liquid materials for polymerization show good surfaces, whereas extrusion processes exhibit the typical line structure. Surface quality also depends on the angle between the building plane and the *x-y* plane. Sintering and 3D printing processes show surface qualities that are proportional to the particle diameter.

2.6.6 Development Potential

In polymerization processes, the surface qualities and the degree of detail reproduction present the greatest potential for development, considering that suitable shaping processes can transpose the advantages of its extremely high accuracy to metal prototypes and series components.

In sintering processes, the laser performance is limited by the realizable diameter of the powder particles, and during sintering, by the ratio of powder particle diameter to laser beam diameter. (The powder particle diameter chosen cannot be smaller than a certain size.) The potential is seen especially in producing metal parts and also plastic and ceramic parts directly. They provide the basis for direct digital production.

In layer and extrusion processes, in particular the anisotropic model properties connected with these processes, are considered a limit to the development. In extrusion processes, the relatively large diameters of their extrusion nozzles also limit the realizable degree of detail reproduction.

2.6.7 Continuous 3D Model Generation

All currently available industrial systems are 2D processes with regard not only to their functionality, but also to their basic physical principle. They take the shaping in one plane element and add the third dimension by stacking up neighboring planes. To differentiate between this process and a real 2D process in one plane, they are referred to as 2½D processes.

Those processes that can also produce continuous contours in the *z* direction, by at least following basic physical principles, and superimpose them onto the *x-y* direction, that is, they are able to reach any spatial point, are called (real) 3D processes. There are two possibilities:

1. Polymerization: The energy necessary for solidification is generated by superimposing two or more partial beams, and by suitable control and interference, any spatial point can be reached (beam interference, two-focus process).
2. Ballistic process: The material is transported by a movable heated nozzle to any spatial point (five-axis CNC), where it is added to the model.

This potential of 3D capability is currently not yet technically applied (see FLM). It is not merely a question of expenditures; the necessary software and suitable control algorithms are not yet available.

The qualitative properties described here are quantified in the appendix and are attributed to materials and machines in the form of technical data.

3 Machines for Rapid Prototyping, Direct Tooling, and Direct Manufacturing

Since 1987, more than 100 machines for direct digital production have been developed and introduced to the market that follow the principle of layer manufacturing using the basic physical principles introduced in Chapter 2.

Out of these machines, the most often used and therefore the most important of the have been selected and will be introduced and discussed according to the main families of additive manufacturing:

1. Polymerization
   - (Laser) Stereolithography (SL)
   - Polymer Printing/Jetting
2. Sintering/Melting
   - (Selective) Laser Sintering ((S)LS)
   - Selective Laser Melting (SLM)
   - Selective Mask Sintering (SMS)
   - Electron Beam Melting (EBM)
3. Layer Laminate Manufacturing (LLM)
   - Paper Lamination
   - Plastic Lamination
4. Extrusion Fused Layer Modeling (FLM)[1]
   - Fused Deposition Modeling (FDM)
5. Powder Binder 3D Printing
   - Three-Dimensional Printing (3DP)
6. Aerosol Printing

This list contains the generic names according to Fig. 2.38 and commonly-used abbreviations. Based on these families, manufacturers have developed different processes and machines, and they have given them their own company or brand names, which will be introduced as well in this chapter.

[1] According to the source, the "M" stands either for "manufacturing" or for "modeling."

The names are often sketchy or even heterogeneous. In many cases the generic name has been derived from the brand names. As an example, even many experts do not know that LOM (Laminated Object Manufacturing) is the registered trademark of the company Helisys[2] and consequently a brand name and legally cannot be used as a generic name.

As long as a process is the only one on the market, nobody pays much attention, but when competitors show up, manufacturers see to it that their trademark is protected. Therefore, brand names and generic names should be kept separate.

This situation becomes more and more complicated because, due to changing strategies or company acquisitions, well-known names are substituted for by new ones.

The industrial fabricated machines for additive manufacturing are presented and described one by one in this chapter, selected according to the active principle (as specified in Fig. 2.38), and then among these, according to the manufacturers.

Individual machines or families of machines made by the same manufacturer will be described. Only the machines that can be used as industrial processes for which upgrades and updates are available, for which professional services are guaranteed, and have arrival times that are comparable to machine tools are mentioned here.

Due to extremely rapid development, machines are also mentioned that are on the edge of commercialization. In addition, at the end of each chapter, processes and machines are mentioned that are expected to be commercialized in the not-too-distant future, that seem to be extremely interesting, or that show thrilling applications or just great potential.

Systems that are gone from the market will not be described in the book but they can be found in one of the preceding editions.

This chapter is structured as follows:

Machine or Brand Name, Manufacturer, Important Reseller

Contains the official name(s), which can be different for the machine and the process and can be subject to change.

Short Description

Provides a quick overview. Points out the special capabilities and fields of application of the machine.

[2] The succeeding company, Cubic Technologies, Carson, CA, acts solely as a LOM service bureau.

Application Level, Development Status

The application level according to Fig. 1.6 and Table 1.1, providing a comparably rough classification. A more detailed assignment would be confusing.

The development status indicates how long the machine (or preceding ones) has been on the market.

History

Throws a spotlight on the history of the company and its products.

Strategy, Partners

Discusses the background of the development under the corporate and technological perspectives and illuminates strategies.

Data Format/Software

Describes which software comes with the machine, its capabilities, and what kind of interfaces are available.

Principle of Layer Generation

Explains the building principle of the machine.

Type of Construction

Describes the specials aspect of the machine's engineering design.

Material/Construction Time/Accuracy

Construction or build time and accuracy are described in the literature in many different ways. The build time usually is specified as the growth per unit time (mm/h) of the part in the z direction. Alternatively, the volume built per unit time, the time to make one layer out of the overall time to make the whole part is used.

The accuracy is seldom specified as an IT value according to ISO 286. Mostly, absolute values in terms of the total dimensions of the part are used, and more and more often, dpi values are provided. Often there is no differentiation between resolution and accuracy.

Surface properties are often described only qualitatively. The use of R_A or R_z values is very rare.

Postprocessing

Postprocessing describes all of the procedures that are necessary to obtain a usable part after the printing process. Postprocessing is a part of the entire AM process and includes manual and automated work as well as investment.

Finishing

Finishing is the name for all nonadditive process steps that do not belong to the AM process but, due to the later application, must be done to obtain a perfect part. It contains subtractive work as well as electrochemical processes, mainly for better surface finish.

Follow-up and Secondary AM Processes

Follow-up processes are non-AM process steps, which are mainly copying processes based on additive masters.

Most of the space here is used to describe the machine, and the emphasis was placed on doing that it neutrally.

The sequence as well as the length of the description does not indicate a valuation.

The user should bear in mind that even machines that are on the market show rapid changes due to technical upgrades and innovation. As a matter of fact, only the newest information and data sheets are valid.

Currently (2016), 3D printers are capable of making end products, but many of the parts are prototypes rather than products or components of them.

Preferably, 3D printers use process-dependent materials that mostly mimic the materials used in nonadditive manufacturing processes but do not lead to identical parts. This is true even if the chemical composition of each is identical, as is the case for parts made from polyamide either by laser sintering or by injection molding.

Today's 3D printers still require a reasonable amount of manual work, mainly for material handling and removal of the parts from the machine.

From a high-volume point of view, the reproducibility of the parts is still not satisfactory, as measured by the requirements of a modern, quality-assured, automated high-volume production of parts with different geometries. *In-process* control is under intensive development and on its way to commercialization. In this chapter, therefore, many machines are described that are prototypers rather than fabricators.

The situation changes rapidly. It can be expected that a couple of machines, which are under development today, will enter the market soon.

Common technical details regarding the transformation of the fundamentals of the layered technology (Section 2.1, "Basic Processes") into the machines presented in the same section will be jointly discussed preceding the description of the entire machine.

3.1 Polymerization: Stereolithography (SL)

Local polymerization as a principle of selective hardening of liquid or paste monomeric precursors occurred in the first 10 years after the launch of additive processes exclusively as laser stereolithography (more precisely, but usually not referred to as, laser scanner stereolithography). Meanwhile, processes have gained importance that work with masks (lamp mask method) or with print heads (polymer printing process). The methods coexist, and each requires customized processes and machines.

The following section is based on the classic (from today's perspective) method that is designated as laser scanner stereolithography, but it also discusses briefly the most common variants for polymerization. It was deliberately described in more detail than the other sections. The stereolithography commercialized in 1987 is still the benchmark for many other additive processes. More important at this point, the detailed description of the production that is suitable for stereolithography parts gives a sense of additive processes as a whole and can be transferred to other methods in many aspects. Therefore this section is recommended for the reader even if he or she is not at first interested in stereolithography.

3.1.1 Machine-Specific Basis

The industrial application of the principle of solidification of liquid or pasty monomers by polymerization (Section 2.3.1.1, "Photopolymerization: Stereolithography (SL)") is called stereolithography. It includes laser scanner stereolithography, lamp mask stereolithography, the polymer printing method, and the polymerization of pastes.

Variants for contouring and solidification of the layers will be presented in subsections.

3.1.1.1 Laser Stereolithography

The company 3D Systems name their stereolithography method *Stereolithography, SLA*; the company EOS called it *Stereography*. The terms are trademarks or registered names of their respective companies.

Laser scanner stereolithography is the ancestor of all industrially offered additive manufacturing processes and is represented with 4500 installed systems worldwide (as of the end of 2006); after the extrusion machines, it has the most industrial applications. The following section focuses on laser stereolithography but shows parallels to related processes where this seems appropriate.

Principle of Layer Generation

Laser stereolithography is based on the point-wise solidification of photosensitive monomers (polymerization) using a laser scanning exposure apparatus (galvo scanner).

Stereolithography machines utilizing the laser-scanner method consist of a container of liquid monomer, the installation space, which is usually also used as a reservoir, a construction platform, which is displaceable in the *z* direction in this container, and a laser-scanner unit, which *writes* the current layer information on the surface of the resin bath. The platform supports the part by support structures. This allows for the production of overhangs, fixes unconnected parts of the model, and ensures the defined building up and subsequent removal of the construction platform. After solidification of a layer, the construction platform is lowered by one layer thickness. Thereafter, a new layer is applied (recoating), and this layer is exposed to the data of the new layer and thus solidified. Then, the process proceeds to the exposure of the following layers. In this way, the part "grows" in layers from bottom to top (Fig. 3.1).

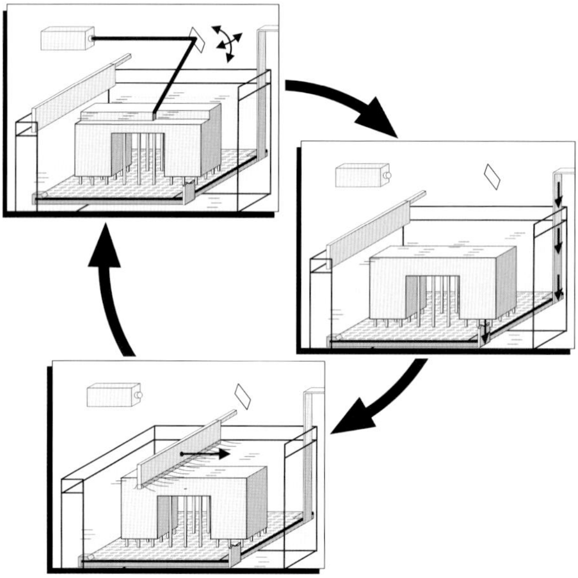

Figure 3.1
Stereolithography (laser scanner) principle

Stereolithography processes try to realize the solidification of a layer with a row of single consolidations, so-called voxels. The geometry of the voxels is given by the energy distribution in the laser beam and the penetration characteristics of the resin. The ideal geometry has the shape of a paraboloid of revolution. In order to achieve the necessary component strength, the laser penetrates both of the voxels in one layer and the two adjacent layers ("overcure") so that the actual penetration depth of the laser is greater than the layer thickness (see Fig. 3.2). The generation of a layer and the toothing with the underlying previous layer takes place simultaneously.

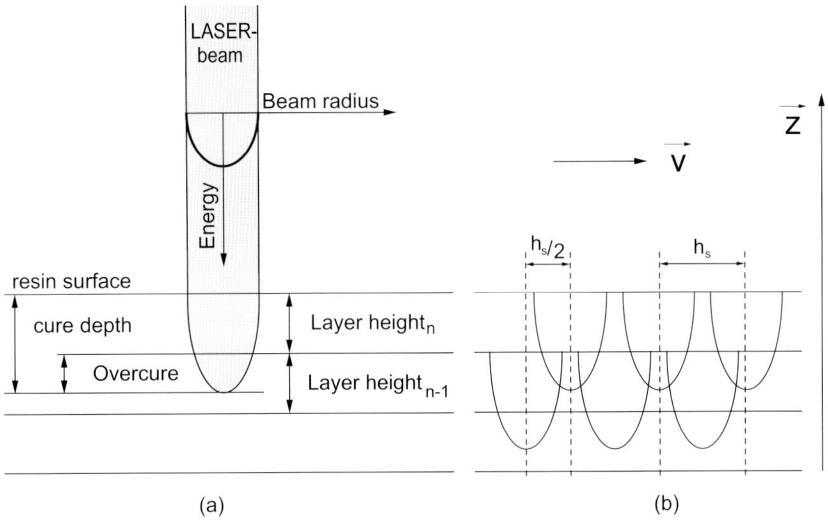

Figure 3.2 Exposure to the laser beam on the resin surface: (a) conditions in the single beam; (b) voxel structure

In practice, the tuning of laser power, beam parameters, scan speed, and material parameters (resin type) determine whether a voxel structure actually arises in the layer or whether virtually continuous paths are written.

For fast and accurate polymerization, each manufacturer uses different exposure strategies. Basically, the areas to be solidified are contoured by boundary curves (borders) and finally hardened inside by appropriate hatching (hatches; see Fig. 3.11). For exact generation of the boundary curves, the beam diameter is compensated. For this purpose, the path of the laser is postponed by half of the amount of the beam diameter of the geometrically exact contour into the component (beam-width compensation or line-width compensation). In addition, the beam diameter is changed in some systems.

As a result of polymerization, the volume of the liquid resin decreases, and the component shrinks. The problem of shrinkage has been substantially mitigated with the change of acrylates to epoxy resins in the second half of the 1990s (linear shrinkage: acrylate = 0.6 % versus epoxy resin = 0.06 %). However, epoxy resins require up to three times higher exposure energy. Methods that use lamps for the polymerization therefore operate today preferably with acrylates. To counteract the effects of shrinkage, there are a number of building strategies that can be used in addition to the optimization of the process parameters. One way is to not continuously connect opposite walls but, as shown in Fig. 3.3(a), periodically generate spaces that counteract the deformation due to internal stresses (retracted, 3D Systems).

Figure 3.3(b) shows the conditions on a real device. The individual hardening tracks do not touch (hatch spacing = 0.26 mm) and have a sufficient distance from the edge (retracted 0.2 mm). The pattern was made using the STAR-Weave technique (3D Systems).

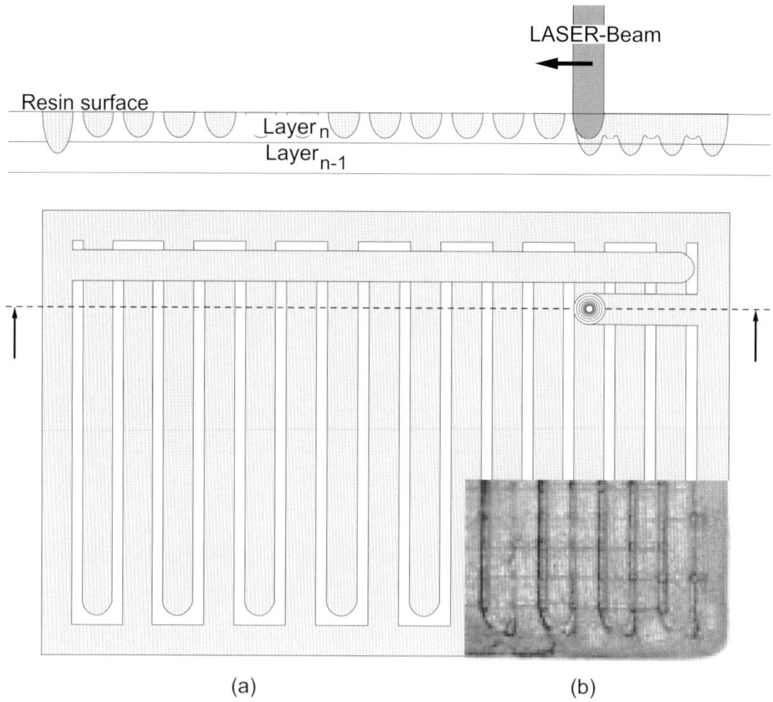

Figure 3.3 Design strategy to reduce internal stresses (retracted): (a) schematic representation; (b) real structures, STAR-Weave (Source: CP GmbH)

Components produced by the laser scanning process have a relatively low strength (green strength) during the construction process. They can therefore be easily deformed by the coating mechanism, internal stresses, or by their own weight. For this reason, protruding elements or cantilevered walls must be lined at a certain overhang or slope angle by supporting structures (Fig. 3.4).

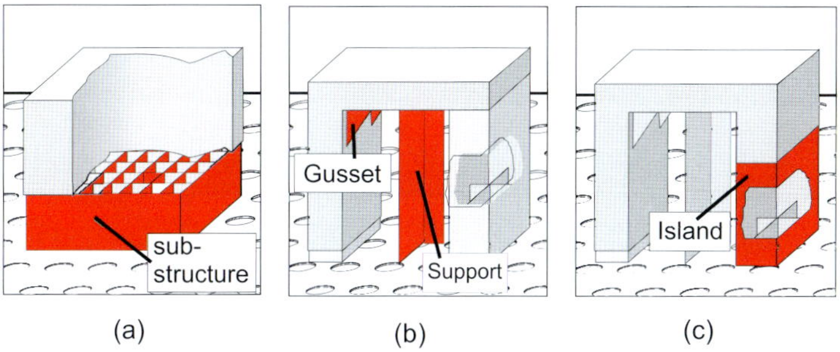

Figure 3.4 Support structures: (a) base; (b) support; (c) "island"

In addition to this function, supports also serve to pull down device areas that want to "roll up" due to shrinkage processes (curl) and to position and fix the component on the build platform.

Until the past few years, supports were considered as volume elements and formed a wall like a square of twelve triangles. They also had special features. The supports in Fig. 3.4(b) are designed as elbow fittings (gussets) to support the particular rectangular geometry branches. Figure 3.4(c) shows a structure that is also called an island. It is used to position and support the component areas that start at later layers and later grow together with the rest of the component. The handles of cups, for example, could not be manufactured without support because otherwise they would start "in the air."

With such support structures, the amount of data grows considerably, especially in STL formulations. Therefore, the triangles (seen from above) were generated only in one line and thus produced without any volume in CAD. But the track gets a volume in the SL process due to the width of the laser tracking. Figure 3.5(a) illustrates this situation. The color of the triangles shows the alternating direction of the normal vectors. Figure 3.5(b) shows the conditions on a real device. Shown are supports that were built with the Fine Point Method (3D Systems). Due to their small diameter, a particularly fine connection can be removed easily and almost without damage to the part.

(a)
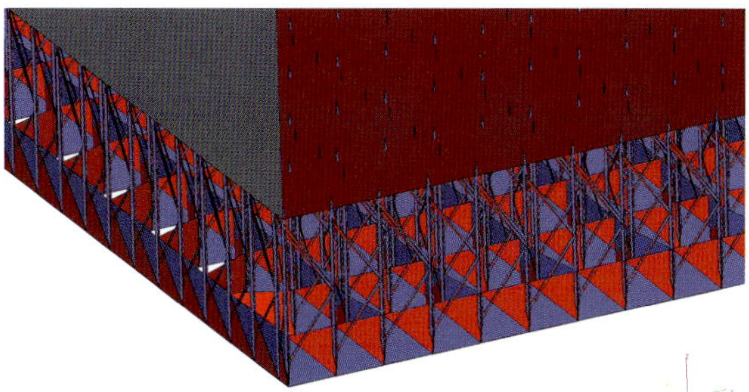

(b)
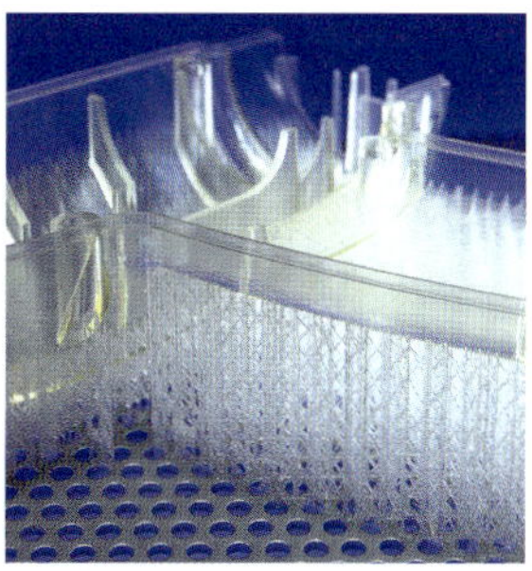

Figure 3.5 Supports: (a) support as a volume element; (b) situation in a real part

Supports are generated (automatically) during the data preparation, but they must be manually removed from the finished part. In some methods, for example FDM or PolyJet, the supports may also be washed out automatically.

Regarding hollow-built stereolithography components, all manufacturers implemented building strategies for the stereolithography processes with a larger cavity volume fraction because of the increasing speed of construction and material savings, but especially to use the models directly as lost forms in the investment casting process (burning of ceramic porcelain molds). In fact, solid walls are designed as spatial trusses, that is, as two thin partition walls, which are interconnected by webs. In addition, so-called outer layers (skins) must be introduced to complete the model or individual model parts up and down (in the z direction).

Hollow-built stereolithography components must have openings through which the noncross-linked monomer may leak. These openings must be closed if the model is going to be used in investment casting so that the ceramic mass does not run into the model; in this way casting defects are avoided.

With the consolidation of the last (top) layer, the building process is complete. The device, which is now fully immersed in the monomer, is moved out of the resin bath upward so that the excess resin can drain and drip back into the resin tank. In terms of economic utilization of the machine, it is advantageous to provide for the component to drip into a separate, preferably heated drain container. Committed users have constructed devices for themselves that support the bleeding of the resin by the spin of a honeycomb related centrifuge process. This is particularly of advantage for hollow-built walls.

Design/Construction

The basic structure of a laser stereolithography machine is shown in Fig. 3.6. The machine illustrated, a 3D Systems SLA-250, dates back to the year 1992. The machine is out of date in this design today, but very well suited to illustrate the basic design.

The design principles as well as the essential components and their arrangement today determine the development of high-performance laser stereolithography machines. The Viper SLA may be regarded as the successor. Its production has now ceased. (For current machines, see Section 3.1.3, "Stereolithography Apparatus (SLA), 3D Systems.")

Materials and consequently *construction times* and *accuracies* are machine- and manufacturer-specific. Therefore, they are not described in general but are explained in the following sections in the course of the descriptions of each installation.

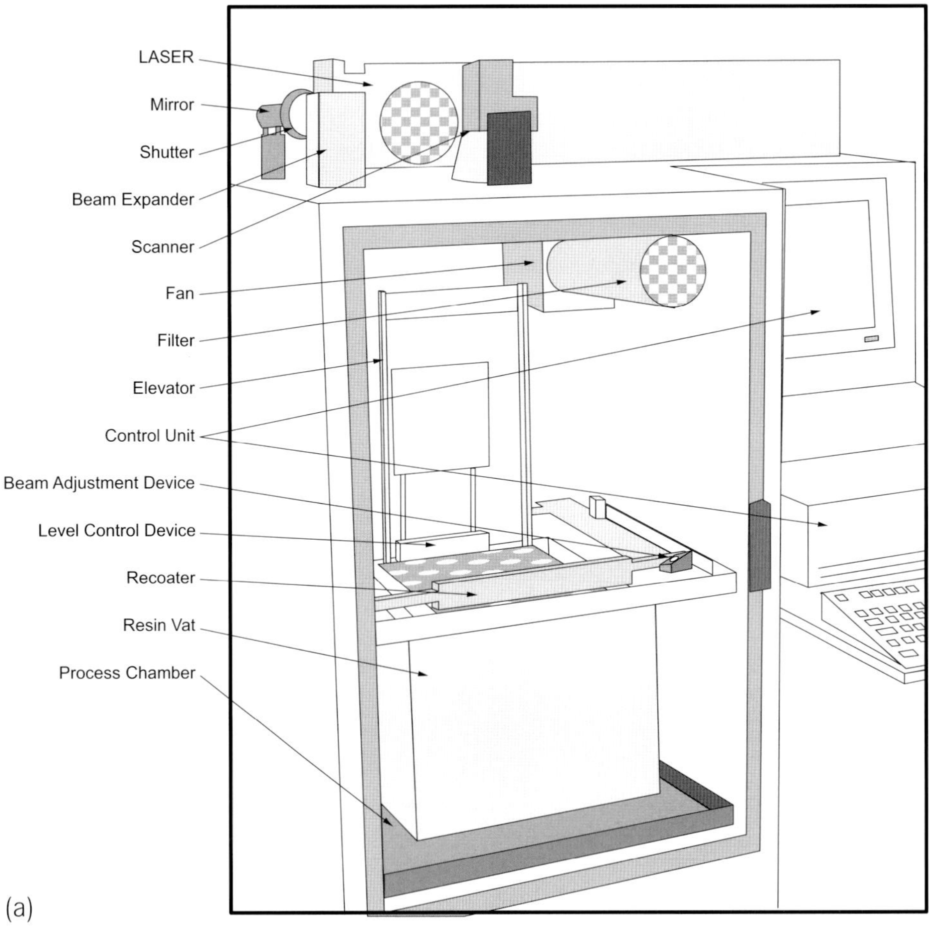

(a)

Figure 3.6 Laser stereolithography machine SLA-250 (3D Systems, 1992). Representation for Illustration of the basic construction: (a) schematic structure;

(b)

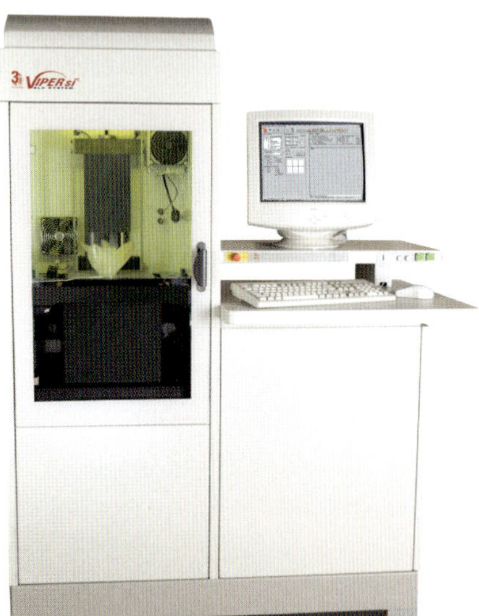

(c)

Figure 3.6 *(continued)*
(b) view; (c) Viper SLA with Nd:YO_4 solid-state lasers

Postprocessing

The procedure of cleaning, removal of the supports, and post-cross-linking of the components is known as postprocessing and is the same for all laser-scanner-based stereolithography processes. The basic procedure is shown schematically in Fig. 3.7.

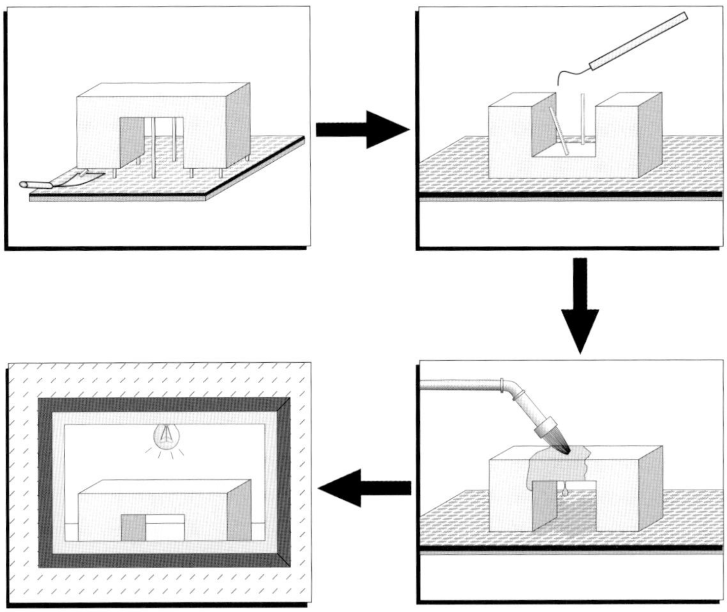

Figure 3.7 Postprocessing for laser stereolithography (laser scanner process)

After the construction process, the models are removed from the build platform and cleaned (acrylates with TPM), and after cleaning the solvent is removed as completely as possible (TPM with isopropanol).

Laser-based stereolithography methods generally do not polymerize the component completely, but solidify it depending on the hardness penetration, up to about 96%. The components are therefore post-cross-linked outside the machine in special ultraviolet (UV) chambers (postcuring oven) so that they are cured or fully polymerized.

In *laser* stereolithography processes, the supports are removed before or after curing, depending on the material, the complexity of the components, and the requirements for accuracy.

Removal of the support is a manual activity that requires a lot of care and is therefore time-consuming.

In terms of the time-optimized production of a component, it is advisable to take care in the generation of restraints so as not to (automatically) generate too many unnecessary posts. Recent support geometries and support strategies, such as the already-mentioned Fine Point supports (3D Systems), have brought considerable relief here. In acrylates it is occasionally useful to remove the component including supports after the curing of the build platform to increase model accuracy. However, difficulties can arise when they are connected with supporting delicate areas of the component.

For postprocessing, mechanical reworking by sand blasting, grinding, milling, and so on, is done as far as it is necessary to achieve the desired process-typical characteristics.

Finishing

In contrast to the process-oriented postprocessing, the finishing is process-independent and includes further treatment of the component, in particular of its surface. This can be done, for example, with paints, fillers, and filling of all kinds, but also by electrodeposition or plasma coating and is in any case not an additive process step.

Equipment and tools for finishing (grinding, sandblasting, painting, and so on) and for subsequent molding processes are not provided by the manufacturers of the equipment in general (occasionally but recommended) and must be selected by the interested user.

Typical Process Sequence Methods

Stereolithography models have the highest detail and the best of all surfaces of the additive manufacturing processes. The disadvantages are the unsatisfactory mechanical and technological properties of the resins, despite the progress made in the recent past, when compared with the characteristics of plastics as they are used in plastic injection molding.

Subsequent processes help to bridge this problem. In order to develop a wider range of materials, the strategy of separation of the properties is frequently used: a detailed geometrically exact model is made quickly by a stereolithography process. The required mechanical, technological, and optical properties are obtained by molding processes. In the manufacture of plastic prototypes, therefore, the process chain of stereolithography followed by vacuum casting has been established as a standard when numbers between 5 and 50 are required and the simulation of construction material by pourable plastic (polyurethane) can be accepted (see Section 5.3.2.1.1, "Vacuum Casting").

In order to obtain metal models, wax models are produced by pouring off from additively manufactured components or are directly additive manufacturing, rendered in a classic investment-casting process. Alternatively, hollow-built stereolithography models are used directly as lost models. Tools are produced by molding (or equivalent watering) from additive manufactured models with metal-resin compounds or low-temperature coatings (see Chapter 5, "Rapid Tooling").

3.1.1.2 Digital Light Processing

Digital light processing (DLP) machines are very similar to the laser stereolithography machines in construction and in operation.

The main difference is in the type of exposure: it uses a digital light processing projector, via a lamp. With the projector, an entire cross section at a time is always exposed. Therefore, the construction time is independent of how much the build platform is occupied with components and also how complex they are. As in laser stereolithography, only one material is used for the component and the support. Therefore, the supports must be added to the record for the component and manually removed from the finished part. The components must be post-cross-linked, purified, and finished. The building materials are acrylates.

"Bottom Up" Construction Types

A variant of the DLP machine builds from the bottom up. The building platform dives down into a resin bath with a glass floor so that between the building platform and the glass bottom just one layer thickness of building material is left. A DLP projector projects the current component's cross section from below through the glass plate onto this layer and solidifies it. The building platform and the layer (partially finished) model are raised by a layer thickness out of the resin bath, new resin flows in, and the next layer is exposed. The process is repeated until the part is finished.

An advantage of this design is that very little resin has to be held in the building space and rapid exchange of the resin is not a problem.

The disadvantage is, as with all solidifying processes by means of lamps, that the space must be carefully protected from ambient light.

Film Transfer Imaging

In contrast to the working out of the resin bath machines, in film transfer imaging (FTI) the building material is assembled as a monomer layer on a transparent film and supplied in a cartridge. For coating, the crystal-clear carrier film is applied thereon with the photosensitive layer between the build platform mounted upside down (with the partially finished component) and a glass plate, and then exposed to light from below. Structures belonging to the component and the supports adhere after the polymerization in the part of the partially finished component (in the case of the first layer of the component onto the build platform), and the parts that are not exposed further are transported with the carrier film. Thereby, a non-exposed portion of the very long film enters the building platform and the next layer is exposed to light, and so on.

The materials are available in several colors. Supports do not have to be removed, and the postprocessing is now virtually automatic.

3.1.1.3 PolyJet and MultiJet Modeling and Paste Polymerization

The methods of the company Objet are called *PolyJet*. 3D Systems calls it MultiJet Modeling (MJM). The machines are commonly referred to as Thermojet printers.

MJM systems consist of a building platform moving in the *x* direction and a print head overhead in the *y-z* direction pushing the building material through multiple parallel nozzles in layers.

The solidification takes place by photopolymerization. Simultaneously with the application of the building material, the layer is hardened by UV lamps, which are moved with the print head. Because lamps have a lower intensity than do lasers, acrylates are used as a building material, which have a correspondingly higher reactivity than do epoxy resins.

A complete layer is always printed. The component and the regions of a layer not corresponding to the component are produced simultaneously by separate nozzle rows. The supports are thus not generated separately. The component and supports are made of different materials. The entire volume of component and support areas are directly completely cured in the construction process. With lamp light, after the printing process, the solidified polymer components come from the printer as a massive block.

The support structures are washed out after the build. Special cleaning equipment, water-jet cabins, or ultrasonic cleaning systems are used. There are a variety of materials having the specific properties, depending on the purpose. Most have wax-shares and are additionally heated in the nozzle area.

Paste Polymerization

Paste polymerization is similar to the MJM process. In contrast, the building material is not printed on the build platform or the uppermost layer of the component, but as a monomer layer applied with a doctor blade to the build platform. The exposure takes place by laser. The pasty material can be provided with ceramic or metallic fillers.

3.1.2 Overview: Polymerization, Stereolithography

Figure 3.8 gives an overview of *the additive manufacturing* methods based on polymerization. It uses the structure of *additive manufacturing methods* according to Fig. 2.38 and assigns them to the industrial process. The classification criterion is the technical implementation of the physical principle that the additive manufacturing process family of polymerization/stereolithography is based on. It is structured based on the concrete machine, as well as on the product and company identification. In each case, the section is indicated that describes the process in more detail.

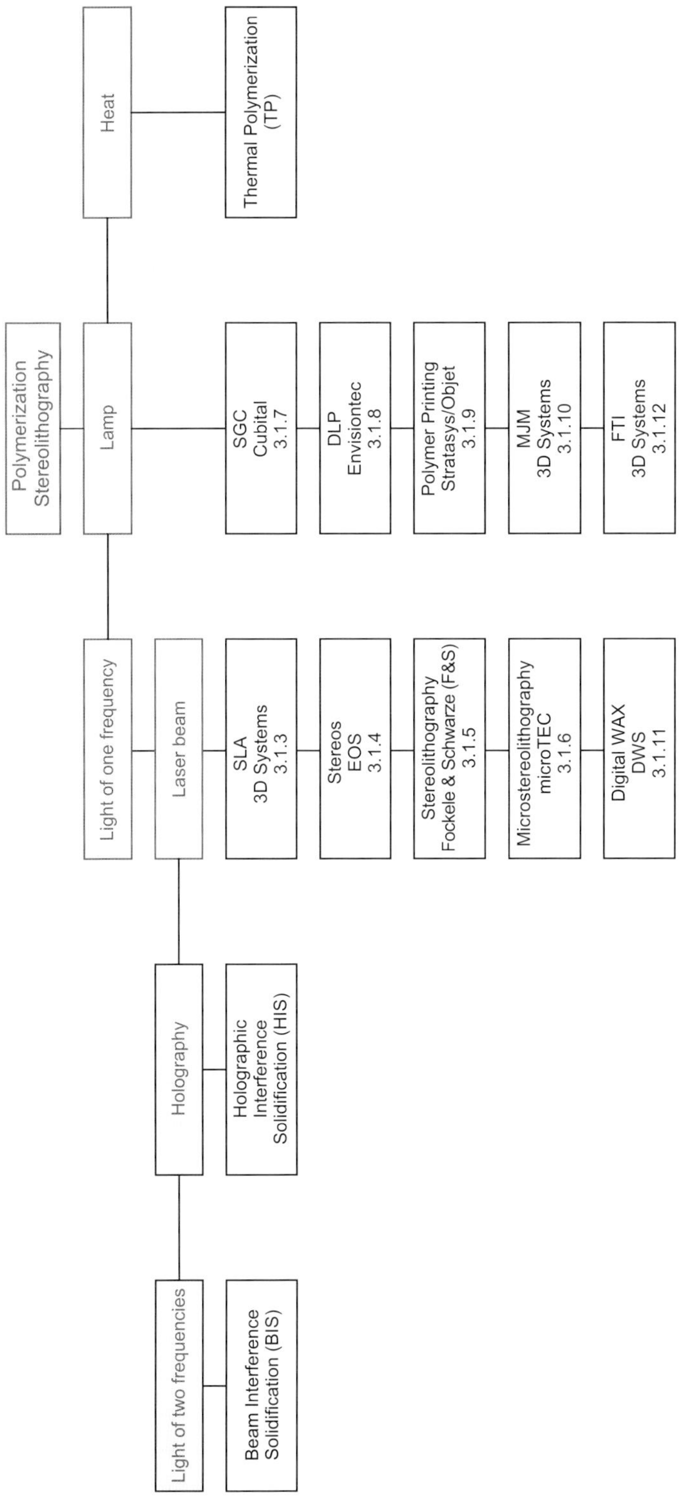

Figure 3.8 Polymerization and stereolithography process, overview in continuation of Fig. 2.38

3.1.3 Stereolithography Apparatus (SLA), 3D Systems

SLA[3] Production Printers: iPro 8000, iPro 8000 MP, iPro 9000, iPro 9000 XL

3D Systems Inc., Rock Hill, SC
3D Systems GmbH, Darmstadt, Germany

Short Description

The stereolithography machines from 3D Systems are classified as production printers and overall counted as stereolithography (SLA) machines. Accordingly, they operate on the laser-scanning method.

3D Systems counts the polymer printers of the 6000 and 700 series, ProJet SD 7000, ProJet HD 7000, ProJet MP 7000, ProJet SD 6000, ProJet HD 6000, and ProJet MP 6000 also as production printers and assigns them to the method of stereolithography (SLA). That is not according to the methodology set out in Section 3.1.1, "Machine-Specific Basis" and in particular, does not permit a conclusion on the additive manufacturing process. Also, it obscures the fact that lamp-based polymer printing methods work with acrylates, but laser-based processes work with epoxy resins.

In the case of stereolithography (SLA) machines that work on the laser-scanner method, the component is created on a platform in a resin bath that is movable in the *z* direction. After contouring and networking of the top layer, the platform and the component with associated supports are lowered by a layer thickness in the resin bath. On the surface of the component a new resin layer is applied, and the process is repeated until the model is completed.

The distance of the supports, the post-cross-linking in the UV oven, and the finishing of the surface are process steps that are indispensable outside the machine for achieving a high-quality model.

Area of Application

Solid imaging, concept modeling, functional prototyping, direct manufacturing, prototype tooling directly from ACES injection molding, indirectly through molding processes (Keltool).

Development Status

The predecessors of the current machines have been commercialized since 1998.

---

[3] SLA stereolighograghy apparatus is a protected designation of 3D Systems Inc., Rock Hill, SC, US.

History

The company 3D Systems was founded in 1986 by Chuck Hull and others. Since the early 1980s, Hull had been working on a method to build a layered model of photocurable monomers at the company MSRP (Ultraviolet Products, CA). In 1984 he was granted a patent on the "Stereolithography Apparatus" and therefore is considered one of the fathers of stereolithography. In November 1987, the first stereolithography machine was unveiled in Detroit (SLA-1) at the trade fair Auto-Fact. Based on this concept, the first commercial stereolithography machine (SLA-250) was presented in 1988. It initiated the rapid development of rapid prototyping systems. Around 1990, the first stereolithography machines arrived in Europe and Germany. The company moved in 2006 from Valencia, California, to Rock Hill, South Carolina. The basic design principle of the SLA-250 (Fig. 3.6) continues to be found in all laser-based stereolithography and laser sintering machines, as well as in most other additive manufacturing systems.

The current SLA family consists of the iPro 8000 SLA and the iPro SLA 9000 (Fig. 3.9) and its variants 8000 and MP 9000 XL.

Figure 3.9
Stereolithography machine
iPro 9000 XL SLA production printer
(Source: 3D Systems)

All other SLA machines are no longer offered. Because there are many machines on the market that have remained in operation for a long time, they are still regarded as current. There is also a growing secondhand market.

The technical data of the current machines are summarized in the appendix section "Additive Manufacturing Systems (Prototypers and Fabricators)." Figure 3.10 shows the previous machine (Viper Pro SLA) and the system consisting of the machine (left), loading and unloading station for a second automatic construction process (middle), and the auto offload cart (right).

Figure 3.10 Viper Pro SLA system: machine (left), loading and unloading station for a second automatic construction process (middle), and auto offload cart (right)
(Source: 3D Systems)

Strategies/Development Partners

The development aims are to offer a family of powerful additive manufacturing machines for the production of concept models, functional prototypes, and production parts for industrially relevant parts with characteristic dimensions of typically a few centimeters to 1.5 meters and in conjunction with suitable molding processes and process chains.

The previous function of the SLA-level model was represented by the Viper SLA and is now represented by the ProJet Professional 3D printer line. The iPro SLA production printer has been optimized as production machinery for availability and productivity in the sense of maximum throughput. For this purpose, they offer replaceable resin tanks (vats) called *resin delivery modules* (RDM).

The iPro 8000 SLA may include one optional container (x, y = 650, 750 and z = 550, 275, or 50 mm or x, y, z = 650, 350, 300 mm^3), the iPro 9000 SLA, two containers or one large container (XL, Fig. 3.9) with the dimensions x, y, z = 1500, 750, 550 mm. Two containers used simultaneously may contain different materials. The machine can build parts simultaneously in the containers. 3D Systems goes a step toward a manufacturing machine with the *auto offload cart*, an optional device for automatically changing the build platform and starting a second building process after the first.

Development partnerships are available with software vendors (Solid Concepts, Materialize, and others). In the field of resins, 3D Systems acquired partnerships with Ciba Geigy Special Chemicals, Vantico, and resulting from Ciba in 2002, Rapid Prototyping Chemicals AG (RPC), Marly, Switzerland.

Data Formats/Software

The ProJet 8000 and 900 series can read STL and SLC files. 3D Systems supplies the parts-preparation software 3DManage 1.23 (2013), which prepares the build files for the control software and the control software 3DPrint. All SLA systems no longer produced, the SLA 500/3500/5000/7000/Viper, run with Lightyear (current as of 2007 was version 1.5) a comprehensive front-end software (preparation software). It works with the control software Buildstation 5.5 Lightyear and includes a number of subroutines that are no longer announced separately today. In the following, these functions are shown because they were valid individual program elements for the SLA-250. Even though the SLA-250 is no longer being built it is used here because of the individual subfunctions that are more apparent than in the consideration of an integrated software system.

The body can be observed by the view program and can be rotated, stretched, mirrored, and moved. The automatic support generator Bridgeworks (originally a stand-alone program of Solid Concepts) is very helpful, especially for standard tasks. For complicated components, experienced users should make use of the possibility of manual intervention, to reduce the construction and thus the construction period and the postprocessing. The current Lightyear package 1.5 (2007, available since 2005) includes several features that automatically detect supports and correct them when they break through the walls or would end in cavities.

The records of supports and components are separately loaded into the section manager Partman and there edited separately. In Partman both the recoating and the component parameters are entered. The recoating parameters include the number of wipings, speed of wiping, speed of the elevator, acceleration of the elevator, wait time between the end of the scan and the beginning of the elevator movement, and wait time between the end of the elevator movement and the beginning of the scan. The component parameters include thickness, edge overlap, overlapping hatch, hatch spacing, depth of hardening, filling gab, filling direction, edge distance, beam compensation, and construction technology. The component parameters are defined individually for each component and for each set of supports. Already-saved component parameters can be read and edited.

After processing in Partman, a slice file is generated from the STL file. In this data set, the geometry of the body is stored in layers. After this, four more records are automatically generated that contain the control data for the SLA. The vector file contains all of the contour data for controlling the laser. The values for the hardening depth are stored in the recoating file, giving rise to the overlapping of the individual levels. The parameter file contains general information, such as the shrinkage factors of the resin or the position of the component on the elevator platform. The Layer file contains information such as the layer thickness and the hatching distances of the individual supports and components. In order to start the construction process, the resin parameters (C_d, D_p) are added to the four data sets.

For the SLA 5000 and 7000, only a so-called BFF file is given, in which all information is summarized. The processing of border and hatch structures is done in parallel during the construction process in the machine.

The machine-oriented part of the software is called Buildstation (since 2005: version 5.5) and corresponds with the current version Lightyear.

Principle of Layer Generation

The procedure of additive manufacturing is described in detail below with reference to the SLA-250. It is similar with restrictions to all stereolithography processes and basically to many additive manufacturing processes that operate on a laser-scanner principle.

For a firm connection between the supports and the building platform, the first layer is cured in the pattern of holes on the platform. Controlled by the data of the vector files, the laser beam cures the layers of the model together with the supports. The speed of the laser beam is in this case a function of the hardening depth, material, and laser beam diameters up to 25 m/s. At the beginning of a new layer, the laser travels along the contour (border) of the component first. Subsequently, the surface between the boundary lines is hatched (hatch). The parts of the layer that form the surface (outer layer) of the component can be filled, in addition to the hatching. The component is hardened stronger with the filling parameters than the hatch. The surface of the component receives greater stability. Figure 3.11 illustrates the differences between the edge, hatch, and fill.

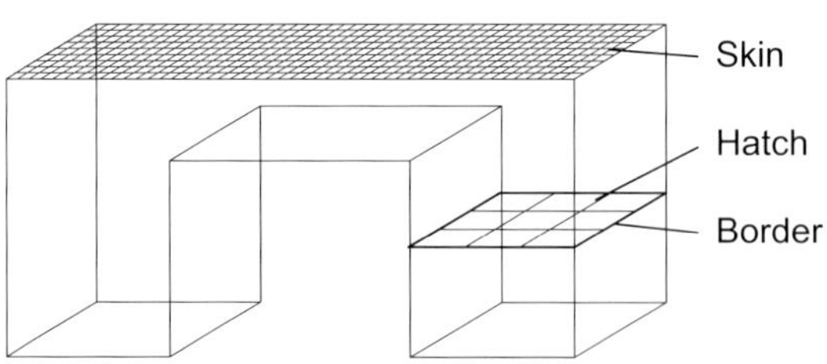

Figure 3.11 Definition of boundary curve (border), area filling (hatch), and outer layer (skin)

Originally each layer was assigned as a border or hatch zone. As the layers got thinner, each layer was no longer provided with hatches, and this was called the *cyclic build style*. The mixing of cyclic and normal building style should be avoided wherever possible.

The hatch spacing is chosen so narrow that only a small portion of the resin remains in a liquid state. The construction process of the iPro 8000/9000 is accelerated by using a very fine laser beam with a diameter of 0.13 mm to trace the contours (border) with a diameter of 0.13 mm. For the filling (hedge), the beam diameter (dual spot) is significantly increased to 0.75 mm.

To connect the individual layers, the hardening depth of the contour and shading is greater than the layer thickness, resulting in an overlap in the range of about 0.03 mm (overcure, Fig. 3.2). Downward-facing surfaces that begin during the construction process on any *z* coordinate are greater due to this fact in the *z* direction. To compensate for this, the machines are therefore equipped with a *z* correction.

Figure 3.12 shows a construction platform with SLA models.

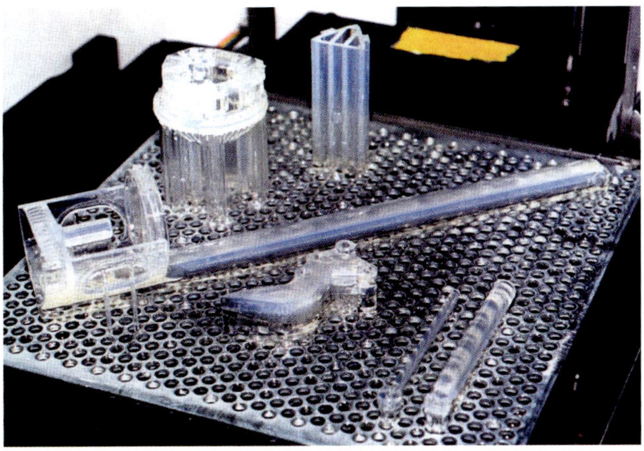

Figure 3.12 SLA construction platform with models

The minimum layer thicknesses are 0.05 mm, the thickest 0.15 mm. All machines can be operated with different architectural styles. By the choice of architectural styles, the operator can optimize between speed (productivity) and accuracy.

After a layer has been cured, the build platform is lowered by a layer thickness, and a new resin layer is applied (*recoating*). This is done with an active recoater, the *Zephyr recoating system,* which is designed to apply material using a vacuum system.

The 3D Systems' patented *Zephyr recoating system* has an open hollow wiper in the bath in which a resin reservoir is installed. A slight negative pressure is placed on the surface of the reservoir so that it is filled with resin from the bath surface. The pressure ratios are adjusted so that when sweeping solid areas of the component, the resin sticks to these structures and is thus applied but is sucked from the free surface so that the resin is filling the reservoir. This system is faster than the passive predecessor because it is generally driven over the bath once. A so-called *Smart Sweep,* limiting the wiping operation in the *y* dimension of the cross section model, additionally accelerates the process. Unlike systems that promote the resin via pumps, the Zephyr system is relatively quick and easy to clean after resin exchange.

The architectural style designated ACES, *accurate clear epoxy solids*, can manufacture 100 % solid epoxy cured components. To achieve this, in particular the hatches are set so closely that they partially overlap.

Full models are especially needed when higher (pressure) strength is required, for example in rapid tooling applications (AIM, ACES injection molding) or if the optical properties are important.

Quick Cast is a special architectural style for producing components with more than 80 % voids (based on a full model). It combines the two important properties for building hollow models: good drainage effects and the highest possible void fraction. The components are preferably used for investment casting. Special resins are used to create smooth walls, and because of their relatively low viscosity, they easily run out of the cavities, such as in the Accura 60 for systems with solid-state lasers.

How the build hatch is based on an algorithm of 0°/60°/120° to the horizontal *x* direction hatch can be seen in Fig. 3.13. In the second step, the hatch pattern (Fig. 3.13(a)) is based on the first shot (Fig. 3.13(b)) and repeated in the *y* direction. The result is a honeycomb structure (Fig. 3.13(c)) in the *z* direction having continuous cavities that allow a simple drainage and the complete emptying of the model of the liquid residual monomers. Quick Cast components and the associated investment-casting components are shown later in Figs. 3.15 and 5.15.

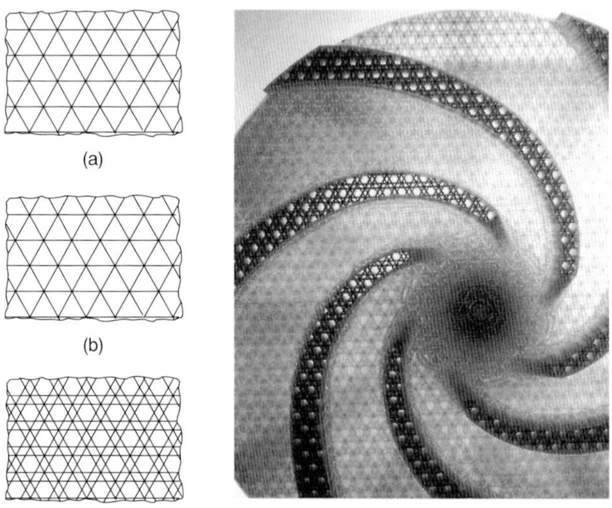

Figure 3.13 (a) to (c) Quick Cast building strategy; (d) practical implementation of the example of a pump impeller (Source: 3D Systems)

Design/Construction

A closed process chamber, where the actual model is generated with the process and control computers and the laser scanner, forms the structural unit on all models. In Fig. 3.6, the upper machine cover of an SLA-250 is removed so that the (purchased) lasers are clearly seen as separate components, such as the beam guide and scanner unit. The construction of the machines from as many commercially available (off the shelf) components as possible and hence an extremely low level of vertical integration is another hallmark of most manufacturers of additive manufacturing systems. The iPro 8000/9000 family has a Nd:YVO_4 solid-state laser (wavelength 354.7 nm) with a power of 1000 mW at the laser head. The lifetime is guaranteed to 10,000 hours. The technical data can currently assume a repetition rate of 10 kHz.

The running of a perforated plate construction platform emerge as the basis from above into the resin bath and can be completely moved out to remove the components. Because of this construction, the resin holder can be changed in all machines upgraded to the latest state.

The SLA 7000 (Fig. 3.14), the predecessor of the iPro 8000/9000 family, has an 800 mW solid-state laser (with 5000 hours of guaranteed lifetime). It has two different beam diameters: from 0.23 to 0.28 mm (standard beam diameter), and 0.685 to 0.838 mm. With the smaller beam diameter, a high accuracy is achieved. The larger is used in order to shorten the time-consuming hatching and thus the construction time, without increasing the scanning speed.

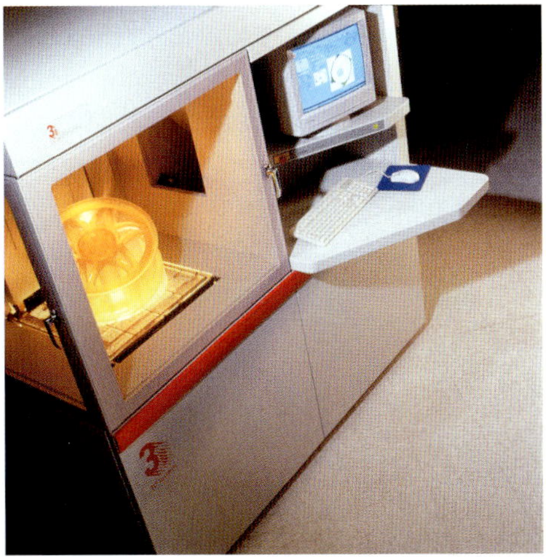

Figure 3.14
Stereolithography machine SLA-7000 (Source: 3D Systems)

The maximum scanning speed may not be increased above the product of the repetition and spot diameter, if a coherent model should be generated. As long as the repetition rate of the laser cannot be increased over 10 to 20 kHz, this design solution is a good way out. The technical data can currently assume a repetition rate of 10 kHz. Material/Construction/Accuracies

The available range of materials is still small compared to the variety of engineering plastics, even though the number and capacity of the resins has increased enormously.

For stereolithography there are the material families Accura and RenShape. All resins are epoxy resins and suitable for systems with a solid-state laser (Nd:YVO_4). Accura 25 (white) simulates the properties of PP (polypropylene), Accura 55 (white) the properties of ABS. Both resins are no longer ivory-translucent, as was long the case with SL resins. Accura 60 is transparent and also simulates the properties of polycarbonate (PC). Accura Bluestone is a nanocomposite (filled resin). It reaches the tensile strength of Accura 60 (up to 68 MPa) but has a maximum of one- to almost four-fold higher modulus of elasticity (up to 11,700 MPa) at a significantly higher thermoforming resistance (up to 66 °C as opposed to up to 55 °C for Accura 60).

Accura Amethyst is a special material for the production of master models and lost forms of detailed components, as they occur preferably in the jewelry and dental industries. The material is rendered directly in the investment casting process or used to make silicone molds. With the help of the silicone molds, the necessary number of lost models can be made for a series by investment casting.

The current 14 Accura materials cannot be described clearly. An overview of their preferable application is the Accura® product selection guide for Stereolithography (SLA®) materials (see appendix or download from the 3D Systems website).

The construction result depends not only on the type of resin, but also on the interaction between laser and resin. The manufacturers themselves and independent third parties provide resins for special applications. With the use of these materials, attention should be paid to the fact that well-tested parameter sets are present for the fusion machine-laser material. Many resins are hygroscopic and therefore should be stored and processed accordingly. Components often do not reach their expected strength if water absorption is excessive.

The results of the construction process are strongly dependent on calibration, as in all rapid prototyping methods. For this purpose, the service has a calibration plate that scans the position of the laser beam by diodes and evaluates it online immediately. This method is very effective. In HR (high resolution) mode, a wall thickness of 0.1 mm can be realized. Different architectural styles allow different layer thicknesses and therefore different speeds and accuracies. With the EXACT style, a layer thickness of 0.076 mm can be reached, with the Quick Cast style 0.101 mm, with the FAST architecture 0.127 mm (which was the original standard), and with the thin-layer architecture 0.025 mm.

In order to join successive surfaces, the hardening depth of the layers is chosen to be greater than the layer thickness (overcure). For free downwardly oriented surfaces (downskin), that would lead to a too-large layer thickness. A *z* correction controls and eliminates this effect.

The structural unit of building space and storage container has the disadvantage that large amounts of expensive resin are in the machine, especially for large stereolithography machines (for the iPro 8000/9000, this may come to be worth about 10,000 €).

The machine should be used continuously for economic reasons because the resin tends to age, and accurate air conditioning is required. Swap bodies, especially at lower volumes, alleviate the problem.

Postprocessing

In these systems, the postprocessing corresponds to the typical stereolithography postprocessing described in Section 3.1.1, "Machine-Specific Basis."

Finishing

All of the typical stereolithography finishing techniques and molding processes are possible (see Section 3.1.1, "Machine-Specific Basis").

Typical Process Sequence Methods

Investment casting components (Section 5.3.3.1, "Investment Casting Process with Additive Manufacturing Process Steps") are obtained when using the Quick Cast style, which prevents shell cracking in hollow structures when burned out of the ceramic molds. The hollow structures of the component have to be completed with top surfaces for use in the investment casting process. To drain the part, corresponding drainage openings should be provided or drilled mechanically. Before the actual investment-casting process, they must be sealed again. Figure 3.15 shows an assembled engine block built with the Quick Cast process and the associated casting. By optimizing the curing parameters, it is important to make sure that relatively thin outer surfaces do not occur.

Forms and tools are obtained directly via AIM (ACES injection molding, Section 5.4.1.1), or indirectly through molding using the 3D Keltool from Course4 Technology (Section 5.3.2.2.5).

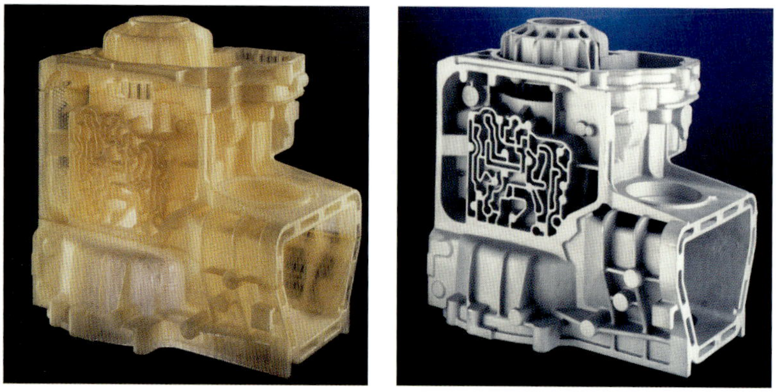

Figure 3.15 Quick Cast building method, cavity model for investment casting, component (Source: Volkswagen AG)

3.1.4 STEREOS, EOS

STEREOS DESKTOP/300/400/600/MAX 600

Electro Optical Systems GmbH (EOS), Krailling, Munich, Germany

Short Description

The STEREOS stereolithography machines of the company EOS work on the laser-scanner principle. The model arises on a platform that, after contouring and mashing of the current top layer, is lowered by one layer into a resin bath so that on the surface of the model a new resin layer can be placed. The process is repeated until the model is finished.

The removal of the supports, the after-mashing in the UV oven, and the finishing of the surface are process steps that, outside of the prototyper, are essential for creating a high-quality model.

Application Field

Concept models, functional prototypers

The manufacturing of molds and tools takes place indirectly by molding processes.

Development Status

It has been commercialized since 1990. The manufacturing and sales of stereolithography machines was stopped in 1997.

History

EOS was founded in 1989 and introduced its first laser stereolithography machine (that was called a laser stereolithography machine), the STEREOS 400, already in 1990. Until 1997 the delivery program was greatly expanded and included a range of stereolithography machines, from a quasi-office-friendly desktop machine to the STEREOS MAX 600 with a Nd:YAG laser and an active recoater.

In 1997 the company EOS stopped the production and sale of stereolithography machines. According to reports, a patent swap with 3D Systems took place. EOS took over all of the patents concerning laser sintering from 3D Systems, and in reverse they handed over all of the stereolithography-relevant patents to 3D Systems. EOS sold their stereolithography section to 3D Systems and retreated completely from the stereolithography market. The company since then has focused on the development and production of sintering machines.

Until the exit of the stereolithography technique, EOS was at one time the worldwide single manufacturer that developed and produced stereolithography and sintering machines.

Because at the time of the termination of the distribution a row of quite new stereolithography machines from EOS were in use, and because 3D Systems promised the service of those machines, those were discussed in detail in the first edition of this book (2003).

3.1.5 Stereolithography, Fockele & Schwarze

FS-Realizer

Fockele & Schwarze (F&S) Stereolithographietechnik GmbH, Bielefeld, Germany

Short Description

The prototyper FS-Realizer from the company Fockele & Schwarze (F&S) works on the laser-scanner method. The model is created on a platform that is sunk by the height of one layer into a resin bath after contouring and mashing of the top layer. On the surface then a new resin layer can be applied. The process is repeated until the completion of the model.

The removal of the supports, the after-mashing in the UV oven, and the finishing of the surface are process steps that, outside of the prototyper, are essential for getting a high-quality model.

Application Field

Concept models, functional prototypers

The manufacturing of molds and tools takes place indirectly by molding processes.

Development Status

Commercialized

History

The company Fockele & Schwarze was founded as a spin-off of the University of Paderborn (Germany) in 1991. It developed its own stereolithography system and launched it in the year 1994 as a laser model system (LMS). In 1996 the FS-Realizer replaced the LMS system.

Sales stopped soon afterward. The machines were used in the in-house service bureau.

3.1.6 Microstereolithography, microTEC

RMPD (Rapid Micro Product Development) mask, stick2, multimat, 3D-CSP

microTEC Gesellschaft für Mikrotechnologie mbH, Duisburg, Germany

Short Description

These are proprietary additive systems and methods based on stereolithography for the manufacturing of prototypes, single parts, and series. The focus lies on the industrial application of micro- and nanotechnology. Parts of one or more different horizontally or vertically stacked polymers (RMPD multimat) are delivered, but also polymers are applied on different substrates such as glass, silicon, or metal (RMPD-stick2). Through a combination of the methods and the possibility to insert parts such as silicon chips (bare dies), microlenses, sensors, or the like into cavities during the construction process, functional microsystems are manufactured (3D-CSP, *chip size packaging*). Different UV-hardening materials are available (sol-gel and others).

Application Field

Functional prototypes, final products, even in large series. The focus is on the direct fabrication of integrated microcomponents.

Development Status

The methods are patented internationally and available at microTEC in the form of proprietary machines. The machines are used for individual production to the client's order. Sales of the machines to end users are currently not provided.

History

The company microTEC was founded in 1996 by Dipl.-Ing. Reiner Götzen and his team in Rummeln close to Duisburg (Germany). Today it operates at two locations (Duisburg, Bad Dürkheim) as contract manufacturers and development partners.

Strategies/Development Partners

The focus of the company is on the production of microparts and microsystems. microTEC thus belongs to those companies that have practiced *rapid manufacturing* since 1996, and they emphasize this by the name of the process: *rapid micro product development.* By using micro-optics, high quantities at high productivity are reached (up to 150,000 parts per hour). The company is involved in European and national development activities and networks. Industrial development partners among others are Diagnostics (Switzerland) and Bürkert (Germany).

Data Formats/Software

The machine takes in geometric information directly from CAD systems by neutral interfaces. Supported are STL, IGES, and other formats. Even microelectronic circuit data (OrCAD) can be directly integrated.

Principle of Layer Generation

The manufacturing of functional micro systems (RPMD) is based on stereolithography. The layer is either polymerized with parallel laser beams (RPMD, Fig. 3.16(a)) or mask supported (*RPMD mask*, Fig. 3.16(b)). In a variant of the method, the combination of several different polymers in horizontal or vertical stratification is possible (*RMPD multimat*). It is also possible to apply polymers on different substrates such as glass, silicon, or metal (*RMPD stick2*).

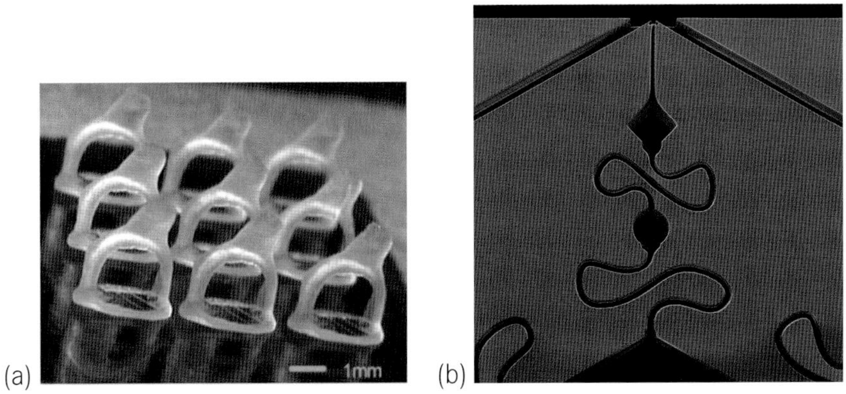

Figure 3.16 Microstereolithography: (a) parallel production, auditory ossicles (RMPD); (b) microfluidic system (RMPD mask) (Source: microTEC)

The process offers the opportunity to insert parts such as silicon chips (bare dies), microlenses, sensors, or the like into cavities during the construction process. In this way, functional microsystems are manufactured (3D-CSP, *chip size packaging*).

The additive process is adjusted for high-precision requirements and integrates the principle of stereolithography. The exposure is done with the use of UV lamps over the whole mask area (up to 14") or point by point with a pulsed excimer laser.

Design/Construction

How the growth of steps of less than 1 µm in the *z* direction is realized is as much a secret as the method that allows the insertion of microelements, such as shafts, to find the exact current *z* coordinate after insertion and to prevent contamination by material not participating in the process. Conceivable–although pure speculation–is a glass plate as a top (or bottom) cover for the construction space that takes care of the exact *z* positioning and for an absolute *x-y* plane. Then there is still the secret of how the resin can be removed from the glass plate after the polymerization. After all, in the beginning of stereolithography, parts were built on glass, when especially precise parts were required. Also in regard to the insertion of parts and because of the necessary cleaning, it is hard to imagine that, as in classic stereolithography, the part lies in the resin bath.

For the RMPD mask, the masks known from microelectronics are used. The machines support the mounting on and the building up of existing structures as well as the automated filling.

Material/Construction Time/Accuracies

Plastics are used, among others, acrylates and epoxy resins as well as sol-gel materials. According to manufacturer information, within one part, different materials can be used, and thus partially different mechanical and technological properties (hard/soft, hydrophobic/hydrophilic) or different colors can be achieved. The materials are developed partly in-house or with partners. Different functional materials are available up to a PEEK-similar material. Selected materials for medical techniques have FDA[4] admission.

The maximum part size (*x-y*) is DIN A4. A solution of 1 µm is reached, aspect ratios of 200 can be achieved, and R_a values of 0.92 nm have been reported.

Postprocessing

There is no information about postprocessing.

[4] Food and Drug Administration (US), which has the authority for the approval of drugs.

Finishing

A special processing enables the production of nanostructured surfaces.

Typical Follow-up Processes

The parts can be metallized.

3.1.7 Solid Ground Curing, Cubital

Solider 4600, 5600

Cubital Ltd., Raanana, Israel
Cubital Europe, Nidderau-Heldenbergen, Germany

Short Description

In the Solider method, the illustration of the layer geometry and the exposure are done by a high-power UV lamp through a mask (type: lamp-mask method). To the actual stereolithography, a second process, the generation of the mask, is added. The models are created without support structures and can be placed even one above the other so that the machine achieves a very high productivity. The models have, particularly in the z direction, a very high accuracy because the layer is milled planar after each exposure process step. Those areas not belonging to the model are filled with wax. In the *postprocessing* outside of the machine the wax must be removed.

Application Fields

Geometrical prototypes, functional prototypes (technical prototypes)

The manufacturing of molds and tools is done indirectly via molding processes. Investment castings arise from lost solid models based on the principle of the lost-wax process in the *SoliCast process.*

Development Status

Commercialized 1991 to approximately 2000

History

The Israeli company Cubital was founded in 1987 and thus belongs to the oldest manufacturers. The machine was presented in 1991. In the year 2001, Cubital handed over all patents to Objet. This was done with the approval of Scites Corp., the biggest investor in both companies (since January 2006, Scailex Corporation Ltd., Tel Aviv). Cubital later stopped the production.

The machines have meanwhile become so rare that they will not be discussed in detail here any further.

3.1.8 Digital Light Processing, Envisiontec

Perfactor[4] Standard, Perfactory Mini, Perfactory Desktop, Ultra

Envisiontec GmbH, Gladbeck, Germany

Short Description

The additive machine family of Envisiontec consists of stereolithography systems that work on the lamp-mask method. The mask is created with a DLP projector (digital light processing technology, Hewlett-Packard), projecting the layer information directly onto the construction plane. The machine family is divided into the Perfactory and the Ultra series. Besides these, there are the bigger ones Xtreme and Xede.

All of the machines work with supports that are created, during the process, out of the same material.

The *Perfactory* family, except the Ultra, has a "head over" arranged stereolithography system. It contains the Standard, Desktop, and Miniseries with 17 variants altogether. In all of them, the projector projects the mask from underneath onto a glass plate, which defines the construction plane, and at the same time, is the bottom of the resin reservoir. In this way, each time a complete layer is exposed and hardened at once. The construction platform is located one layer of thickness above. After each exposure, it is moved up one layer and "draws" the partly finished part head-first layer by layer out of the material container. This contains so much material that the recoating at the raising is ensured. The system therefore is especially very suitable for fast material changes.

The Ultra (in the US it is sold under the label ZBuilder Ultra by the sales partner Z Corp) in 2009 was presented for the purpose of manufacturing big parts after the Perfactory principle, but it does not work "head over" because of the greater part weights.

Perfactory Xede and Perfactory Xtreme, which were developed from the Vanquish, work with the same DLP exposure system but are designed for the manufacturing of bigger parts (up to 444.5 × 355.6 × 500 mm). They contain, as in the laser-scanner stereolithography machines, a resin bath and an inner movable construction platform.

Application Fields/Development Status

Concept models, functional prototypes, series production

The Perfactory was presented as a prototyper in 2001 and was distributed since the end of 2002; currently it is in its fourth generation (*Perfactor*[4]). The *Vanquish*⁺ was presented at the end of 2004 as the *Vanquish FC (Flash Cure)* and the *Desktop* in 2006.

History

The company was founded in August 1999 in the context of a technology screening of Degussa GmbH (earlier: Hüls AG, Marl) as a start-up of Dr. *Hendrik John*. In 2004, American private investors were brought in. The largest shareholder and CEO of Envisiontec is currently (2013) Al Siblani.

Strategies

Envisiontec sees itself as supplier of *rapid prototyping* and *rapid manufacturing* systems and sees all of its activities as being under the CI description, "computer-aided modeling devices" (CAMOD).

The *Perfactory* was designed especially for the end user as a personal fabricator, from which also comes the name: Personal Factory. It is optimized for the manufacture of small and intricate parts and thus is recommended especially also for applications in jewelry and in the medical field. There are the variants "Desktop," "Standard," and "Mini." The *Perfactory Mini* is optimized for smaller parts but higher resolution.

All machines create the part from voxels with the dimensions of 16 × 16 × 15 μm in the *x*, *y*, and *z* directions. The different machines result from a good balance between resolution (accuracy) and construction speed. A dynamic adjustment of the layer thickness between 15 μm and 150 μm and of the focal length of the projector enables an *x-y* resolution between 16 and 69 μm. The large number (of a total of 17 current variants) follows the branch-specific requirements, for example, for the dental field or jewelry manufacturing.

The *Perfactory* can be installed quickly and easily and requires no additional installation after a change of location. A material change can automatically as well as manually be realized very quickly. By design, very little material must be stored. The few moving parts facilitate the maintenance.

The *Desktop* is smaller and designed as a *low-cost* machine. It focuses particularly on the markets of the jewelry and dental fields.

With the *Vanquish*⁺, the Perfactory principle is applied to larger construction spaces. The principle of movability and the small amounts of material in the construction space had to be given up for that reason.

The strategy to focus on interesting part markets such as the medical field, and here, *tissue engineering* gained the committment of the company in the development of the BIOplotter, which was developed in cooperation with the Freiburg Materials Research Center (FMF), of the Albert-Ludwigs University of Freiburg.

A distribution agreement with Agile Manufacturing provides access to the Canadian market and through this to the US market since April 2004.

Data Format/Software

Envisiontec offers for all machines an autonomic, running on the machine, easy to handle *rapid prototyping* software package called *Suite*, which accepts STL or CLI files. The software can be used in *auto mode* or in *expert mode*. In the *automatic mode*, the system works fully automatically after the transfer of the STL file. In the *expert mode*, the process parameters can be set manually, by reworking the 3D data set with RP software.

The machines can be integrated into an Ethernet network.

Principle of Layer Generation

Perfactory: To drive the DLP projector, the construction data are converted into a bitmap format and projected as a mask onto the construction plane by the mirror unit. Each layer is assigned to one picture mask, in which the exposed parts are transparent and the other areas are nontransparent. The nonsolidified material remains in the construction space.

Each layer adheres after the exposure on the previous one (the first one at the construction platform), but because of a special coating, not onto the glass. After each exposure, the platform is drawn up out of the construction space (in the z direction) to the height of one layer. Because of the suction of the immersed areas of the part and because of gravity, the material flows in between the previously exposed layer and the glass plate and constitutes the material for the next layer. As the construction plane (glass plate) is fixed in space, the model grows "head over" layer by layer out of the construction space.

The *Ultra* works on the same principle from bottom to top, just like a classic laser-scanner stereolithography machine. The part, which is much heavier than in the Perfactory, is created on a platform that is moved down after each layer. The free resin surface is covered by a transparent plate that defines the projection plane and ensures the surface quality. It also enables the suction effect that is necessary for the re-exposure. A roll ensures the evenness of the layer and prevents air inclusions.

Design/Construction

Perfactory: The machine consists of a barrel-shaped base, a top-opened glass tub as construction space, and a construction platform positioned above, precisely movable in the *z* direction (Fig. 3.17).

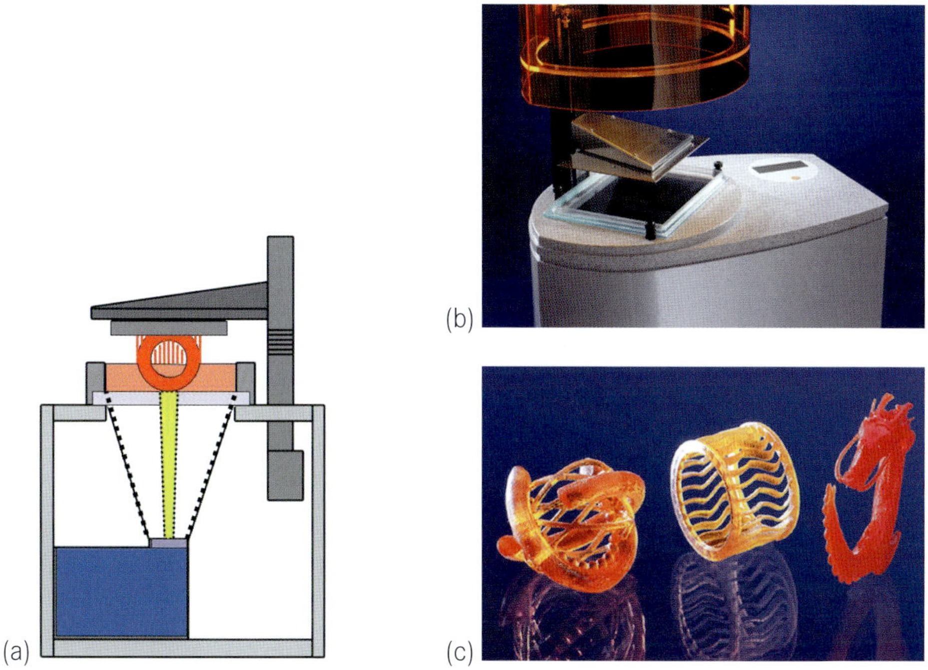

Figure 3.17 Envisiontec Perfactory: (a) principle; (b) machine; (c) part (Source, b, c: Envisiontec)

The base contains the DLP projector, the control computer (displayed at the top), and the storage for the material. The glass bottom of the construction space is the projection plane for the exposure. Because the projector contains the whole exposure system, the machine has only a few movable parts.

The DLP projector sends out the mask as a pixel graphic. The construction field size results from the optic used. This is determined by a zoom lens (Perfactory Standard), with three different factory-installed lenses with fixed focal lengths (Perfactory Standard UV), or with three changeable fixed focal lengths (Perfactory Mini). The determined amount of pixels also defines the accuracy. The material is delivered in a sealed container and brought to the construction space via peristaltic pump. The container and hose can thus be easily and quickly changed without cleaning and thus without contamination of the construction material.

The material can be filled into the construction space without using the storage container, which enables experiments with very small amounts of material.

As a result of the design, the material piles up on the bottom of the construction space. This always forms the layer and therefore is completely used up, so there will never be an accumulation of old material.

The construction space and the whole area of the construction platform are covered by a transparent, UV-absorbing, and lightly yellow-toned cover, which prevents an external exposure and together with the oval shape of the base defines the characteristic design of the machine.

The *Desktop* is a variant reduced to the minimum necessary in a classic housing.

The Ultra (Fig. 3.18(a)) has a resin bath into which dips a construction platform movable in the *z* direction. The DLP projection unit is situated in the head.

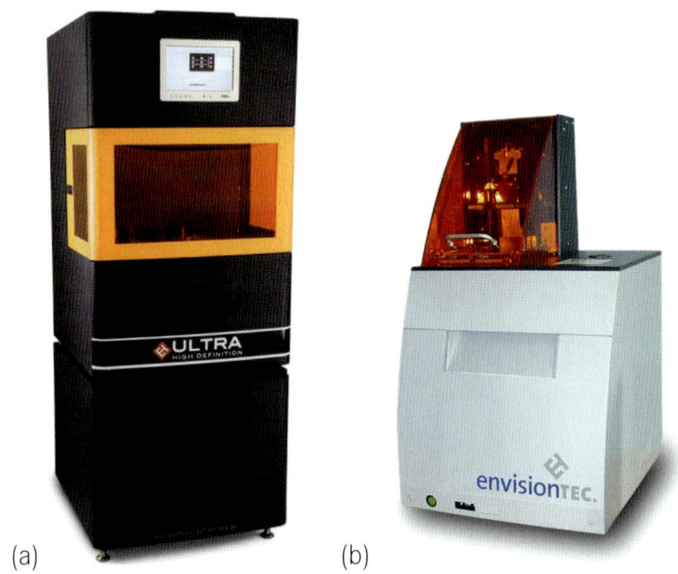

Figure 3.18 Envisiontec: a) Ultra; b) Desktop (Source: Envisiontec)

Because in the Ultra the same, very UV-reactive materials are used, its construction space is also covered by a yellow, UV-absorbing panel.

Material/Construction Time/Accuracies

A part of the strategy of the company is the development of proprietary materials. All materials can be used with all machines.

Perfactory R5 and Y8 are reddish (this is a characteristic of the Perfactory parts) methacrylate. R5 can be seen as a standard material. Its properties are comparable with those of polypropylene (PP). Y8 is a material that is optimized for higher flexibility.

PIC100 is for the investment casting, after the melting process, of optimized material for the manufacture of cast models, especially for the jewelry industry.

E-Shell 100 is a biocompatible photopolymer (class 6) that was developed for the target market "shells for hearing aids," but which is suitable for other biomedical applications. Formerly it was colored beige, but today it is also available in different colors and in transparent. E-Shell 200 is a biocompatible material of the class IIA with higher bending strength.

Nano Cure RC25, a resin with up to 80 wt% filled with ceramic particles, was presented in 2006. It connects higher strength and higher temperature stability.

The construction speed of the Perfactory and the Vanquish⁺ is around 10 to 20 mm/h, and depending on the specification, up to 40 mm/h in the z direction. The Desktop is up to 7 mm/h.

The DLP projector creates an x-y pattern with 1400 × 1050 pixels in the SXGA⁺ graphic format.[5] In the new *Enhanced Resolution Module* (ERM) mode, a virtual doubling of the resolution in x and y can be achieved by a subpixel shift, thus 2800 × 2100 pixels. The module that only concerns the projector can also be upgraded in older machines. The *Desktop* has a resolution of 1024 × 768 pixels in the XGA mode.

Experience has shown that the accuracy is at ±1 pixel, and in the ERM mode, therefore 1/2 pixel.

The construction field size, and at a given number of pixels, the resolution and the layer thickness, are within certain limits regarded as parameters that enable an optimization up to a maximum accuracy or to a maximum construction progress.

Perfactory Standard: The construction volume is, depending on the optics, x, y, z = 120, 90, 230 mm^3 to x, y, z = 190, 142, 230 mm^3. This results in a projection size per pixel between 86 µm and 136 µm for the Perfactory Standard and between 43 µm and 86 µm for the Perfactory Standard in ERM mode. The *Perfactory Desktop* has a construction volume of x, y, z = 40, 30, 100 mm^3.

The layer thickness can be varied between 50 and 150 µm. Envisiontec calls that the *dynamic voxel thickness*. The *Desktop* is preset at a layer thickness of 38 µm.

The Perfactory Mini machine delivers a great solution in a smaller construction space. This can either be used to get finer details or to increase productivity. The construction volume is, at different optics, between x, y, z = 45, 34, 230 mm^3 and x, y, z = 110, 82, 230 mm^3. The projection size per pixel is then between 16 and

[5] SXGA, Super Extended Graphics Array, refers to an array of 1280 × 1024 image points in the field of computer graphics, and SXGA⁺ normally consists of 1400 × 1050 image points.

39 µm. This layer thickness can be varied between 15 and 59 µm (dynamic voxel thickness).

The Vanquish⁺ has a construction space, depending on the (zoom) lens, of x, y, z = 300, 254, 381 mm^3 up to x, y, z = 215, 162.5, 381 mm^3 and a projection size per pixel between 76 and 117 µm. The layer thickness can be varied between 25 and 200 µm (dynamic voxel thickness). In addition, it can be built in the Standard and in the Enhanced Resolution Mode. This results in values for the standard resolution of 0.246 and 0.157 mm, and for the enhanced mode, 0.144 mm and 0.093 mm. This corresponds to the best accuracy of ±46 µm.

Postprocessing

The supports must be removed and the part must be cleaned. An advantage can be seen that the parts can be cleaned with alcohol-free solvents.

Typical Follow-up Processes

All of the applicable follow-up processes for stereolithography parts are possible, especially vacuum casting (for parts of the material R5) and investment casting (for parts of the material PIC100).

3.1.9 Polymer Printing, Stratasys/Objet

Objet Geometries was taken over in 2012 by Stratasys. The machines are therefore branded with Stratasys, but have kept their Objet machine description. The Objet machines were integrated into the Stratasys Design Series Precision, including

Objet 24, 30 Pro, Eden 260V, 350V, 500V, 260 Connex, 350 Connex, 500 Connex, 1000.

The Stratasys Idea Series includes the Professional desktop 3D printers

Mojo, uPrint SE, uPrint SE Plus and FF.

The Stratasys Production Series includes the

Fortus FDM Series 250mc, 360mc, 400mc, and 900mc.

These machines are extrusion machines based on the FD method. They are described in Section 3.5, “Extrusion Process: Fused Layer Modeling.”

Stratasys Ltd., Eden Prairie, MN (Objet Geometries Ltd., Rehovot, Israel)

Short Description

Stratasys Objet encompasses a family of additive manufacturing machines using the Objet polymer printing process *PolyJet*, which uses a multinozzle print head to print liquid monomer directly on the component and then polymerize by high-intensity lamps. The machines are designed to be intuitive and suitable as office equipment, but because of their large space requirements, the Connex Objet 500 and 1000 should not be in an office. There are different and also flexible materials. The support material is washed away, leaving no contact points, which is an advantage for internal hollow spaces that can be cleaned only with difficulty.

Application

Concept models, functional prototypes, and components from different plastics

Development Status

Commercialized since the end of 1999

History

Objet was founded by Scitex Corporation (since January 2006, Scialex Corp. Ltd., Tel Aviv, Israel) and private investors in 1998. The first machine, the Objet Quadra, was introduced in late 1999 and the Quadra pace in November 2001. Then, almost at yearly intervals, the Eden330[6] (2003) and the Eden500V were introduced. To complement the machine series to smaller design spaces, the Eden250 (May 2006) was introduced on the RapidTech in Erfurt, Shanghai, and Birmingham. The Objet 1000 was introduced on the EuroMold 2012. Objet in 2012 was acquired by Stratasys, but the name Objet continued as a product name. From 2003 to early 2007, *Stratasys* took over distribution of the Eden machines in North America, but then abandoned it again.

Strategies/Development Partners

This machine family transfers the expertise in the production of large-format printers for the graphics industry to the production of three-dimensional components. Accordingly, the company develops the hardware, software, and materials itself. The design series consists of easy-to-use office-compatible machines that are designed to work as network printers. All machines can (with restrictions on the Eden250/260) handle all materials and use the same software. With Objet 1000 for large and with Objet24 for small components, Objet covers the entire range of macroscopic components.

Data Formats/Software

Using their own *Objet Studio software,* data sets from 3D CAD programs are taken in, automatically sliced, and provided with supports. The positioning in the building chamber is done manually with a "drag and drop" function, but the software

---

[6] In the US, known as Eden333.

makes a proposal for the orientation in space and positioning on the build platform. It accepts STL and SLC data sets. The machine controller also has a module to harmonize the wear of the nozzles.

Principle of Layer Generation

Each layer is generated by simultaneous printing of the building and the support material. The print head consists of eight (Eden260/350/350V/500V) or four (Eden250) multinozzle elements that apply half each of the building or support material. Simultaneous to the printing, the layer is cured by using two UV lamps, each mounted on the right and left of the print head and moving with it. The building chamber is passed through in the *x* direction in strips of 65 mm each, which are then transferred several times in the *y* direction, depending on the machine. The entire *x-y* plane is printed in every case so that after completion of the building process, a block of support material with the component located therein is removed from the machine. The minimum layer thickness is 16 microns and is very thin. This results in a very good model accuracy in the *z* direction. But thin layers are also required because then the drainage of the still-liquid material between coating and solidifying is substantially reduced. This is also a reason for the relatively low *z* dimension, particularly for the Eden500V.

Design/Construction

The Eden family builds on two substantially externally identical platforms. The smaller, also called *office-based*, leads to the Eden250/260. The larger, the *professional*, leads to the Eden350, 350V, and 500V. The 350 is the evolution of the Eden330. The technical core of all of the machines is the printing and exposure unit, which is the same for all, and the *z* mechanism for thin layers. The machine (with the exception of the Eden260/350) can be switched from high quality (HQ) to high-speed (HSR) mode, whereby the layer thickness is increased from 16 to 30 microns. The print head consists of eight individually replaceable segments (SHR, single head replacement) and the Eden250 of four. Each half of them transports building or support material. The traverse with the mounted print and exposure unit moves in strips, each 65 mm in the *x* direction completely over the component. The machines are stand-alone systems.

The building chambers vary from x, y, z = 260, 260, 200 mm^3 (258, 250, 205 mm^3) in the Eden250 (Eden260) to x, y, z = 350, 350, 200 mm^3 (Eden350/350V), to x, y, z = 500, 400, 200 mm^3 in the Eden500V. The accuracies are equal for the 250/260 and the 350/350V/500V if the same layer thicknesses are set.

The V variants have a material volume with up to four cartridges of 3.6 kg. The cartridges are leak-proof and are easy to change and switch for unattended long-term operation automatically. The V types have a high-speed mode, which builds much faster at approximately twice the thickness (30 microns) and with equal *x-y* resolution.

Material/Construction Time/Accuracy

There are three families of materials: standard plastic simulation, engineering plastic simulation, and medical. Simulation indicates that the materials behave similar to the simulated thermoplastic, but they are not identical to those.

Standard plastic includes the Vero materials that are used on all Objet machines (with the exception of the special dental machine Objet 30 Orthodesk, for which there are special materials). The Vero family enables essentially similar properties for producing white (VeroWhite), blue (VeroBlue), or black (VeroBlack) components.

Tango is a flexible material with elongations at break of about 50%. The black material, called TangoBlack, has a Shore A hardness of 61, the gray, TangoGray, one of 75. Tango materials are only suitable in the Eden and Connex machines.

The support material FullCure-705 is the same for all. It solidifies as a jelly-like material that can be removed manually by a water jet and washed out of difficult-to-access areas such as inner channels.

The material is supplied in 2 and 3.6 kg cartridges, which are integrated with all machines in an automated material management system, which switches automatically when a cartridge is empty.

Specifically for the production of hearing aid shells, there is the VeroDent material that can be processed only on the Objet 3D printers Dental and Hearing Aid. The transparent resins are available in colorless, pink, and skin colors but are the same in their other properties (see Figs. 3.19 and 3.20).

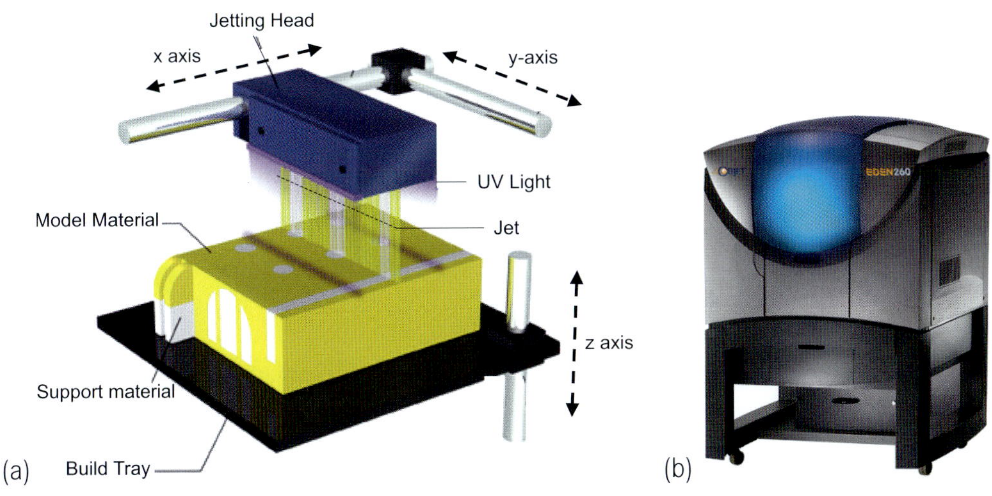

Figure 3.19 PolyJet: (a) principle; (b) Machine Eden260 (Source: Stratasys/Objet)

Figure 3.20 PolyJet component (Source: Objet)

The Connex machines can process two materials simultaneously and in different mixing ratios, so it can simulate two-component (for example, hard and soft) injection-molding components.

Depending on the parameter settings, a building speed of up to 10 mm/h is achieved.

The resolution in the *z* direction is 16 microns (alternatively and in some machines as standard, 28 microns) and in the *x* direction 600 dpi (corresponding to about 42 microns). In the *y* direction, the machines achieve a resolution of 300 dpi to 600 dpi.

The accuracy is about ± 1 pixel. A comprehensive study with the previous machine Quadra tempo certifies that the PolyJet method achieves excellent accuracy and very good surface quality [Grimm, 02].

Postprocessing

The component leaves the machine in a cured condition. The supports are easily removed manually by a water jet or washed out. Soaking in a sodium hydroxide solution is advantageous.

Typical Process Sequence Method

All process-typical subsequent proceedings, including vacuum casting, apply.

3.1.10 Multijet Modeling (MJM), ProJet, 3D Systems

Professional Printer, ProJet 3510, ProJet 5000

3D Systems Inc., Rock Hill, SC
3D Systems GmbH, Darmstadt, Germany

Short Description

The ProJet model series 3510 and 5000 work on the multijet modeling method (MJM method). They use photosensitive materials that are extruded and solidified immediately after the application of UV lamps. Therefore, they are stereolithography systems. The print heads have several piezoelectric nozzle rows arranged in a line. If the thermally assisted extrusion is moved into the foreground, one can consider the MJM method as an FLM method.

Application

Concept models, functional prototypes from investment casting

Development Status

Commercialized since 2002

History

3D Systems presented in 1997 the Actua 2100, a *concept modeler*, which was based on the *multijet modeling* process (MJM). The successor was the wax Sprinter ThermoJet (see first edition, Section 3.3.4.5, "Multijet Modeling (MJM), 3D Systems," pages 176–178). The InVision (2002) complemented the classic MJM method with a polymerization with light (stereo). With the unified designation of 3D Systems machines as ProJets, the model series 3510 and 5000 were assigned to the current generation of machines. For more on the history of 3D Systems, see Section 3.1.3, "Stereolithography Apparatus (SLA), 3D Systems."

Strategies/Development Partners

The MJM systems are referred to as 3D printers to underscore that they can be operated in CAD networks just as easily as today's 2D printer. The machine is consequently designated by the manufacturer as a *solid object printer*. It is not a 3D printer within the meaning of the MIT patent of 3D printing.

Data Formats/Software

With the help of the *InVision print client* software, the machine reads STL data and prepares the building data. The software allows editing of the input data, although this is not necessary according to the manufacturer's instructions for the actual print process. The user can visualize the CAD file with *preview* and look at different perspectives. It can copy files, resize, rotate, flip, and move. There is a switchable measurement function between inches and millimeters. The software also has an autocorrection device for incorrect STL data. Several STL files can be built next to each other on a platform and summarized as one job.

Principle of Layer Generation

Using each of multiple print heads, small drops of a thermoplastic acrylic photopolymer are applied as a building material and a waxy thermoplastic as the support material in layers of 0.04 mm thickness. Each layer is then smoothed using a roll called a *planarizer.* Then, the whole construction platform moves under a UV light unit that completely cures the building material, and the process starts again with the order of the following layer.

Design/Construction

In the 3500 series, there are several variants that are fitted to specific and industry-specific requirements. As a basic version of the 3510er series, the ProJet 3510SD can be considered. This has, like all other variants of this line, a space of 298 × 185 × 203 mm^3 and has a weight of 323 kg. Differences between the machines are, in particular, the possible layer thickness and the maximum resolution. The layer thicknesses can vary between 16 and 32 microns in maximum and the resolution between 375 × 375 × 790 dpi and 750 × 750 × 1600 dpi. The additions to the name of the variants give a clue to the application in which it can be used. Variants with a "CP" in the name are suitable for the RealWax method, in which a model is created from wax, and then it is melted out in an investment casting process.

In order to ensure the required accuracy for this indirect method, the provided variants are offered with a special thin layer thickness and a high resolution. For medical and dental production, a DP or MP is appended to the name of the variants. These variants do not provide as high a resolution as do the investment casting versions. But they are suitable for specific materials that are necessary in the dental field. Because a detailed description goes beyond the constraints of this book, it will be referenced to the data sheets in the appendix, A2.19.

The *ProJet 5000* is offered in only one variant. This has a space of 550 × 393 × 300 mm^3 and a weight of 538 kg. Depending on the selected print mode, different resolutions and layer thicknesses are set. It will work with a layer thickness of 64 microns and a resolution of 375 × 375 × 395 dpi in high-speed mode. In UHD mode (ultrahigh definition), the layer thickness is reduced to 29 microns and the resolution enhanced to 750 × 750 × 790 dpi. Using these values, it is readily apparent that a desired reduction in building time is at the expense of the accuracy of the model, see Figs. 3.21 and 3.22.

(a) (b)

Figure 3.21 ProJet 3510: (a) machine of HDplus variant; (b) component (Source: 3D Systems)

Figure 3.22 ProJet 5000 (Source: 3D Systems)

Material/Construction Time/Accuracy

For the variants of the 3510er series, there is a wide range of different materials. These vary especially depending on the application field of the machine. In the following, some materials are exemplified and explained.

VisiJet X: This material is a plastic material, which should have properties similar to ABS after the injection-molding process, according to the manufacturer.

VisiJet Crystal Plastic: This is a clear plastic material, which is very similar to standard design plastics. These properties make this material ideal for rapid prototyping, but also for rapid tooling and rapid manufacturing applications.

VisiJet Prowax: This material is a casting wax that can be melted out. It is suitable for medium-sized and large casting processes that are working with a lost form.

VisiJet Hi-Cast: This material is suitable for the highest resolution models with very fine details, according to the manufacturer. Corresponding parameter sets are only available from the ProJet 3510CPX, because this machine supports resolutions that exploit the potential of the material. Because there are no differences in melting point from the Prowax material, it seems likely that these two materials differ mainly in the viscosity.

VisiJet® Dentcast Dental Wax-up Material: This is a special material for the production of dental prostheses. According to the manufacturer, very smooth surfaces arise that are ideal for the production of crowns, bridges, and implants.

VisiJet® Stoneplast Dental Model Material: Stoneplast Dental is suitable for the economical production of dental models from digital data.

For some of these materials, 3D Systems has obtained a certification according to USP Class VI so that use in the medical and dental sectors is possible. The certified materials include the VisiJet® Crystal Plastic, VisiJet® Stoneplast Plastic, VisiJet® Clear Plastic, VisiJet® EX200 Plastic, and VisiJet® MP200 Plastic materials.

Postprocessing

The components are removed from the supports in a cleaning bath, the *InVision Finisher,* and can then be inserted. If the components are used for medical applications, the instructions of 3D Systems must be observed. This results in an increased postprocessing effort in certain circumstances because certain cleaning procedures must be strictly adhered to in order to get the USP certification.

Typical Process Sequence Method

The typical process sequence method depends strongly on the machine used and the planned field of application of the models. The "CP" models create master models for investment casting; as a result here a process of casting is necessarily followed. Concept models that are produced can often be used directly after the building process and require no further finishing work. In the case of MP and DP printers, the necessary finishing work depends on the selected printing process. There is the possibility of printing models with a chalky or with a smooth surface.

3.1.11 Digital Wax

DIGITALWAX

008J, 009J, 028J, 028J+, 029J, 029J+, 030J, dfab, 008D, 020D, 028D, 029D,030D, 029X, 030X

DWS S. r. l, Zanè (VI), Vicenza, Italy

Short Description

This is a stereo system using the laser-scanner process for producing fine melt-out components, especially for dental technology (D) and the jewelry industry (J), as well as design and product development (X).

Application

Duration models for the production of wax parts and lost forms and models for investment casting

Development Status

Commercialized since September 2003

History

DWS, Digital Wax Systems, was founded in Vicenza, Italy, in 2007. The predecessor company, *NextEngine*, is a manufacturer of scanners and (was) a distributor of Meiko in Japan, Europe, and the Middle East. Everything indicates that the offered machines are developed and built by DWS.

Strategies/Development Partners

The focus is on industry solutions for the jewelry industry (J), dental technology (D), as well as for design and product development (X), consisting of equipment, software, and aid equipment such as intraoral scanners. The DIGITALWAX series includes many machines whose sizes are finely tuned. NextEngine also sells stereolithography systems from Meiko (LC-510, LCV 700, and LCV 810) and from Autostrade (E-darts). The partners probably are the technology provider. Meiko is a company that focused on the Japanese jewelry industry. Meiko had built small stereolithography machines since the early 1990s and has developed a CAD program, JCAD, fitted to the jewelry industry. Autostrade has offered a stereolithography system since 1998 that builds upside-down like the Perfactory of Envisiontec. The strategy is underscored with the new DFAB, which is oriented as a desktop version in dental clinics, the closely related chairside version supports the strategy with the final restoration done in only one session.

New jewelry designs are also implemented, such as the direct casting of stones in the wax blanks (stone-in-place casting).

Data Formats/Software

The machine accepts STL and SLC data. The machine control works with the control software DigitalWax.

Principle of Layer Generation

The machine works according to the laser-scanner stereolithography process. The component is built overhead and exposed from below through a transparent plate. With the use of a solid-state laser, no post-cross-linking is necessary.

Design/Construction

In contrast to the similar Envisiontec principle, no mask is used, instead a laser with an *x*-*y* plotter method is used, which is designated as a scanner. The structure largely corresponds to the E-darts series of Autostrade. The construction building space is $x, y, z = 200, 200, 200\ mm^3$ in the DIGITALWAX 020 and $x, y, z = 200, 200, 60\ mm^3$ in the DIGITALWAX 010. The DIGITALWAX 010plus and 015 are not mentioned in the publications and on the website of the company. The external appearance is equal to the compact design of the desktop 010 and 020 (casing dimensions of about $x, y, z = 600, 600, 600\ mm^3$).

The DIGITALWAX 030 is a huge stand-alone machine with dimensions of $x, y, z = 1600, 650, 1900\ mm^3$. The building chamber dimensions correspond to the smaller DIGITALWAX 010. A solid-state laser called BluEdge (with a wavelength of 360 and 480 nm) is used. The machine is equipped for autonomous long-term operation and has its own air conditioning. See Fig. 3.23.

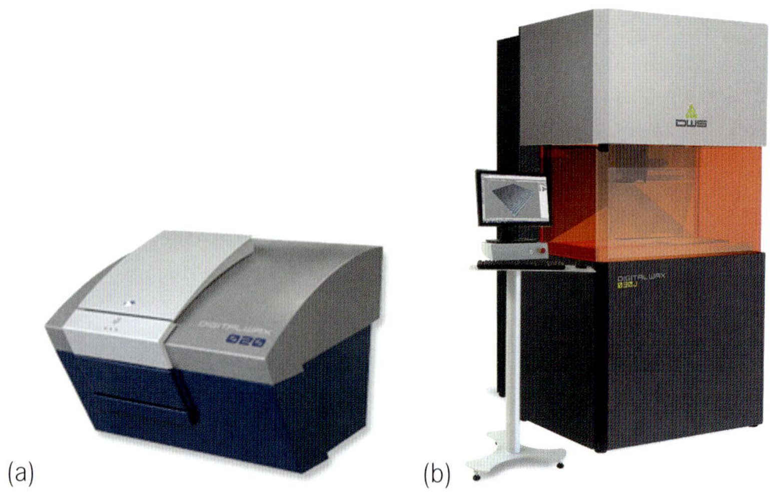

Figure 3.23 DWS: (a) DIGITALWAX 020J; (b) DIGITALWAX 030J
(Source: NextEngine)

Material/Construction Time/Accuracy

Proprietary polymerizable waxes are used. The model can be machined and polished by hand and is completely melted out of the lost form. Another family of materials is designed as a molding material for wax castings or for the casting of plastics. The layer thickness is 0.001 to 0.10 mm. However, the resolution of the hardware is also indicated in all directions as x, y, z = 0.01 mm.

Postprocessing

The components leave the machine ready for the subsequent processes. Postcuring is necessary.

Typical Process Sequence Method

The parts are either used directly or as investment-casting molds for the production of wax models or plastic components.

3.1.12 Film Transfer Imaging, 3D Systems

ProJet Personal 3D Printers: ProJet 1000 Personal 3D Printer, ProJet 1500 Personal Color 3D Printer

3D Systems, Rock Hill, SC

Short Description

This is a cheap desktop solid imaging system based on the film transfer imaging (FTI) process for the production of plastic components for the stereolithography process.

Application

Solid imaging and production of concept models and functional prototypes

Development Status

Commercialized (since November 2001, ProJet 1500; November 2012, ProJet 1000)

History

For information on 3D Systems, see Section 3.1.3, "Stereolithography Apparatus (SLA), 3D Systems." The ProJet™ Personal 3D Printer family is the further development of the V-Flash FTI 230 Desktop Modeler.

Strategies/Development Partners

The ProJet machines are based on the FTI process. They are compact, inexpensive, and low maintenance. They are designed as low-cost, fast but accurate machinery. Because of the material cartridges, the conditioned storage of material in the machine is bypassed and also allow the occasional operation. The components do not require postprocessing. Strategically, the mass market is targeted. The high-quality components are silhouetted against the fabber. In the background of the development is the manufacturer's experience in the areas of materials development and processing of photopolymers, especially of pastes, as well as in dealing with films. The low machine price is offset by higher material prices. This is a benefit for occasional users. The machine does not require an installation. See Fig. 3.24.

Since 2007 there has been a development partnership with Drewe Otoplastic, which has led to the development of a production system for hearing aid shells, the V-Flash HA 230 Production System. In this consortium, Drewe takes over the materials development.

Figure 3.24 ProJet 1500 Personal Color 3D Printer, machine. External view (left), process (right) (Source: 3D Systems)

Data Formats/Software

The machine is network-capable. It handles error-free STL data and reads CTL[7] data.

Principle of Layer Generation

The building material, applied as a layer to a carrier film, is pulled over the building plane made of glass and positioned so that the material shows in the direction of the component (upward) and the foil in the direction of the DLP exposer. The exposure takes place from below by the glass plate and the carrier film. The materials belonging to the model (including the supports) are exposed and solidified and adhere to the pressed-top (or on the semifinished component) building platform. The unexposed material adhere to the film, and can be removed with the solidified material and support structures. The carrier film with the deposited material is further transported to a space dimension and repositioned, and the next layer is exposed. After solidification of the last layer, the component is removed with the building platform out of the machine. The used film with the material remains is transported back into the cartridge and replaced with it.

[7] CTL is a file format for an ActiveX control file of Microsoft Visual Studio.

Design/Construction

The compact machines (x, y, z = 555, 914, 724 mm^3) have a space of x, y, z = 171, 228, 203 mm^3 (ProJet 1500) or x, y, z = 171, 203, 178 mm^3 (ProJet 1000). The machine is exposed with a DLP module in the variant of a patent-pending Flash Curing[8] system. From this probably the name V-Flash is derived. All material handling, including ensuring the exact layer thickness and the calculation of the available amount of material, is transferred to the material cartridge. The machine has neither a material reservoir (except the cartridge) nor a coating system. From the cartridge, a transparent film comes from the width of the building space (about 180 mm) with a polymer layer applied there in the thickness of a layer. The layer thickness is 102 microns. It can be increased by the ProJet 1500 in the so-called high-speed mode to 152 microns.

Material/Construction Time/Accuracy

The building material is a photopolymer of the acrylate type. The manufacturer promises components of stereolithography quality specifically in terms of accuracy, resolution, and durability. The VisiJet FTI materials are delivered in 2 kg cartridges and returned with the waste (carrier film and support material). They are available in the colors red, gray, blue, and black.

The building progress is, according to the manufacturer, 12.7 mm/h (0.5 in./hour) and in the high speed mode, 20.32 mm/h (0.8 in./hour) and is thus higher than comparable machines.

The smallest representable detail amounts to 0.254 mm, and the thinnest freestanding wall must not be less than 0.64 mm. The resolution in the building plane (x, y) will specify 1024 × 768 dpi.

Postprocessing

Finished components leave the machine. They need to be cleaned in a part washer and postcured in a part curing unit by means of UV radiation. Both are offered by the manufacturer, but they can hardly operate in an office.

Typical Process Sequence Method

Because of the concept used, no follow-up procedures are provided. Molding processes such as vacuum casting and perhaps even the use as a cast model are possible.

[8] US Patent Application 20050253308

3.1.13 Other Polymerization Processes

3.1.13.1 Paste Polymerization, OptoForm

Opto 15/400, 40/400, 90/800

OptoForm LLC, Rock Hill, SC

Short Description

The OptoForm process is a variant of laser stereolithography, with pastes used in place of liquid monomers. The material is applied with a squeegee. The process allows for the direct production of composite parts.

This interesting machine has been running since 2002 in selected research centers in test mode, but it has not been commercialized. It is therefore not included in this edition.

■ 3.2 Sintering/Selective Sintering: Melting in the Powder Bed

The additive manufacturing technology "sintering" is known as laser sintering because of the use of lasers when it was developed in the 1990s. To separate the process from the industrial sintering processes, the name "selective laser sintering" is more suitable because the solidification of the material is limited to a selected area. Since 2001, when a machine that uses an electron beam instead of a laser beam (Arcam) was introduced, and since 2005, one that works with an infrared source (Speed Part AB), the generic name "selective sintering"[9] or "sintering" is correct, so for the following chapter the term "sintering" is used. In combination with the processing of metal, the term "melting" also has its relevance. Sometimes the wording "beam melting" is used for general purposes.

3.2.1 Machine-Specific Basic Principles

Principle of Layer Generation

In the laser sintering process (see also Section 2.3.2.1 and [Noe, 97]) particles of usually 50 to 100 µm, which are closely packed together to form a powder bed, are slightly pressed if this is necessary for the process. They are then slightly melted locally by a laser beam, solidified by cooling due to heat conduction, and are

[9] "Selective sintering" in the context "Selective Laser Sintering," is a protected name of 3D Systems.

thereby joined together to form a firm layer (see Fig. 3.27). The unmelted particles stay in the part as support material and are removed after the end of the process. By lowering this layer and recoating it with powder is analogous to the first layer, the second layer is solidified and connected with the first. Figure 3.25 shows the principle of the process on the basis of laser sintering.[10] Processes with other energy sources work analogously.

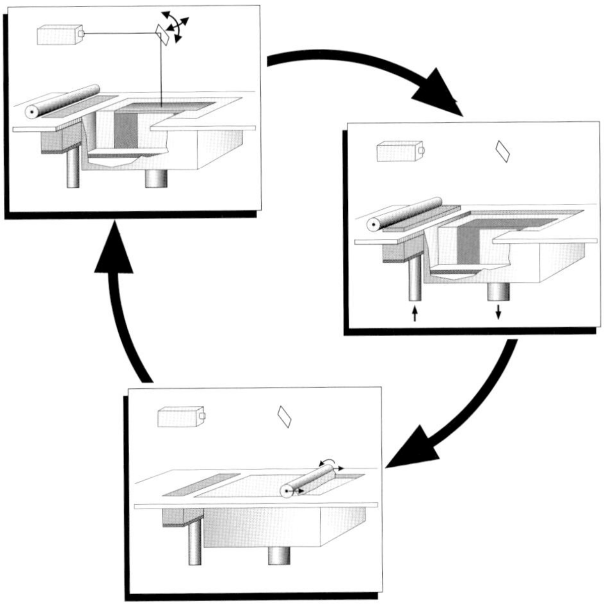

Figure 3.25
Principle of laser sintering and melting

Design

The process chamber offers the building room, which can be enhanced by lowering the bottom of the chamber by one layer height. It is important for the technical functioning of this process to construct the process chamber in such a way that it can be preheated up to very nearly the melting temperature of the sinter material. This is the requirement for the quasi-isothermic sintering of plastics. The energy source (laser or electron beam, infrared lamp) therefore only needs to add a small differential energy for the sintering. The process temperature must be kept as steady as possible and within narrow tolerances (a few degrees). Furthermore, oxidation of the material must be avoided, which is usually achieved by inerting the machine. This is done by generating a nitrogen atmosphere inside the machine (0.1 % to 3.5 % residual oxygen, depending on the material).

The laser sintering process always works without supports because the unsintered powder stays in the bed and supports the model. Depending on the geometry of the

[10] A 3D animation can be found at *http://www.rtejournal.de/filme/SLS-RTe.wmv/view*.

model and the material being used, it has proved to be useful to build a platform (base) as well and to build on this base. Metal processes reduce tension by the use of support structures. In addition or alternatively, firm understructures can be constructed if required using specialized construction strategies. Starting with about half laser performance and double scanning speed, and by adjusting the laser performance as well as the scanning speed to the optimal parameters within several layers, structures with low distortion are generated that favor a reliable model construction.

To maintain an even temperature field, it is useful in some cases to place loose parts deliberately near the model or to put a grid around the component.

Metal processes work with support structures to position the parts at the platform and to avoid deformation.

Some users suggest positioning the model, especially long drawn-out parts, under the so-called Kodak angle relative to the cylinder's longitudinal axis. It is assumed that especially distortion-free components can be generated under this angle. Some sources mention an angle of about 15°, others about 10°.

In practice, nearly all models are positioned in the machine without a base according to the criteria of accuracy and economy. When the build parameters are carefully matched, and especially when the temperature fields are kept even, the results are excellent.

To achieve models without defects, it is necessary to adjust and control the temperature and the temperature distribution in the build chamber very carefully. Temperature differences of a few degrees can cause useless components that are either poorly sintered or deflect due to excessive heat and even change color, depending on the material. Therefore, temperature control plays an important role. Current machines have very precise temperature control.

Material

Plastic materials are best suited to laser sintering owing to their relatively low melting temperatures and their low heat conduction properties that facilitate a local limitation of the sintering process. The process can, however, be used in principle with all materials that can be incipiently melted and sinter-fused under thermal treatment and that solidify again after cooling. The term thermoplast, usually used only for plastic material, must therefore be understood as a generalization.

Particles of 20 µm up to about 100 µm diameter are used in sintering machines. By sifting, fractions are achieved with grains that are usually 90% below the given diameter. Correspondingly rough surfaces are the result. This is especially the case with amorphous materials (polystyrene, polycarbonate), which tend to melt only incipiently and therefore essentially retain their form. Crystalline materials (polyamide) that become completely liquid form smoother surfaces owing to their surface tension. They tend to form dimples, however, so rough surfaces result here as well.

Because the sintering process is not a pressurized process, the achievable density is lower than with injection-molded materials. This applies especially to amorphous materials that have relatively large hollow spaces and therefore reach densities of only 60 to 85%. These properties, which act negatively on the load-carrying capacity of the parts, are beneficial toward parts that are melted out in a precision-casting process. Amorphous materials are more easily melted out and are not as susceptible to *"shell cracking."* Crystalline materials can easily reach densities that are very near those of injection-molded parts.

There are differences also in the shrinkage of crystalline and amorphous materials. Crystalline materials alter their volume during the melting process quite significantly and therefore tend to shrink. Amorphous materials are considerably less prone to shrinkage, depending on the density achieved. The shrinkage usually leads to curling, the tendency of the models to "roll up" in the direction of the heat source. This tendency can in principle be controlled by shrinkage factors, which should already be taken into account during the generation of the construction data. Because these factors are empirical values, they cannot be defined with sufficient accuracy right from the start.

Theoretically, there is no limit on how often thermoplastics could be melted and solidified again. But the materials could not be used for further build processes even if they have only served as support material. They receive thermal damage due to the temperatures in the sintering process. Therefore, new powder must be used for every build process. As a guideline, a 0.5:0.5 up to a 1:1 mixture of old to new powder is considered. The exact conditioning is a matter of experience.

The sintering process is today used for plastics, metals, ceramics, and resin-bonded sands. Metal parts are fully melted and therefore become dense.

Construction Time/Accuracies

Construction times and accuracies depend on the geometry and therefore on the positioning of the component in the construction chamber.

The build time in the *x-y* direction is much higher than that in the *z* direction. Typical build times are approximately up to 10 mm/h. The accuracies are limited by the laser beam diameter. Beam diameters of about 0.4 mm result in accuracies of ±0.15 up to 0.2 mm and the finest details such as freestanding walls of 0.5 mm thickness minimum.

Due to the heat conduction of the component into the surrounding powder, particles of the surrounding powder are baked onto the contour of the part. Even small faults of the temperature field lead to a reduction of the precision and surface quality. Therefore the model accuracy depends on an exact process control.

Postprocessing

For plastics, after the construction process, the compound is completely enclosed in a powder cake. Most sintering processes take place at a temperature of between 170 and 200 °C, especially when plastic materials are sintered. To ensure a uniform cooling, it is important that after the last layer an additional layer of powder several centimeters thick is applied. After the compound has cooled completely, which may take several hours owing to the poor heat conduction, the powder cake is carefully removed from the exterior (Fig. 3.26). If changeable build chamber modules are used, the removal and cleaning process takes place half-automatically in screens and with high pressure outside the machine. Although, in theory, the compound is merely embedded in a loose powder cake and only needs to be taken out and the powder blown off, in practice it is wise to proceed carefully. First, the compound can be easily damaged because its position is not exactly known; second, there are, depending on the temperature control, slightly sintered areas around the model (fleeces) that have to be removed with great care by using special tools. Therefore, patience and skill are required in cleaning sintering models, especially those with internal hollow spaces, drillings, and fine details. To exacerbate the situation, the models and powder have the same color.

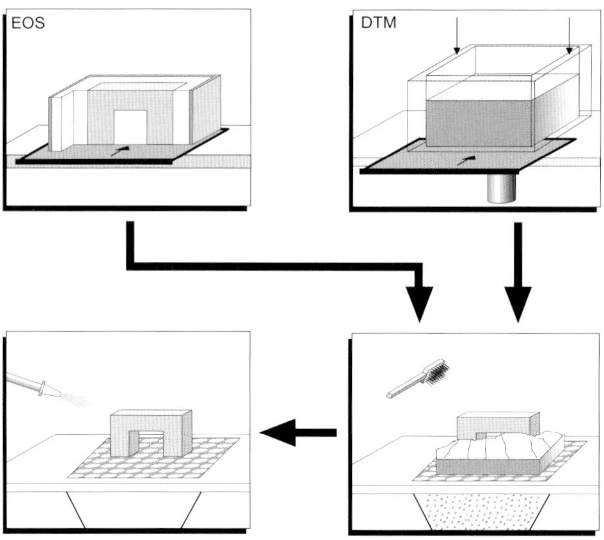

Figure 3.26
Laser sintering, postprocessing principle

For metal processes, the postprocessing is easier because the process chamber and the powder stay nearly cold. The problem is that they use support structures, which have to be removed. Direct ceramic sintering machines work with high preheating temperatures.

After the parts are cleaned in the postprocessing, the surface is treated further by manual polishing or sand blasting. Parts of models or broken-off parts of models can be glued on with *cyanoacrylate* glues or with epoxy resins. To finish the models, cylinders filled with polishing material can also be used. It has to be taken into account that the type of the abrasive used determines the amount of material removed. There is a high risk of rounding sharp-edged corners. Because sintered models are generally porous, all infiltrating surface sealings could generally be used. This includes all kinds of hard wax, epoxy resins, and also primers on an enamel base.

Every kind of postprocessing changes the geometry. This is important because the additive manufacturing processes produce parts directly from the 3D data sets.

Follow-up Processes

Laser sintering processes are preferably used for functional prototypes (functional parts) or for direct production (rapid manufacturing). Therefore, the direct application is used more often than casting processes. Vacuum casting is possible in general but requires that the surfaces be exceedingly well finished.

Metal parts could be achieved either by direct selective laser melting (SLM) or by precision casting on the basis of additive manufactured master models. It is especially interesting to use polystyrene or polycarbonate models directly for precision casting. This procedure is successful only if an appropriately careful wax impregnation and surface preparation are applied. The model itself must be preheated up to approximately the temperature of liquid wax (between 190 and 210 °C). The impregnation can take up to 30 min; if critical spots have to be reworked manually, it could take significantly longer. Finally, the classic precision casting process is started. The thorough baking should be done with sufficient oxygen so that the chemical reaction is supported. A ventilator is probably helpful.

In areas with thick walls, ash can accumulate that is difficult to remove later. Such material accumulations can even cause local shell cracking. A number of precision-casting foundries, however, have in cooperation with the model makers, managed to control this process so reliably that for every polycarbonate model presented, a casting model is returned. The advent of CastForm PS should solve these problems once and for all.

Following the basic principles of laser sintering, machines have been designed that differ especially in the details, the target applications, and consequently in the materials used. At first some of the constructive solutions were devised owing to the patent situation, but later they prevailed and blazed the way forward.

3.2.2 Overview: Sintering and Melting

In Fig. 3.27 a structural overview of the sintering and melting processes is given. Referring to the generic definitions, it contains the industrial processes and machines and the product and manufacturer definitions.

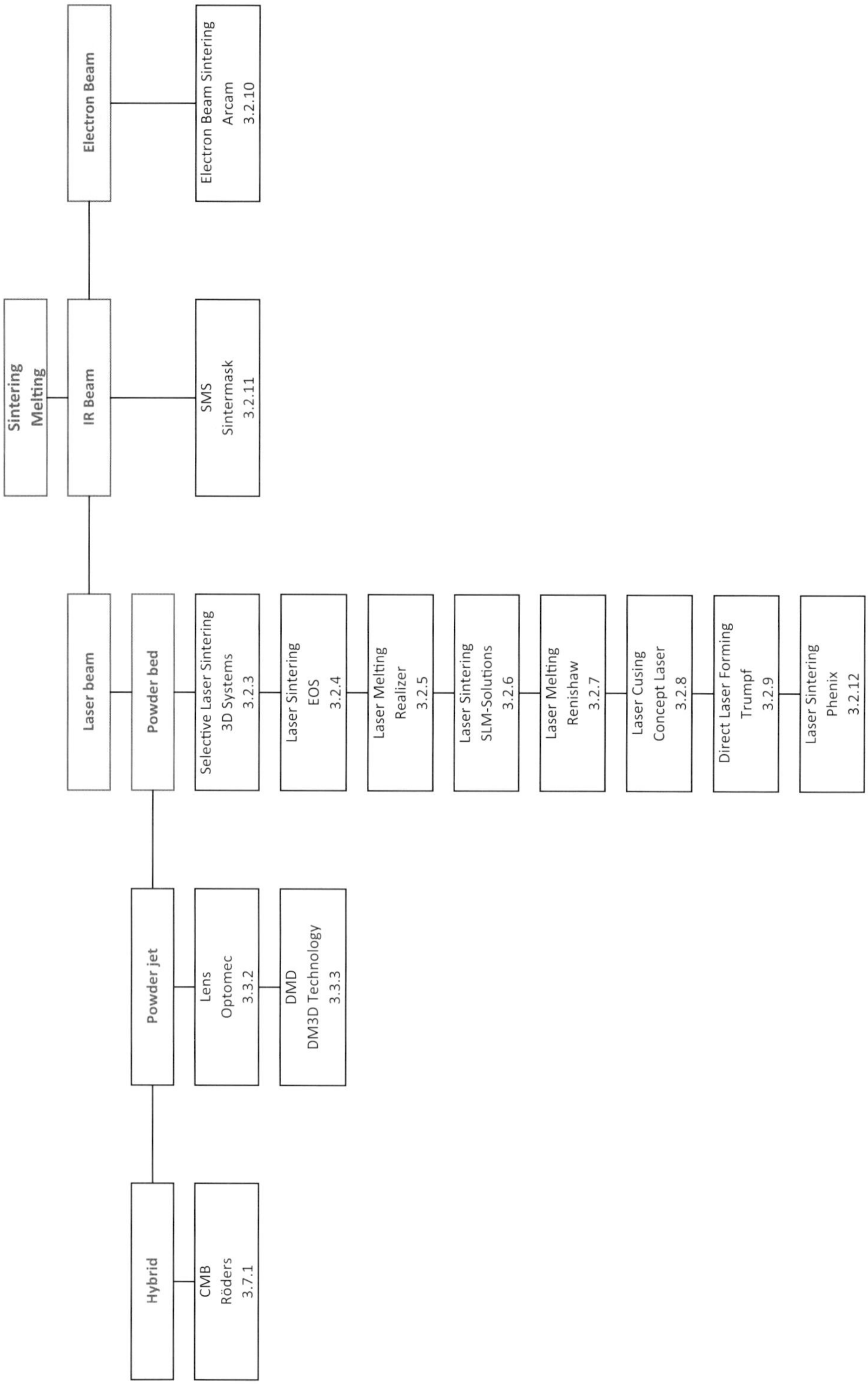

Figure 3.27 Sintering and melting processes, continuation of Fig. 2.38

Included are all processes that work with a powder bed (Section 3.2, "Sintering/Selective Sintering: Melting in the Powder Bed") and those that freely build up with a powder nozzle (Section 3.3, "Coating: Melting with the Powder Nozzle").

The sections that describe the processes more precisely are given in the figure.

3.2.3 Selective Laser Sintering, 3D Systems/DTM

Production 3D printers:
sPro60SD, sPro60HD, sPro140, sPro230, sPro125
DirectMetal, sPro250 DirectMetal

3D Systems Inc., Rock Hill, SC

Short Description

This is a family of prototypers for the sintering of plastics, metals, and sands. The machines work on the laser-scanner principle. A CO_2 laser scans the surface of a powder bed and incipiently melts the particles that form the model after solidification. The process is repeated after every layer. Owing to the process technique, no supports are necessary.

Range of Application

Plastics: Geometrical prototypes, functional prototypes (technical prototypes)

Metals: Technical prototypes, direct tooling (series parts via tooling)

Sands: Series parts

Development Status

Commercialized by DTM in 1992, since 2001 as 3D Systems

History

Sintering activities by 3D Systems go back to the DTM Corporation of Austin, TX. The DTM Corporation was established in 1987 by Carl Decker as a branch of B. F. Goodrich, and at the beginning of 1999 it sold to a group of investors. The laser sintering process is called selective laser sintering (SLS) by DTM and is protected by patents (Carl Decker). The first machine, the Sinterstation 125, was introduced in 1989–1990. For some time it was used in the service office (from 1990 onward). The first industrial machine for SLS, the Sinterstation 2000, was developed simultaneously. Marketing began in the United States in 1992 and in Europe in 1993. The largely identical Sinterstation 2500 with a rectangular construction chamber followed in 1997; the Sinterstation 2500^{plus} was introduced in 1998. In 2001, DTM was taken over by 3D Systems. After the takeover, the Vanguard (2001) was developed, which is based on the 2500^{plus} hardware with some specific improvements such as scanners, heating, and control. In addition, the software was upgraded. The actual machine generation is called the ProX and sPro series. There is one line of plastic printers for selective laser sintering (SLS) and one metal line for selective laser melting (SLM).

Development Partners/Strategies

3D/DTM wants to be able to use every kind of material in one prototyper. Software and material packages have been developed that are matched to one another. Therefore, the user has the freedom to work with different materials in the same machine. The costs for the software/material packages, the stockage, the time to convert the machine, and the necessary training lead to a practical use of only one material in the machines. The actual machines do not work on a one-machine principle, so there is one metal and one plastic series.

Development partners of 3D Systems are the University of Texas and the Institute for Rapid Product Development at the University of Applied Science, St. Gallen, Switzerland.

At the moment, 3D Systems targets the market of additive manufacturing. They call it *advanced direct manufacturing* (ADM). Therefore, the sPro line was developed as a manufacturing system, which includes the machine with changeable building chamber modules and supplemental modules for the powder handling and the cleaning of the building chamber. This is accompanied by the development of materials for direct manufacturing, such as the fire retardant material DuraForm FR.

This strategy is underlined by the Manufacturing Center, and the affiliate 3D Systems University at its headquarters in Rock Hill (South Carolina).

Another strategic application is metal mold inserts for injection molding and die-casting tools.

Early in 2004, a long-standing patent dispute between 3D Systems and EOS GmbH was settled and a patent exchange was agreed upon.

Data Formats/Software

The machine reads error-free and complete STL data. Neutral interface formats can also be read by means of an additional module.

The standard program system supplied by the producer is similar to the MAGICS RP Software from Materialize and supports the most important operations. Optionally, there is the SinterScan software for better surfaces and the RealMonitor software with an SPC (statistical process control) module, which provides important system information during the process and support its evaluation.

All in all, the machine control is organized in such a way that the computer calculates one layer at a time while the machine is constructing the preceding one. This process is effective as long as the calculation of one layer takes no longer than the construction of the preceding layer. Otherwise the machine has to wait for the slice algorithm. This can easily happen with extensive data records and complicated geometries such as, for example, the cranial base of a medical cranium model.

Principle of Layer Generation

The scan strategy consists of a raster scanning of the relevant cross section (raster technique). This also results by necessity in stair steps in the construction plane, which further diminish the quality of the surface. The stair-stepping effect manifests itself when the model is generated without any additional borders. To make the stair-stepping effect less noticeable, the scan direction is turned after each layer generation, thereby spreading the steps more evenly over the model. If this should be avoided (especially for thin walls), scan strategies that run along the contours first and then raster scan the internal areas are preferable. To avoid thermal imbalances, the scanning takes place alternately in the $+y$ and $-y$ direction, that is, from the "front" or from the "back." The sPro60 in the variant HD + HS (high speed) is offered with a HS Celerity BDS (beam delivery system) and a 70 W CO_2 laser, thus building much faster than the basic version. Digital scanning systems, as in the sPro, support complex scanning strategies. The optional SinterScan software optimizes the building process by ensuring identical component properties in the x and y directions.

System Type/Construction

The process chamber of the *sPro* consists of two reservoirs and two overflow containers, and between them is fitted the build chamber with the powder bed. The containers are firmly installed and form a structural unit. In the previous models, they were located below the building plane. Since the introduction of the Vanguard and also in the HiQ, the containers are fitted above the building plane. This allows for an automatic refilling, as it is installed in the sPro. The material is applied to the powder bed by a roller rotating against the transport direction. Excess material enters the opposite material cylinder or the overflow container. The opposite material containers allow a bidirectional coating and therefore a faster repetition of the coating process. The building chamber contains a moveable bottom, which is lowered by one layer height after every finished layer. Figure 3.28 shows the principle of the construction of a 3D sintering station. It is identical to the Sinterstation HiQ (see Fig. 3.29), except for the overhead powder containers.

The sPro represents a modular manufacturing system (see Section 6.3.3.1, Fig. 6.20). It consists of the actual machine with a nitrogen generator and condenser, an external thermostated and interchangeable building module (rapid change module,

RCM) with temperature control (offline thermal station, OTS), the unpacking BOS (breakout station), and the powder management system. This is fed by transport and storage containers (intelligent powder cartridges, IPC). The powder management system mixes new and unused (but has been in the process) powder and feeds it to the machine. It also ensures a uniform powder quality. Due to the preheating and the cooling in the RCM outside the machine, loss by these processes is minimized, and the machine is not blocked for the building process at these times. In the following, the SLS models are now first explained in detail, and then the SLM series is described, see Figs. 3.28–3.31.

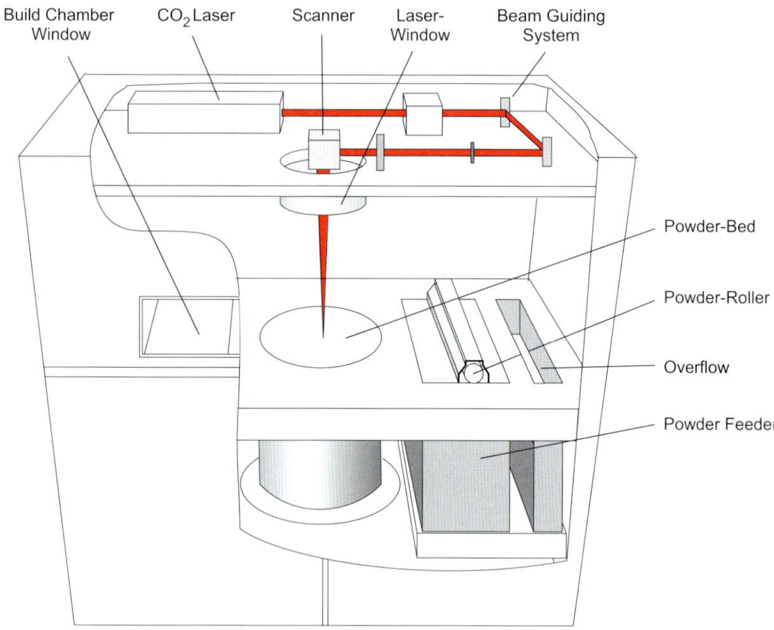

Figure 3.28 Construction of a 3D sintering machine, principle

Figure 3.29 Sinterstation HiQ (Source: 3D Systems)

Figure 3.30 sPro 60 (Source: 3D Systems)

The sPro 60 is offered in three different variations. There is the SD, the HD Base, and the HD-HS version. They all have the same build chamber dimensions of 381 × 330 × 457 mm^3 and a flexible layer height between 0.08 and 0.15 mm. There are differences in the scanning speed and the laser power. The speed for the SD version is 5 m/s, for the HD Base 6 m/s, and the HD-HS version can use 6 m/s and also 12 m/s. All versions use a CO_2 laser with a laser power of 30 W for the SD and HD Base versions and 70 W for the HD-HS version.

The sPro 140 (Fig. 3.31) and the sPro 230 are available in a Base and an HS version. The numbers show the powder volume that they can contain. For the sPro 140 this is 140 liter, and for the sPro 230, 230 liter. The dimensions in the x and y directions for both machines are 550 × 550 mm^2. The building height is 460 mm for the sPro 140 and for the sPro 230, 750 mm. The differences are the same for the Base and the HS version. Thus the scanning speed is 10 m/s in the Base version and the laser power is 70 W. The HS version offers a laser with 200 W power and a scanning speed of 15 m/s.

The entire machine is preheated to a temperature of about 4 °C lower than the melting temperature of the applied construction material. The machine is then flooded with nitrogen. A continuous nitrogen stream through the powder bed supports an even temperature distribution in the building space.

There are two different metal machines (SLM). One is the sPro 125, and the other is the sPro 250. The sPro 125 offers a build chamber of 125 × 125 × 125 mm^3. The laser power depends on the chosen configuration and can differ between 100 and 200 W, with a spot diameter of 35 µm at the powder surface.

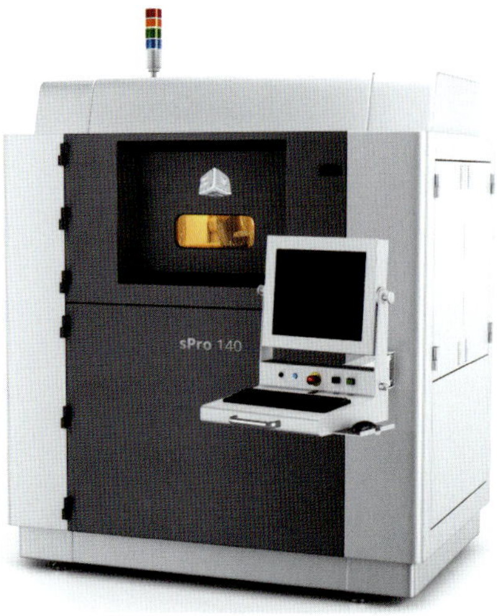

Figure 3.31
sPro 140
(Source: 3D Systems)

The sPro 250 has a build chamber of x, y, z = 250, 250, 320 mm^3. In comparison to the sPro 125, it offers a higher laser power, which could be 200 W or 400 W, and the laser spot has a size of 70 µm at the powder surface. See Fig. 3.32.

Figure 3.32
sPro 250
(Source: 3D Systems)

Both machines can build layers with a height between 20 µm and 100 µm, and the building speed is between 5 and 20 cm^3/h. The scanning speed could be 1000 mm/s at maximum.

The SLM machines do not heat up the whole chamber. Only the platform is heated.

To reach the build chamber, the laser beam has to cross a special glass window, which is made of zinc-selenite. To clean the machine, the window can be removed. To cool the glass and to avoid pollution of it, a constant stream of the entering nitrogen is flushed around it. The nitrogen is heated by this stream around the glass, but also by the preheating of the chamber. The user of the machine can watch the process through a big window, which shields the process from the surroundings.

To avoid deformation of the model, in addition to correctly orienting the model in the build chamber, careful preheating of the machine is necessary. An optimal heating across the powder bed and of the upper walls of the build chamber lead in conjunction with a fast regulation to uniform temperature fields.

To permanently ensure optimum recoating, it is advisable to specifically roughen the heated coating roller at longer intervals (dressing the roller).

The supply of nitrogen is carried out by a (built-in) air separation unit. This ensures that the nitrogen does not go out during the construction process. If it did, the process would be aborted and the component would be useless.

Before removing the model, the entire machine must be cooled at the sintering station HQ. This can take several hours to complete, depending on the material. Accordingly, the entire machine must be warmed up again for the next construction process. In order to remove the components, the entire machine must be suspended to the ambient oxygen-containing atmosphere. Therefore, a rebuilding process requires a new inerting. The cooling and inerting must be added to the construction period and are included in the calculation. In the sPro 125 and sPro 250, only the RCM module is changed. The machine is then immediately ready for use.

The calibration of the machine is done by exposing a special film with markers. The evaluation is carried out by the manufacturer in the US. The procedure is lengthy.

Material/Construction Time/Accuracy

The family of materials for laser sintering is called the Form LS materials. It includes plastics, which are called DuraForm, metals, which are called LaserForm, and CastForm, which is used to produce lost forms for investment casting.

Sand mold Si (silicate-based) and sand mold ZR (on zirconium base) are no longer offered. The same is true for waxes that were available for some time with the introduction of the first machine in 1992. Both are indications that the machine can handle a much wider range of materials than is currently offered commercially.

It should be noted that the usable build room could be reduced in accordance with the material used, due to prescribed minimum distances from the wall.

Material – Plastic Materials

DuraForm PA, GF: Under the name DuraForm, two polyamides (approximately equivalent to PA 11) are available for all laser sintering printers, either as a pure nylon (DuraForm PA) or as a glass-filled one under the name DuraForm GF. The grain size is typically 90 % < 50 µm and is equivalent to the former fine nylon. Glass fibers cannot be used because of the process, but glass globules may be mixed into the powder. They impede extension, thereby increasing the mechanical stability. The SEM photo (Fig. 3.33) shows that, because of the accumulations of globules, internal hollow spaces are formed that influence the dynamic properties of the material; test results are therefore not necessarily applicable to series parts.

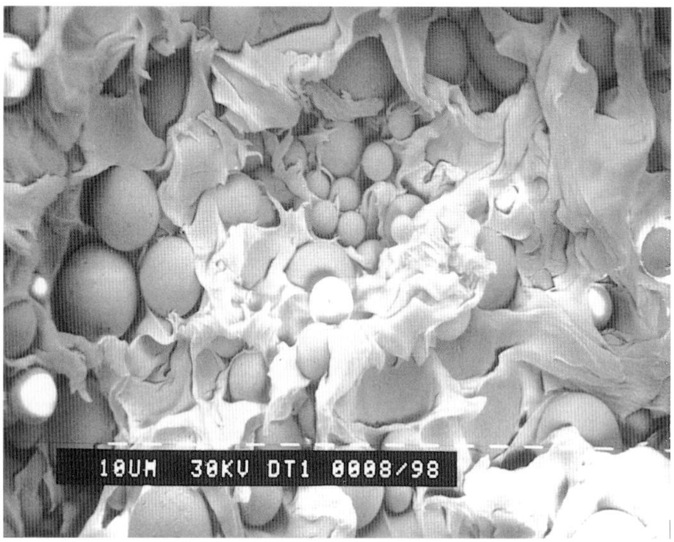

Figure 3.33 SEM photo of glass-filled polyamide (Source: Conrads/FH Aachen)

In 2007, Boehler[11] and Levi introduced a new material, a fiber-reinforced material named Polyamide 12 [Boehler, 07], which is currently referred to as 06-430-3. It is expected to be called V6. In this material, there are no fibers embedded in the matrix of the PA, but the fillers are shaped like suppositories (Fig. 3.34). If these fillers have a length of 300 µm and an aspect ratio of around six, they can be handled in the process. In comparison to unreinforced PA, it has a tensile strength that is around 15 % higher and is also stiffer (around one-third of the elongation of unreinforced PA, but one-fourth of glass-filled polyamides). The tensile modulus is around

[11] The Valspar (Switzerland) Corporation AG, www.valspareurope.com.

20% higher than that of glass-filled materials, and the heat resistance (HDT) is at least 10% higher than the values for glass- and aluminum-filled materials. That means it could reach around 165 °C. The material started in a beta phase in 2007.

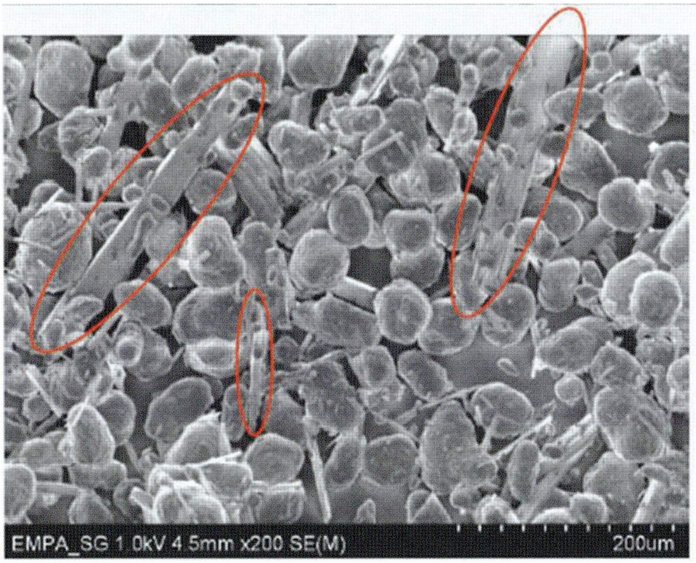

Figure 3.34 SEM-26 photo of reinforced polyamide (Source: Levy/FHSG)

Sintered parts are porous, so they have to be infiltrated (with epoxy resin) if they need to be pressure-tight.

DuraForm Flex: For elastic models, a thermoplastic elastomer is available that is known as DuraForm Flex. It is used for the simulation of rubber-elastic materials. The same Shore hardness series properties such as tightness and locking capability can be tested at room temperature. The behavior of the materials differs considerably between low temperatures (< – 30 °C) and high temperatures (> 50 °C), as well as over long periods of time. In addition, the flexible material Somos 201 is still available.

CastForm PS: CastForm PS is used as a material for the production of fusible models for investment casting. It is a polystyrene-based material with a low density that could be used for aluminum and titanium investment casting, but also for low-melting alloys. CastForm PS is optimized for the standard casting process. It is autoclavable, softens at about 80 °C, and therefore avoids the sprinkling of the shell (shell cracking). It has a low water absorption and a low residual ash content of 0.02 %.

Duraform EX: Duraform EX is a high-strength and temperature-stable material that has approximately the properties of polypropylene or ABS plastic. It is typically used to reflect functional details like snap hooks or film hinges. It was developed as a material for direct production and is available for the current machines.

DuraForm AF: DuraForm AF is a polyamide-aluminum mixture in a ratio of about 40:60 (PA:AL), which gives parts the appearance of die-cast aluminum components. It is used for the production of Al-geometry prototypes and for the direct production of components that should have an appearance of aluminum (see also the discussion above about material 06-430-3 and DuraFormGF).

DuraForm FR: DuraForm FR is a flame-protected plastic, developed for aerospace components. The material is aimed at the direct production of parts. It has passed the FAA[12] tests regarding flammability and the production of smoke and toxic gases. The mechanical and technological properties are similar to those of DuraForm PA. DuraForm FR is available only for the sintering Station Pro.

Material – Metals

DTM has developed a multistep indirect metal process (RapidTool) on the basis of polymer-coated metal powders (RapidSteel 3.0) for the production of metal models and molds. This process is based on the knowledge that all existing sintering machines can sinter any material if it is embedded in plastic. On this basis, a multistep metal process–the so-called indirect metal process–was developed.

First the model is sintered in a normal sintering process from plastic-coated metal powder. The model achieves a "green stability" (green product) owing to the plastic coatings of the metal particles that sinter fuse with one another.

This model is then resintered in a reducing hydrogen atmosphere inside a special high-temperature oven, thereby also expelling the plastic material. This is especially important because in this way oxide skins, which would considerably lower the wettability of the rather loosely sintered metal particles (brown part), are avoided.

In a third process step, a low-melting metal, usually copper, is infiltrated into the hollow spaces of the porous model. This is done in the same oven if the infiltrate has a corresponding melting temperature. The infiltrate should merely be placed right next to the model to enable the infiltration to take place by capillary forces.

Figure 3.35(a) shows such a brown part before infiltration and the copper cylinders used for infiltration. Figure 3.35(b) shows the result after the infiltration has taken place.

[12] FAA (Federal Aviation Administration): US federal agency responsible for aviation safety in the US. FAA rules are often internationally recognized or accepted.

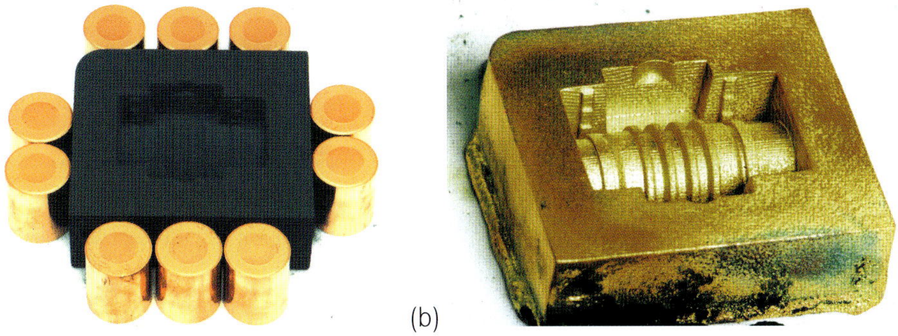

(a) (b)

Figure 3.35 (a) brown part with copper cylinders used for infiltration (before infiltration) (b) part (after infiltration) (Source: DTM)

This is a relatively complicated process that demands a great deal of experience and additional investment. Furthermore, the indirect metal process has the paramount disadvantage that, owing to the multiple heat treatments via green product and brown part, and the shrinkage factors that cannot be disregarded, it becomes so complicated that it is impossible to control the parts satisfactory without empirical values. In addition, this process is very time consuming.

Copper polyamide (PA) is a material that is simple to process but also softer and is used preferably for molds and tools for the casting of soft materials.

Construction Time/Accurancies

Construction times and accuracies typical for laser sintering are achievable (see also Section 3.2.1). The absolute accuracy of the Sinterstation HiQ is given as 90% smaller than ±0.10 mm.

For metal part tolerances and standard parameters, the accuracy is from ±0.10 up to ±0.50 mm. In the literature, surface qualities of R_A = 8–10 µm and accuracies up to ±0.05 mm for the tooling-build style for tool inserts are reported [Mareco, 98]. Freestanding walls and perpendicular overhangs could reach a broad-height ratio of 10 up to at most 15 for a wall thickness higher than 1 mm.

In the HS (High Speed) version, the build speed is raised by a factor of 1.85, and the scanning speed is doubled to 10.0 m/s. This is achieved by a 50 W laser and the Celerity Beam Delivery System (BDS) with HS optics. A technology called Vector Bloom Elimination, which avoids the undesired beam expansion especially at the end of the scanning vectors and at high deflections by intervention in the laser power control, is integrated. This improved scanning strategy leads in combination with the higher laser power to higher build speeds. The speed of a vertical build process is significantly raised from 7 mm/h up to nearly 15 mm/h.

In the standard mode, a build process speed of 18 mm/h in the vertical direction could be reached by higher layer heights and correspondingly worse details, but only with a slightly worse accuracy.

Additional build jobs could be loaded in the machine by the software during the build process. Technically this is possible because the machine slices the actual layer immediately before it is processed (slice on the fly). Therefore the productivity is raised, and the production time, crucial for the user, is reduced because the build process starts on the fly.

Postprocessing

The postprocessing is identical to the postprocessing typical for laser sintering, as described in Section 3.2.1, "Machine-Specific Basic Principles."

After the actual construction process is completed, the entire machine must be cooled down, which can take several hours. Only then should the whole cake with the embedded component be taken out. This could be done by moving the whole powder cake in a frame with the same dimension as the build chamber and taking it out with the frame. After the model has cooled down sufficiently, it can be broken out of the powder cake and cleaned by brushing and sandblasting.

In the manufacturing system of the SinterStation Pro, the quick-building module is removed from the machine and inserted in the BOS (break out station). Therefore, the cooling does not block the machine. In the unpacking, the components are removed automatically, and the unsintered powder is screened and transferred to the integrated recycling station.

For metal processes, a debinding and postsintering with subsequent infiltration is necessary. The manufacturer provides the LaserForm oven for this. The shrinkage factors, which needed to be considered for the multiple heat treatments, can only be evaluated by good experience. Even so, the process time is reduced to currently five days with parallel sintering and infiltration, and the whole process is quite time consuming. In contrast to the selective melting method, the heat input is uniform in the entire component, which counteracts distortions.

Typical Follow-up Process

All follow-up processes typical for laser sintering can be applied (see also Section 3.2.1). When using sintering models for vacuum casting, it has to be taken into account, however, that the surfaces are not as smooth as those achieved by stereolithography, and a finishing process could be necessary.

3.2.4 Laser Sintering, EOS

EOSINT P: Formiga P110, P395, P760, P800
EOSINT M: M290
EOSINT S: S750

Electro Optical Systems GmbH (EOS), Krailingen-Munich, Germany

Short Description

These are additive manufacturing machines for the direct melting of plastics (P), metals (M), and sands (S). The machines work by the laser-scanner mechanism. A CO_2 laser scans the surface of a powder bed and incipiently melts the particles that form the models after solidification. After the lowering of the build platform and the recoating, the process repeats until the part is finished. The process requires no additional supports.

Specially optimized prototypers exist for the processing of plastics, metals, and sands.

Range of Application

Plastics: (EOSINT P) geometrical prototypes, functional prototypes (technical prototypes), series parts by means of precision casting

Metals: (EOSINT M) technical prototypes, direct tooling (series parts by means of tooling)

Sands: (EOSINT S) series parts (castings)

Development Status

Commercialized since 1994 (P) and since 1995 (M, S)

History

EOS was founded in 1989 by Dr. Langer as part of a project with BMW AG for the development of a stereolithography machine. Shortly after developing its first stereolithography machine in 1991, EOS presented a sintering machine for sintering plastics on which the later EOSINT model range was based. The EOSINT P series are machines for sintering polystyrene and polyamide. It was first presented in 1994 as the P350. This machine was followed by the P360, P380, P385 (2005), and the P385 (2006). The current version is the P396. The EOSINT series also includes the largest laser sintering machine, the double-laser machine P700 (1999), which is the P760 in the current version. The EOSINT M series was developed for the direct sintering of metals and metal alloys and was founded by the M160 prototype. It was followed by the EOSINT M250/M250^{xtended}, M270 (2004), M280 (2010), and M290. The P350 was developed into the S350 as a prototype of an installation for sintering molding sands and later into the EOSINT S700 for the production of (preferably) cores for sand castings.

Until EOS gave up the stereolithography business, it was the only company worldwide that built both stereolithography and sintering machines at the same time.

Development Partners/Strategies

EOS relies on the optimization of a machine-material system and on an accordingly specialized prototyper. The production line P is optimized for the processing of plastics, M for metals, and S for sands. The performance of the laser, the scanner, and the scan strategy match the material in hand. Details such as nitrogen generators are brought into play only if the material requires them. EOS is continually searching for development partnerships with material producers such as Electrolux (Finland) and with software suppliers. In early 2004, a long-standing patent dispute with 3D Systems was settled amicably. According to reports, 3D Systems will use the double-laser technology from EOS on the basis of this agreement. Since the end of 2005, there has been a cooperation with 3D-Micromac, Chemnitz, for the application of the microlaser sintering as part of a technology and service center.

EOS is focusing on the market of rapid manufacturing and therefore is actively aiming at the field of "e-manufacturing," which is also protected as a trademark. The machines could be upgraded to manufacturing solutions by the use of additional modules, such as for powder handling or unpacking (Fig. 3.36).

Figure 3.36 Plastic laser sintering machine EOSINT P395[13], manufacturing system with powder management, unpacking station, and thermo case (Source: EOS)

[13] The picture shows the processing machine, P385, which is, except for the label, nearly the same as the P395.

The smallest machine, the P110 Formiga (in Portuguese, “the ant”), is completely integrated in this strategy. It is only a little bit slower than the P395 and can build parts down to a wall thickness of 0.4 mm. It is also compatible with the system equipment, and due to its size it can be carried through a normal door (1067 mm width). Therefore it can be set up in normal rooms, and the installation effort is low.

Data Formats/Software

The machines are able to read complete STL and contour data (CLI, common layer interface). There are also optional interfaces for standard data types. The software is published with the name EOS RP Tools. It also covers the data-preparation software EOSPACE (V3.0, 2005) for the optimized, automatic placement of parts in the build chamber. Alternatively, also available are the software packages Materialise Magics and Deskart Expert Series. The EOS processing software is PSW (V3.2, 2005). It contains a lot of features for optimizing part accuracy, part quality, and build speed, as well as an optimized system productivity and ease of use.

Principle of Layer Generation

The principle of layer generation is identical to those processes typical for laser sintering, as described in Section 3.2.1. The machines work on the laser-scanner principle. Especially for the metal process, the skin and core method, which was originally developed for the stereolithography process, was modified in a way that for outer and inner structures different, optimal parameters could be applied.

The scan strategy is adjusted to the respective material.

System Type/Construction

EOSINT P – Plastic sintering machines

The EOS P machines work on the laser-scanner principle. They use a CO_2 laser with a power of 30 W (P110), 50 W (P395), and 2 × 50 W (P760). For higher quality structures, an F-theta lens is used. Available is also a dual-focus module, which can change the beam diameter during the process. By using this technology, it is possible to build the contour with a fine laser beam diameter but a slower build-up speed and the inner hatches with a bigger beam diameter and therefore a higher build-up rate.

The compact EOSINT P110 Formiga has a rectangular build room with a size of x, y, z = 200, 250, 330 mm^3. It builds layers with a height of 0.06 mm, 0.1 mm, and 0.12 mm. The build chamber of the EOSINT P395 (Fig. 3.37) is around x, y, z = 340, 340, 620 mm^3 with a maximum height of 660 mm. The EOSINT P760 offers a remarkably larger build chamber with a size of x, y, z = 700, 380, 580 mm^3. It works simultaneously with two laser-scanner modules. The EOS P 800 is the larger version of the EOS P 395 (x, y, z = 700, 380, 560 mm^3).

The recoating system for the plastic machine is schematically shown in Fig. 3.37. It consists of a channel-like, open powder feeder converging in the direction of the powder bed. This feeder wipes across the powder bed and thereby applies the necessary powder. To optimize the powder flow of the plastic powder, the whole unit is vibrating. The powder feeding system is designed in a way that the whole powder is precompressed.

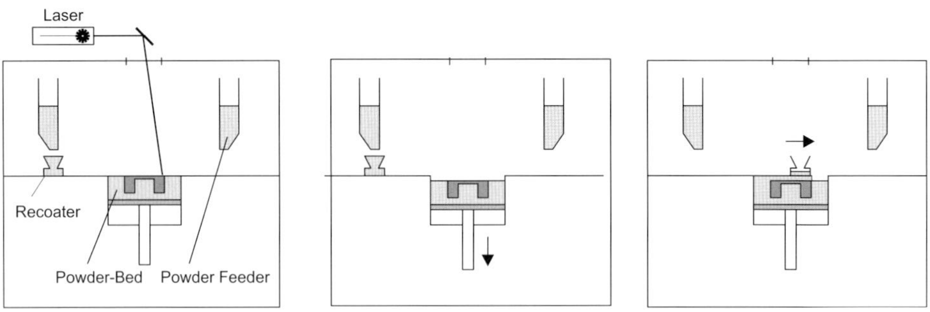

Figure 3.37 Recoating system EOSINT P (schematic)

The machine has an online calibration for continuous monitoring and correction of the laser beam during the construction process.

The walls, which are needed for the inclusion of the powder bed, are built up during the process. This requires build time and material, but it enables the removal of the whole model, which is embedded in the supporting powder. Because of this constructive step, there is no need for a complete cooldown process in the machine, which could take hours because of the poor thermal conductivity of plastics. Instead of this, the model, which is embedded in a powder cake, can be removed from the machine while it is hot, and an externally controlled cooldown takes place. Therefore, the reloading times between two build processes can be shortened. Optionally, a portable tank is offered, which makes building the enclosure walls unnecessary. The EOS machines feature an integrated nitrogen generator, which operates the gas supplies. This and an automatic powder management system expand the machine into a (semimanual) additive manufacturing system (see Fig. 3.36).

The EOSINT P760 offers such a wide build chamber that most parts that exist in the automotive industry, especially in engine construction, fit into it. To maintain the productivity, two simultaneously working 50 W CO_2 lasers are used. The laser beam sources and other technical data correspond largely to the valid values for the P395.

The layer heights vary between 0.1 and 0.15 mm.

To calibrate the machine, an optical scanner calibration system guides the laser beam over a thermosensitive paper. The result is then analyzed outside of the machine using a video image processing system that proposes a calibration. Depending on the grid spacing (typically 5 mm), the calibration in the machine takes between 15 and 45 minutes. Together with the evaluation and subsequent calibration, the process takes a few hours if it can be done in-house. The evaluation at the machine manufacturer accordingly takes longer.

EOSINT M - Metal sintering machines

The EOSINT M290 is a machine for the melting of metal powders. This technology is called direct metal laser sintering (DMLS) by EOS. There are specialized applications in material and construction for the production of molds and mold inserts (the DirectTool process) and for the direct production of metal components (the DirectPart process), see Fig. 3.38.

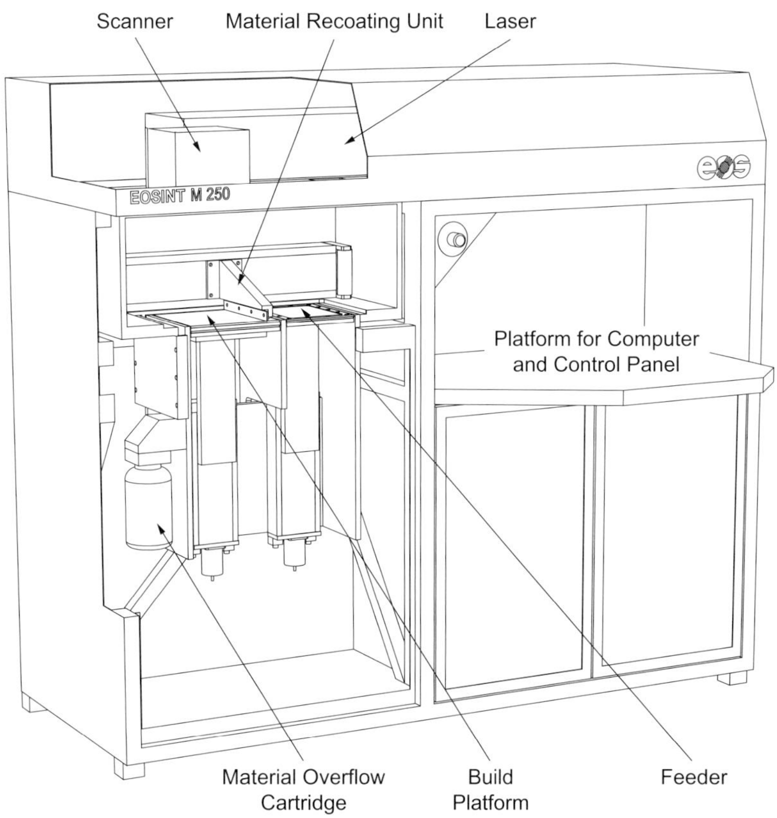

Figure 3.38 EOSINT M sintering machine for metals, schematic construction (Source: EOS)

The software allows not only the transfer of the geometry data, but also the transmission of reference points that enable postprocessing with conventional tool machines safely and effectively.

Design details include the fully enclosed building space of x, y, z = 250, 250, 325 mm³. The reservoir is below the build platform and opens into the building plane. The recoating system takes the material from there, distributes it on the build plane, and carries the material, which is not needed for the process, to the overflow tank. The system works unidirectionally and therefore strokes twice over the build plane. The recoating system is designed as an asymmetric blade. The mirrors of the scanner system are cooled to avoid deformation caused by heat and to increase the accuracy. The machine has a gas-tight process chamber, which allows effective working with protective gases. The EOSINT M290 is equipped with a 400 W Yb (ytterbium) fiber laser, which combines the highest beam quality, high beam intensity, and high dynamics. It also offers a variable focus (0.1 to 0.5 mm beam diameter), see Fig. 3.39.

One of the previous machines, the closely related EOSINT M250^{xtended}, has a build chamber that is 15 mm smaller in the z direction and a 200 W CO_2 laser with a fixed beam diameter of 0.4 mm.

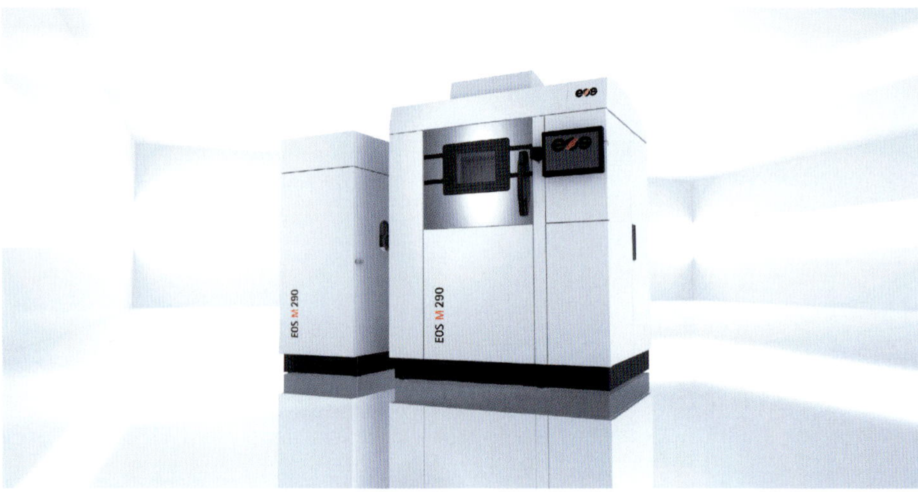

Figure 3.39 EOSINT M290 sintering machine for metals (Source: EOS)

The parts are built on a ground plate. It offers special clamping and reference elements and thereby determines reference points and planes, which are reproducible from the CAD data. These clamping and reference elements could be used for subsequent measuring or cutting operations as a bracket.

EOSINT S - Sand sintering machines

The EOSINT S750 (Fig. 3.40) is developed for the direct sintering of polymer-bonded molding sands, especially for the manufacturing of complex cores. The development partner of the process called DirectCast was Dr. Wendt (ACTech, Freiberg in Saxony, Germany). Development goals are a larger build chamber fitting the requirements of the engine design industry, a higher build-up rate suitable for the larger models, and at least a partial compensation for the higher efforts by the elimination of all features unnecessary for sand sintering. The S760 has a double-laser system consisting of two 100 W CO_2 lasers with dual focus.

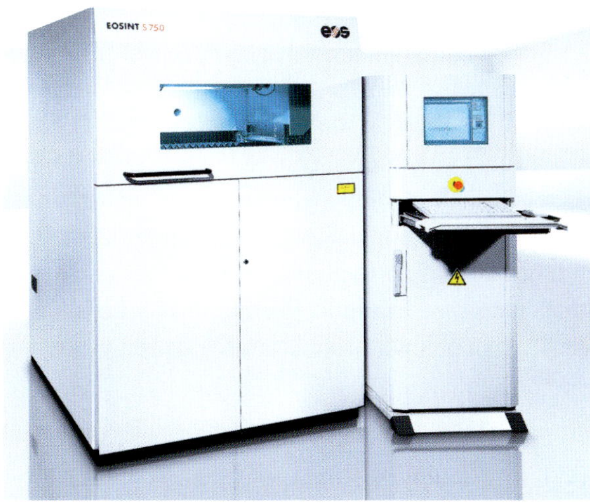

Figure 3.40 EOSINT S750 sintering machine for molding sands (Source: EOS)

The sintering of polymer-bonded molding sands follows the principle of polycondensation, not the one of liquid sintering.

The machine offers a build chamber of x, y, z = 700, 380, 580 mm^3 and therefore can build up molds and cores for typical engine components like crankshafts, engine blocks, and cylinder heads as one part. With a standard layer height of 0.2 mm and a scanning speed of 3 m/s for both lasers, a throughput of approximately 2.5 liters of sand per hour is reached. The continuous charge of molding sands is ensured by a pneumatic conveying system.

Material/Construction Time/Accuracies

EOS is selling self-developed materials, which are documented in detail. But they are also involved in the development of materials for the customer. Naturally, the results of such developments are communicated in less detail.

Materials – Plastics

These are unfilled polyamides (PA 2200, PA 3200 GF, PrimePart). In the EOS machines, normally fine polyamides are used, which are similar to PA 12. They are available without a glass filling as PA 2200 and with glass filling as PA 3200 GF. The grain sizes are about 60 μm. The strengths of the materials reach up to 90 % of the strengths of injection-molded parts made out of basically the same materials. Therefore, the components are not only great as functional models and for the simulation of injection-molded parts, but also for direct production. This is why sintering is used in the context of "e-manufacturing" (see Chapter 6). The polyamide PA 2200 is certified as biocompatible.

PrimePart is a fine nylon, which is designed for thinner layers and better detail. It is referred to as a high-end material; in particular this means that an optimized balance between tensile strength and elasticity is provided.

Filled Polyamides (AluMide, CarbonMide, Flame-Retardant Polyamide). AluMide is an aluminum powder filled polyamide. This results in a metallic look with interesting color and surface effects, especially when the components are polished after the building process. AluMide has approximately the modulus of the PA 3200 GF and the tensile strength of the PA 2200 but at 170 °C and around 40 °C higher temperature resistance. EOS made, for example, the handles of the EOSINT M270 with this material and underlines in its own production the strategy of e-manufacturing.

CarbonMide (2005) is a polyamide that is filled with carbon spheres. This combines a higher tensile strength with a lower weight and an interesting black appearance.

In 2005, the world's first flame-retardant material for laser sintering, a flame-retardant nylon, was introduced. The polyamide material based on PA 3200 complies with the relevant aerospace requirements, the JAR/FAR 25[14], Appendix F, for flammability (Part 1) and smoke density (Part 5), the ABD0031/AITM 3.0005 (2) (Airbus), and the BSS 7239 (Boeing) for smoke density in the class wall thickness of 2 mm. In addition, for a wall thickness of 2 mm, the classification V0, according to UL 94[15] of the directive, has been reached. With these classifications, EOS has qualified a flame-retardant material for in-flight applications.

With the high-temperature material EOS PEEK HP3, a high-performance polymer for the EOS P 395 and the EOS P800 is offered.

Polystyrene (PS 2500, PrimeCast 101). As fusible materials for investment casting, the amorphous polystyrene powder PS 2500 and the more recent PrimeCast 101 are available. With PrimeCast 101, thinner layers can be created and thus map

[14] The JAR (Joint Aviation Requirements) of the JAA are the European equivalent of the FAR (Federal Aviation Regulations) of the FAA. The JAA (Joint Aviation Authorities) is an association of the civil aviation authorities of 34 European countries. The FAA is the Federal Aviation Administration of the US.

[15] UL (Underwriters Laboratories): an organization for verification and certification of product safety. UL 94 regulates flame propagation.

finer details. Polystyrene can be processed at relatively low temperatures, which favors a lower shrinkage and higher model accuracy. The construction process is also much faster than with polyamide. For the investment casting process, the component must be sealed, for example by overgrowing.

Materials – Metals

For the sintering of metal powders, two standard material groups exist: *DirectMetal* and *DirectSteel*. In *DirectMetal*, powders with low-melting components are melted locally and serve after solidification as a binder for the higher-melting components. In the *DirectSteel* process, one-component powders or alloys with nearly the same melting point are directly sintered.

DirectMetal 20 is a fine nickel-bronze powder with a grain size of approximately 20 microns. Some components expand as a result of the sintering process and mostly compensate for the shrinkage, and the parts have a high dimensional accuracy. The heat conduction is higher than that of steel, but the mechanical-technological properties are worse in general.

Depending on the application and the injection-molding material, the inserts for injection molding made of this material are infiltrated either with low-melting alloys, or today more commonly with epoxy. Of course, the molds have to be polished before use, like conventional molds today. Experience shows that the material can be easily processed both by machining and by electrical discharge machining (EDM) (Serbin already documented good experience with the previous material [Serbin, 95]). The shapes may be used for the production of small and pilot series of several hundred to several thousand parts made of different plastics, such as polyethylene (PE), polypropylene (PP), ABS, or even glass-filled nylon.

DirectSteel H20 is a 20 micron steel-based powder. It was already introduced in 1998 (as DirectSteel 50 V1) as the first pure steel powder usable in commercial machines. The information of the manufacturer that the material is "without plastic binder," has a melting temperature of "700 °C" and a tensile strength of up to 1100 MPa, suggests that by a proper optimization of the alloy components (Cr, Ni, Mo, Si, V, C), a steel with a relatively low melting material was developed whose properties correspond roughly with cast steel. The thermal conductivity of $\lambda = 25$ W/m · K (at 23 °C), is significantly lower than that of steel ($\lambda = 45$ W/m · K) and still falls significantly with the temperature. The critical step is to sinter a dense component (density = 7.8 g/cm^3) without extensive follow-up processes and thereby avoid the resulting deformation. A surface hardness of 40 HRC could be reached. To increase the build-up rate in accordance with the necessary surface quality, different parameter sets for layer heights of 40 and 60 µm are available. Freestanding walls of a wall thickness down to 0.28 mm are possible. Infiltration is neither necessary nor due to the high density reasonably possible. The material can be machined and wire-cut.

Stainless Steel 17-4 is a stainless steel that corresponds to the European material number 1.4543. It is characterized by a very high corrosion resistance, good mechanical properties, and an excellent ductility without further treatment. This material is recommended for the production of components (DirectPart). With the default parameters, layer thicknesses of 20 microns can be made. The material melts completely. The components are therefore fully dense and have nearly isotropic properties. To speed up the building process, the sheath core method can be used. The components can be nonadditively processed, including by welding, microblasting, polishing, or coating.

Cobalt Chrome MP1 is a very fine-grained powder mixture based on cobalt, chromium, and molybdenum. It is characterized by very good mechanical properties as well as corrosion and temperature resistance. It corresponds to the chemical composition of the UNS R31538 CoCrMo alloy, is nickel-free (<0.1% nickel content), and is sterilizable. It is therefore particularly suitable for biomedical applications. The laser-sintered components completely fulfill the requirements of ISO 5832-4 and ASTM F75 for cast CoCrMo implant alloys, as well as, except for the strain, the requirements of ISO 5832-12 and ASTM F1537 for forged CoCrMo implant alloys. By hot isostatic pressing (HIP), the elongation can be increased, and it then also meets this standard.

CoCr alloys are used for highly stressed components, such as in aircraft engines and engine construction.

Cobalt Chrome SP2 is a version for dentistry. It meets, in addition to the properties of MP1, the special requirements of dental restorations, which are faced with dental ceramic.

Titanium Ti64 is a prealloyed Ti6Al4V alloy. It is a TiAl alloy for structural and engine parts from the aerospace industry, chassis parts, and also for medical implants. The material is characterized by very good mechanical properties and corrosion resistance combined with a low weight. It is biocompatible. The ELI version (extra-low interstitial) has a particularly high purity.

Maraging Steel MS1 is a martensitic hardening steel. Its chemical composition corresponds to the US classification 18 Maraging 300, the European 1.2709, and the German X3NiCoMoTi 18-9-5. It is characterized by the highest strength with good ductility. It can be very well postprocessed by machining. By a subsequent thermal treatment, a hardness up to 55 HRC can be reached. It is used as a tool steel for injection and die-casting molds, but also for highly loaded machine parts.

Nickel Alloy IN625 is a nickel alloy powder that is heat and corrosion resistant. The alloy is characterized by very high tensile, creep, and fracture resistance. In addition, a subsequent heat treatment of the material is possible. Because of its high corrosion resistance, it is especially suitable for marine applications, where a high resistance to pitting, crevice corrosion, and stress corrosion cracking is necessary.

Nickel Alloy IN718 is also a nickel alloy, which corresponds in its composition to Inconel™ 718. Due to its aging resistance, tensile strength, high-temperature strength, and breaking strength, which are preserved up to a temperature of about 700 °C, this alloy is particularly suitable for series products in the aerospace industry. Other applications are in the fields of the chemical, petroleum, and natural gas industries.

AlSiMg10 is a cast aluminum alloy, which is used for thin-walled components with complex geometries. A high strength and hardness is achieved by the percentage of silicon and manganese in the alloy. Because of the mechanical properties and low density, the alloy is used for applications in motor sports, aerospace, pneumatics, automotive engineering, and several other industries.

Through a process called microblasting, the surface quality can be improved to about R_z = 20 microns. With conventional tool-making polishing methods, surface qualities of R_z = 3 microns can be realized.

The perforated plate used in the DMLS process can be clamped directly onto machine tools. The powder that participated in the build process but was not processed for the part can be fully reused.

Materials – Sands

The range of materials includes silica and zircon sands for applications in the higher temperature range. They can be directly used for sand casting and deliver properties that are identical to series parts. It has to be noted that not only cores are necessary for building the sand molds. The binder content is approximately 4 to 6% (mass fraction). Customized materials are available.

Ceramics 5.2 is an aluminum silicate sand that is coated with a phenolic resin (synthetic mullite). The material is suitable for the additive manufacturing of complex sand cores and molds for all kinds of foundry applications. Because of its high heat capacity and low thermal expansion, the main application of this ceramic sand is in high-temperature casting.

Quartz 4.2/Quartz 5.7 is a phenolic-resin-coated quartz sand suitable for additive manufacturing fabrication of complex sand cores and sand molds for all casting applications.

Construction Time/Accuracy

For the laser sintering, typical build-up times and accuracies for the laser sintering can be achieved. As a measure for the build-up time, a build process rate in the *z* direction of 24 mm/h (P110 Formiga) and the very good rate of 35 mm/h (P395, P730) are given. With the DirectMetal, material wall thicknesses down to 0.6 mm are safely realizable. With a build-up rate of 2 up to 20 mm^3/s, the build speed in the *z* direction is from one up to a few millimeters per hour.

Postprocessing

For the plastic sintered models, the postprocessing is identical to the postprocessing typical for laser sintering, as described in Section 3.2.1.

EOS distributes an unpacking station, that can be connected to the powder management system.

The metal models lie in the nearly cold powder bed after the sintering process is finished so that only the powder needs to be moved into the powder container after the construction chamber is raised. The construction platform can then be easily unscrewed from the elevator. The supports must be removed manually, depending on the material. The platform needs to be smoothed afterward and should be completely polished after several construction processes. In the DirectTool process combined with DirectSteel material, it could be directly used as a tool carrier.

The sand cores must be postprocessed several hours in the oven. To avoid deformation, they are often placed on glass balls.

3.2.5 Laser Melting, Realizer GmbH

Selective laser melting (SLM):
Realizer SLM50, Realizer SLM100, Realizer SLM250

Realizer GmbH, Borchen (Paderborn), Germany

Short Description

This is an additive manufacturing machine for the direct melting of pure metal or ceramics powder. The machine works on the laser-scanner principle.

Application

Metals functional parts, tool inserts

Ceramics: tool inserts

Development Status

Commercialized since 2002

History

Dr. Fockele and Dr. Schwarze founded a company in 1990 for the development of a stereolithography system, and in 1992, the F&S Stereolithography GmbH. Since 1995 they have developed the SLM process, and in 1998 they presented the SLM Realizer. In 2005 the development efforts in the company Realizer GmbH were controlled and supervised by Dr. Fockele, who currently manages the production and marketing of the systems.

Strategies/Development Partners

F&S and the Fraunhofer Institute for Lasertechnique (FhG-ILT, Aachen) together developed the basic technology, *laser melting*. One of the primary objectives was the processing of different commercial-practice metal powders. There is a strategic partnership with the Institute for Microprocessing technology of the Research Centre Karlsruhe (IMVT-FZK) for the application of the technology on microparts.

Data Format/Software

There is a combined control software for the STL and SLM Realizer. It enables the reading of STL and slice data, the automatic support creation, simple CAD functions, like the generation of multiple offset structures and the generation of shell models out of slice data and the component placement. The software takes over the complete machine control including the compensation of the jet range.

Principle of Layer Generation

The laser-scanner system basically works as described in Section 3.2.1, "Machine-Specific Basic Principles." Layers of 200 to 100 μm are produced.

Construction

Realizer SLM250. The Realizer SLM250 is the common basis of the technology. The Realizer 250 is equipped with an infrared-emitting fiber laser with 200 W, 400 W, or 600 W. The beam diameter is 30 to 250 μm. The building space dimensions are x, y, z = 250, 250, 300 mm. The machine can be inerted. See Fig. 3.41.

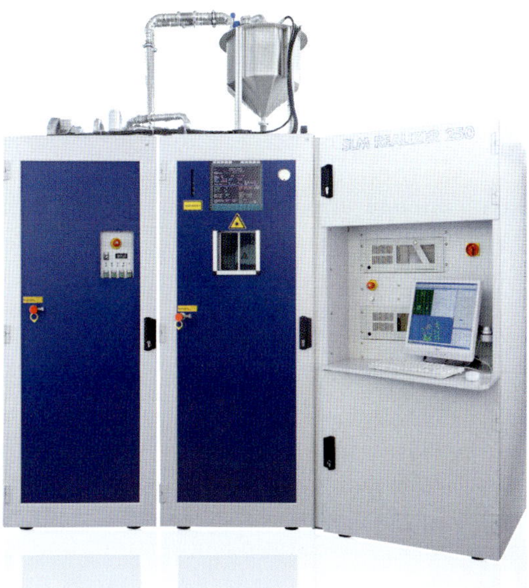

Figure 3.41
Realizer SLM 250
(Source: Realizer GmbH)

Realizer SLM100. The Realizer SLM100 is a smaller machine from Realizer GmbH that is optimized for building intricate parts. The circular building space has a diameter of 125 mm and a maximum height of 80 mm. It is completely capsulated and enables working under shielding gas and vacuum. Also high precision powders can be processed without a risk. The machine includes a very fast digital scanner. See Fig. 3.42.

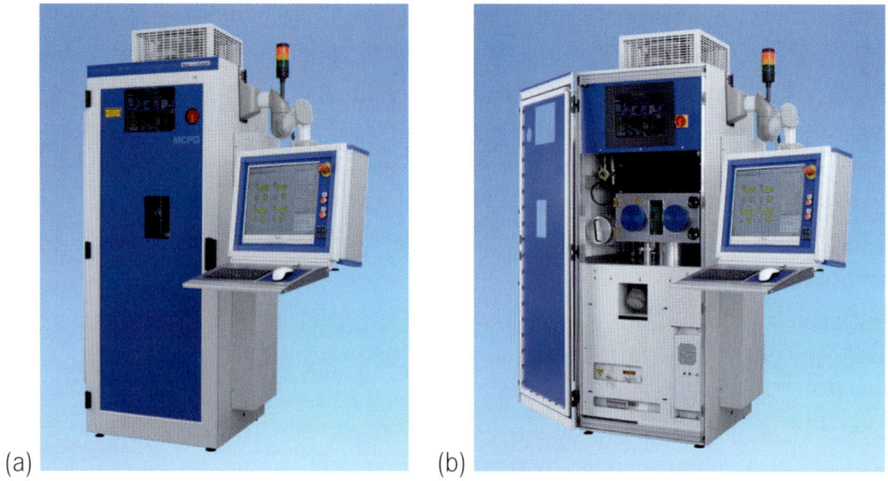

Figure 3.42 Realizer SLM 100: (a) outside view; (b) view into the building chamber (Source: Realizer GmbH)

Realizer SLM50. The Realizer SLM50 is the worldwide first SLM desktop machine. Like the Realizer SLM100, it has a circular building space with a platform diameter of 70 mm and a maximum height of 40 mm. The laser has a power of maximum 120 W and it can be focused on a diameter of between 10 and 20 µm, which depends on the lens used. The SLM50 was specially designed for applications in the dental and jewelry industries. See Fig. 3.43.

In the development of the machine, the manufacturer pays special attention to the processing of materials for these industries, like cobalt-chrome or gold. A further specialty of the system is the integrated stereomicroscope. This enables diverse applications of the SLM50 in research and universities.

All of the machines supply components with a density of 100 %.

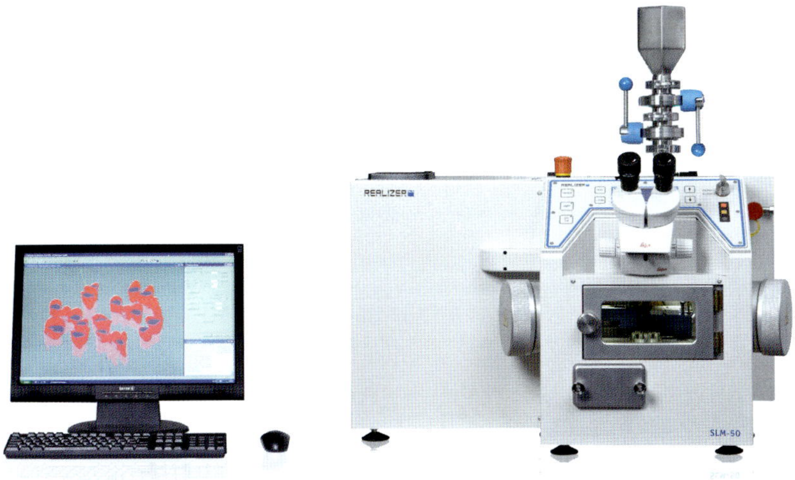

Figure 3.43 Realizer SLM50 desktop machine (Source: Realizer GmbH)

Material/Construction Time/Accuracy

The materials that are used are customary powders with a diameter of 10 to 45 µm. They must have almost round grains (gas atomized) to be able to roll and to enable a good coating. The powders must not contain any binder material and need to have a constant grain diameter. All customary powders can be used, but the advice is to work with the recommendations of the manufacturer at first and to gain experience.

Layers can be built in an area of 20 to 100 µm. But when focusing on the effort of the postprocessing and the details, layers of small thickness are recommended. The building speed is given as 7 to 30 cm^3/h.

Based on the manufacturer's details, stainless steels (316L), tool steels (1.2344), cobalt-chrome steels, and *titanium* can be processed. For the tool-steel 1.2344, a density of > 99.8 % and a hardness of >52 HRC can be reached. The yield strength is around 500 N/mm^2, the tensile strength around 600 N/mm^2, and the fracture elongation reaches around 30 %. The results are scattered depending on where the samples were taken, and in what direction of the layer or perpendicular to it. At the end of 2006, the first parts of aluminum were presented. The limits of the process are walls around 80 µm, which correspond to 50 µm layers. Therefore the process must be optimized precisely.

Postprocessing

No typical follow-up processes like infiltration or thermal treatment are necessary. The component can be treated like all other components of the same material. Metal parts can be machined, eroded, welded, or thermally treated. It is possible to polish them until they achieve a mirror-like gloss.

3.2.6 Laser Sintering, SLM Solutions

Selective laser melting (SLM):
SLM® 125 HL, SLM® 250 HL, SLM® 280 HL, SLM® 500 HL

SLM Solutions GmbH, Lübeck, Germany

Short Description

SLM Solutions GmbH offers laser melting applications that work with the SLM process for the direct production of finished products. According to their statements, their machines were the first ones that were able to work with reactive aluminum powder. In addition to that, they were the first company that merchandised titanium implants made with the laser sintering process.

Application

Finished products made out of metal

Development Status

Founded as MCP-HEK Tooling GmbH in 2007, renamed SLM Solutions in 2010

History

The primary distribution partner of F&S, MCP-HEK GmbH, formed HEK Tooling GmbH in 2007 for the further development of SLM technology. This was renamed MTT Technologie GmbH in the following year. The activities of the MTT Technology Group, which consisted of MTT Technologie GmbH and MTT Technologies Ltd., were split in the year 2011. MTT Technologies Ltd. moved to Renishaw, and MTT Technologie GmbH was renamed SLM Solutions GmbH.

Data Formats/Software

The standardized software for the machines is Magics AutoFab of Materialise.

Principle of Layer Generation

One must differentiate between several processes of layer generation. The standard version builds up the layers with the conventional laser sintering process. Regarding bigger machine types, there is the additional possibility of using the so-called double-beam technology, which uses two different lasers. One of them has less power and a small focus diameter for producing barrier layers; the other one uses a higher power and a top-hat profile. The more powerful laser is used for the inner areas of a contour. Because of this differentiation between the lasers, SLM Solutions calls the process the shell-core strategy. See Fig. 3.44.

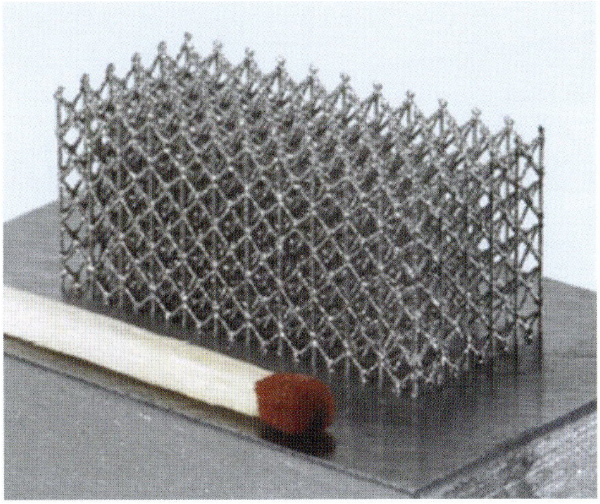

Figure 3.44 Selective laser melting, SLM, microstructure (Source: Realizer GmbH)

Building/Construction

SLM 125 HL: The SLM 125 HL works with a 100 W or a 200 W fiber laser and offers a building space of the size x, y, z = 125, 125, 125 mm^3. It is designed for research and development issues and the production of smaller models.

SLM 250 HL: The SLM 250 HL has a building space of 250 × 250 × 250 mm^3 and works with either a 200 W laser power or a 400 W laser power. It is possible to apply laser sintering structures on already preproduced parts. This is a particular advantage for producing hybrid models, which must meet different demands.

SLM 280 HL: The SLM 280 HL offers a building space of 280 × 280 × 350 mm^3. The fiber laser can work with 400 W or 1000 W of power. One distinctive feature of this machine is the possibility to use the double-beam technology. Therefore, a second laser is applied inside the machine. The combination consists of a 400 W laser with a focus diameter of 100 µm and a 1000 W laser, which uses a top-hat profile and a focus diameter of 700 µm. With a special software control, the 400 W laser fabricates the outer contour with a fine focus diameter, and the stronger laser produces the inner areas of the contour. This method is supposed to optimize the profitability and the building rate of the machine.

SLM 500 HL: The SLM 500 HL has a building space of maximum 500 × 280 × 325 mm^3. Overall, it works with four fiber lasers, which are divided into two units with one 400 W laser and one 1000 W laser. Therefore, the maximum power of the whole application is 2.8 kW. Both units can work in parallel or independently to produce a series of smaller models. Just like the SLM 280 HL, the SLM 500 HL uses the shell-core strategy, which results in a very high building rate. See Fig. 3.45.

Figure 3.45
SLM® 500 HL laser sintering machine
(Source: SLM Solutions)

Materials

SLM Solutions exclusively uses metal materials. The usable materials are titanium, cobalt-chrome, tool steel and stainless steel, aluminum, and Inconel. The particle sizes can vary between 10 and 30 µm.

3.2.7 Laser Melting, Renishaw Ltd.

Selective laser melting (SLM):

Laser Melting System AM125, AM250

Renishaw Inc., Gloucestershire, United Kingdom

Short Description

The company Renishaw offers laser sintering systems for the direct manufacturing of end products. The company focuses on the aerospace industries, high-tech mechanical engineering, and on the electrical industry. The machines are designed for processing with metallic powder materials.

Application

End products made of different metals

Development Status

Originally MTT Technologies Limited. Renishaw took over the laser sintering technique in 2011.

History

The company Renishaw was founded in 1973 by David McMurtry and John Deer and is still supervised by them. First Renishaw mainly produced measuring systems for machine tools.

With the takeover of MTT Technologies Limited in 2011, Renishaw gained competencies in the area of laser melting and offers systems from MTT registered in their own name. The process works with the laser sintering process; see the description in Section 3.2.1, "Machine-Specific Basic Principles."

Data Format/Software

Renishaw offers software packages for their machines from the two software suppliers Materialise and Marcam Autofab.

Principle of Layer Generation

The machines built by Renishaw work on the classic laser sintering principle. They generate thicknesses of layers between 20 and 100 µm. During the running process, it is possible to switch the powder collection chamber to a powder storage container. This enables very economical work.

Construction

At this time, Renishaw offers two different systems: the AM250 and the AM125.

AM250: The AM250 is equipped with an ytterbium-fiber laser where one can choose a power of 200 W or 400 W. The diameter of the laser focus on the powder surface is 70 µm. The maximum build volume is 245 × 245 × 300 mm^3. Upon request, the z axis can be extended to 360 mm. The scanning speed in the x-y direction is 2000 mm/s, and the positioning speed is 7000 mm/s. The build-up rate varies between 5 cm^3 and 20 cm^3 depending on the material, density, and geometry. See Fig. 3.46.

AM125: The AM125 is a smaller machine that has a building space of x, y, z = 120, 120, 125 mm^3. The laser has a power of 100 or 200 W. The diameter of the focus on the powder surface is 35 µm. The speed of building is the same as in the AM250. See Fig. 3.47.

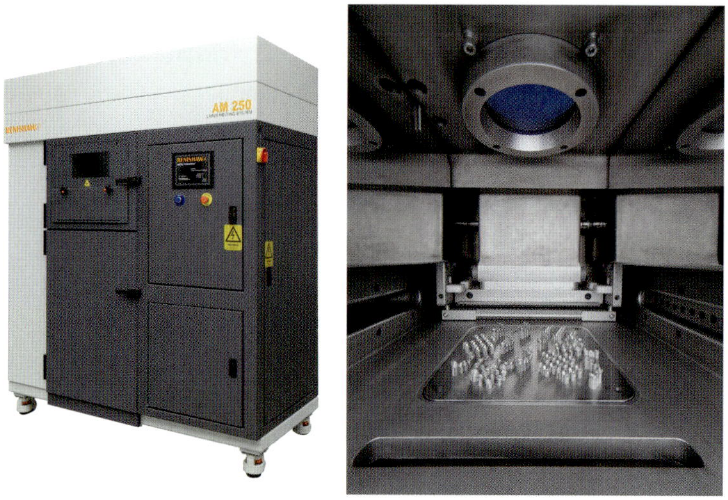

Figure 3.46 Renishaw AM250 laser sintering machine (left), build chamber of the AM series (right) (Source: Renishaw)

Figure 3.47 Renishaw AM125 laser sintering machine (Source: Renishaw)

Materials

With the systems of Renishaw, many different materials can be processed. These include stainless steels 316L and 17-4PH, tool steel H13, the aluminum alloys Al-Si-12 and Al-Si-10, the titanium alloys CP, Ti6Al4V, and Ti6Al7Nb, cobalt-chrome (ASTM 75), and Inconel 718 and 625. Parameters for other materials are under development.

3.2.8 Laser Cusing, Concept Laser

LaserCUSING®
Mlab cusing, M1 cusing, M2 cusing, Xline 1000R

Concept Laser GmbH, Lichtenfels, Germany

Short Description

These are additive manufacturing machines for the production of metal components using the laser-scanning methods by the laser cusing referred melting process.

Range of Application

Metals: functional prototypes, final products as metal components for the automotive, aerospace, medical, dental, jewelry, and engineering fields, as well as tools and tool inserts for plastic injection molding or metal casting

History

Concept Laser Ltd. was founded in 2000 as an autonomous subsidiary of the Hoffman group. Since then, the company has reported an annual growth of 20%.

Development Status

Concept Laser has four commercialized types of machines on the market: Mlab cusing, M1 cusing, M2 cusing, and Xline 1000R.

Strategies/Development Partners

Hofmann is a traditional tool-making company and thus the natural development partner of Concept Laser.

Concept Laser also maintains development agreements with universities and institutes.

On the material side, in the dental sector they partner with DENTAURUM, in the field of jewelry with Legor and Hilderbrandt. They offer standard machines and customized machine concepts for metal laser melting. The high quality requirements, level of experience, and references from Concept Laser stand for process-safe and cost-effective solutions that prove their efficiency in the daily production and are primarily focused on unit cost reductions.

Data Formats/Software

The machines have the machine software LCM+, which edits the final data for the building job.

Principle of layer generation

The laser cusing machines principally work like the sinter system described in Section 3.2.1, "Machine-Specific Basic Principles." The designation "cusing" is a combination of the terms concept and fusing. It points to a liquid phase sintering. The material is melted completely in the process, and dense components can be produced. The manufacturer states that the warp is reduced significantly by their own patented exposure system.

System Type/Construction

Mlab cusing

The machine has a 100 W fiber laser, and the focus diameter is 30 microns. The Mlab cusing is equipped with easily replaceable space modules. There are spaces available in following sizes: x, y = 50 × 50 mm^2, 70 × 70 mm^2, and 90 × 90 mm^2. The height is in each case z = 80 mm.

The Mlab cusing is a low-cost starter model oriented toward the production of intricate metal components such as those that occur, for example, in the dental and jewelry fields. In addition to dental materials and precious metal alloys, reactive materials such as titanium can also be processed with the Mlab cusing.

Figure 3.48
Mlab cusing
(Source: Concept Laser)

Here, the material handling is performed by means of a flexible handling carriage. The Mlab cusing consists of a space and a component-handling station, in which also the powder material is filled in the building module. The module is pulled out of the space for an ergonomic setup of the machine. Furthermore, it is possible to store the material, for example for precious metals such as gold; the building module can be uncoupled in seconds and stored safely in a safe. If it necessary to change the material, it can be carried out like the change of the component modules as needed. A secure processing of titanium alloys and pure titanium is ensured by the use of the so-called handling carriage. The building module can be pulled out simply with the handling carriage. This is then flooded with argon. Through glove ports, the intervention by the operator takes place in the build chamber in order to perform the loading operation or the removal of components. Once finished, the building module is pulled back into the machine, and finally the handling carriage is undocked.

M1 cusing

The M1 cusing is normally equipped with a 200 W or 100 W fiber laser. The focus diameter is between 70 and 200 microns. The building space size of the M1 cusing measures x, y, z = 250, 250, 280 mm^3. The exposure is via a galvo scanner and a flat-field lens. The machine consists of the installation space (right) and a handling station (on the left), in which the building material is also filled in the building module (Fig. 3.49). The building module, where the building process takes place, is reciprocally moved into the space or in the handling station. The access to the handling station is pushed back. Thus, the handling station for the setup process is easy to access with a crane, such as with heavy parts or tooling inserts.

Figure 3.49
M1 cusing
(Source: Concept Laser)

The M1 cusing is a machine for processing nonreactive materials. Its ergonomic machine architecture is appealing, especially in the fields of tooling and dentistry, because of the safe and ergonomically designed component removal outside the process chamber. A zero-point clamping system, an insertion device (allows the melting of geometries on long, thin hybrid subfields), and a QM powder sieve to ensure the quality of powder are available as options, as well as the quality monitoring modules QM laser, QM coating, QM atmosphere, and QM documentation.

M2 cusing

The M2 cusing is normally equipped with a 200 W or 400 W fiber laser. The focus diameter is between 70 and 200 microns. The building space size of M2 cusing measures x, y, z = 250, 250, 280 mm³.

The exposure is via a galvo scanner and a flat-field lens. The machine consists of the actual space and a handling station, in which the building material is also filled in the building module. The building module, where the building process takes place, is reciprocally moved into the space or into the handling station. Two gas connections allow the inert gas supply and, as may be necessary, the change. In the machine, a completely sealed powder circuit is realized, so the reactive powder is not coming into contact with atmospheric oxygen, even with the operator. The residual oxygen is constantly controlled so that components can be manufactured with reproducible properties and the provisions for explosion and fire protection can be strictly adhered to. This system is designed according to ATEX guidelines and suitable for the processing of reactive powders. It has a fully sealed space that is under inert gas.

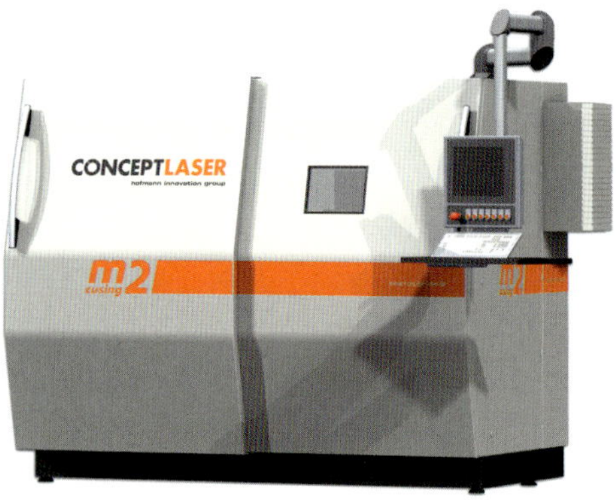

Figure 3.50
M2 cusing
(Source: Concept Laser)

Preferably, components made of aluminum and titanium are produced. Furthermore, secure powder handling and preservation of the powder quality is guaranteed by a sealed powder circuit. The M2 cusing is especially appealing to the medical, aerospace, and automotive industries and in mechanical engineering. A zero-point clamping system, a QM powder sieve to ensure powder quality, and the quality monitoring modules melt pool QM, QM laser, coating QM, QM, and QM atmosphere documentation are available as an option for the M2 cusing.

Xline 1000R

The Xline is normally equipped with a 1000 W fiber laser. Through the use of a 1 kW fiber laser system, the layer thickness and track widths can be increased by several factors such that, compared with conventional systems with typically up to 400 W laser power, the productivity can increase tenfold. The focus diameter is between 100 and 500 microns. The building space size of the Xline 1000R measures x, y, z = 630, 400, 500 mm^3.

The system concept of the Xline 1000R is designed according to the European ATEX guidelines to ensure the safe handling of reactive materials such as aluminum or titanium. The process takes place in the process chamber with the exclusion of oxygen. To carry out the powder handling and setup process, the build chamber is moved into the handling station. A risk-free powder handling of reactive materials such as aluminum and titanium is enabled.

Furthermore, for the first time a metal laser melting system with automated powder feed, even in the construction process, was developed. This means that powder handling by an operator is no longer necessary. The sieving station and containers needed to prepare and store the large amount of powder are an integral part of the system. With the use of the sealed system, contact can be avoided with oxygen, see Fig. 3.51.

Figure 3.51
Xline 1000R
(Source: Concept Laser)

In addition, the Xline 1000R was equipped with a rotating mechanism for reciprocal use of two building modules. Thus it is possible that the optics and laser unit crucial for productivity can operate continuously so that maximum efficiency is achieved. The usual downtimes for setting up the system or for powder recycling are completely eliminated. On the process side, the additive manufacturing building process takes place. With the rotary mechanism, the build chamber is moved with a finished construction job to the process side of the handling side. Furthermore, the building chamber of the Xline 1000R has integrated heating of the building panels.

Material/Construction Time/Accuracies

The process provides fully dense parts of commercially available materials, though optimized by the manufacturer for the process. There are various one-component powders that have been optimized for selected applications. They allow layer thicknesses to be generated between 20 and 50 microns. The grain size varies accordingly.

The range of materials includes stainless and tool steels, aluminum and titanium alloys, nickel-based super alloys, cobalt-chromium alloys, or precious metals such as gold or silver alloys. CL 20ES (1.4404) is a stainless steel for the production of acid- and corrosion-resistant components or tool inserts preferably for preproduction tools. The tensile strength is 650 N/mm^2, and the hardness is 20 HRC. CL 50WS (1.2709) is a hot-work steel, which is more tenacious than 1.2343, has a tensile strength of about 1100 N/mm^2, and has a hardness of 35 to 40 HRC. The main application is the production of tooling inserts for the series injection molding of plastic parts.

With the CL 91RW, a martensitic chromium steel is available that is preferably used for the production of inserts for plastic molds. It is particularly resistant to corrosive plastics. In tools and molds, it is also widely used because it is very easy to polish. There is also a nickel-based alloy called CL 100NB that corresponds to Inconel 718. For processing under inert gas (M2), the materials aluminum CL 30AL (AlSi 12), CL 31AL (AlSi10Mg), and titanium (Ti6Al4V ELI) are available in the specification CL 40TI ELI.

For the dental market, the materials remanium® star CL and CL rematitan® of the cooperative partner Dentaurum are suitable. Remanium star CL is a certified Co-CrW alloy according to CE0483 for the manufacture of metallic dentures. The tensile strength is according to the recommended heat treatment at 1030 MPa, the bond strength according to ISO 9693 at 40 MPa.

Rematitan CL is suitable for the production of metallic dentures and implant designs. The material is a CE0483 certified titanium alloy Type 4. The tensile strength is 1005 MPa, the bond strength according to ISO 9693 at 37 MPa.

CL 80CU is a bronze alloy that is great for the additive manufacturing process due to its melting behavior and is usually used in jewelry work for the production of master patterns. The material is easy to polish and has a higher hardness than silver (hardness 171 ± 7 HV0.2). Gold and silver alloys are offered by partners Hilderbrand & Cie and Legor SRL.

In their processing, depending on the type of machine, an inert gas consumption of about 0.6 to 34 l/min is expected. Metal powder that was used for building process but is not part of the finished component can be fully reused. In the M2 cusing they are sucked off in the handling station under inert gas and stored in a powder container in the machine or on an external powder shelf. The building progress rate is about 2 to 100 cm^3/h, depending on the machine type. The wide range shows the strong geometry, material, and laser dependency.

The accuracies are usually at ± 50 microns. This does not replace the fine finishing passes, but the rough machining and optionally also the semifinishing milling. The component properties have die-cast qualities, whereas the surfaces correspond more to sand casting.

Postprocessing

No further process-related postprocessing such as infiltrating or sintering is required. The part can be handled for rework like nonadditive manufactured components of the same material, for example by wire cutting, milling, grinding, or polishing.

3.2.9 Direct Laser Forming, TRUMPF

TrumaForm LF130, TrumaForm LF250

TRUMPF Laser GmbH & Co. KG, Ditzingen, Germany

Short Description

Direct laser forming generates metal components by melting binder-free powder layer by layer. A disk laser ensures a high performance and a good beam quality. Two building spaces, which are separated and used alternatively, enable the machining of sensitive materials and provide a high productivity.

The machine DMD 505, also known by the name TrumaForm, works with the direct metal deposition (DMD) process. In this process, powders of different metals and alloys are applied on a workpiece by powder nozzles and then locally melted (see Section 3.3.3, “Direct Metal Deposition (DMD), DM3D Technology (TRUMPF)”).

Application

Functional prototypes, end products

Development Status

Commercialized since 2003

History

The company TRUMPF (founded in 1923) is a manufacturer of machine tools for sheet metal forming and material processing. They also produce laser beam sources and laser-supported processing facilities for the industry. Other branches of the company also deal with electronics and medical technology. By working with laser-welding applications and laser-coating systems in the 1990s, they acquired special knowledge in repairing badly damaged components. The processes of direct metal deposition (DMD) and direct laser forming represent a systematization of the single solutions according to the main areas and their practical implementation. The two main areas are *building in the powder bed* and the *free application by powder nozzles*. The TrumaForm *Direct Laser Forming Machine* was awarded the MWP Award 2004 for the “Best Physio/Chemical Machine Tool” by the English field-related magazine *Metal Working Production*.

Strategies/Development Partners

In the laser technique section, TRUMPF put their trust strategically in complete systems for whole fields of application. Because of their modular concept, they can be adapted to the customer's individual needs. Corresponding to the company history, there is a focus on the machine technology. The development of a machine that generates components out of a powder bed opens up a new activity. But because of their competence in the areas of laser and laser marking systems, it fits the company's portfolio very well. The process of laser melting was developed in the Fraunhofer Institute of Laser Technique (FhG-ILT) in Aachen, Germany.

After successful trials, the machine was pulled off the market in 2006 and is no longer mentioned in the company presentations. A few (modified) machines are still running, but they will not be mentioned in this book.

3.2.10 Electron Beam Melting

Arcam Q10, Arcam A2

Arcam AB, Mölndal/Gothenburg, Sweden

Short Description

This is an additive manufacturing machine for sintering pure metal powder. The surface areas of a powder bed of the component contour are selectively melted with an electron beam in vacuum, which forms a layer after solidification (*electron beam melting*, EBM). The component is created layer by layer according to the respective recoating.

Application

Functional prototypes, end products

Development Status

Commercialized since 2001

History

In 1993 *Ralf Larson* applied for a patent for the direct sintering of metal powders. The process was developed in cooperation with Chalmers University of Technology in Gothenburg, Sweden. When the patent was granted in 1997, Larson and a venture capital fund founded the company Arcam to commercialize the technology.

Ralf Larson is also the inventor of *selective mask sintering* with infrared radiators and is the founder of the company Speed Part (for the history of the company Speed Part, see Section 3.2.11, "Selective Mask Sintering (SMS), Sintermask"). He runs the development company Larson Brothers Co., with his brother Ove, which forms the technological and entrepreneurial background of both companies.

Strategies/Development Partners

Arcam has a strategic focus on the direct production of components made of high-load metal, especially in aerospace and medical technology by free-form fabrication (FFF). This is also why they teamed up in a strategic partnership with the California medical device manufacturer DiSanto Technology in February 2013. Arcam products have been distributed exclusively in North America by Stratasys since January 2006.

Data Formats/Software

The Arcam EBM Control software needs slice files (SLC), which are derived from STL data. To import the CAD data, to process the STL data, to prepare individual construction processes, and to slice the STL data sets, Arcam recommends the RP software Magics (Materialise). Via the software, prepared building strategies can be selected or the operator can design his or her own. With their help, support structures that are very fine and therefore particularly easy to remove might also be constructed. They are barely noticeable and are therefore also called *invisible supports*.

The support assembly is loaded together with the component data in the *Arcam Build Assembler* and is merged together there into a final building data set. In this data set, a structure can be assigned for each partial data set. Massive (solid), sintered (porous), or a spatial framework (grid) structures can be chosen so that the component can be better adapted to meet requirements. In addition, different building strategies can be assigned to the individual data sets.

Principle of Layer Generation

Pure metal powder is applied to the surface of the powder bed and melted by means of an electron beam according to the contour of the current layer. There is one difference here compared to other methods, in the possibility of reaching several melting points of the layer. This is achieved by a special control of the electron beam. Due to heat conduction in the partially finished part and the surrounding powder, the melt solidifies and forms a solid layer. The process runs under vacuum. Arcam calls this process *CAD to Metal*. Characteristic is the complete melting of the powder in a vacuum, accompanied by a very accurate temperature control.

System Type/Construction

The Arcam system is a combination of a vacuum-resistant and therefore correspondingly massive build chamber and an approximately equal-size control unit. The building chamber has a movable base. It is coated by an overhead material deposit on a kind of filling shoe. The machine has a standard electron beam source of about 4 kW. In a common vacuum-resistant chamber, it is positioned above the building space. In the top portion, the electrons are emitted and accelerated, focused, and diverted by means of coils along the way. The entire exposure unit has no moving parts such as scanners or plotters. Focusing is done solely by the magnetic field and requires no lenses. The system is quite insensitive in this respect. The electron beam can be spread (diffracted) at a speed of up to 1000 m/s. For the melting process, scanning speeds of around 20 m/s are actually used. The power reserves are sufficient to overlay the melting process with a heat treatment process in the form of a defined pre-, post-, or simultaneous treatment. Because the penetration depth of the electron beams (in a high vacuum and in steel) can be up to 300 mm, a defined heating of the component is realized. If Arcam makes use of it, it is not found in the publications of the company. With a laser beam, this effect is basically impossible to achieve because the beam is absorbed from the surface and carries the absorbed energy by thermal conduction into the component. The results are higher temperature gradients and much greater penetration depths.

An integrated cooling device is optionally available.

Arcam Q10: The Arcam Q10 is the third generation of the EBM machines. It was developed specifically for the production of orthopedic implants and can produce components with a maximum size of x, y, z = 200, 200, 180 mm^3. It has an electron beam with a power of up to 3 kW and a minimum beam diameter of 100 microns. See Fig. 3.52.

Arcam A2: The A2 has been developed for applications in aerospace engineering, as well as general industry. It provides a variable laser power of 50–3500 W and a focusing ability of the beam between 0.2 mm and 1 mm.

The A2 is available in three different versions. The standard version A2 has an exchangeable building chamber, for which two different forms of spaces are available. There is a rectangular space with dimensions x, y, z = 200, 200, 350 mm^3 or a cylindrical space with a diameter up to 300 mm, and it allows a building height up to 200 mm. See Fig. 3.53.

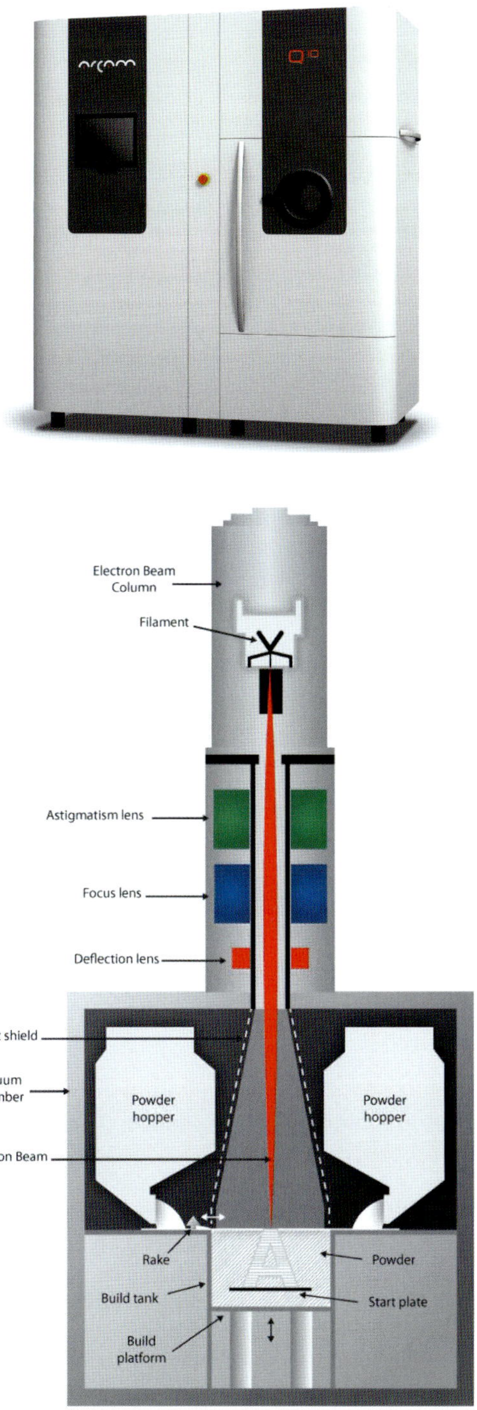

Figure 3.52 Arcam Q10 (top), section through the electron beam unit (bottom) (Source: Arcam)

Figure 3.53 Arcam A2 (left), EBM build chamber of the Arcam A1 (right) (Source: Arcam)

The A2X version is optimized for the processing of high-temperature materials and offers a space of x, y, z = 200 × 200 × 380 mm^3.

The A2XX is based on the same concept as the A2, but it was equipped with a larger cylindrical space diameter of 350 mm × 380 mm.

Material/Construction Time/Accuracies

In the machine, practically all technically relevant materials can be melted. Arcam has a number of materials qualified for the process, but it also supports customized developments.

There are three families of materials: steel, titanium materials, and CoCr alloys.

Arcam 200 is a low-alloy steel, preferably for the production of prototypes. It is soft and can be easily machined. It achieved a tensile strength of about 290 N/mm^2 and a hardness up to 220 HB.

Arcam H13 is a high-alloy tool steel corresponding to 1.2344 (DIN), which is mainly used in tools for injection molding and die casting. The tensile strength is 1400 N/mm^2, the hardness up to 52 HRC.

Ti6Al4V (Grade 5) is the most common industrial titanium alloy for highly stressed components, preferably the aerospace industry. The tensile strength is above 970 MPa, the 0.2% yield strength above 910 MPa.

Ti6Al4V ELI (extra-low interstitial) is a particularly pure alloy in which the melting impurities and enclosed elements (especially C, O, N) were minimized. It is particularly suitable because of its purity for medical applications. Because of its generally better mechanical properties, and mainly because of its low temperature resistance, it is also used in other industries.

Arcam ASTM F75 CoCr is an interesting alloy, especially for medical applications. It exceeds the ASTM[16] specification F75 not only with regard to the tensile strength and the yield strength, but also in terms of hardness and elongation at break.

In addition, a range of materials have been and are being evaluated, such as superalloys (Inconel 718, 625, and 620), stainless steel, and beryllium alloys (AlBeMet).

The layer thicknesses are, depending on the material, from 0.05 to 0.2 mm. From this and from the beam data, the resulting rates of the molten bath are from 0.3 to 0.5 mm/s. A precision of ±0.4 mm is specified. Depending on the material, a building speed of 30 mm/h can be achieved in the *z* direction.

Postprocessing

The final component is preferably taken out with the optionally available handling in the powder bed of the machine and cooled definably in a cooling chamber.

Finishing and Process-Typical Sequence Method

The components can be finished with all of the classic manufacturing processes, preferably by machining or by erosion.

3.2.11 Selective Mask Sintering (SMS), Sintermask

Sinter mask technology (SMS)[17]: Pollux 32[18], Zorro, Additive Fabrication System

Sintermask GmbH, Eichenbühl 10, 92331 Lupburg, Germany

Short Description

The Speed Part is an additive manufacturing machine for sintering proprietary plastic powder. It works with the mask process. An infrared light heats the surface of the powder through the mask and melts the exposed particles, which form a layer of the model after hardening. The process will be repeated after every layer after the next varnishing. There are no supports necessary during this process.

Application

Concept models, functional prototypes

[16] American Society for Testing and Materials

[17] The process is sometimes called infrared mask sintering (IMS).

[18] The machine is called the Speed Part RP3 at the actual homepage, but it was announced as Speed Part Do in 2005.

Development Status

Commercialized since 2004

History

Speed Part was founded by Ralf Larson and a risk capital fund in 2000. Ralf Larson had already developed a plastic sintering process in 1996, which used an infrared light instead of a laser. He and his brother Ove owned the Larson Brothers Co. In 1989 they developed among other things one of the first rapid prototyping machines and founded SPARX AB. There they produced the *Hot Plot*, a widely used manual layer laminate machine. It works with Styrofoam plates contoured with a hot wire, and its products were show-and-tell models. Larson also developed electron beam sintering (Section 3.2.10, "Electron Beam Sintering, Arcam") and co-founded the company Arcam.

The Speed Part RP3, now called the Pollux 32, was operated as a beta-test machine from 2006 to 2008, but there was no serial production. An alternative name for the machine was Speed Part Do. The company's name Speed Part AB was changed to Sintermask AB in 2007. In 2008 there was the acquisition through the German company FIT GmbH, and the transfer to the newly founded company Sintermask followed in 2009. The production of the beta machine was cancelled right in the beginning of 2009. The Zorro is presented as the replacement machine with planned commercial use.

Strategies/Development Partners

Selective mask sintering (SMS) was developed as a competitor for laser beam sintering. An advantage is that there is no laser-scanning device and no inertization of the arrangement. Another advantage is the high building velocity of 700 layers per hour and the independence of building time considering different geometries. The available material is optimized for vacuum forming; other materials for other applications have not been announced.

Data Formats/Software

Front-end software reads CLI data and includes all necessary functions for printing preparation and positioning of the models in the building space, and furthermore it pilots the machine. If STl files have to be read, the Materialise Slice Module is used for slicing. The Materialise Magics program can also be included. Several machines can be examined and telecommanded by a PC. A *Fault Tracking System* makes a remote diagnosis possible.

Principle of Layer Generation

The machine consists of a device for making the mask, a lighting device, and the building space. The contour of each layer will be positioned and connected to a glass plate electrostatically. The masks in the Pollux 32 were generated by an electron beam printer. After that, the mask drives above the building space, which is filled with powder. Several infrared lights heat and melt the powder at exactly the positions of the gaps in the mask, the positions where the light can get through the mask. The heat gets conducted into the surrounding powder and the already built layers, which causes the current layer to cooldown. The bottom of the building space is deepened by one layer thickness for the next layer and new powder gets spread out of the powder container. Then the plate is removed from the building space, emptied, and cleaned. After that step, the process starts again from the beginning until all layers are lightened. This method needs no support structures.

Building/Construction

The machine consists of a device for making the mask, the lighting device, and the building space. The lighting device is positioned behind the building space, which causes the machine to be much longer than wide with a greater width in the back than in the front. The building space size is x, y, z = 300, 200, 800 mm^3. The mask generator is designed similar to a copier. The sections that have to be lighted are electrostatically charged and varnished with powder that is impermeable to the infrared radiation. Depending on the size of the models, it is possible to choose the full- or half-sized mask, which reduces the varnishing time and the consumption of the toner. The mask generator and the infrared radiator are already optimized purchased parts. Another option is an air blower device and a handling system for the material. The infrared system's input power is 10 kW.

Material/Construction Time/Accuracy

Speed Part uses a self-developed, glass-filled polyamide powder, VT3, which is optimized for building tools for the vacuum-forming process (Section 5.4.1.2, "Deep Drawing or Thermoforming"). It is also called the *Speed Part Toolmaker Powder*. With it, the models achieve a higher porosity, so there is no need to drill any drains for the vacuum-forming process for plastic foils. The melting point is 180 °C.

The *Speed Part Partmaker Powder* is designed for generating prototypes and models. It is also a polyamide, but without infill. The duration of generating one layer is about 10 to 20 s. The lighting time of one layer is independent of the complexity of the model thanks to the mask process. The building time is dependent on the absolute size of the model because the mask can be lit in whole or in half. A build time of 10 s per layer and a layer thickness of 0.1 mm means a building velocity of 36 mm per hour in the z direction. Doubling the time for a layer to 20 s reduces the building velocity by half, to 18 mm per hour. The declared resolution is 15,000 × 10,000 dpi, see Fig. 3.54.

Figure 3.54
Selective mask sintering. The Speed Part RP3 is identical to the Pollux 32 (Source: Speed Part)

Postprocessing

Easy cleaning by brushing or blowing

Typical Follow-up Processes

The models are optimized for the vacuum forming process. After sealing and finishing, it is also possible to use the vacuum casting process.

3.2.12 Laser Sintering, Phenix

PXL, PXM, PXS, PXM and PXS Dental

Phenix Systems, Clermont-Ferrand, France; since 2013, 3D Systems

Short Description

This is an additive manufacturing machine for direct, one-step sintering of commercial metal and ceramic powder. The machine works with the laser-scanning process. There are no supports needed.

Application

Products are made out of ceramic or metal.

Development Status

Commercialized since 2001

History

Phenix Systems was founded on July 20, 2000, by Francois Reymondet, Patrick Teulet, and Jean Marie Gaillard. The first prototype was built in 1996. The market entry with the Phenix 900, which is also called a prototype, retrospectively, began in 2001. The current Phenix PXM 250 is the subsequent machine. The company was taken over by 3D Systems in June 2013.

Strategies/Development Partners

The strategic aim is the direct production of metal and ceramic models out of commercial powder with the same machine. The machine is a stand-alone solution with integrated software for the parameterized usage of the machine's building data and for the improvement of the material-dependent building data. This software was developed by the company C4W.

Data Formats/Software

There is a proprietary (but free and independent of the machine) CAD software given: *Phenix Processing*. With this software it is possible to edit all CAD and rapid prototyping operations in addition to creating the STL data. Phenix Processing assigns the data to the operating software *Phenix Control*. This software includes several modules for the creation of layer information and automatic sequence control for building the layers, for heating the building space, and its gas management and for operating the laser and the scanner. The machine accepts IGES, STEP, and STL data. There is specialized software available for the PXM Dental and the PXS Dental for the creation of dental products out of CoCr steel, and it works with STL data.

Principle of Layer Generation

The additive manufacturing process corresponds to the classic laser sintering process. The powder gets spread out by a cylinder out of the container over the building platform, compacted, and selectively sintered by the laser and scanner device.

Building/Construction

The PXL has a fiber laser with 500 W power and a building space of 250 × 250 × 300 mm. The processable materials are metals and ceramics. This machine uses an automated material loading and recycling process.

The PXM and PXM Dental machines use a 300 W fiber laser, and there is a smaller building space of $x, y, z = 140 \times 140 \times 100$ mm^3. The difference between the standard machine and the PXM Dental can be found in the software and the workable materials. The software packages delivered with the PXM are Phenix Processing and Phenix Manufacturing. The dental version uses Phenix Dental, and only cobalt-chrome material is possible, in contrast to the standard version.

In the PXS and PXS Dental machines, there is a 50 W fiber laser used in the smallest version of the PX series. The building space is 100 × 100 × 80 mm³. As already stated about the PXM series, the differences between the PXS dental and the PXS standard version can be found in the software and the workable materials.

The machine has a fully sealed process chamber, which can be heated up to 900 °C and flooded with process gases (Fig. 3.55, PM250). Therefore it is also called an oven. The powder stock is arranged inside the process chamber, and the driving step motors of the moveable building and containment platform are positioned outside. Fiber lasers with 50 or 100 W are coupled into a laser head and light the layer through a laser window. The varnishing cylinder spreads out the powder above the building platform and compacts it at the same time.

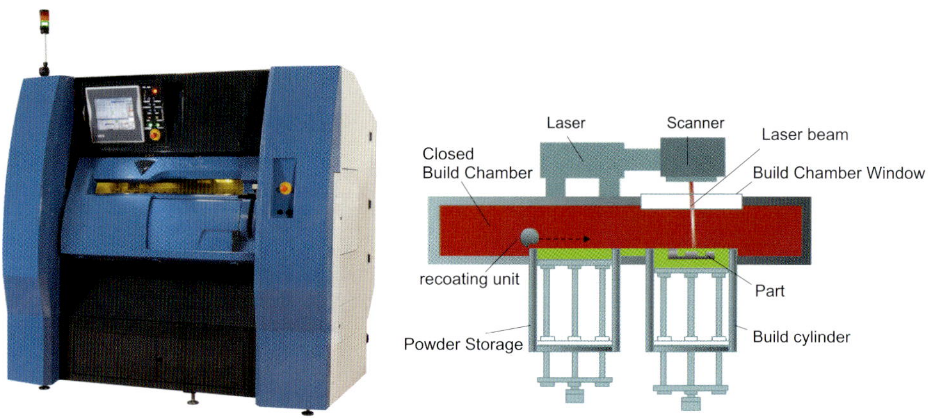

Figure 3.55 Laser sintering, Phenix machine PXL (left), section through the closed process chamber (right) (Source: Phenix)

The PM100 has a 50 W fiber laser and builds on a platform of 10 mm thickness. It is optimized for the creation of dental crowns and dental prostheses. The building height is not stated but it is unlikely that it is higher than 20 mm.

Material/Construction Time/Accuracy

The PXL and the PM100 are configured for different materials of ceramic or metal. The fabrication can be optimized through different parameters. The PXM Dental and the PXS Dental both use a chrome-cobalt powder after ISO 9693 and ISO 16744. They have an accuracy of about ± 20 µm. The PX standard version also has a declared accuracy of ± 20 µm. The resolution of the minimum possible details are about 100 µm in the *x* and *y* axes and about 20 µm in the *z* axis.

Postprocessing/Finishing

Metals get polished, and ceramics get resintered. The models can be treated based on their materials.

Typical Follow-up Processes

None

■ 3.3 Coating: Melting with the Powder Nozzle

Already in the 1970s, thin-walled, shell-type components for highly stressed nuclear power machine components made of nickel-based alloys were produced by deposition welding. The process was called *shaping welding*. After it was possible to control the residual stresses with the aid of lasers, which were insurmountable procedural obstacles until then, and given the growing importance of rapid prototyping, the Americans have taken up the activities and further developed *shape welding*.

Simultaneously, in material processing, the laser coating process led to the control of multilayer layers with higher thicknesses. The step of building being controlled by any bounded geometry was no longer great.

3.3.1 Process Principle

Processes that put powder through a nozzle in a local molten bath generated by a laser beam are called *laser coating* or *laser generating*, and in German *laser cladding* is also used. The term laser generating comes from the laser material processing where it describes the multilayer coating with the laser, in contrast to single-layer coating. In the context of additive manufacturing processes, this term has the disadvantage that it also represents the generic term for all layer manufacturing processes with lasers.

In most processes, the material is applied via freely programmable nozzle systems that simultaneously locally melt the material with the help of the laser beam. There is a variant that has arisen from deposition welding. There, the material is supplied in the form of wires; compare Section 3.7.1, "Controlled Metal Buildup." The layer formation is done by solidification of the melt due to heat conduction in the component. In contrast to the sintering process, a powder bed does not exist.

Metal and ceramics are used as materials. The processes are single-stage, and thermal post treatments are possible and often necessary.

The building material and the energy that is necessary for layer formation is introduced only by the moving laser and nozzle unit, which is why the process is fully 3D capable. It is not bound to generate cross sections one after the other to complete the component. In contrast to processes that make the layer-wise application of the material and the contouring of the layer one after the other, the material can therefore be applied to any substructures or even to components to be repaired.

The process principle is shown in Fig. 3.56.

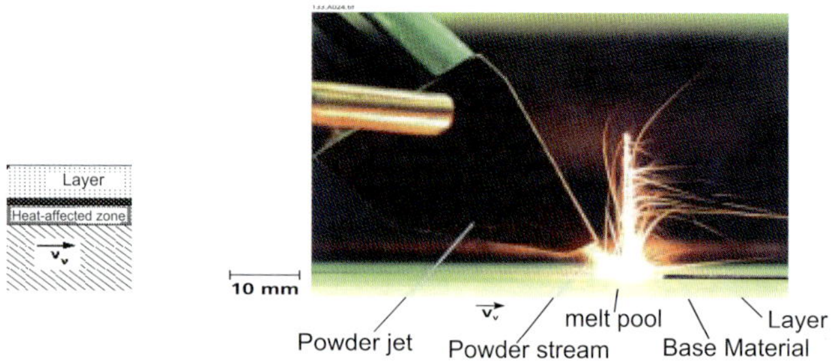

Figure 3.56 Single-layer coating with laser radiation (Source: FhG-ILT, Aachen)

The material is guided with the aid of an inert gas stream (such as argon) and a powder nozzle aimed at the processing point. Three-dimensional components result from overlaying many layers.

In the process, the filler wire and a thin edge layer of the substrate or the partially finished component are melted so that a melted metallurgical bond is generated. The degree of dilution can be set by parameters. Generally it leads to a dense (real: 99.6 and higher), fine-grained microstructure without cracks, voids, or pores; see Fig. 3.57.

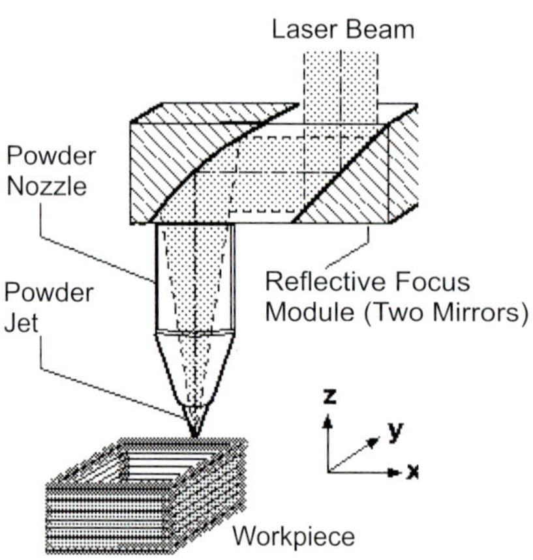

Figure 3.57 Generation with laser radiation (Source: FhG-ILT, Aachen)

The most important process parameters are the laser power P_L, the feed speed v_S, the powder mass flow m_P, and the power density distribution. Typical values for P_L are between 300 W and 5000 W, for v_S, between 300 and 1200 mm/min, and for m_P, 3–8 g/min.

As laser beam sources, CO_2, Nd:YAG, and diode lasers are used because of the good beam quality with a small focus diameter. Increasingly, laser disks are used too, but especially fiber lasers.

3.3.1.1 Concepts of Powder Nozzles

Mainly pneumatic powder-conveying nozzles are used. The gas then is used for transporting the powder and functions as the inert gas. Preferably argon and helium are used.

In principle, two different powder nozzle concepts are suitable for coating with laser radiation: the off-axis or outer axial and the coaxial nozzle.

Off-axis nozzles are positioned outside the beam axis and aligned so that the powder and the inert gas flow in the melting range are available. The inner contour of the nozzle is generally cylindrical. At the outlet of the gas in the atmosphere, it forms a free beam that splits open after a distance of about five diameters and reduces the effective amount of powder due to turbulence. In order to achieve a high powder efficiency, a maximum working distance of approximately 10 to 15 mm is required. Typical diameters of nozzles are in the range of about 2 to 3 mm. At such low distances, high heat loads occur at the nozzle tip. To avoid damage or clogging, the nozzles have an integrated water cooling system. The nozzle material is copper because of its good thermal conductivity. A disadvantage is that the feed directional of the nozzle affects the results of the process in an off-axis nozzle. Forehand and backhand processing sections alternate, depending on the geometry.

The powder is fed coaxially to the laser beam in a coaxial nozzle. Therefore, a directional dependence does not occur. Continuous and discrete coaxial nozzles are separated.

In a continuous coaxial nozzle, the powder is evenly distributed around the circumference and conveyed through an annular gap around the laser beam. Problems include the relatively complicated internal construction of coaxial jets. This can lead to powder congestion inside them, and consequently, there is periodic powder throw-off. When difficult to access and tight positions are generated, the relatively wide, and due to the annular gap, laterally extended nozzle tip leads to problems. A continuous coaxial nozzle cannot be adapted to different focusing optics or the rotationally symmetric deviant power density distributions.

Discrete coaxial nozzles or multijet nozzles lead the powder over several single nozzles, distributed around the laser beam. Often three individual nozzles are positioned 120° apart. Because of the relatively steep inclination of the individual nozzles, about 75° to the horizontal, and the possibility of a separate adjustment, the three subbeams can be successfully focused with a depth of several millimeters.

3.3.1.2 Process Monitoring and Control

Typical deviations from the nominal layer thickness are in the range of 10 to 100 microns and are added at several hundred layers up to deviations in the millimeter range. Therefore, apart from a homogeneous powder density distribution, a powder supply that is adjusted correctly and conveyed without temporal fluctuations has to be guaranteed in the powder gas jet. Finally, a good process is defined only when a distance control for a reproducible layer construction is provided.

Depending on the component geometry, local temperature peaks may arise and could lead to construction defects. This can be avoided by temperature control. In the molten bath, a temperature window from 1400 to 1700 °C is detected by a photodiode and is used as a signal for the laser power control. References: [Argawalw, 91], [Carter, 91], [Haferkamp, 95], [King, 94], and [Mariques, 91]

3.3.2 Laser-Engineered Net Shaping (LENS), Optomec

LENS MR-7, LENS 450, LENS 850-R

Optomec Inc., Albuquerque, NM
Neotech Services MTP, Nürnberg, Germany

Short Description

In laser-engineered net shaping (LENS), process powders are applied from different metals or alloys via powder nozzles locally on a workpiece and melted locally using a laser. In this way, material is generated in rows and layers, and so a component can be constructed or repaired. The process corresponds to laser generation.

Application

Functional prototypes, final products in metal

Development Status

Commercialized since May 1998

History

Optomec was founded in 1983 by Thomas Swann. The company supplies optomechanical systems to research institutions and universities and as an accessory for rapid prototyping systems. The LENS technology was invented and developed by David Keicherin in Sandia National Laboratories (SNL). He joined as a partner at Optomec in 1997.

Strategies/Development Partners

LENS is a registered trademark of Sandia National Laboratories (SNL) and Sandia Corporation. In 1997 Optomec took over the LENS process as a license from SNL and developed an industrially applicable system. The strategic goal is the direct production of metal functional prototypes and identical components from a variety of metallic and ceramic materials. There is a development partnership with SNL.

Optomec tracked their overall strategy to become a manufacturer of additive manufacturing systems for the processing of high-performance materials. Accordingly, the metallurgy stands in the foreground.

For the production of microelectronic components, Optomec has mesoscale aerosol printing under the name Maskless Material Deposition (M3D), (see Section 3.6.6, “Maskless Mesoscale Material Deposition (M3D), Optomec”).

Data Format/Software

The LENS system can read error-free STL data and CAD data via the usual neutral interfaces. The system software contains all elements that are necessary for the preparation and control of the additive manufacturing process. The system is operating optionally with SolidWorks rapid prototyping software. The system provides a modular assembly according to user requirements, such as additional axes, and process monitoring, which can be integrated into the machine control.

Principle of Layer Generation

The powder is introduced through several nozzles that are distributed around the laser beam in the focus area of the laser, melted there, and applied to the molten component. The solidification after cooling occurs by heat conduction into the component. Cyclone powder feeders are used as a standard.

The model is created line by line (raster process), wherein the line structure from layer to layer is rotated by 90°. The advanced scanning strategies are not known. The LENS 850-R has a fiber-coupled 1, 2, or 4 W fiber laser.

The building process takes place in a fully closed process chamber. It can be viewed with a video camera. The trial chamber is flooded, depending on the material, with a suitable inert gas, normally argon. The system has a complex gas management system. A recirculation allows an exact flow in the process chamber. It is connected to a process gas cleaning system that constantly monitors the oxygen content, among others.

Design/Construction

The company Optomec provides complete systems consisting of software and hardware for converting the CAD data and to control the process. It consists of a process chamber sealed against environmental influences and working under inert gas, the externally positioned laser, which is connected to the chamber via an optical fiber, the powder conveying and coating unit, and the build platform.

In the LENS 850-R (Fig. 3.58), the dimensions of the process chamber are x, y, z = 900, 1500, 900 mm^3. The substrate is clamped onto a rotary table that can be moved in the x and y directions. The x-y contour of the component is provided by a positioning system with a brushless servo drive. The laser head of the coating LENS 850-R is moved by an x-y portal system in the building plane and the entire portal in the z axis. The system has up to seven axes. Two powder feeders allow the adjustment of a material gradient. A front feed palletizing system supports the handling of large parts and bigger series.

Figure 3.58 LENS 850-R, view (left), process chamber (right) (Source: Optomec)

A visual inspection system and a molten pool monitoring and control is integrated. The LENS 450 has a smaller construction space of x, y, z = 100, 100, 100 mm^3 and a 400 W fiber laser. It is provided for small parts in small quantities. The MR-7 has a construction space of x, y, z = 300, 300, 300 mm^3 and a 500 W, 1 kW, or 2 kW fiber laser. Most features of the 850-R are at least optionally available for the MR-7.

Material/Construction Time/Accuracies

According to the manufacturer, fully dense components can arise from a variety of materials, including stainless steel, titanium, and special alloys, which are intended to have the physical and technological properties of standard parts.

The manufacturers claim that due to the laser process a grain refinement of the resulting structure occurs and causes both the tensile strength and ductility of the parts produced with LENS to exceed the characteristics of the relevant materials [Kreicher, 98]. So with the LENS process the steel 316S reaches a tensile strength of 500 MPa, compared to 243 MPa from nonadditive processing. This trend is also evident in Inconel 625 (584/403 MPa) and Ti6Al4V (973/834 MPa), in which the elongation at break values are at least 10% higher than the comparative values, and the expected hardening thus does not occur. The manufacturer specifies that 0.08 mm details can be resolved and identified a tolerance of ± 1/10 mm relative to the finished size. The LENS process has the fundamental advantage that components made of different materials can be built. It is obviously possible to provide the component with supports made of copper, both externally and in terms of internal cavities. More information about it does not exist.

Postprocessing

Optionally removing the supports, polishing

Typical Follow-up Processes

None

3.3.3 Direct Metal Deposition (DMD), DM3D Technology (TRUMPF)

TrumaForm DMD 505, POM 5050 Direct Metal Fabricator, DMD 105

TRUMPF Laser GmbH & Co. KG, Ditzingen, Germany
DM3D Technology, LLC, Auburn Hills, MI

The TrumaForm DMD 505 and the POM 5050 Direct Metal Fabricator were identical machines that arose from the partnership between TRUMPF and POM. Therefore, they are considered together. Meanwhile, TRUMPF no longer has the DMD machines in their program, but provides the same or similar custom configurations available on request. POM has focused its activities on the DM3D technology and offers service or customized machine concepts. A distribution of configured machines does not exist.

Short Description

The systems use the direct metal deposition (DMD) method. Powders of different metals and alloys are applied over powder nozzles locally on a workpiece. It is simultaneously melted with a laser and solidified by heat conduction in a solid layer. In this way, a component in rows and layers of material is generated and constructed or repaired. The process corresponds to a regulated automated laser generation.

There is another system family called Trumaform, including the LF130 and LF250, which also produce metallic components, but uses the method of direct laser melting in the powder bed, called the TRUMPF *direct laser forming*.

Application

Rapid manufacturing

Development Status

Commercialized since 2003

History

TRUMPF is primarily known as a manufacturer of machine tools for sheet metal and material processing based on laser sources and laser-assisted machining equipment for the industry. TRUMPF preferably provides systems that consist of beam sources and handling equipment. For application fields with appropriate market relevance, special solutions such as pipe welding systems are offered. According to the company's history, the machine technology stands in the foreground (compare Section 3.2.9, "Direct Laser Forming, TRUMPF").

POM was founded in 1998 as *Precision Optical Manufacturing*, which is technology for automated laser cladding in a closed loop machine; POM was implemented and marketed after many years of preparation by the founder, Dr. Mazumder. The first machine, POM 2400 (the name comes from the 2.4 kW CO_2 laser from TRUMPF), was introduced in 1999. The machine was mainly used in the company's service center. In 2002, POM presented the DMD 5000 (with a 5 kW laser), the second generation. DM3D took over POM in January 2013, including all assets at the Auburn Hills site. It is headed by interim CEO Roger Parsons.

Strategies/Development Partner

In 2002, a strategic alliance to develop and commercialize the DMD technology was agreed upon between POM and TRUMPF. The goal of the agreement was the sharing of the DMD patent, which includes the continuous and automatic laser application controlled via integrated optical sensors with a powder nozzle laser system. In 2003, the POM DMD 5050 (which is also called the DMD 505, Fig. 3.59) and the TrumaForm DMD 505 followed. The two machines are identical, if one leaves aside the numerous possibilities for customers to customize the machine. In Germany only the TRUMPF machine is marketed.

POM has also presented a smaller machine, the DMD 105.

The machines are no longer associated with a configuration and marketed as a kind of catalog product, but form the basis for services or customized solutions.

The description is based on the old machine names, because technical data are present. They also form a good basis for individual concepts.

The term *direct metal fabricator* emphasizes that final products are manufactured. The machines are mainly used for repair or the upgrading of heavy-duty tools. In addition to armoring, thermally conductive alloys are also used as well as built up so-called bimetallic systems that allow fuel in casting tools. This allows a cycle time reduction while increasing service life.

Figure 3.59 TRUMPF DMD 505/POM DMD 505 direct metal fabricator (Source: TRUMPF)

Data Format/Software

The system works with the TRUMPF 3D CAM programming system ToPs 800, and also with the 3D CNC data of the respective component. For components, which need to be repaired, these datasets usually are available, as they are machined in most cases before (and after) the repair process. Otherwise the programming is possible by a teach-process right at the machine. Basically the control of the tool path (nozzle) is track-oriented in any case. The machine can be accessed over the internet.

Principle of Layer Generation

The powder material is applied by a powder feeder coaxial with the laser beam on the workpiece surface and completely melted. The weld pool solidifies by heat conduction into the component. Each layer is applied, path-controlled, line by line. Therefore, no stair-step effect. The individual tracks have a width of about 0.2 to about 1 mm and a corresponding height. To form a layer, they have defined overlaps. The material application is monitored by the geometry of the weld pool by three CCD cameras in 3D and controlled in real-time closed loop.

As it is preferably attached to existing geometries or substrates, in general the support structures are omitted.

Design/Construction

The system is based on a TRUMPF Laser Cell, a modular five-axis machining system for laser material processing. Three linear axes (A, B, C), with a traverse of (A = x, B = y, C = z) 2000, 750, and 1000 mm and two axes of rotation that can be pivoted at (B) to ±45° and at (C) to ±190°, form a five-axis motion system. CO_2 lasers are used at 3.2 to 6 kW (default is 5 kW) and focus diameters are around 0.2 mm. The machining head with the optics of the concentric powder nozzle and sensor depends on the two axes of rotation along the z axis. The result is a very easily accessible machining area. The machine is completely enclosed, but it can be operated with the doors open and fed on three sides. This allows the processing of components that are larger than the machining chamber. The processing can take place under protective gas. The machine is not completely closed (sealed) and therefore does not allow editing in a circulating inert gas atmosphere. The use of up to four powder conveyors is also provided by software. Powder materials can be mixed directly in the machine.

The DMD 105 is equipped with a fiber-coupled diode laser of 1 kW, has a five-axis handling system, and has a processing space of x, y, z = 300, 300, 300 mm^3. It is designed for machining of smaller parts, such as the repair of gas turbine blades.

Material/Construction Time/Accuracies

Commercially available single-component metal powder without a binder in a typical grain size from 40 to 105 microns can be used (maximum 160 microns). Typical materials are super alloys and the Ni-Co-Cr-Al system, Ni and Co alloys, tool steels such as 1.2344, and WC hard coatings. Due to the relatively low heat input, in particular, crack-sensitive materials such as nickel-based alloys (like Inconel) can be handled excellently. Typical material application rates are 20 to 60 (90) cm^3/h (maximum up to 150 cm^3/h). The layer thicknesses are usually several tenths of a millimeter to a few millimeters, with a machining allowance of about 0.2 mm to be considered for finishing by milling. Thicker layers are not used, less for technical reasons than for economic reasons. In the machine a surface quality of 175 microns can be achieved; see Fig. 3.60.

Heat treatment can also be done in the machine with the (defocused) laser.

Figure 3.60 Direct metal fabrication, partially finished component with mechanical repair coating (Source: TRUMPF)

Postprocessing

Process-specific postprocessing is not required. The components can be heat treated.

Typical Follow-up Processes

The components are usually finished by machining or erosion.

3.4 Layer Laminate Manufacturing (LLM)

3.4.1 Overview of Layer Laminate Manufacturing

Figure 3.61 shows a structured overview of layer laminate manufacturing. Based on the generic names (Fig. 2.38), it contains the industrial processes and machines discussed hereafter, as well as their product and manufacturer names.

Software-based methods are included, which are either completely hardware independent, or essentially for the purpose of marketing, have been adjusted to a specific hardware. In each case, the sections are indicated in which the procedures are described in detail.

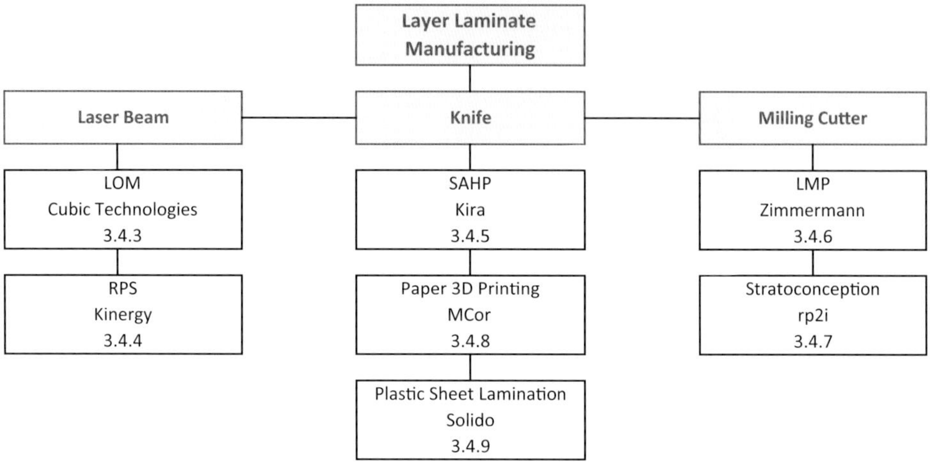

Figure 3.61 Layer laminate manufacturing; overview in continuation of Fig. 2.38

3.4.2 Machine-Specific Basics

"Layer laminate manufacturing" (LLM) denotes a method family in which the layers are individually contoured from prefabricated sheets or plates using a laser, knife, or cutter and simultaneously or subsequently are connected (usually in the form of bonding) by joining. They are often described as laminated object manufacturing (LOM). However, LOM is the product name and a registered trademark of Helisys Inc., and respectively its successor, Cubic Technologies (Section 3.4.3). It is therefore not usable as a generic term.

Many approaches lead to nonautomated processes and are therefore not strictly *additive manufacturing* processes.

Principle of Layer Generation

The LLM process is the direct implementation of the theory of the layer method by a machine. A CO_2 laser, a blade, or generally a parting tool cuts the contour corresponding to the respective layer information (see Fig. 3.62). The 3D part is created in the z direction by joining the following layers to each other. Strictly speaking, the LLM process is a combination of a subtractive (cut) and an additive (connect) method. Papers, plastics, metals, and ceramics are used in sheet form. Paper, wood, and plastic are either laminated with polyethylene adhesive, or the adhesive is applied during the layer generation. Metals are welded at the contours or surface bonded by diffusion welding.

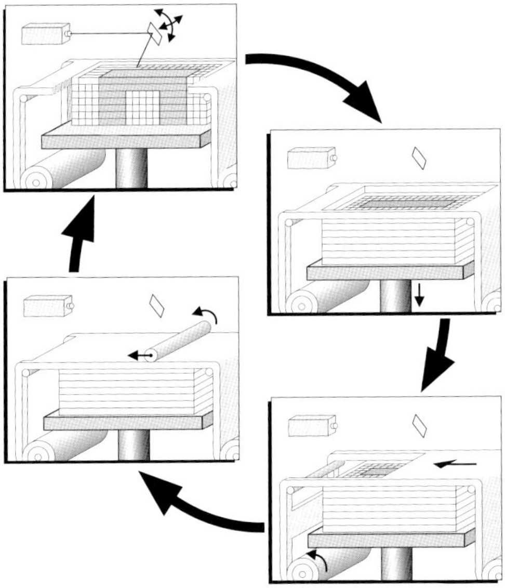

Figure 3.62
Layer laminate manufacturing (LLM), in the variant of laminated object manufacturing (LOM) principle[19]

As a result of the construction principles, in which the individual layers are only contoured (cut out) but do not have to be filled, the LLM method compared to the raster scanning method is more advantageous the greater the surface of a layer is in relation to its application. The method therefore has particular advantages when massive models must be built with a large-volume model. Intricate structures are inadequate and expensive to produce. Construction strategies for the production of hollow structures, such as "Quick Cast," "sheath core," and "Fast Sculp" are only possible if their cleaning is ensured.

Supports are not needed because the component associated material that is separated usually remains inside of the part and supports it. For later and better extraction, they are cut into small squares (hatched), so this results in cubes remaining in the part that must be removed after its removal from the machine. There are also variants of the process in which the parts not belonging to the final part are removed after each layer or directly removed as chips. A variation of this is to fill the resulting voids with wax, for example, for stabilizing the part, so that the wax can be melted out after completion.

The same is true within limits for heavily undercut model parts. For this reason, internal bores that are even longer and more difficult to access, especially if they run spatially, are difficult to clean and therefore are not arbitrarily finely representable.

[19] A 3D animation can be found here: www.rtejournal.de/archiv/index_html/filme.

Design

The basic principle was picked up by several manufacturers that developed different method families out of it.

In the very simple *(partially) manual layer laminate manufacturing*, the contour is usually cut layer by layer with a knife cutting plotter or with a hot-wire cutter plotter from relatively thick plastic sheets. The model contours are then released manually and glued to the model. The processes are inexpensive, and because of the relatively large thicknesses of polystyrene, allow the building of large, but very rough, models. A disadvantage is the dependence of the model quality on the technical skills of the operator.

Automatic layer methods, which cut the contour out of paper sheets with lasers or knives, have the largest market share. These include the LOM machines from Helisys and the very closely related prototypers of Kinergy (Zippy). The material not related to the model is incised, but remains in the layer and is removed after completion. If paper is used, the models are similar to wood. Other materials such as plastics, ceramic composites, and metals are available or announced. While papers and plastics can be bonded with relative ease, metals must be welded and ceramics must be sintered. The postprocessing, the removal of the unused internal material, is complicated in every case.

Layer-milling processes use plates of up to 20 mm thickness. The materials are wood or modeling materials. The contours are formed by milling in the *x-y* but especially also in the *z* direction. Therefore, parts can be produced that have no more gradations. Representatives are the layer milling process (LMP) (F. Zimmermann, Germany) and the Strato Concept PC method (Charlyrobot, France). Stratified object manufacturing (SOM), in which the model is not necessarily divided into layers, but into any undercut-free parts models. The "Millit," a software tool for the segmentation of 3D objects into undercut-free layers, belongs to the same family. Millit works with STL data. In both cases, the layers are milled and then joined to the model. Methods that cut the individual contours with lasers from steel plates are schematically also assigned to the layer milling processes.

Materials/Construction Times/Accuracies

These are machinery- and manufacturer-specific. This is discussed in the descriptions of the individual machines in the following chapters. Basically, most components have anisotropic material properties because the layers and the connecting layers are different in their properties. Exceptions are the diffusion welding of metals and the sintering of ceramics, both of which lead to isotropic components.

Postprocessing

After the construction process, in the paper and plastic processes, the entire block with the model lying therein will be removed from the machine. The frame and parts not included in the model are removed (Fig. 3.63). For finishing, especially in paper models, the immediate fixation of the model with varnish takes place so that the highly hygroscopic parts on freestanding corners and edges are not flaky like a puff pastry (delamination). The surfaces can, for example, as in wood, be reworked, especially by grinding.

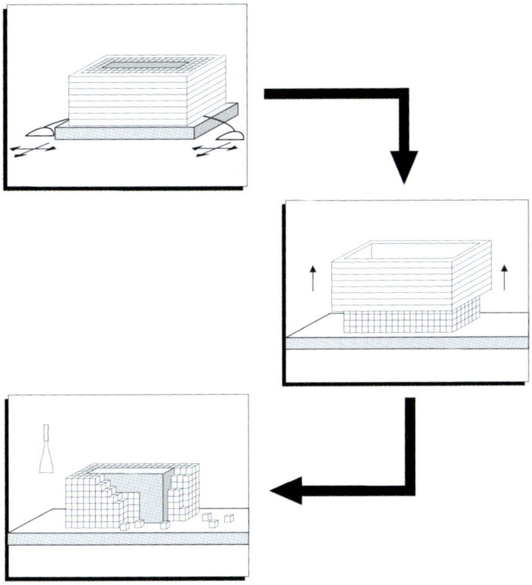

Figure 3.63
Postprocessing in layer laminate manufacturing (LLM) in the variant of laminated object manufacturing (LOM); principle

Metal models are released from the unused material layer by layer. Figure 3.64 shows an LLM workpiece before and after unpacking.

Figure 3.64 Layer laminate manufacturing (LLM) in the variant of laminated object manufacturing (LOM), part before and after the "unpacking" (postprocessing)

Typical Follow-up Processes

LLM parts of paper are generally suitable for all molding processes and follow-up techniques in which the part is seen as the master for investment casting and vacuum casting but also for sand mold processes.

In the direct burning of the parts in the investment casting process, the high ash contents are problematic. It is therefore recommended to choose the process chain over wax models. For this purpose, the LLM parts are manufactured as divisible forms in which the wax is injected.

LLM parts are well suited for sand casting because of their high compressive strength and are accepted by foundries as an alternative to the tried and tested wood models because of their wood-like structure and properties. After the building, they must be sanded, sealed, and painted. LLM components are ideal supplements for laser-sintered sand cores for rapid assembly of boxes with complex shapes.

LLM parts are, because of their mechanical and thermal stability after appropriate pretreatment, suitable to be used directly in thermoforming processes. Because of their temperature resistance, polycarbonate, ABS, propylene (PP), and similar plastics can be used. It is also advantageous that the parts can be built very large and double-layer thicknesses may be used. Finally, in combination with the sand casting process, the production tools can be made on the basis of the prototypes.

After careful finishing, LLM paper parts can be used in vacuum casting.

3.4.3 Laminated Object Manufacturing (LOM), Cubic Technologies

LOM 1015plus, LOM 2030H

Cubic Technologies Inc., Carson, CA
RTC Rapid Technologies GmbH, Hofheim, Germany (materials and services)

Short Description

These are prototypers based on layer laminate manufacturing, which at Helisys, is called laminated object manufacturing (LOM, a protected name). Layer by layer an adhesive-coated film is adhered to a building platform or on the partially completed model and contoured by the use of a CO_2 laser plotter. All not-inherent regions of the model are removed manually after completion. Basically all materials that are available as film can be used. Paper sheets are standard. In addition, plastics and ceramics are available, but they are not very common.

The LOM machines currently do not play a major role and are no longer sold. The description was left in this book because it reflects the common basic principle of all later developed machines from other manufacturers.

Application Field

Concept models, functional prototypes. The method is used to advantage when large, solid models are required, and possibly also with large steps in the wall thickness.

Development Status

Commercialized since 1991

History

Helisys was founded in 1985, even before 3D Systems, by Michael Feygin and thus (except for Denken, J., 1985) is the oldest company that offers prototypers. The LOM process was patented in 1988. However, sales did not begin until 1991. The machines have changed little in their basic concept and also only a little externally until today, but they were improved in numerous details. Thus the LOM 1015 was followed by the LOM 1015plus and the LOM 2030 by the LOM 2030E and the LOM 2030H. The range of materials has been extended to plastic and ceramic composite films. It is probably due to the first incoming market success, in the form of Kinergy (1996) from Singapore and Kira (1994) from Japan, when basically related systems huddled to the European market. Around 1995, the business started to go bad. Helisys closed in 2000 and continued with an identical product range in the form of Cubic Technologies Inc., which was also founded by Feygin.

Strategies/Development Partners

The American company Cubic Technologies has commercialized the LLM method and provides two technically essentially identical machines of different sizes for the fully automatic production of three-dimensional models of films under the name LOM. The machines are in price, based in a market segment in which they are in direct competition with stereolithography, laser sintering, and other (full-size) rapid prototyping methods. The development focuses on new materials. Development partnerships exist with the University of Dayton in Ohio (LOM models with curved surfaces). Currently there are no sales in Germany, only service and material supplies.

File Formats/Software

The single input file format is the STL format. The software of the LOM system allows preparatory operations with limited CAD functionalities with graphic support on the screen. Data errors cannot be corrected, but the program is very forgiving. The “slicing” is done online with LOMSlice. In complex models, it can occur that the machine has to wait for the slice data. The accuracy of the parts in the z direction is increased by the fact that the current height is permanently measured as a basis for the calculation of the following layer. However, the measurement is made only at one point.

The software does not make any beam width compensation. The calibration of the system is achieved by the measurement of a fixed test specimen. The actual values are entered into a program which then automatically calculates certain correction factors. In the next software version, an explicit facility for beam width compensation is to be implemented.

Principle of Layer Generation

The material is inserted into the machine in a roll form and drawn over the construction platform. There it is laminated using a heated laminating roller onto the construction platform or onto already contoured model parts. The laser then cuts the desired contour. In addition, each layer is provided with an exterior frame that is equal for all layers through which the layer and thus the model are separated from the paper roll. The material of the frame also serves to further transport the material to roll it up on the opposite side and to position the next layer on the construction platform after the contouring of a layer.

For the construction to begin, the first layer of paper must be attached with double-sided tape onto the build platform. On this first paper layer, a base is built up, which means a cut into blocks of 25 × 25 mm, and depending on the model size, a 3 to 8 mm high paper block. The base prevents warping at the start of construction (curl), which can be several millimeters and depend on the direction of rolling. Then the actual construction process starts.

When the first layers of the part shall be reviewed, a period of about 1 to 2 hours is necessary for starting the machine, including the first inspection of the part. The speed with which the roller is transported can be set by software. In the manufacture of large parts with a low ridge width of the paper frame and a high forwarding speed, the risk of a paper crack is a given. But this is detected by a following tapping roller, and the machine is automatically stopped. A startup is possible immediately after a paper crack without loss of quality. Because the paper cools down with time or absorbs moisture and swells, depending on the ambient conditions, at a later restart of the process, the risk of structural defects in the form of an offset cannot be excluded.

Construction Process

Each new paper layer is pressed by a roller heated to about 330 °C. For that, the roller melts a thermally activated adhesive onto it, and thus ensures the connection of the new layer with the partially completed model. For large parts, the temperature of the roller and the pressure should be slightly higher than for small parts.

One to four paper layers can be processed simultaneously. With increasing thickness of the layers, the speed of construction increases, but the accuracy decreases and the discoloration of the edges by the laser increases.

Both the actual model contour and the entire maximum space are cut out as a frame in each layer. The rest of the paper remains as a cutout in the framework of the paper web and ensures that the new layer can be adjusted by means of the winding roller paper. The disadvantage of this design is that, depending on the machine size width, 250 mm or 10 mm basically remains unused and must be disposed of.

Design

Cubic Technologies offers two essentially technically identical machines of different sizes, the LOM 1015plus and the LOM 2030H. The 2030H has a construction space of x, y, z = 813, 559, 508 mm and the 1015plus has the dimensions of x, y, z = 381, 254, 356 mm. Both machines are compact stand-alone solutions, where all necessary elements are integrated. On the user side, a generously sized suction device, a power supply, and depending on the type of laser, a water cooler must be provided. Because of the large cover, the construction space itself is easily accessible. Large doors in the lower frame ease the access to the paper mechanism. The machines require, apart from the on-site air supply and the exhaust system, no major installation expense and must always be run automatically.

The contouring is performed with a CO_2 laser (2030H: 50 W; 1015plus: 25 W) above a plotter mechanism. Figure 3.65 shows a view into the construction space. The plotter mechanism consists of a cross head movable in the x direction (paper lengthwise) and a laser head movable in the y direction (paper-width direction), which is currently in the top-right corner of the construction area. The laser beam sweeps in from the right, above a movable mirror mechanism covered by a folding bellows. The boundary of the actual construction area is clearly visible. The paper has just lifted from the plane of construction and is transported to the left so that a new layer reaches the construction space. The rolling or unwinding mechanism of the paper is situated below the actual construction platform.

In order to ease the removal of unneeded parts that initially remain in the model, these parts are also cut into small squares, resulting in cuboids in the z direction. The grid spacing in the x-y direction is adjustable via software to between about 3 and 5 mm. Horizontal surfaces can be automatically scanned more finely. For this purpose, a grid spacing of about 1 mm is usually chosen. Removing the remaining cuboids from horizontal surfaces is otherwise hardly possible.

After contouring of the last layer, the entire model is taken as a block from the machine and manually released from the parts not required for the model.

The construction of nested movable parts in one construction process is possible if the parts are already assembled as STL files into one another. The parts are movable when the laser passes over the manufactured parts without play twice, so that the necessary gap is created.

Figure 3.65
View into the construction space of a LOM machine
(Source: Cubic Technologies)

Material/Construction Time/Accuracies

Usually the part is manufactured in a "one-layer operation" with a thin (0.1 mm) paper. When utilizing the maximum design space, the construction progress of complex geometries is only a little more than 2 mm/h. For large raster intervals and simple geometries, at most 12 mm/h progress can be achieved. If very intricate parts are built and consequently a fine grid must be scanned (for the rest of the material), the parts tend to be much more expensive. The maximum contouring speed is 456 mm/s.

The accuracy, measured over the entire space, is usually ±0.25 mm. The repeatability of the laser gantry is ±0.05 mm, and the beam diameter is from 0.203 to 0.254 mm. Although wall thicknesses of 1 mm can be built, one should refrain from wall thicknesses of less than 1.2 mm, mainly because of the consequent rework. In order to build parts with thin walls, the roller pressure must be relatively high. The parts otherwise tend to delaminate, not only during demolding but also during the construction process.

As materials, polyethylene-coated paper sheets (LOMPaper), polyester films (LOMPlastic), and glass fiber reinforced composites (LOMComposite) are available. LOMPaper is available in different widths as standard (LPS) in 0.1 mm thickness and as high performance (LPH) in 0.1 and 0.2 mm thickness. LOMComposite (LGF) consists of differently oriented but nonwoven glass fibers. The adhesive layer consists of thermally activatable epoxy, the upper outer layer of acrylate. LOMComposite is 10 times stronger than paper, resistant to moisture, and withstands higher temperatures, so it can be used for special shapes and functional parts. LOMPlastic (LXP) is a polyester film having 0.127 mm thickness, that is, the same as a paper coated with polyethylene copolymer as a thermally activatable adhesive. It is to be used without modification in the machine. The accuracies range from ± 0.25 to ± 0.4 mm and are dependent on the geometry of the part.

The top of the parts is, in the normal case, relatively brown, due to the effects of the laser beam, and is imprinted with strong marks of the basic screen. The bottom, however, is relatively white and has hardly any grid marks on it.

Depending on the complexity of the part, a percentage of the material is not used for the model itself, but for the surrounding frame and the internal support structures. Therefore, for parts with housing character, well over 90% of the construction material can be waste. As a result, the relatively cheap price of paper is relativized. Since LLM procedures always have access to all laser-cutable materials, this aspect may be important in the future in the use of more expensive materials.

Postprocessing

For paper, the finished parts are alternatively separated by a wire, similar to a cake wire, directly from the perforated metal construction platform, then melted from below with an iron, or removed or pocketed from the back of the plate with solvent, and then through the holes, the tape is dissolved.

For demolding of the part, the surrounding frame is completely lifted. This may require the assistance of a spatula and gouge. Subsequently, the model and the surrounding supportive cubes are removed, and the model is exposed step by step in this way.

After cleaning, the parts immediately have to be coated, preferably with Zapon paint, otherwise they will absorb moisture and grow, up to 1% within one to three days. The coating also seals the parts from leaking glue and protects against delamination.

The rework can also be done by milling, turning, and so on, as long as it is well painted and the tools used are sharp. The success during milling depends on the direction of rotation of the cutter according to circumstances. Also, drilling is possible, but if necessary, one must prevent the drill from pressing the layers apart. Usually this is done by clamping the parts transverse to the direction of drilling.

The expense of the finishing increases significantly when the part has thin walls.

Metal parts must be diffusion welded, layer soldered, or otherwise joined; ceramic parts are sintered.

Typical Follow-up Processes

After careful postprocessing, finishing is possible with all of the traditional machining and molding processes, as suited to the respective material (see Section 3.4.2).

3.4.4 Rapid Prototyping Systems (RPS), Kinergy

ZIPPY I, ZIPPY II, SW

Kinergy Precision Engineering, Co. Ltd., Nantong, PR China

Short Description

The prototypers of the ZIPPY family work on the layer laminate manufacturing (LLM) method. They are largely similar in structure and in function to the LOM prototypers after the Helisys LOM principle. The model is created in layers of paper or plastic film. The material is laminated on the model and then contoured by the laser.

Application Field

Concept models

Development Status

Commercialized

Strategies/Development Partners

The main business of Kinergy (Singapore) is the contract manufacturing of electronic parts and components, including tooling. Kinergy also builds mechanically robust prototypers, and offered with the ZIPPY II in 1999, the largest space of an additive manufacturing machine. Revised versions of the ZIPPY II have since not been introduced. In 1999, Kinergy Mechatronics in Shanghai was founded. It is a wholly owned subsidiary of Kinergy and acts as a marketing and service center for the entire company. It builds and sells prototypes (rapid prototyping service, RPS) using machines manufactured by Kinergy RPS. The sale of the machines does not seem to be as strained since then. Kinergy is also active as a system developer. In 2003, in cooperation with the National University of Singapore, a laser sintering machine for metal or plastic was developed.

The machines are currently not in their business program and are therefore no longer included in this edition. All typical follow-up procedures for LLM described in Section 3.4.2 can be used.

3.4.5 Selective Adhesive and Hot Press Process (SAHP), Kira

SolidCenter PLT-A3, PLT-A4, PLT-20 Katana

KIRA Corporation, Hazu-gun, Japan
KIRA Europe GmbH, Erkrath, Germany

Short Description

These are paper layer methods based on a (Xerox) copier (selective adhesive and hot press process, SAHP). The toner is thermally activated, thus enabling the bonding of the individual layers. The contouring is performed after bonding the layer on the model of the current block by a cutting plotter.

The machines are currently not in their business program and are therefore no longer included in this edition.

3.4.6 Layer Milling Process (LMP), Zimmermann

CNC layer milling center LMC (Zimmermann), WorkNC-LMP Module (S.E.S.C.O.I.)

F. Zimmermann GmbH, Denkendorf, Germany

Short Description

This is a fully automatic layer milling center, consisting of the software module WorkNC-LMP (S.E.S.C.O.I.) and the Layer Milling Centre, or LMC (Zimmermann), for the production of parts with large cavities and steep walls. The Layer Milling Centre consists essentially of an overhead-operating, high-speed model milling machine, which contours the model layer by layer, undercut-free, without gradations, and at approximately constant cutting conditions, a disk magazine, and an adhesive unit.

Application Field

Function prototypes, final products (forms)

As the machines are no longer sold by the company Zimmermann, a detailed description is omitted here.

3.4.7 Stratoconception, rp2i

Stratoconcept VR Software, Stratoconcept STE-Series, STO-Series milling machine rp2i milling machine, Charlyrobot milling machine

Stratoconcept, Saint-Dié des Vosges, France Charlyrobot, Cruseilles/Cernex, France

Short description:

The Stratoconception software slices a 3D CAD model into equal layers ("la strate," in French, means the layer) and determines their contours. These are arranged nested on a plate with the corresponding layer thickness, contoured by milling, and after this processing assembled into the part. Because they can be turned before joining, undercuts are possible.

Application Field

Function prototypes, technical prototypes

Because the machines of the company Stratoconception are not available on the market anymore, a detailed description was omitted here.

3.4.8 Paper 3D Printing, MCor

MCor Matrix 300, MCor IRIS

MCor Technologies Ltd., Dunleer, Co. Louth, Ireland

Short Description

This is a layer laminate process using plain paper (A4, 80 gr, and standard letter office) and contouring by knife. The parts have the color of the paper (matrix 300). The IRIS is a full-color machine that produces the color by printing on the paper prior to the 3D process.

Application Field

Concept models

Development Status

Commercialized since 2007

History

The company was founded in 2003 by Dr. Conor MacCormack and Fintan MacCormack.

Strategies/Development Partners

MCor has realized its strategy with the Matrix (Fig. 3.66(a)) to provide a simple and inexpensive, both to purchase and to run 3D printer. An important contribution is the development of a paper layer process without pretreatment of the paper and without an expensive pressure system.

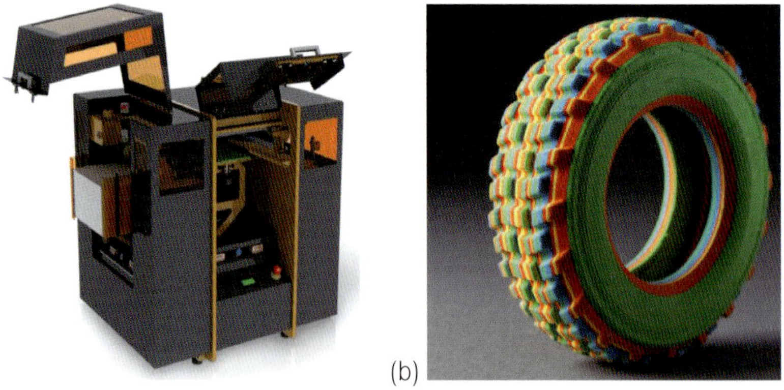

(a) (b)

Figure 3.66 Paper 3D printing; (a) the MCor Matrix 300 machine, (b) a colored part made of precolored paper (Source: MCor)

The first colored parts that were created out of colored paper (Fig. 3.66(b)) were not only visually interesting, but were also the experimentation and development platform for the full-color machine IRIS.

MCor is expanding its strategic approach to color parts with the IRIS. From the basic process, a full-color printer was developed using the project partner Staples. “Staples is a leading manufacturer of office supplies and office solutions and 3D printing is a logical extension of our activities,” says Oscar Pakasi, director of Staples Printing Systems.

File Formats/Software

The machines work with STL files. The SliceIT software cuts the 3D CAD data in the specified layer thickness and supports the positioning in the construction space. For the colored parts, the individual layers are provided before slicing with the color information using the software ColourIT. The software is compatible with STL, WRL, OBJ, 3DS, FBX, DAE, and PLY. Afterwards the file is exported to the WRL file format to the SliceIT program.

Principle of Layer Generation

The core of the layer generation principle is the application of the adhesive, which consists of microdroplets and therefore prevents paper curling. The layers are first glued and then contoured with the cutting blade. Cuts away from the contour toward the paper edge and a lower adhesive application outside the contour facilitate later removal of the excess paper.

Design

The matrix consists of a paper supply container, a positioning system, the coating unit for the application of adhesive, and the cutting plotter.

The IRIS (Fig. 3.67(a)) is a standard matrix machine in which a standard 2D Epson color printer is further integrated. Using this printer, the marginal areas of each layer (about 5 mm wide) are printed on normal paper, and the inner areas are left to white or left in the color of the paper itself. The color printer delivers 1 million colors as CMYK at a resolution of 5760 × 1440 × 508 dpi. For this purpose, it uses four cartridges, including black.

The printer prints each sheet two-sided. That, and an ink that soaks the paper very well, has the effect that the edge regions are colored with defined lasting colors.

The entire set of preprinted sheets is stored in a memory, from where sheet by sheet the 3D part production is supplied.

A bar code, supported by plain text, enables and facilitates checking whether the correct sheet is handled in the correct sequence. Missing or incorrect layers are detected, the machine stops, and then there is a corresponding reprint.

Accuracies

The layer thickness is 0.1 or 0.19 mm at .ply color data sets.

Depending on the slice quality, in spite of the good color quality, the triangle faces are clearly visible, which results in corresponding color fringes. A part is shown in Fig. 3.67(b).

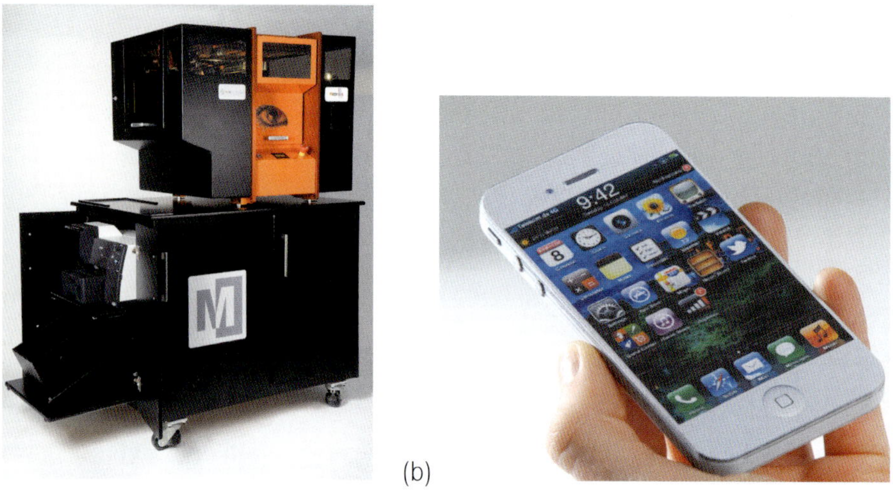

(a) (b)

Figure 3.67 Colored paper 3D printing: (a) MCor IRIS machine, (b) fully colored part made of precolored paper (Source: MCor)

Postprocessing

Postprocessing includes manually freeing the part from the surrounding paper, light sanding, and a colorless coating.

Finishing

All finishing methods suitable for paper can be applied in principle.

Typical Follow-up Processes

None

3.4.9 Plastic Sheet Lamination, Solido

S3D-300 Pro
Solido Ltd.
encee CAD/CAM Systeme GmbH, Kümmersbruck, Germany

Short Description

This is a desktop concept modeler according to the layer laminate process for the production of plastic models with PVC films.

Application Fields

Solid imaging and manufacturing of concept models

Development Status

Commercialized as a 3D Systems product since 2005

History

For information about 3D Systems, see Section 3.1.3. The LD 3D Printer was developed by Solidimension and initially marketed exclusively. Since April 2005, it was manufactured, first made without any technical changes, for 3D Systems from Solidimension and marketed as an LD 3D printer. A further variant that is structurally identical is the Graphtec XD 700. In July 2006, 3D Systems additionally introduced a revised and 30% faster variant without changing the name, but has since run out the product. The manufacturer is Solido.

Strategies/Development Partners

The SD 300 is a representative of the so-called office printer, which strategically has the future mass market in its sights. But because of the personal printer and fabber, it comes increasingly in danger of becoming extinct. Although the removal of the excess material must be made manually, the LD 3D printer can be used in the office because it requires no tools and no solvents.

File Formats/Software

The included software handles STL data and sends it to the machine after processing. A user-friendly Windows interface allows the positioning of the parts in the construction space. Small gaps in the STL data are automatically closed. Large data sets can be divided. A calculator for the construction time and a print job manager to manage the print jobs are integrated. One can assign priorities, as is usual in a conventional network printer.

Principle of Layer Generation

The construction material is drawn from the roll over the construction platform and there positioned and fixed to the self-adhesive base plate (first layer) or on the previous layer. Then a large area of adhesive is applied. The contour of the layer to be built is processed by a *demasking fluid* so that the parts of a layer not associated with the model are not glued. The film is cut off, and a new layer of the roller automatically pulled onto the construction platform and partially glued there. Then the part contour is cut with a cutting plotter. All material remains in the part. Then the process begins anew, until the last layer is contoured.

Design

The compact machine (x, y, z = 460, 770, 420 mm) enables the production of models with the maximum dimensions of x, y, z = 160, 210, 135 mm. It consists of an unwinding mechanism for construction material, a cutter for the film, coating nozzles for the adhesive and demasking fluid, and a cutting plotter with replaceable blades. The expendable materials will be delivered as a kit. No installation is necessary. See Fig. 3.68.

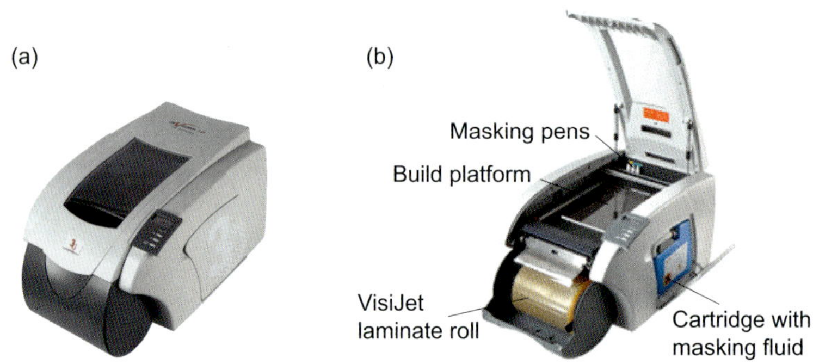

Figure 3.68 Solido SD 300 pro lamination printer, machine: (a) exterior view; (b) view when the machine is open (Source: Solido (3D Systems))

Material/Construction Time/Accuracies

The construction material is a hard PVC. It is inserted into the machine as a roll material. The roll has a diameter of 205 mm (new) and a length of 175 m. The film has a width of 215 mm and a thickness of 0.15 mm. The material is available in the colors amber, red, and blue.

The process results in a layer thickness of 0.165 to 0.168 mm. The accuracy is given in the x-y direction at different sources with ±0.20 to ±0.25 mm, in the z direction with ±1.5%, but at least plus or minus one layer thickness, for example ±0.165 mm.

Postprocessing, Finishing

Out of the machine one gets a cuboid with the width of the film, which completely circumscribes the part. The model is contained inside, embedded in the bonded, not associated to the model layer material. This material was mutually bonded during masking on the sides of each second layer on a narrow strip in a way that it can be separated without tools like a concertina from the model. The model freed from waste is ready for use. It can be treated with glue and painted.

Typical Follow-up Processes

Because of the nature of the concept, no follow-up procedures are provided. Molding by vacuum casting is possible.

3.4.10 Other Layer Laminate Methods

3.4.10.1 Parts of Metal Foils: Laminated Metal Prototyping

The principle of the production of parts from metal foils is colloquially called *metal foil LOM*. It corresponds to the principle of the layer milling process.

It is preferably applied to the construction of tools. The part is designed in CAD and divided into the necessary number of contour cuts. The contours are nested on a metal sheet and cut out using a laser.

At the TU (Technology University) Ilmenau, in the course of basic work already in the 1990s, diffusion welding has been qualified as a joining method, and production-ready tool inserts have been presented [Sändig, 01]. The work was later taken up under the MELATO project at TU Dresden [Techel, 04]. A commercialized system is not available in Germany.

Tschopp Technical Engineering, Ramlinsburg, Switzerland, offers services based on metal foil tools and also runs in-house development. The metal foils are bonded with an epoxy resin called *backlack*. The interconnection layers are about 4 to 6 microns thick.

The company Fast4M Tooling in Troy, Michigan acts only as a service provider.

The hybrid method *ultrasonic consolidation* from the company Solidica, see Section 3.7.2, also belongs to the metal foil methods.

3.4.10.1.1 Metal Foil Tools, Weihbrecht

A very simple implementation of the principle has been established by the company Weihbrecht in Wolpertshausen as a proprietary system, providing services (see Section 5.4.3.4.2).

The CAD model of the additive manufactured die is provided by means of proprietary software on the bottom with addendums in the form of T-shaped mounting slots and clearance holes and virtually cut into 1 mm thick foils crosswise to the channel axis in CAD. The foils are cut with the laser and clamped mechanically on a massive scale by means of tie rods. The edges are not reworked, but covered by a thermoforming sheet, which gives a sufficient quality for prototypes. The final die is mounted on a press. The method provides fast small quantities of thermoformed sheets with topologies that are not very strong.

■ 3.5 Extrusion: Fused Layer Modeling (FLM)

Extrusion processes continuously press a heated, and therefore viscous, thermoplastic material through one or more nozzles and lay it in a defined way on a substrate. The energy in the material is enough to start the melting of the substrate so that after solidification a permanent connection is produced. To create contours, corresponding curved strands are deposited in the building plane, the length of those can be very different depending on the geometry. Surfaces are produced from strands that are placed close together. In support structures, the strands shrink down to a (technical) point.

Processes that do not extrude continuously, but discontinuously apply the melted material (drop on demand) or shoot it onto the component, are attributed to the extrusion process, although they could be regarded as printing processes, as sometimes happens in the literature.

3.5.1 Overview of Extrusion Processes

Figure 3.69 gives a structured overview of the extrusion processes. Based on the generic name (Fig. 2.38), it contains the industrial processes and machines discussed hereafter, as well as their product and company names.

In each case, the sections are indicated in which the processes are described in detail.

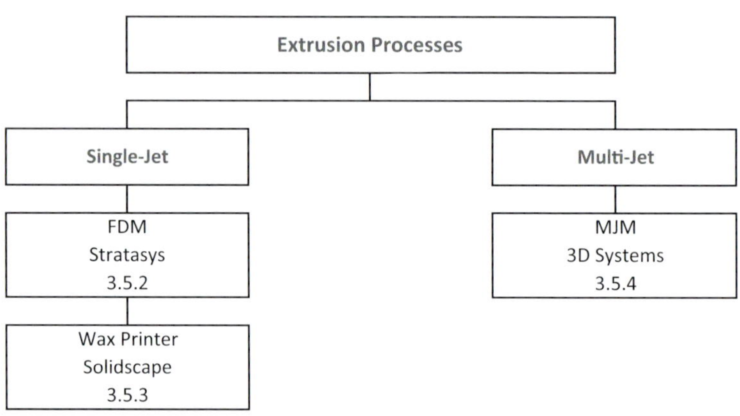

Figure 3.69 Fused layer modeling extrusion processes, overview in continuance of Fig. 2.38

3.5.2 Fused Deposition Modeling (FDM), Stratasys

3D Printers

Idea series:	Mojo 3D Printer, uPrint (SE, SE Plus)
Design series, performance:	Dimension 1200es, Dimension Elite
Production series:	Fortus (259mc, 360mc, 400mc, 900mc)

Stratasys Inc., Eden Prairie, MN
alphaCAM GmbH, Schorndorf, Germany

Short Description

The FDM machines from Stratasys melt prefabricated wire-shaped thermoplastics and lead the viscous melt through a heated nozzle specifically to the component. The stratification is done by solidification due to heat conduction in the partially finished component. The range of materials includes plastics such as polycarbonate and ABS, but also high-performance thermoplastic plastics such as polyphenylsulfones (PPSFS). The design principle can be seen in Fig. 3.70.

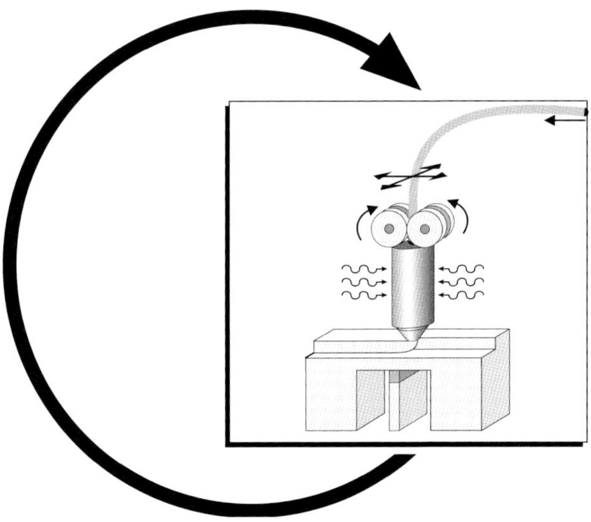

Figure 3.70
Fused deposition modeling (FDM) scheme[20]

Application Areas

Conceptual models, function prototypes, end products, series-identical components of investment castings, tooling of molding processes

Development Status

Commercialized since 1991

[20] A 3D simulation can be found at www.rtejournal.de/filme.

History

Stratasys was founded in 1988 and is one of the oldest manufacturers of additive manufacturing systems. The FDM 1000/1500/1600 prototypers were sold from 1991 on and were systematically developed to larger build spaces and sped-up processes (FDM 1650/2000/8000). With the Prodigy, the redevelopment of the entire series has begun, which today comprises a total of four machine families tailored to particular applications properties. In 1999 the FDM Quantum claimed, with x, y, z = 600, 500, 600 mm^3, the largest space of all prototypers for plastic processing. The current successor is the Fortus 900mc. The product range was expanded with the acquisition of Objet. There was a restructuring of the offering and a general designation as a printer. The classification of the printer depends on the area of use or application. Stratasys divided the printers into the Idea Series, a design series, and a production series. The predecessor of the Dimension, the Genisys, was the first prototyper to be used as a concept modeler and therefore founded a new class of machines that were also called 3D printers.

It should be noted that the Objet machines are incorporated in the printer family Design Series. Therefore, a system that is produced by Stratasys, is not always working by the FDM process anymore.

Strategy/Development Partners

Stratasys has laser sintering as a competitive process, and they designed the machines consistently based on the processing space, near-series material, speed, and easy handling. Preferably the focus is on big, thin-walled, and directly applicable functional prototypes from near-series plastic (nominally series-identical). The focus in development is currently on the material.

Starting with the Genisys, since 1997 Stratasys has developed an office-ready variant that is based on the FDM principle and continues to this day with the Mojo. The machines are called “3D printers”[21] and were marketed under the name *Dimension*, but currently they are incorporated into the program of Stratasys.

Data Format/Software

The process is prepared and controlled by a single, easy-to-use but comprehensive software that is called Insight.[22] It can read complete and error-free STL data. Automatically the data will be sliced and support structures are provided. The program calculates the extrusion path and helps to optimize the construction process. It also includes an assessment of the construction and a module for the optimization of

[21] The reason for this name was discussed in Chapter 1. It is easy to inform laypeople the analogy of three-dimensional printer with the two-dimensional printer. Confusing is the fact that the machines of the Dimension series have nothing in common with the 3D printing process patented by MIT [Sachs, 93].

[22] Insight is a protected designation of Stratasys. The name is very common and is used among other things for a program of the company TSI for evaluation of laser-based measurements as a protected trademark.

the entire construction process. "Insight" replaces the previous QuickSlice, and the generator supports the "Support Works" existing software package. The program internally transfers the construction data to the machine control. This is done with the SSL format (Stratasys System Language). It can also be read in SLC files.

Principle of Layer Generation

In this case, the thermoplastics and waxes are supplied in the form of wire on rolls or cartridges, are melted partially, and are extruded (see Fig. 3.70). The nozzle head is guided by an *x-y* plotter in the building plane. After coating a building plane, the base plate with the model is lowered by a layer thickness, and the process begins anew with the next layer.

The raw material is heated in an electrically heated nozzle head to just below melting temperature (about 68 °C for investment casting wax and about 270 °C for ABS plastic) and joins the viscous material to the previous layer (or the first layer on the base plate). It is melted slightly, cools on contact because of heat conduction, and solidifies. The distance between the previous layer and the extrusion head, as well as the volume flow of the semiliquid material, are coordinated so that gauges result from 0.254 to 2.54 mm and layer thicknesses from 0.127 to 0.330 mm. The nozzle head gap is set to the previous layer to about half of the nozzle diameter. This results in a slightly oval fluid cross section whose main dimension is referred to as the row width (RW). This extruded strand width must always be greater than or equal to the layer thickness. Optimal layers are cross sections with a width-to-height ratio between 3.5 and 6. By this "crimping" of the layer, solid structures and relatively smooth surfaces arise as a function of surface tension and viscosity (as a function of temperature). Clearly visible sausage-like structures are thereby weakened in their effect, see Fig. 3.71.

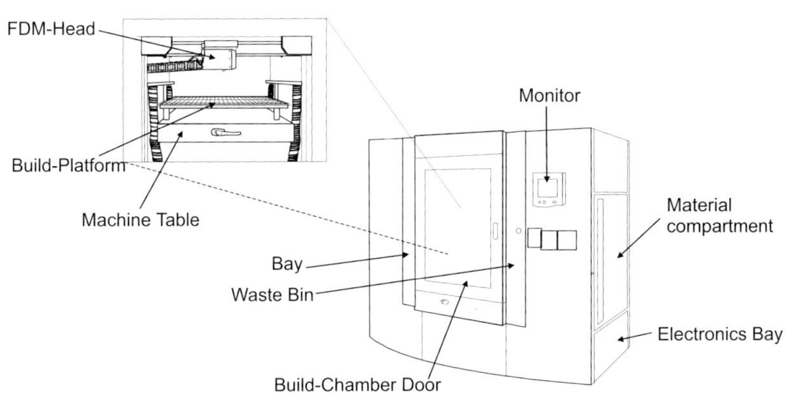

Figure 3.71 FDM Maxxum, overview with detailed view of processing chambers (Source: Stratasys)

Although the material solidifies very quickly, supports are necessary for projecting sections and for the model as a base. The fully automatically generated columns are made of a more brittle material than the component. Therefore, the supports can be quickly removed without damaging the model and without tools by breaking by hand. This *Breakaway Support System* (BASS) method is available for the polycarbonate families PC, PC-ISO, and polyphenylsulfones (PPSF). A support material called *Water Works* (WW) can be removed automatically in a washing station. This support material is for ABS, ABSi, and PC-ABS and continues to spread, despite more and more safety requirements.

The Fortus 900mc is, with a space of 914 × 610 × 914 mm^3, the largest FDM machine. Figure 3.71 shows a diagram with a detail of the design space. It comes from the previous model Maxxum, but this is irrelevant to representing the basic principle.

The entire component has to be supported with supports. The line spacing for the generation of restraints is half the track width of the next component generated. The angles are specified by the x axis in the clockwise direction. The zero point of the system lies in the left front corner platform.

The nozzle head moves in the components usually below a main direction of 45°, wherein the main direction from layer to layer is changed by ±90° in each case. Since the supports are generated under half of this angle, this angle is usually at about 22.5°. All angles can be changed in the software.

The starting point of each layer can be moved automatically by the software. It is visible and builds up if it is always in the same location as seam on. However, it can be relatively easily abraded. The construction parameters are adjustable in the z direction and on individual regions of each component.

During the building, on the outer sides of the component, a radius is created on the outer corners. The radius is created automatically. In practice, the components must always be provided with a borderline. The borderline is composed of circular elements that arise outside a radius for geometric reasons. Inner contours are automatically always sharply displayed (Fig. 3.72).

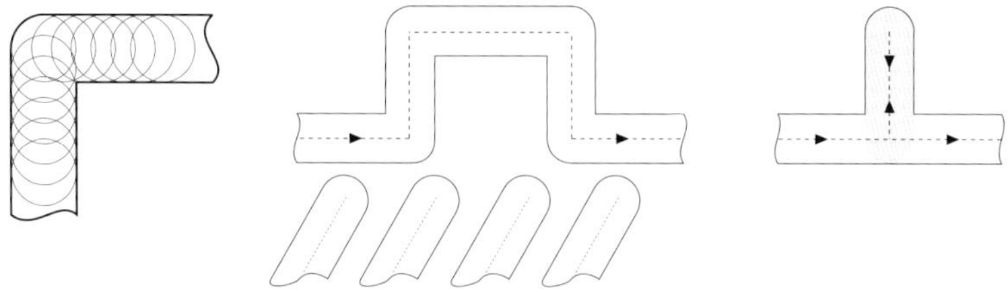

(a) (b) (c)

Figure 3.72 Fused deposition modeling: (a) corner figure; (b) intricate geometric detail; (c) fine webs

The problem is the zone in which the geometry is very delicate. The border is inset by half the row width in the component, similar to the beam compensation in stereolithography, creating areas that cannot be fully passed by a border (Fig. 3.72(b)). The vectors are only calculated in the borderlines, therefore thin structures are not be displayed. This can be changed manually by breaking the dataset (island) and calculating it with parameters.

The worst case forms a wall, which terminates outside of the component, and the wall thickness is track width. The ideal line in the middle would be crossed double on the way there and back, resulting in an unclean surface (Fig. 3.72(b)). The construction process is not interrupted in this case. Figure 3.73 shows an FDM functional model of the material ABS.

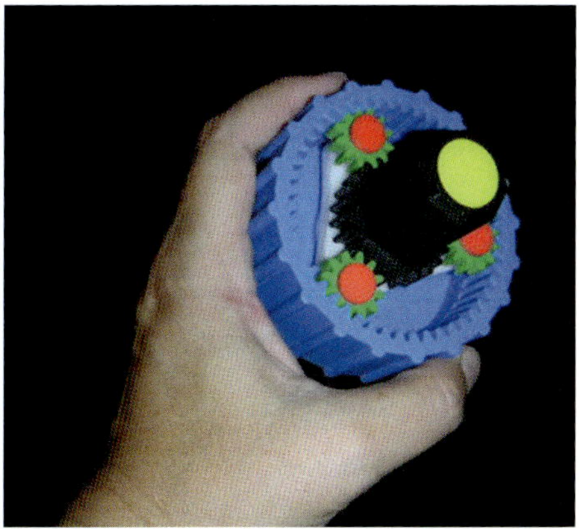

Figure 3.73
Fused deposition modeling, colored ABS functional model (Source: Stratasys)

Type of Construction

The nine model series of the FDM technology family are available in other versions, some of which are equipped only to process certain materials or combinations of materials. An overview is given in Table 3.1.

Table 3.1 Model and Model Variants of the FDM Family

<table>
<tr><th rowspan="3">Machine</th><th rowspan="3">Space (mm^3)</th><th colspan="6">Material</th><th rowspan="3">Material combination</th><th rowspan="3">Support material</th><th rowspan="3">Precision</th></tr>
<tr><th>ABS</th><th>ABSi</th><th>PC-ABS</th><th>PC</th><th>PCI-SO</th><th>PPSF</th></tr>
<tr><th colspan="6">Layer height (mm)</th></tr>
<tr><td>Prodigy Plus</td><td>203/203/305</td><td>0.178
0.254
0.330</td><td></td><td></td><td></td><td></td><td></td><td>ABS</td><td rowspan="7">WW
for ABS, ABSi, and PCABS

BASS
for all PC and PPSF</td><td>n/a</td></tr>
<tr><td>Vantage i</td><td rowspan="3">355/254/254</td><td rowspan="4">0.127
0.178
0.254</td><td></td><td></td><td rowspan="5">0.178
0.254</td><td></td><td></td><td>Single material
ABS or PC</td><td rowspan="6">±0.127 mm
up to 127 mm
length

±0.0381 mm
per mm
above 127 mm
length</td></tr>
<tr><td>Vantage X</td><td rowspan="3">0.127
0.178
0.254</td><td rowspan="3">0.127
0.178
0.254</td><td rowspan="4">0.178
0.254</td><td></td><td>Multiple materials

ABS, ABSi, PCABS or PC, PCABS, PCISO</td></tr>
<tr><td>Vantage S</td><td></td><td rowspan="2">Multiple materials

ABS, ABSi, PCABS, PCISO, PC</td></tr>
<tr><td>Vantage SE</td><td rowspan="2">406/355/406</td><td></td></tr>
<tr><td>Titan</td><td>0.127
0.254</td><td>0.127
0.254</td><td>0.127
0.254</td><td>0.254</td><td>All materials</td></tr>
<tr><td>Maxum</td><td>600/500/600</td><td>0.127
0.178
0.254</td><td rowspan="4">0.178
0.254</td><td></td><td></td><td></td><td></td><td>ABS and ABSi</td></tr>
<tr><td>Mojo</td><td>127/127/127</td><td>0.178</td><td></td><td></td><td></td><td></td><td>P430 ABSplus</td><td rowspan="3">Soluble SR-30</td><td rowspan="3"></td></tr>
<tr><td>uPrint SE</td><td>203/152/152</td><td>0.254</td><td></td><td></td><td></td><td></td><td>ABSplus</td></tr>
<tr><td>uPrint SE plus</td><td>203/203/152</td><td>0.254
0.330</td><td></td><td></td><td></td><td></td><td>ABSplus</td></tr>
</table>

Table 3.1 *(continued)* Model and Model Variants of the FDM Family

<table>
<tr><th rowspan="3">Machine</th><th rowspan="3">Space (mm³)</th><th colspan="6">Material</th><th rowspan="3">Material combination</th><th rowspan="3">Support material</th><th rowspan="3">Precision</th></tr>
<tr><th>ABS</th><th>ABSi</th><th>PC-ABS</th><th>PC</th><th>PCI-SO</th><th>PPSF</th></tr>
<tr><th colspan="6">Layer height (mm)</th></tr>
<tr><td>Fortus 250 mc</td><td>254/254/305</td><td>0.178
0.254
0.330</td><td rowspan="2">0.178
0.254</td><td></td><td></td><td></td><td></td><td>ABSplus-P430</td><td>Soluble</td><td>±0.241</td></tr>
<tr><td>Fortus 360 mc</td><td>406/355/406</td><td rowspan="2">0.127
0.178
0.254
0.330</td><td rowspan="2">0.127
0.178
0.254
0.330</td><td rowspan="2">0.127
0.178
0.254
0.330</td><td></td><td></td><td>Multiple materials

ABS, PC-ABS, PC</td><td>Soluble or break-away for PC

Soluble for ABS-M30 and PC-ABS</td><td>±0.127

±0.0015 mm/mm</td></tr>
<tr><td>Fortus 400 mc</td><td>406/355/406</td><td>0.127
0.178
0.254
0.330</td><td>0.127
0.178
0.254
0.330</td><td>0.127
0.178
0.254
0.330</td><td>All materials</td><td>Soluble for most materials

Break-away for PC-ISO, ULTEM, and PPSF

Soluble or break-away for PC</td><td>±0.127

±0.0015 mm/mm</td></tr>
<tr><td>Fortus 900 mc</td><td>914/610/914</td><td>0.178
0.254
0.330</td><td>0.178
0.254
0.330</td><td>0.178
0.254
0.330</td><td>0.178
0.254
0.330</td><td>0.178
0.254
0.330</td><td>0.178
0.254
0.330</td><td>All materials</td><td>Soluble for most materials

Break-away for PC-ISO, ULTEM, and PPSF

Soluble or break-away for PC</td><td>±0.09

±0.0015 mm/mm</td></tr>
</table>

All machines from Stratasys are compact stand-alone systems and are designed as a “plug-in” system. They do not have a laser and therefore have a low energy consumption, and they do not require cooling water. According to the manufacturer, there are no irritating triggers for humans or even toxic vapors that need to be vacuumed. An operation in an office is therefore possible.

There are different materials available. Each cannot be used in any machine (Table 3.1). The large machines combine several cartridges, each about half a liter, which are automatically switched in long-term operation. For each material a separate nozzle head must be used. The nozzle heads differ in the drive rollers, the gears used, and probably also in the heating elements used, but especially in the nozzle diameters. If the material or the nozzle head is changed, a test item must be created before the construction process begins. As a test part, a cube is used that is 20 × 20 mm wide and about 4 mm high.

The Idea series includes the Mojo 3D printer with a space of 127 × 127 × 127 mm^3. This printer has two nozzles and can create support structures with appropriate material (SR-30). The layer thickness is specified as 0.178 mm.

This class also includes the uPrint SE (Fig. 3.74, top left). The space is equipped with 203 × 152 × 152 mm^3 (in the plus version, 203 × 203 × 152 mm^3). It is larger than the Mojo 3D printer, however has the uPrint SE only through one nozzle. The layer thickness is 0.254 mm (plus with 0.330 mm). It is larger than the layer height of the Mojo. This class is designed as a desktop machine. The printer should efficiently, quickly, and easily create conceptual ideas and first drafts.

The design performance series includes the Dimension printer (Fig. 3.75). The Dimension is, in terms of the usable material, ease of use, and its moderate price, a development of a concept modeler. The layer thicknesses are 0.254 or 0.330 mm and 0.178 mm in the Elite. The precision is specified by the manufacturer as better than ±0.3 mm. This is both sufficient and realistic for a concept model. For the Elite, the accuracy should be increased ±0.2 to ±0.25 mm. The machine name is the space volume in cubic inches. The dimensions of the installation space of the Dimension Elite are x, y, z = 203, 203, 305 mm^3, which is in contrast to the Dimension 1200 with x, y, z = 254, 254, 305 mm^3.

(a)

(b)

(c)

Figure 3.74 FDM uPrint SE (a), Fortus 250mc (b), and Fortus 900mc (c) (Source: Stratasys)

Figure 3.75 Dimension: overview in process space (left), exterior view 768 BST (middle), and 1200 SST (right) (Source: Gebhardt/Stratasys)

More models can be built side by side. The Dimension 1200 is available as a BST and an SST version. The Breakaway Support Technology (BST) is a more brittle supporting material and can be removed (canceled) manually, while the Water Works method (Soluble Support Technology, SST) washes the support structures in a special cleaning machine. Elite is a Dimension 1200 SST, which processes the thinner material (ABSplus) and thus provides higher load-bearing components.

The Production Series includes the large machine called Fortus. In this group there are four different versions: Fortus 250mc, Fortus 360mc, 400mc, and the largest, the Fortus 900mc. The Fortus 250mc is the smallest full FDM machine (Fig. 3.74(b) and Table 3.1). It has, like all of the others, two nozzle heads and can therefore simultaneously handle construction and support material. In the x, y, z = 203, 203, 305 mm space, ABS can be used in the colors white, black, blue, yellow, red, and green. The layer thicknesses can be selected in three steps: fine = 0.178 mm, standard = 0.254 mm, and draft = 0.33 mm. In this class the machines differ primarily in the physical size, but also in the choice of materials and the available layer thicknesses.

For example, the Fortus 900mc has the biggest space with 914 × 610 × 914 mm^3 (Fig. 3.74(c)). Also, the layer thicknesses are adjustable in four stages from 0.127 to 0.330 mm. The Fortus 900mc can be equipped with material like ABS, ABSi, PC-ABS, PC, PC-ISO, ULTEM® 9085, and PPSF/PPSU. Also, the choice of supporting material is greatest on this printer.

Material/Construction Time/Accuracy

There are different materials available, but all materials cannot be used in any machine (see Table 3.1). In addition to ABS and ABSi, there is a PC-ABS blend, polycarbonate (PC), PC-ISO, and polyphenylsulfones (PPSF), a high-performance plastic from the family of polyarylsulfones.

The elastomer (E20) and the investment casting wax (ICW 06) are no longer in the program.

The standard material *ABS* is a white, relatively high-impact plastic material. It has strength properties similar to nylon. The heater nozzle of the support is driven at 265 °C, the heating nozzle for the component material at 270 °C. The ambient temperature is set at from 50 to 80 °C. If the temperatures are set too low, there is the risk of delamination. If the temperature is set too high, the material is too fluid for the construction process, tends to oxidize, and turns brown. Moreover, there is a risk of distortion in the components.

In using ABS, currently the nozzle heads are blown out after each layer because ABS is hygroscopic. If the water evaporates upon the heating of the material, there are dropouts in the coating.

ABSi is an ABS with increased impact strength. It is translucent and available in several colors.

Polycarbonate (PC) has a lower strength than ABS (about 10 %), but it is less brittle (the bending strength is almost two times higher than ABS) and considerably more temperature resistant (about 125 °C compared to 90 °C for ABS). It is therefore an ideal material for functional prototypes and small series.

PC-ABS combines the good mechanical properties of ABS with the flexibility of PC.

PC-ISO is a sterilizable material for medical applications and meets the relevant standards ISO 10993-1 and USP Class VI.[23] PC-ISO can be sterilized by gamma radiation or ethylene oxide.

Polyphenylsulfone (PPSPH) is the first high-performance plastic in additive manufacturing systems. With comparable mechanical properties, it has a nearly two times higher temperature resistance than does ABS (> 200 °C). With PPSPH, the machine runs not only thicker layers (0.254 mm), but it is also much slower. The supports should be removed from the hot component.

Investment casting wax (ICW 06) corresponds to industrial investment casting waxes, and it can be used easily in all known investment casting processes. This satisfies the important criteria of a low ash content (0.0075 %) and a low thermal expansion coefficient, so "shell cracking," which means cracking of the ceramic shell, can be reliably avoided.

For all machines and materials, accuracies of ± 0.127 mm for components are accessible to a characteristic length of 127 mm. About ± 0.381 mm, elsewhere also 0.1 % (but not less than 0.1 mm) can be specified. In the Prodigy Plus, poorer accuracy can be expected. There is no information shipped from the factory.

Ultem 9085 is a high-performance plastic that can be used in the application in the rapid prototyping process. It is a material with very good flow properties for thin and lightweight components.

Postprocessing

The removal of the breakaway supports (BASS) requires no machining. But the contact points props/model should be grinded. The Water Works (WW) supporting material can be washed out. This is done in a kind of washing machine with the addition of chemicals.

---

[23] The United States Pharmacopeia (USP) defines quality standards for prescription and over-the-counter medicines, food supplements, and other health care products, including materials for the machinery used for this purpose.

Typical Follow-up Processes

With appropriate tuning of the prototype materials used from the available range of materials, any known postprocessing techniques in combination with the FDM process can generally be used. In individual cases, it must be checked whether the recoverable detail, the achievable accuracy, and the resulting surfaces correspond to the respective requirements.

Wax models are designed for the classic investment casting process.

3.5.3 Wax Printers, Solidscape

Wax Printers
3Z Pro, 3Z Studio, 3Z LAB, 3Z EDU

Solidscape Inc., Merrimack, NH
Horbach GmbH, Idar-Oberstein, Germany

Short Description

The T66 Benchtop and the T612 Benchtop are additive manufacturing machines for the production of small, very precise wax models, preferably for investment casting. In a drop-on-demand (DOD) principle, the working printing head drops molten material of the type hard wax. A second low-melting wax is applied as a support material. By milling each slice,[24] generally very accurate models arise that are suitable for investment casting of precision parts. The machines are not concept modelers, even if they appear so outwardly.

Application Fields

Geometry prototypes (wax) are functional prototypes to produce small series of metal parts in the investment casting process. Variants of the 3Z machines are aimed at the jewelry, dental technology, precision engineering, and miniaturized model markets.

Development Status

Since 2002, the T66 (BT2, 2007) and since 2003, the T612 have been commercialized. The R66 and D66 were introduced in 2007. The current machines belong to the 3Z series.

[24] This process variant is strictly meaning a combined additive-subtractive method.

History

Sanders Prototype Inc. was founded in 1994 by Royden C. Sanders Jr. The first model, the ModelMaker mm6-PRO, was commercialized in 1994 (and produced until 1998), and the ModelMaker II in 1997. It was finally replaced in 2002 by the T66. In parallel, the PatternMaster was made in 2000 and replaced by the T612 in 2003. The R66/D66 underlines, with basically the same technology, the trend of industry solutions (2007).

The development for the ModelMaker was done between 1992 and 1994 by Sanders Design Inc. (SDI). In the course of corporate law disputes, the company was renamed in September 2000 as Solidscape Inc. Solidscape is not to be confused with the still-active Sanders SDI.

Solidscape was acquired by Stratasys in 2011, but it operates in its own market.

Strategy/Development Partners

The T series has been developed with the aim of producing wax models for small and very precise components as required by investment casting. Target markets include in particular the jewelry industry, medical technology, and the toy industry. The T series is marketed as a production system, while the technically largely identical but much smaller variants R66 and D66 were developed for the target markets of jewelers and dental technicians. They are at a 30% lower price level (about $35,000 in 2007), appealing to the jewelry retailer offering his or her own designs.

Data Format/Software

The 3Z accepts STL and SLC data sets.

The manufacturer-supplied program *ModelWorks* reads the data and allows visualization, orientation, and positioning. It contains simple CAD functionality and automatically calculates the supports. It generates the layers and makes use of the ability to automatically adjust the layer thicknesses. ModelWorks generates the construction data, transmits it to the machine, and automatically controls the construction process.

Principle of Layer Generation

The heated inkjet print head material is melted and spun as microdroplets onto the model structure. Upon impact, the droplet solidifies onto the model due to the heat conduction. Due to the impact energy, the droplet deforms, so the deposited material webs have a height-to-width ratio of about 0.6 liters. It can realize up to 0.013 mm thin layers. The areas not inherent in the model are supplemented by a second print head with a low-melting wax, which acts as a support material. After applying each layer, the surface is milled flat so that very high accuracies can be achieved in the *z* direction. For this a cooling of the material is not necessary.

Type of Construction

The machine consists of a building platform and a traverse that moves the two print heads with a high positioning accuracy of 0.006 mm in the x and y directions. On the same x rail a rolling cutter is mounted, which is connected to a vacuum unit. The build platform is positioned very finely in the z direction (0.0032 mm). The usable space of the 3Z Pro measures x, y, z = 152.4, 152.4, 152.4 mm^3, and the Studio 3Z, the 3Z Edu, and the 3Z Lab have a space of x, y, z = 152.4, 152.4, 50.8 mm^3. See Fig. 3.76.

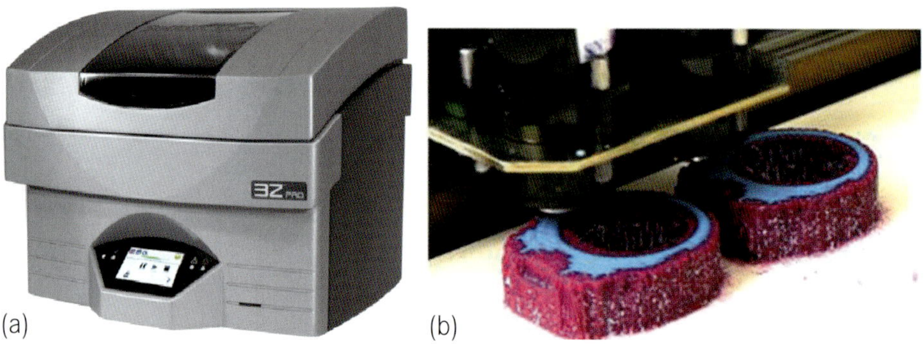

(a) (b)

Figure 3.76 Solidscape: (a) 3Z Pro; (b) wax model (blue) in support material (red) (Source: Solidscape)

The machine is designed for automatic operation and includes a material supply for 24 hours as well as a device for automatic verification of nozzle function.

The print head emits beads having a diameter of about 0.0075 mm. They consist of molten material and result in line plotted of a web of about 0.010 mm wide and 0.006 mm high. Such a line structure can be built at a speed of 30.5 mm/s (which corresponds to a coverage of about 0.11 mm^3/min).

Material/Construction Time/Accuracy

Originally the materials were a green *ProtoBuild,* a hard wax blend, and a red *ProtoSupport,* a support material. The building materials were meanwhile called *BlueCast* and are called today *InduraCast* and *InduraFill,* with the supporting material *InduraBase.*

The materials are designed according to the type of waxes and for investment casting with lost forms. The building material has a substantially higher melting point (92 to 106 °C) than does the backing material (50 to 72 °C). The current materials are specified by the manufacturer to be 200% stronger than their predecessors.

The construction time is relatively long and can be shortened if the layer thickness is increased up to the maximum of 0.076 mm, but there the resolution is impaired. The finest layer thickness is 0.013 mm.

The accuracy is given as 0.025 mm to 76 mm (which is half the distance in the *x* or *y* direction) and, as a result of the milling of each layer, as 0.013 mm to the entire *z* length of 229 mm.

The 3Z machines have a resolution of $x, y, z = 5000 \times 5000 \times 8000$ dpi. The accuracy is given as 25.4 microns per inch in all three spatial coordinates. The layer thickness is 12.7 microns.

Postprocessing

The supports are removed with a solvent. In general, additional postprocessing is not necessary.

Typical Follow-up Processes

The wax models are particularly suitable for investment casting, but also for vacuum casting and other molding processes.

3.5.4 Multijet Modeling (MJM), ThermoJet, 3D Systems

Solid object printer, ThermoJet 3D printer

3D Systems Inc., Rock Hill, SC

Short Description

The ThermoJet 3D printer uses the multijet modeling (MJM) method, so it works, albeit intermittently, on the extrusion principle. As the successor to the Actua (1997), it has proven to be an additive manufacturing machine for the direct production of display models and especially of wax components that are used in investment casting. The building and supporting materials are identical. The greatest disadvantage is that the financial cost for the removal of the supports is high and additionally, visible marks remain on the bottom of the models.

With the InVision series (2003) and its successors, the ProJet 3510 and 5000, these disadvantages have been overcome in which a modified construction material was also polymerized by UV rays and thus solidified. The unpolymerized supports can be removed by heat. Because the solidification of the component is ultimately accomplished by addition polymerization, the method is allocated to the stereolithography process in Section 3.1, "Polymerization: Stereolithography (SL)."

The ThermoJet is no longer built. However, a number of machines are still in operation. Many of them are used in the field of production of lost patterns for investment casting.

3.6 Three-Dimensional Printing (3DP)

3.6.1 Overview: 3D Printing

Figure 3.77 gives a structured overview of the established 3D printing methods. Starting with the generic descriptions (Fig. 2.38), it contains the industrial processes and machines discussed below, as well as their product and producer identifications.

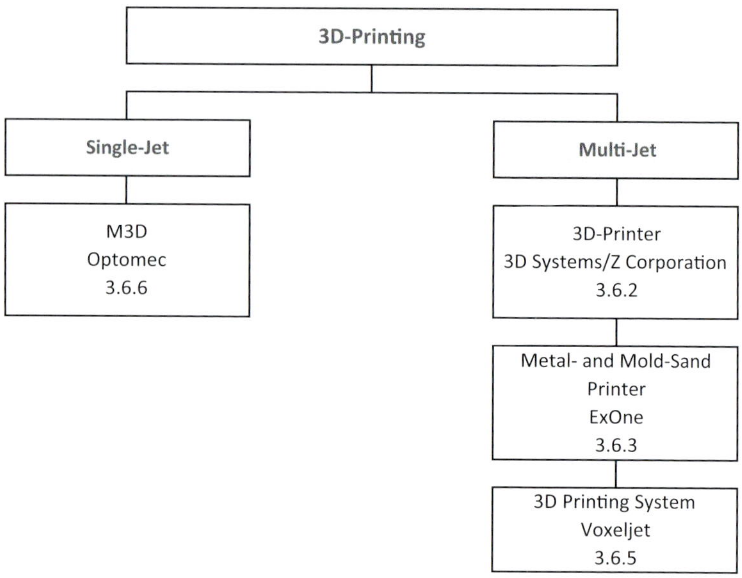

Figure 3.77 3D printing processes[25], overview in continuance of Fig. 2.38

The aerosol print process was included, although it is not a process printed in a powder bed; instead the material and the binder are carried in the drop.

The sections describing the processes in more detail are given in Fig. 3.77.

3.6.2 3D Printer, 3D Systems, and Z Corporation

3D Printers

ProJet® 160C Personal 3D Printer
ProJet® 260C Personal 3D Printer

[25] A 3D simulation can be found at www.rtejournal.de/filme.

ProJet® 360C Professional 3D Printer
ProJet® 460Plus Professional 3D Printer
ProJet® 660Pro Professional 3D Printer
ProJet® 860 Professional 3D Printer

3D Systems Inc., Rock Hill, SC

Short Description

3D Systems' 3D printer family is designed for the fabrication of conceptual models and precision casting of components based on a 3D printing process. The part emerges from the injection of a binder liquid into the surface of a powder bed. Z Corp is the only producer who provides machines that can continuously manufacture colored parts. The Z Corp process is trimmed for speed and the simplest of handling. The degree of detail reproduction is low, without a special infiltration the models are not resistant to mechanical stress, and the material is environmentally acceptable.

A modified process named Z-Cast was particularly developed for the manufacturing of precision casting models.

Range of Application

Solid images, concept models

History

The Z Corp was established at the end of 1994 by Marina Hatsopoulos and Walter Bornhorst. The development of a prototyper started after an agreement was signed with MIT concerning the rights to the special use of the 3DP master patent. The Z402 was introduced in 1997. The 402C, the first additive manufacturing machine that was able to build colored parts, followed in 1999; the 406 with an upgraded color system was introduced in 2002, and the 310 as an enhanced monochrome machine was introduced in 2005. From the significantly larger 806 (2001) followed the 810. The small series was completed with a color machine, the 510. The 450 was introduced as an especially inexpensive color machine in 2007. The 810 is not mentioned any longer on the current home page (since 2007). In 2012, Z Corp was taken over by 3D Systems, and the products were integrated into their 3D portfolio. Ever since then, they are placed in the class of professional printers and identified as ProJet.

Development Partners/Strategies

The machine is based on the MIT 3D printing (3DP) patent. It closely follows the concept of a 3D printer. Installation and handling are extremely simple and can be done without prior training. With the current ZPrinter 450, Z Corp takes a huge step toward the promised office machine: the machine can be run in an office environment, and it possesses a nearly closed powder circuit.

A partnership exists with Huntsman for the development of new materials.

Data Formats/Software

The producer-supplied software *ZPrint* adopts the geometry information as an error-free, buildable file in the formats STL, VRML(II), PLY, and 3DS (3D Studio). By means of VRML, FEM and scan information can also be imported. The PLY format can also be exported.

The *Zprint software* generates the slice and machine control files and visualizes and automatically monitors the construction process. The software also supports 3D views, markings, and the scaling and interlacing of parts. The data sets can be rotated and tilted.

Other CAD and rapid prototyping functionalities are not provided. The ZPrinter 450 is easy to handle through a one-button menu control.

The preparation of the geometric data and the STL conversion need to take place outside the machine. It is expected that the software will be standardized.

Principle of Layer Generation

A colored, water-based liquid is injected into a powder bed of cellulose powder by means of an *inkjet* print head, resulting in local solidification and thereby generating the elements of a new layer and joining it with the preceding one. Powder that is not wetted stays in the construction chamber and supports the model. The models must be infiltrated with wax or epoxy resin because otherwise they are not resistant to mechanical stress.

System Type/Construction

The machine consists of a construction chamber with three chambers, above which a coating and plotter mechanism are installed. Two chambers, the powder supply container and the actual construction chamber, have movable bases. The base of the supply cylinder is raised, and a distribution roller takes up a certain amount of powder and, moving across the construction chamber, distributes it evenly. Surplus powder passes to the overflow container. The recoating process takes only a few seconds (two, according to the producer).

After recoating, the layer is contoured by injecting a colored water-based liquid. The print head is an inkjet from Hewlett-Packard with 304 parallel nozzles. The Z310 machine contains one, the Z406 four, and the Z810 six print heads. The Z450 contains two newly developed three-color single print heads of type HP57 and one single head for clear binder, and the Z510 contains four print heads.

The wetted powder particles are firmly linked with one another, forming a physical layer. The unwetted powder stays in the construction chamber and supports the model. The liquid accounts for about 10 % of the model volume. After the solidification of the layer, the base of the construction chamber is lowered by one layer thickness, the base of the supply cylinder is raised accordingly, and the process starts again.

The most important data of the machines are assembled in Table 3.2.

All machines read error-free STL, VRML, PLY, 3DS, FBX, and ZPR data.

Table 3.2 Overview of 3D System's 3D Printers

Machine	Monochrome/ Color	Minimum detail resolution	Build chamber	Number of jets	Number of print heads	Build rate
ProJet		mm	mm^3			mm/h
160	monochrome	0.4	236 × 185 × 127	304	1	20
260 C	64 colors	0.4	236 × 185 × 127	604	2	20
360	monochrome	0.15	203 × 254 × 203	304	1	20
460 Plus	color, CMY 2.8 mill.	0.15	203 × 254 × 203	604	2	23
660 Pro	color, CMYK	0.1	254 × 381 × 203	1520	5	28
860 Pro	>6 mill.	0.1	508 × 381 × 229	1529	5	5-15

The disadvantages of 3D printing are the high content of manual work, especially through the powder handling, and the danger of damaging the part while unpacking and cleaning it, are overcome since the Z450 by means of a nearly closed powder system. The building material is immediately vacuumed and waste powder is recycled. After the building process, the part is cleaned in the construction chamber, which is transformed into a screening machine through a combined vibration and vacuum attachment. The final cleanup through microbeams also takes place in the machine. For this purpose, intrusion holes with beam gloves are integrated into the chassis. A very simple change of the print head and the addition of the binder (ZB59) by means of cartridges with one liter of volume complete the picture of an office-suitable machine.

Material/Construction Time/Accuracies

All of the machines build with the VisiJet® PXL™ powder and with a layer thickness of 0.1 mm. The parts receive their mechanical properties through infiltration. For that purpose, an epoxy infiltrate for functional models with snap fits and integral hinges as well as cellulose-based elastic infiltrate for flexible parts (according to the producer, rubbery-elastic) are available. Solidity specifications are tested as a precaution because of the spreading dependent on the infiltration. According to the producer, the machine builds five to ten times faster than other additive manufacturing machines (monochrome). Facilitating the speed is a very fast slice process. The coating takes only a few seconds and is also very fast. The print head passes over only those sections of the construction chamber that contain a model. In this way the construction time is optimized. Because the print head always passes over the entire area of the construction section, the construction time is not influenced by the complexity of the geometry.

There is the additional effect of particles, which really lie outside the contour, being "glued on" by capillary activity and, similar to the sintering process, giving the appearance of a "fleece." The models are therefore relatively inaccurate, with a rough surface and a porous structure, but they are suitable for "show and tell" applications.

Postprocessing

The models are taken out of the powder bed after the process, and the supporting powder is simply removed by a vacuum cleaner. They can then be carefully removed from the construction chamber. Slight sand blasting is an advantage. This can easily be done with a *depowdering unit*. The parts (except for the Z-Cast ceramic models) have to be impregnated immediately after the building process. Thereby the accuracy is influenced. As an infiltrate, the producer offers wax, two qualities of epoxy resins, elastic urethanes, and an adhesive of the cyanoacrylate type. For the usage of the infiltrates there are extensive references and for the wax a special dipping station. The infiltration makes relative the time advantages and the usage of the machine in the office.

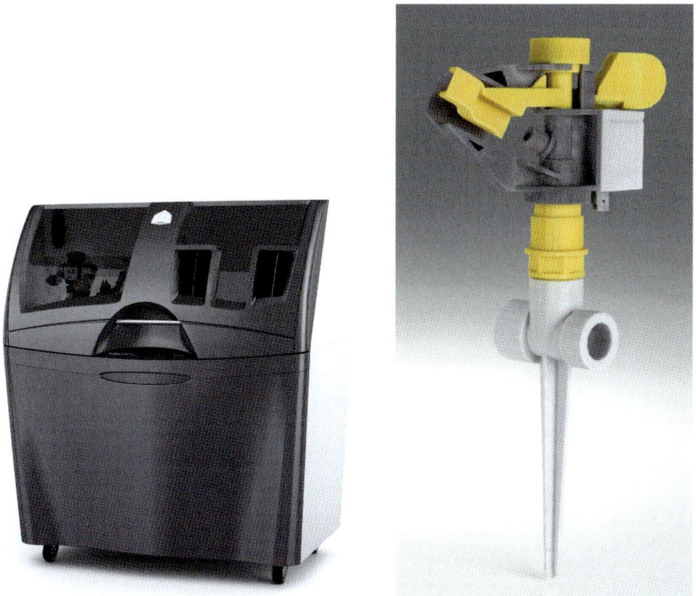

Figure 3.78 3D printing, ProJet 460 Plus (left), colored model (right); (Source: Z Corp)

Typical Follow-up Processes

Concept models do not in principle require molding processes. If the models are to be used for precision casting or–after epoxy resin infiltration–as functional models, then it has to be kept in mind that they are in fact concept models, and the corresponding properties must be accepted.

3.6.3 Metal and Molding Sand Printer, ExOne

Molding Sand Line: S-Max, S-Print
Metalline: M-Print, M-Lab

ExOne, North Huntingdon, PA
ExOne Europe, ExOne GmbH, Augsburg, Germany

Short Description

Prometal promotes two product lines: the metal line, *direct metal printing*, for the manufacturing of metal parts, and the molding sand line, *direct core* and *mold making machines*, for the manufacturing of sand cores and molds for foundries.

The metal line contains additive manufacturing machines for the manufacturing of functional devices, molds, and tools made of steel powder according to the 3DP process (licensed from MIT). In the machine, a green part is generated. After the building process, it is debindered in a *densification furnace* and infiltrated into a dense part with metal.

The sand line has additive manufacturing machines for the manufacturing of complex molds and cores made of foundry sand according to the 3DP process. The parts leave the machine with the favored final strength.

Currently (2013), ExOne offers two of the biggest printers for 3D printing of sand and metal worldwide. They feature a building chamber with dimensions from $1800 \times 1000 \times 700$ mm^3 for sand and $780 \times 400 \times 400$ mm^3 for metal.

Range of Application

Direct manufacturing of parts made of sand and metal.

Development Status

Commercialized since 1997

History

Prometal was founded in 1996 as a division of the Extrude Hone Corp. The development of a rapid tooling system was a new branch of business for a company that had specialized in machines and processes for surface finishing. The RTS 300 was launched in 1997, the R10 in 2001, and the R2 in 2003. At the beginning of 2005, Extrude Hone was sold to ExOne and continued as the Prometal Division. There is also the German subcompany Prometal GmbH in Remscheid and its subcompany, Prometal RCT (Rapid Casting Technology) GmbH in Augsburg. The last one emerged from the company Generis (founded in 1999), which especially practiced the development of 3D printing of sand molds and for that purpose used a sublicense of the MIT license of Soligen. Now all divisions trade under ExOne.

Strategy/Development Partners

Metal line: The target of the developers is the manufacturing of functional devices and tools of steel, ceramics, and other high-melting materials suitable for series production. Development partnerships with MIT and IMAGEWARE, together with the core competence of the company, the surface treatment of metals, form the basis for this strategy.

With the development of small and very precise machines (R1), the jewelry and dental markets are occupied strategically.

Sand line: The target market for machines for the manufacturing of sand molds and cores is the foundry industry. The facilities are offered as manufacturing systems, which allow an automatic material supply and a semiautomatic unpacking. Therefore they can be easily integrated into the foundry operation.

Data Formats/Software

The machines accept geometry information in the STL and SLC formats and feature an automatic slicing program. A partnership with IMAGEWARE assumes the availability of powerful rapid prototyping software.

3.6.3.1 Metal Line: Direct Metal Printer

M-Print, M-Lab

Principle of Layer Generation

The component is built layer by layer by injecting the liquid binder in the surface of the powder bed with the help of a multinozzle print head. The nozzles have a diameter of 46 µm. They continuously emit the binder stream. An oscillating piezo-element interrupts the stream (60 kHz) and forms drops out of it with a diameter of about 80 µm. Before hitting the surface, every drop passes through an electrostatic field. In this field the drop is being charged in order to position it precisely.

At first in the machine a model called a *green part* is generated. It does not get its green strength by thermal melting like in the sintering process, but by injecting the binder in the metal or ceramic powder. The advantage over the sintering process of multicomponent powders is that a separation of the building material and binder segregation in the powder can be excluded. Furthermore, the process of the machine runs under environmental conditions and thus is relatively cold, so there is minimal thermal distortion.

The binder ensures that the components are kept together. Via irradiation with a high-energy lamp, every layer is dried and solidified so that a transportable green part is generated. The components that are ready to use are produced by debinding and sintering on 60% of the desired density. Afterward, it is infiltrated and there is postprocessing like polishing or blasting. Newer binder systems not only ensure that the component is kept together; it also establishes the physical and technological characteristics of the component by additives. With the help of locally different binders, locally different material properties can be reached (graded materials).

Construction

There are two production lines, the M-Print and the M-Lab. The smaller M-Lab is designed as a direct metal research and development printer with the target group of jewelers, laboratories, and education. The M-Print is conceived as a production machine.

The M-Lab machine has a base frame that is designed as a portal. It includes a stock of powder and binder and the control system. The mechanism of coating is integrated into the portal. The portal also carries a lamp that is used as a heater to dry the binder. The process chamber is arranged below the portal and includes a movable bottom, one in the building space and one in the stock space. In the coating process it can be lowered or lifted layer by layer. The R2 has a building space and a powder feed system that is exchangeable.

The maximum size of the components for the M-Lab is $x, y, z = 40 \times 60 \times 358\ \text{mm}^3$. For the M-Print it is $780 \times 400 \times 400\ \text{mm}^3$. To enhance the building speed of the M-Print, the number of nozzles has been increased in the several development steps from 8 to 32 and then to 96. With that, the productivity (measuring unit: production rate) has been increased tenfold. The manufacturer is offering a *densification furnace* to burn out the binder and for the infiltration. They also supply a stripping station to recover the unused powder.

Figure 3.79 shows the M-Lab, and Fig. 3.80 shows a component.

Figure 3.79
The ExOne metal and ceramic printer M-Lab, externally identical to the dental system RX-D (Imagen); (Source: ExOne)

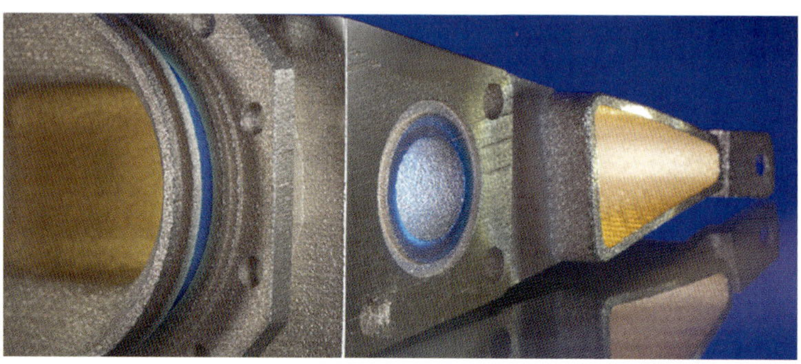

Figure 3.80 Prometal part "engine adapter"; (Source: IFAM, Bremen)

Interesting are also hollow spheres with a diameter of 1.2 mm made of 316L, which were developed at the IFAM in Bremen, Germany.

The process itself is cold, but the drying and the postprocessing include several thermal steps that lead up to 1230 °C and fundamentally bring distortion with them. But they are good to handle because they change very slowly and do not produce high local temperature peaks. The reproducible shrinkage is around 1.5 %.

The printing speed is about 1.5 m/s. The building speed for the R2 is given as 30 to 90 s, and for the R1 as 90 s per layer.

The simultaneous process of sintering, debinding, and infiltration takes about 10 hours. The component in Fig. 3.80 (deviations x, y, z = 170, 65, 90 mm^3) has a building time of 13 h, a debinding and sintering time of 11 h, and an infiltration time of 11 h too.

Typical thicknesses of layers for the standard powder are between 100 and 170 µm. The M-Lab in *fine resolution mode* builds layers of 50 µm. The single drops can be placed with an accuracy of 25 µm. Component tolerances up to ±0.02 % can be reached.

Postprocessing

The component leaves the machine as a green part. It is debindered for one hour in a *densification furnace* at around 600 °C and is sintered for 1.5 hours at around 1230 °C. Because of the required slow heatup and cooldown, the process takes between seven and 10 hours. In another thermal oven process, the infiltration follows in two temperature steps (1100 °C and 900 °C) to a density of 100 %. Alternatively, the components are directly sintered at high temperature.

The surface must be finished mechanically.

Typical Follow-up Processes

The target of the process is the manufacturing of tools or the direct use of the components. Follow-up processes are not provided.

3.6.3.2 Molding Sand Line: Direct Core and Mold-Making Machine

S-Print, S-Max

Principle of Layer Generation

The component is built layer by layer by injecting the liquid binder, called the activator, with a multinozzle print head into the surface of the powder bed consisting of premixed molding sand and hardener. It is to be expected that the print process complies with the Prometal process. The binder causes the direct hardening of the component in the machine.

Construction

The S-Max is the direct follower of the Generis GS 1500. The S-Print is based on an identical technique, but it has a smaller building space.

The material is applied by a vibrating coater. The material is conveyed out of the automatic mixer (S15MX sand mixer) through a vacuum line (S-15). The component is generated in an exchangeable building space that includes the building platform. This space of the S-Print is called the job box (JB). For both big machines there is an unloading station (S-15 UL respectively, S-Print UL, unloading station), which is equipped with a suction system. The S-Max has an automatic transportation system on a rail system. The empty building space of the S-Max weights 125 kg and must be transported with external lifting aids. Both systems are production facilities for the foundry industry. All machines must be placed in a foundry industry environment because of the transport of the sand cores.

The dimensions of the building spaces of the S-Max are 1800 × 1000 × 700 mm^3 and of the S-Print x, y, z = 750 × 380 × 400 mm^3.

Material/Construction Time/Accuracy

Some materials are molding and forming sands based on a furan resin. The building material consists of the sand and the hardener (Activator FA001). The hardener is mixed with the sand in the mixer by a screw conveyor. The sand is offered, depending on the necessary ventilation, in grain sizes of 140 µm and 185 µm. With the help of the print heads, the binder based on a furan resin is injected into the cross section and selectively solidified by the hardener. The component contains a very small portion of resin (<2 %) because not all of the sand is mixed with the artificial-resin binder, so it has better ventilation. See Figs. 3.81 and 3.82 for the ExOne molding sand printer and system.

Figure 3.81
ExOne molding sand printer S-Max (Source: ExOne)

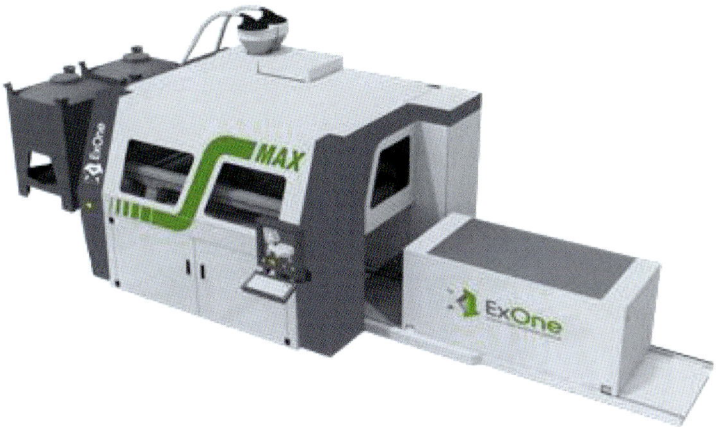

Figure 3.82 ExOne molding sand system for the S-Max: blender (left), unpacking station (right) (Source: ExOne)

Ferrous materials, aluminum, or copper-based alloys can be molded. For the casting of magnesium there is a magnesium inhibitor that can be added by an optional dosing device in the mixing station.

The building speed for the S-Max is given as 14 to 20 mm/h. For the S-Print there is given a building speed of 28 mm/h and a building volume of 7500 mm^3/h. The thicknesses of the layers are 200 to 250 µm.

Postprocessing

The component leaves the machine ready to use.

Typical Follow-up Processes

The process aim is the direct use of the components as casts or thus elements for the foundry. Follow-up processes are not necessary.

3.6.4 Direct Shell Production Casting (DSPC), Soligen

DSPC-1

Soligen Technologies Inc., Northridge, CA

Short Description

These are processes and prototypers based on the 3D printing process producing ceramic molds for precision casting by injecting liquid binder into a ceramic powder.

The process, known as direct shell production casting (DSPC), includes the entire process chain, starting with the transmission of CAD data up to the ceramic mold ready for baking. Series-identical molds are produced, and the process is technically suitable for production (rapid manufacturing).

Range of Application

Technical prototypes, rapid manufacturing

Development Status

Commercialized since 1993, but the prototyper itself is no longer sold (1999). The company offers casting parts through the service bureau *Parts Now* at www.partsnow.com.

History

Soligen was one of the first commercial users of the 3D printing technology introduced in 1991. Several of the machines were sold from 1993 onward. The first industrially usable machine was introduced in May 1994 and was presented on the GIFA in Germany and elsewhere in Europe. Although the producer announced a market introduction for 1995, no industrial machines were sold. Although the rapid prototyping process is robust, it seems that so much foundry technical know-how was required that the marketing was terminated and a service that sold molded parts was established.

General Remarks/Development Partners/Strategies

Soligen adopted the process of *three-dimensional printing (3DP)* that had been developed by MIT and developed a machine to produce ceramic molds directly for precision casting processes. This process, known as *direct shell production casting (DSPC)*, consists of software and hardware and should therefore be considered as an independent solution. The possibility to produce clusters and therefore several parts simultaneously qualifies this process as a miniseries production process. A general advantage is that the classic multistepped precision-casting process, including the model generation, is reduced to two steps: the automatic generation of the shell and the baking. The prototypes are absolutely identical to those of the subsequent series parts with respect to the material as well as their mechanical-technological properties.

Data Formats/Software

The machine requires CAD data via neutral interfaces as the basis for the optimization of foundry technology.

The producer-supplied software generates a one-piece mold on the basis of the CAD data for the required mold model, while observing all of the technical rules of casting. The basis of the process is a CAD solid of the mold (positive) subsequently produced. The geometry of the ceramic mold (negative) is calculated by the so-called *shell design unit (SDU)*, a combination of hardware and software, while observing all accepted casting techniques. These include, for example, scaling to compensate for shrinkage, casting angles, cast-ons, and risers and monitoring the contour for details that are awkward to realize in foundry techniques, such as breakthroughs or pocket holes. After a mold is defined, a nest of molds of any size can be cast within the restraints of the construction chamber with the aid of the software, depending on the geometry of the parts. The necessary foundry technical geometries such as feeders, channels, cast-ons, and risers are stored in a program library. A special software option allows the casting process, including casting simulation and the calculation of metal volumes and weights, to be observed and assessed.

The data gathered in this way are read into the machine.

Principle of Layer Generation

First, the powder coating head applies a thin layer (0.12 to 0.18 mm) of aluminum powder (corundum) onto the construction plane. To do this, a portal above the construction chamber is traversed that helps to distribute the powder evenly over the construction plane. With the aid of a roller that moves over the powder bed, the powder is evened out and slightly precompressed. The dimensions of the work space are x, y, z = 355, 355, and 455 mm^3.

The mold is produced in the DSPC-1 machine in layers. A print head, basically similar to the known design of inkjet printers (speed < 1.6 m/s), injects jelly-like silica (colloidal silica compounds, silica gel) into all contour areas that are to have a firm consistency later. The affected powder particles are thereby compounded with one another as well as with the preceding layer. After each layer, the piston of the construction chamber is lowered by one layer thickness, and the next layer is added.

The cycle is repeated until the complete mold is finished. It is presintered in the machine and then taken out of the powder bed. See Fig. 3.83.

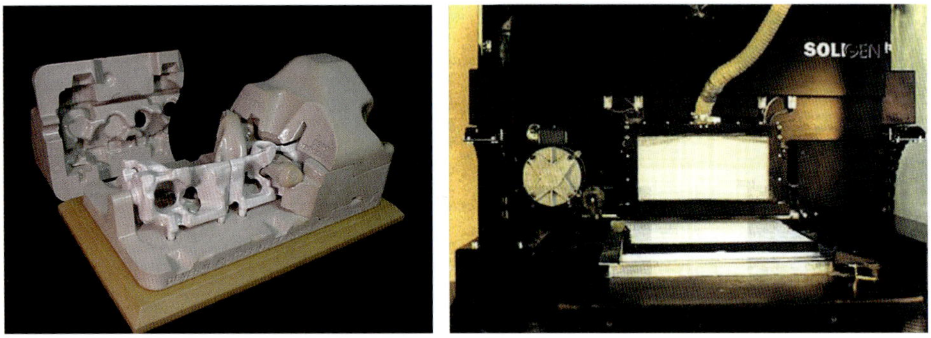

Figure 3.83 Direct shell production casting for the manufacturing of a ceramic mold according to the 3DP process: ceramic mold with core, sliced (left), view into the building chamber (right); (Source: Soligen)

System Type/Construction

The machine has a construction chamber with a movable base. A movable portal above the base contains a print head that contains the necessary amount of material and a roller that precompresses the material.

The advantage of this process is that it permits great freedom of design; owing to its ability to produce internal ceramic parts, it also opens up possibilities of casting hollow components.

Material/Construction Time/Accuracies

Any pourable material can be used, including, for example, titanium, magnesium steel, or superalloys such as Inconel.

A disadvantage is that internal areas, the surfaces of the molds, are hardly accessible after the mold is finished, and therefore they cannot be appropriately finished.

Very thin walls of up to 0.2 mm thickness can be realized. According to the producer, accuracies of up to ±0.05 mm are achieved, thereby meeting the casting tolerances of GT9 to GT12.

In contrast to castings from stereolithography and laser sintering models that, in principle, allow only one casting per model, the DSPC process is significantly faster. Its greatest economical advantage is seen on those solitary occasions when only a few mold prototypes are needed.

Postprocessing

In areas where no binder appears, the ceramic particles are not firmly attached and can be removed with a brush after the building process. The mold is burnt out. Thereby the binder is driven out and an appropriately thick ceramic shape is generated.

Typical Follow-up Processes

The follow-up procedure is similar to that of classic precision-casting processes. The mold is baked (sintered). Afterward, any surplus powder that is not firmly attached can finally be removed. In the last step, the mold is filled with liquid metal. After this has cooled down, the ceramic mold is destroyed, and the model is removed. The surfaces may be polished if necessary.

The process is meant to be used directly for precision casting. Therefore, there are no doubling techniques.

3.6.5 3D Printing System, Voxeljet

3D printing system: VX 200, VX 500, VX 800, VX 1000, VXC 800, VX 4000

Voxeljet Technology GmbH, Augsburg, Germany

Short Description

This is an additive manufacturing printing process based on the MIT 3D printing (3DP) patent. The parts are generated by means of injection of a binder into a plastic powder. After the manufacturing process the parts are ready for use.

Range of Application

Concept models, end products out of plastics

Development Status

Commercialized

History

Voxeljet Technology GmbH emanated from the company Generis GmbH in 2004. Generis GmbH was established by Dr. Ingo Ederer, Prof. Dr. Joachim Heinzl, and Rainer Höchsmann and has been promoting the 3D printing process based on the MIT patents for the application of the manufacturing of sand molds up to the Generis 1500 (today the Prometal S-Max). In the course of a cooperation agreement with the company Extrude Hone (today ExOne), the sand activities of Generis were transferred to the company Prometal. At their facility in Augsburg, Germany, Prometal RCT (Rapid Casting Technology) was established as a subcompany of Prometal GmbH in Remscheid (Section 3.6.3, "Metal and Molding Sand Printer, ExOne"). Afterward, Voxeljet put the emphasis of technology development on the subject of printing and allotting and on a modification of the GS 1500 for the printing of plastics, the VXC 800. There is also a smaller machine, the VX 200.

Development Partners/Strategies

Voxeljet promoted the 3D printing technique, specializing in the application of molding sand and plastics, and developed their own solutions. Smart cooperation politics ensure the collaboration with the licensees of the MIT basic patents (Soligen, Z-Corp) and opens up worldwide commercialization.

A close cooperation exists with TU Munich. Voxeljet also works as a service provider.

At the end of 2012, Voxeljet introduced the worldwide first continuously working printer and made a huge step into the direction of additive manufacturing.

Data Formats/Software

The machine accepts STL files.

The Principle of Layer Generation

A plastic powder layer is applied with a kind of squeegee-type coater. A piezo print head with 640 individually controllable nozzles extrudes the liquid binder according to the layer contour into the powder. Through the powder's reaction with the binder, the powder immediately hardens in the machine. The environmental powder supports the part. User-defined interlacing, even three-dimensional, is possible. The process occurs at room temperature and at surrounding atmosphere. See Fig. 3.84.

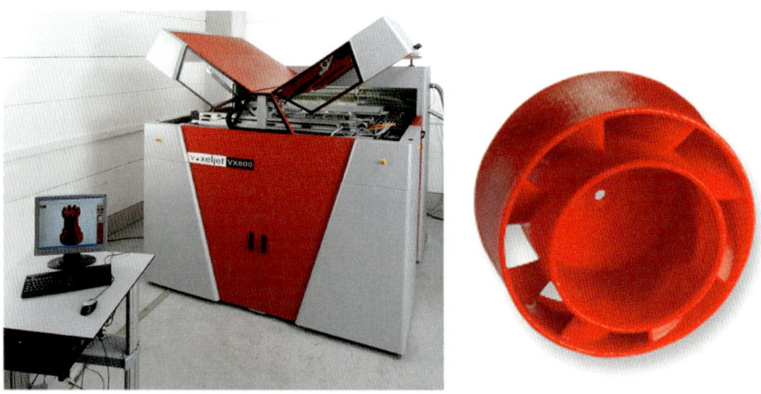

Figure 3.84 3D printing system by Voxeljet: VX 800 machine (left), part "swirl insert" made of epoxy-infiltrated PMMA (right); (Source: Voxeljet)

System Type/Construction

The machine was developed out of Generis GS 1500 and resembles it in terms of composition and in many details. The construction chamber is a changeable basin that can be moved to the unpacking station over a special rail system. Changeable basins ensure a continuous working. The coating system and the portal with the print head move in the x direction over the building platform. The print head also moves in the y direction. The construction chamber of the VX 800 measures x, y, $z = 2400 \times 2800 \times 2000$ mm^3. The machine contains an automatic material handling system. Powder that has not been used can be used again completely. The VX 200, VX 500, and VX 1000 form a machine family.

The VXC 800 works continuously with a printing process that occurs at the beginning of a conveyor belt, the belt is also the continuously moving building chamber boundary. At the end of the conveyor belt, the continuous unpacking process occurs. The so-defined building space has a width of 850 mm, a height of 500 mm, and is a theoretically not limited in its length. The resolution adds up to 600 dpi, and the layer thickness lies between 150 and 400 µm.

Voxeljet's largest machine is the VX 4000. It has a building chamber of $4000 \times 2000 \times 1000$ mm^3 and is designed for the manufacturing of sand casting molds. One special feature is the wide print head. It possesses 26,560 nozzles and can print one complete layer with only two crossings. Through the use of different building platforms, a continuous building process is ensured. Similar to the smaller machines, the VX 4000 features automatic material handling. Because of the size of its building chamber and the wide print head, the machine is suited for the fabrication of big cores as well as for the production of series. See Figs. 3.85 and 3.86.

Figure 3.85
Voxeljet VXC 800 (Source: Voxeljet)

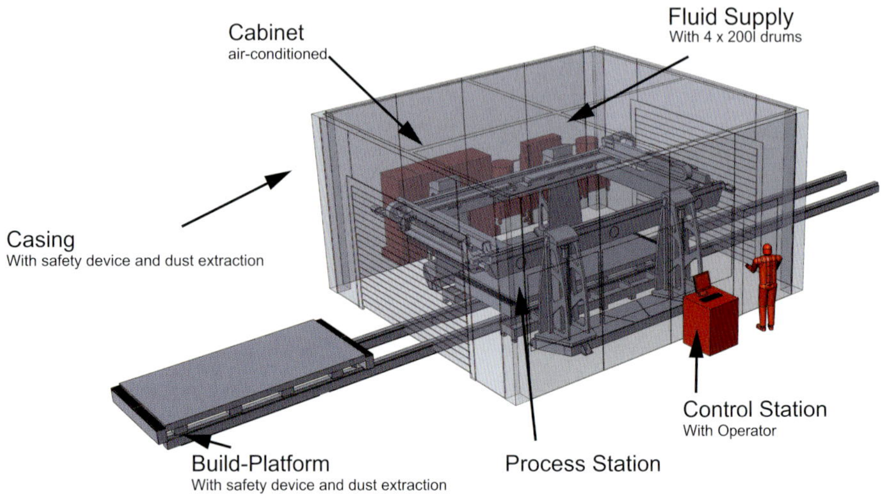

Figure 3.86 Voxeljet VX 4000 machine layout (Source Voxeljet)

Material/Construction Time/Accuracies

The building material is a modified and therefore proprietary PMMA (polymethyl methacrylate), which can also be described as acrylic glass. The binder, also proprietary, is of a dissolver type. The parts harden inside the machine but have some remaining porosity. They can be infiltrated. The infiltration with epoxy resin on plastic parts leads to solidities of 19 MPa at a breaking elongation of 2.9 %. The modulus of elasticity adds up to 1700 MPa. The producer declares a glass transition temperature of 65 °C. The infiltration with wax leads to meltable forms for cast metal. The grain size of the PMMA adds up to 45 µm (primary particle size d50). The standard value of the layer thickness amounts to 0.15 mm. The building progress at 0.1 mm layer thickness is about 24 mm in the z direction and can be doubled at doubled layer thickness. Thereby one can assimilate about 190 liters of building volume in ten hours. The example twist application with x, y, z = 206, 206, 117 mm^3 was built 12 times in eleven hours. The resolution of the print head adds up to 250 dpi in the x and y directions and 100 dpi in the z direction. The accuracy is declared to be 0.3 % of the characteristic part dimension, but at least with ± 0.1 mm.

Postprocessing

Postprocessing involves blowing of the loose powder and if necessary infiltration with epoxy resin or wax.

Typical Follow-up Processes

All of the material-specific finishing and follow-ups can be applied.

3.6.6 Maskless Mesoscale Material Deposition (M3D), Optomec

Aerosol Printing[26]
Optomec M3D
Optomec Inc., Albuquerque, NM

Short Description

This machine is designed for the direct application of nanoscaled particles on substrates. The particles are transported in a steamed fluid (aerosols) and applied in a defined way by a nozzle. Nearly all materials can be processed. Most of the time, there is no postprocessing needed.

Application

Functional prototypes

Development Status

Commercialized since 2004

History

Optomec was founded by Thomas Swann in 1983. The company produces optomechanical systems for research centers and universities and as add-on parts for rapid prototyping applications (also see Section 3.3.2, "Laser-Engineered Net Shaping (LENS), Optomec").

Strategy/Development Partners

The focus of the company's strategy is the market of microstructuralizing. The processes should be improved, accelerated, and get cheaper because the often-used masks are not needed in this process. It is also possible to print on the surfaces of bumpy or spatially curved substrates.

Optomec also developed the LENS process for the direct production of metal models made out of common powders (see Section 3.3.2, "Laser-Engineered Net Shaping (LENS), Optomec").

Data Formats/Software

The machine adopts the layer information as a virtual mask in DXF format. Any preparations have to be done outside the machine.

[26] Compare with Section 2.3.3.1 "Aerosol Printing"; the process is also called aerosol jetting.

Principle of Layer Generation

A flow of aerosols with a diameter between 1 and 5 µm gets piloted precisely onto the substrate by using a nozzle system. This flow contains particles up to a size of 200 nm in diameter, which build up a solid coat after hitting the surface of the substrate with the now-evaporating aerosol. Regarding the chosen material combination, there might be a postprocessing necessary, for example with a laser beam. This process has its benefits in 2D-oriented models, those that are much bigger in the x and y directions than in the z direction. In addition to that, it is often necessary to resinter the several layers, which is a disadvantage in printing large 3D models. There is the option to combine the process with a laser material dissipation.

Building/Construction

The machine consists of a print head (deposition head) and an ultrasonic atomizer, in which the aerosol, consisting of the vehicle fluid and the suspended building material, gets prepared and supplied. Inside the print head there is a flow of gas, which contains and guides the vaporized aerosol. This is called *aerodynamic focusing* by the manufacturer. The print head is moved by a three-axis portal system above the building area (x, y = 305, 305 mm^3). A shutter controls the amount and makes it possible to print about 10,800 drops/hour. The machine is a stand-alone solution and is controlled manually. A video camera supports the navigation. The optional available laser is a 100 mW Nd:YAG laser with a wavelength of 532 nm (frequency doubled), a focus diameter about 10 µm, and a residence time of 100 ms.

Material/Construction Time/Accuracy

Normally, all materials can be processed that can be made into particles of a size less than 200 nm and that can be carried by the vehicle fluid. These are mainly plastics, metals, and ceramics. In addition, it is said that cell structures are also possible. The viscosity spectrum is between 1 and 1000 cp.

Layer thicknesses down to 75 µm and lane widths under 10 µm up to 150 µm are realizable, but the standard is around 25 µm. The example given in Fig. 3.87 shows a connection with a height of 200 µm between the integrated electric circuit and the gold contact of a smart card.

Building rates are declared to be in a dimension around 0.0135 cm^2/h by the manufacturer, and the driving velocity can be up to 15 mm/s.

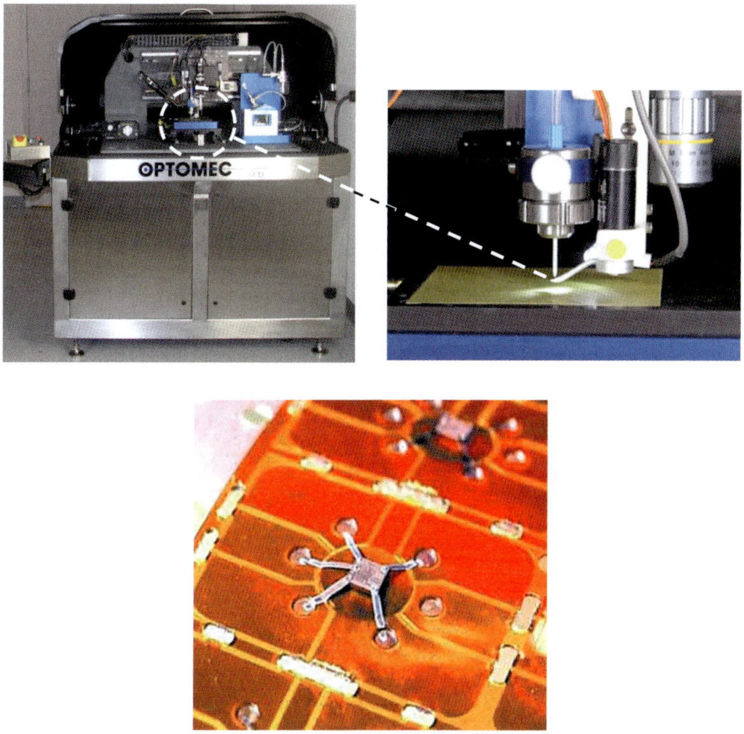

Figure 3.87 Aerosol printing, M3D machine with detail of the print head; Part: "smart card connection"; (Source: Optomec/eppic-faraday)

Postprocessing

Regarding the different material combinations, the vehicle fluid has to be removed and the applied material has to be postprocessed. Often metals have to be locally resintered, which is done by a laser most of the time.

Typical Follow-up Processes

None

■ 3.7 Hybrid Processes

In the analysis of the additive processes, it is found that, for many methods, additional subtractive processes are necessary. For example, the LLM methods use knives or milling tools to contour the added layer. Many additive processes can therefore be seen as hybrid processes. The distinction between the additive and hybrid methods is therefore arbitrary to a certain degree.

Hybrid processes as meant in this section are processes that use at least one additive and one subtractive process step. A characteristic is that a conventionally operated machine, preferably a milling machine, is upgraded to an additive manufacturing machine by adding additive manufacturing hardware modules.

Milling machines have been proven in hybrid systems because two of their advantages, good surface quality and continuous *z* contouring of the parts, compensate for the two major disadvantages of today's additive manufacturing processes.

3.7.1 Controlled Metal Buildup (CMB)

CMB: controlled metal buildup, supplementary module for the HSC Cutter Machine

Röders GmbH & Co., Soltau, Germany

Short Description

Controlled metal buildup (CMB) is a selective laser buildup weld with wire. By means of the CMB process, thin welding beads are placed close together on a workpiece in such a way that thin layers are generated. For this purpose, common filler wire is melted on in layers by a diode laser. Because of the small melting pool, the warping remains low. This process is integrated into the high-speed cutter machine of Röders as a supplementary module so that after each application the definition of the layer can be cut. This process generates layers with parameters, as is common in tool making. The CMB process was primarily developed for fast and economic repairs and alterations to tools and forging dies.

Range of Application

Functional prototypes, technical prototypes, rapid tooling

The molds are suited for injection molding, pressure casting, and forging dies.

Development Status

Commercialized

Development Partners/Strategies

CMB was developed in a close cooperation between Rödersand the Fraunhofer Institut für Produktionstechnologie in Aachen (FhG-IPT) (Fraunhofer Institute of Production Technology, Aachen, Germany). The process is designed as a retrofitting supplement for Röders' high-speed cutting machine and is used preferably for the alteration of tools and their repair, but it is also considered as an alternative for the production of tool inserts, especially if a few fine details would require a high amount of cutting in a conventional process.

Data Formats/Software

The machine reads complete and error-free STL files. The manufacturer supplies software that calculates the track geometry and the control data for the laser and wire advancement, analogous to the commercial CIM programs that were designed for milling.

Principle of Layer Generation

Using a diode laser, the tool surface is incipiently melted point by point while a common welding wire is dipped into the molten mass by means of a wire feeding unit; material is applied in this way. A layer is generated by laying several welding beads next to one another. The low heat application by the laser together with the small melting bath ensure that the tool distortion is practically unmeasurable. After application, the contour of each layer is cut to the defined measurements by means of a high-speed cutter. Highly accurate, 100% dense bodied, new tool areas are thereby generated. Although the process was developed for alterations and repairs, completely new forms can be constructed with it as well (Fig. 3.88).

Figure 3.88
Spray casting mold completely manufactured by the CMB process by means of laser surface cladding and HSC milling, hardness 52 HRC (Source: Röders)

System Type/Construction

The CMB unit can be retrofitted in the common Röders high-speed cutter. For this purpose, a movable slide is fitted in front of the cutting spindle on the *z* axis. This carries the diode laser and the wire feeding unit. The slide is movable to enable it to be pulled out of the work area during the cutting. The HSC milling machine is thereby not restricted in its performance and can be used at any time for HSC cutting as well as for the CMB process. The regulations for laser safety require a complete encasing of the entire installation (laser safety Class 1). The glass panels are therefore replaced by metal sheets, except for one laser-safe observation window.

Material/Construction Time/Accuracies

The process resembles the classic laser buildup welding. There is a relatively wide choice of material because, in general, all common weldable wires can be used. The area welded on is 100% dense. According to producer information, reproducible pore-free surfaces can be achieved. The hardness is > 50 HRC. Because of the refinishing work with the miller, the accuracy of the applied area is equal to the typical machine accuracy of the HSC machine.

Postprocessing

None

Typical Follow-up Processes

None

3.7.2 Laminating and Ultrasonic Welding: Ultrasonic Consolidation, Solidica

SonicLayer R200, SonicLayer 4000, SonicLayer 7200 UAM system
(72″ × 72″ × 36″)

Fabrisonic/Solidica Inc., Ann Arbor, MI

Short Description

This is a milling machine with an integrated ultrasonic welding device for buildup welding and contouring (trimming) of preproduced aluminum sheets. Aluminum tools can be created, especially for wax spraying in precision casting, in addition to prototypes made out of aluminum. With this process, it is possible to establish inner cavities, channels, and structures like honeycombs.

Range of Application

Functional prototypes, models

Development Status

Commercialized since 2001

History

Solidica was founded in the year 2000 by Dawn White with a specialization in the development and production of sensors. In 2005, the company got an order to develop a mechanically embedded RF (radio frequency) sensor for military operations of the US Marines. Solidica received the technology prize *2005 Best New Product* award from the Ultrasonic Industry Association (UIA, Dayton, Ohio) for the development of the ultrasonic consolidation (UC) technology. Fabrisonic was founded for merchandising as a joint venture with Solidica EWI, a research organization in Ohio.

Strategy/Development Partners

The combined additive manufacturing process of the Form-ation milling process and the *ultrasonic consolidation* buildup welding device (called UC process) has been developed by Solidica for the production of their own sensors. The core of the sensor technology is the integration of different subprocesses, for example, microfluidics, electronics, and piezoelectrics, in fully encased metal housings. There are applications for sensors in an aggressive environment, for example in aerospace engineering or in seafaring, but for implantation, too. The company is a member of the *National Center for Manufacturing Sciences* (NCMS), a consortium of well-known US companies for the combined development of high technology in the field of machining technologies. There is a cooperation with the University Of Michigan College Of Engineering. In cooperation with the Michigan Molecular Institute and the Saginaw Valley State University (SVSU), Michigan, the processing of stainless steel will be developed.

Data Formats/Software

The Sonic Layer technology integrates the strapwise buildup welding of one layer and its milling. The so-called *RpCam* software imports 3D-CAD data, for example with the standard interface IGES, but also STL or NC data, and transfers them as an internal data bundle to the machine controls (ISO compatible G-Code). The slicing program creates layers with a thickness of 0.1 mm. They are adapted to aluminum sheets, but generally they are variable.

Principle of Layer Generation

The UC technology consists of two basic steps: *tape and trim* and *part finishing*. The *tape and trim* process begins with the positioning and unrolling of the aluminum belt on the part's bottom. After that, the vibrating sonotrode presses the belt against the part's bottom. First of all, the surface gets cleaned with a high frequency and a low amplitude, and then comes the welding with a low frequency and a high amplitude. The connection occurs because of the material flow in the barrier layer between 10 to 20 µm. The belt gets cut off with a little brim and the next lane will be welded on, just until the whole layer is covered. The blowoff of the sections marks the end of the taping process. Trimming means the cutting of the outer contour with a 2D end milling cutter and the after all blowoff.

The tape and trim process proceeds layer after layer, until the model reaches a height of 3 to 6 mm. During the finishing, a standard milling process, the desired contour and surface accuracy will be reached. Both steps alternate until the part is ready.

If inner cavities grow, it gets tougher to apply the belts. Therefore, a polymer material, which is able to be washed out, is filled inside the cavities during the building process.

The building process can be interrupted at any time, for example, to place something additional into the model. It is possible to continue the process, even after long breaks, without any loss of quality.

Building/Construction

The Solidica Form-ation machine is a three-axis milling machine with an integrated ultrasonic welding device, including a material container. The milling head and the welding application are mounted next to each other on top of a moving device (in the x direction) and above a fixer plate. That is why the machine is much wider than would be necessary for the single milling process. The welding device is about 500 mm in width and includes a coil with the aluminum belt, an unwinding and pressing device, and the ultrasonic head itself, the sonotrode for welding the single lanes together with the partly ready model. The welding velocity is 2.5 m/min. The milling process does not use any lubrication. The machine has a blowoff device for transporting the belt sections and the chips.

The machines mainly differ in the size of their building spaces. SonicLayer R200 is designed for experiments and research (x, y, z = 500, 300, 150 mm^3); the SonicLayer 4000 (x, y, z = 1000, 600, 600 mm^3) correlates the transonic Form-ation and the SonicLayer 7200 UAM system (x, y, z = 1800, 1800, 900 mm^3) and is a production machine. The electric input power is between 4.5 and 9 kW. The R200 gets supplied manually; the others have an automated conveyment of the aluminum belt.

Material/Construction Time/Accuracy

The building material is an aluminum belt with a thickness of 0.1 mm and a width of 25 mm. Indeed, the thickness is synchronized with the contouring (trimming), but it does not influence the accuracy because this gets reached by the subsequent finishing on the milling machine. Therefore, the accuracy is declared to be ±0.05 mm.

Figure 3.89 Ultrasonic consolidation machine: SonicLayer 7200, honeycomb structure (semifinished, open and finished, closed); (Source: Fabrisonic)

A steel material (stainless steel) is being developed at the moment.

The building speed is declared as being independent of the geometry and is 12.5 mm/h.

The UC license only applies for wax injection tools and prototypes. Tool cavities should not include thin freestanding peaks or walls. Undercuts cannot be produced without step effects. Solidica also uses the machine for the production of some of their own sensors.

Figure 3.89 (right) shows a full metal honeycomb structure, a construction without any glue spots or connection layers, open and closed. It is an example out of the CASM (Computer Assisted Satellite Manufacturing) project of the Utah State University. Structures with integrated sensors, electronic parts, batteries, and so on can be produced. It is also possible to integrate multimaterial composites.

Postprocessing

The part is ready to use when it leaves the machine. If the cavities are filled with the supporting polymer, which is able to be washed out, a cleaning process is necessary.

Typical Follow-up Processes

None

Literature: [Domack, 05]

3.8 Summary Evaluation of Rapid Prototyping Processes

The desire of the user to have a simple-to-use and meaningful process at their disposal that assists in choosing the most suitable rapid prototyping processes for any specific requirement in a fast and reliable way is as old as rapid prototyping itself. This becomes even more acute as more new processes are introduced.

Although it was still possible 10 years ago, when the first edition of this book was published, to make direct comparisons between industrially usable machines on the market at that time (because they produced parts of approximately the same size with comparable degrees of detail within comparable price categories), this is in general no longer possible today, nor would it make sense to do so. To compare the construction chamber of an FDM Maxxum with a microstereolithography installation, or the costs of material for stereolithography resins with those of materials used for the Z-Corp machine, would be meaningless.

The following paragraph will try to provide a sharpened perception toward the questions of which process is usable under what circumstances, which alternative processes would also fulfill the requirements, and which processes in special cases are less suitable.

For this purpose, we first take a look at the basic properties of *additive manufacturing processes* compared to conventional processes (Section 3.8.1, "Characteristic Properties of AM Processes Compared to Conventional Processes"). On this note, we put emphasis on the criteria of accuracy and surface quality (Sections 3.8.2, "Accuracy" and 3.8.3, "Surfaces"). Thereafter, we take a look at benchmark parts and user parts.

Attempts are often made to carve out the properties of particular processes by the means of benchmark tests and user parts. The benefit is discussed in Section 3.8.4, "Benchmark Tests and User Parts."

Focusing on the important application of rapid prototyping, the industry-suitable processes are categorized in Chapter 4, "Rapid Prototyping," according to the model classes and therefore their preferred fields of application.

3.8.1 Characteristic Properties of AM Processes Compared to Conventional Processes

The evaluation of the potentials of *additive manufacturing* is easier when comparing it to the characteristic properties of better-known conventional nonadditive processes.

As criteria there are

- materials,
- tools,
- part structure,
- CAD compatibility,
- accuracies, and
- the influence of manual work.

Materials

Conventional processes work with materials in terms of semifinished products, which are chosen as being suitable for the mechanical-technological properties of the product. This is called *product-oriented material selection.*

Additive manufacturing processes work with few materials, which are especially optimized for the additive manufacturing process and therefore less for the product. So the favored mechanical-technological properties can only be approximately reached. This is called *process-oriented material selection.*

Tools

Conventional processes work with different tools, which are optimized for the specific subtask, and eventually manufactured tools, which eventually have to be changed frequently during the process. This is called *product-oriented tool selection.*

Additive manufacturing processes work “tool-less,” which means that they do not use any tools that are optimized for the particular part. The “tool” is an element that generates and construes the layer. This element is neither changed during the manufacturing nor varying from part to part. This is called *process-oriented tool selection.*

Part Structure

Conventionally manufactured parts are usually rigged or compiled out of several pieces or elements consisting of different materials for manufacturing reasons. This is called a *process-oriented part structure.*

Additive manufacturing parts can be optimized and built out of one piece for the specific function because of their nearly unlimited geometric freedom. If they are rigged or compiled out of several pieces, this is a result of the limited building chamber space or is done to simulate a serial assembly, not due to manufacturing or constructive requirements. This is called a *function-oriented structure* of the parts.

CAD Compatibility

The CAD models of conventionally manufactured parts are implemented into editing programs for tool machines by appropriate programs called NC modules. Thereby machine-specific issues are so substantial that in general the removal of a program has to occur on the machine (or with its controls) for which it was programmed. This is called *machine-dependent programming.*

The data for the manufacturing of additive produced parts are derived from standardized STL data out of the CAD data set. Based on STL data sets, all known additive manufacturing machines can be controlled nowadays. So the selection of an ideal process can be ensured without newly generated data sets. This is called a *part-oriented* and *machine-independent programming.*

Accuracies

Nowadays the leading conventional processes correspond to the maximum reachable machine-specific accuracies. They are defined as *state of the art.*

Additive manufacturing processes mostly feature these accuracies only regarding the contouring and are therefore two-dimensional. In particular, the layer principle leads to inaccuracies in the third dimension.

Conventional processes today are handled so well regarding tools and the machine that a defined accuracy is reached, which accords to the machine accuracy.

Additive manufacturing processes do not reach the state of the art of conventional machines. Furthermore, the majority of them are heavily dependent on calibration. Ideal results are under high requirements only reachable if the model is attached first and afterward the resulting deviations are used for the calibration of the machine. With experience these calibrations can be reduced to a minimum.

Influence of Manual Work

Accurate conventional processes, such as high-velocity milling or eroding, create surfaces that either do not need to be refinished at all or need refinishing that is basically limited to surface effects. Therefore, a measurable deviation is not caused.

Additive manufacturing parts feature process-caused steps, which can be easily flattened in relatively soft materials such as plastics. In this way, excellent surfaces can be reached manually. Nevertheless deviations are caused. This can be problematic if a charge of very accurate parts is refinished manually by different coworkers.

These explanations show that conventional parts have advantages over additive manufacturing processes, especially regarding accuracy, material properties, reproducibility, and, depending on the geometry, also the building speed. This applies especially when the geometries are relatively simple. Conversely, here is one of the most important demands for the successful application of additive manufacturing processes:

Additive manufacturing processes are used most beneficially when the parts are complex and needed very quickly in small quantities.

During this discussion, one must not forget that additive manufacturing processes always have one advantage, which is especially brought to bear in product development:

Additive manufacturing processes ensure a consistent and equal and binding data set. Therefore, they are an integrated component of product development strategies as rapid prototyping processes, and as rapid manufacturing the sustainable basis for the overall life cycle of a product.

3.8.2 Accuracy

When considering model accuracy, we should differentiate between machine accuracy and process accuracy:

- Machine accuracy is the achievable accuracy under ideal circumstances, as specified in the producer information. It is listed in the tables in the appendix.
- Process accuracy is the accuracy resulting from taking the entire process chain into account.

For the user, the accuracy of the resulting model is more important. This results as process accuracy from error propagation over the entire process chain. Machine accuracy is thereby only one element. Errors in the data model, the processing of imperfect data, and errors arising during data transfer, during model preparation, and during postprocessing and finishing must also be considered.

The sum of all these influences cannot generally be expressed in numbers. Therefore, the influences will be discussed in an attempt to give the user a sense of those aspects that are most important for achieving a high-quality model.

The achievable *machine accuracy* is limited by three main influences:

1. The machine and its constructional features,
2. The principle of layer generation (the process), and
3. The material.

These aspects in relation to single prototypers have been discussed previously for each type of machine. In the following paragraph, these prototypers are compared with one another.

Machine: Machine accuracy depends decisively on the dimensions of the sphere of action the form-giving element has (laser beam diameter, droplet diameter, nozzle diameter, and so on) and on the method for controlling this element (scanner, plotter). The active diameter of the form-giving element defines the basic accuracy, while the scanner/plotter unit can introduce additional inaccuracies into the system.

Principle of layer generation: When using plotter units that move the form-giving element on an *x-y* slide across the construction plane, it can be assumed that any inaccuracy they cause in the construction does not result from their positioning. If anything, only dynamic effects such as curves, corners, and reversals should be taken into account.

Unlike plotters, scanner units cause additional angle errors so that varying accuracies may result depending on their position on the construction plane. In addition, and this applies especially to fast scanner units, dynamic effects (overshooting) must be taken into account that are often also geometrically dependent (shadowing).

Scanner units are faster not only at contouring, but they are also much faster at positioning than plotter units. This is important for the implementation of construction strategies in which alternating model areas are exposed to avoid warpings. Because warpings, which result in model inaccuracies, also lead to unsatisfactory models, they are tantamount to inaccuracies caused by the machine.

Material: The reproduction and the mechanical-technical properties of modeling material, together with the basics of model making and the considerations of specific prototypers, have been discussed in detail. For a comparable discussion, please refer to Section 2.6. The achievable model accuracies are inseparably connected to the material properties. Therefore, the material properties must also be taken into account when calculating the achievable accuracies.

To assess the influences of single process elements on the *process accuracy,* the most important elements of preprocessing, the additive manufacturing construction process, and postprocessing are the following:

Preprocessing

- quality of CAD modeling
- loss of quality caused by interfaces and data transfer
- support structures and positioning the part in the construction chamber

The aspects of preprocessing that determine accuracy were discussed in detail in Section 2.2. The need to standardize the data transfer and the interfaces should only be reiterated here. In doing so, one can avoid having to deal with problems in prototyping that basically are not prototyping problems at all.

Additive Manufacturing Construction Process

- construction parameter
- material influences

Particularly in Sections 3.1 and 3.2, the additive manufacturing construction processes were described in detail using the examples of stereolithography and laser sintering.

In addition, a specific problem regarding prototyping is to be noted here: in most technical applications, the models have a dominant direction in which it is especially important to keep to the required accuracies, while for other directions the required accuracies are of secondary importance.

Postprocessing

- number of process steps
- quality of postprocessing and finishing

All elements have been described in detail in Chapter 2 and in the relevant machine-specific sections of this chapter and are therefore not discussed here again.

The Human Factor

It is especially important to be aware of the human factor in the rapid prototyping process chain. Descriptions of rapid prototyping processes often give the impression that these are model-making processes that run automatically without human supervision.

All currently used processes are highly dependent on experienced, skilled labor. This is valid for all process steps:

- Manual interventions are necessary when preparing the CAD data for the construction process and, if necessary, for the supports.
- Optimal adjustment of the machine is necessary. A model may have to be built until it has reached an assessable size, enabling the machine to be calibrated accordingly.
- In two-step processes (postcuring), the influences of this second step must be taken into account. In a stereolithography process, for example, it should be considered whether the supports are best removed before or after the postcuring.
- The careful handling and the expert use of the material are important. Careful handling includes proper storage conditions for the material (moisture, light) and also correct recycling of reuseable material. In fact, most materials used for stereolithography and sintering go through a certain aging process if they have been in the machine several times even without having been directly used in the process. Certain quantities of new material should be added so that the desired material properties are maintained.
- Optimally position the model in the construction chamber with regard to required accuracies, taking into account the varying properties of the material when used in different building directions, and the economic usage of the construction chamber.
- The significance of the human factor is obvious in the areas of model cleaning and finishing processes such as polishing, drilling, impregnating, varnishing,

and so forth. Here, manual skill is required, and an appropriately equipped workshop is also absolutely essential.

3.8.3 Surfaces

The largest disadvantages of rapid prototyping processes compared to conventional machining processes are the significantly rougher surfaces and the stepped surfaces in the z direction. Another cause for these disadvantages is the grain size or the diameter of precast filaments, depending on the particular process.

The surface quality for machinable treatment in the industrial area for treatment centers is higher than 0.4 µm (x, y, z = 600, 500, 500 mm^3). In daily operation, qualities for rotating of 1 µm and for milling up to 1.5 µm are accepted.

For additive manufacturing parts, accuracies of about 1/10 mm up to 5/100 mm are declared.

Whereas these qualitative statements are considered to be exact, quantitative statements are relatively rare. Therefore, surface qualities were measured, for example, on an excavator arm. Those same parts were examined after they had been made by a stereolithography process (3D Systems, Acrylate) and by selective laser sintering (3D Systems/DTM, nylon). Each time, a measurement was made in the horizontal direction and another in the vertical direction at right angles to the first [Van Crüchten, 95].

The average peak-to-valley height R_A was established, which is defined as the arithmetical mean of all the values of the roughness profile within the measured length. A roughness profile (R profile) for the measured stretch results. This is the high-frequency part of the surface profile, resulting within the measured stretch after the elimination of the gradient of the compensation lines and after the waviness has been filtered out.

Owing to their layered construction, the surfaces of additively manufacturing built components are structured regularly. For this reason, the maximum wave depth W_T, representing a characteristic value of the waviness profile, is specified in addition to the specification of the average peak-to-valley height R_A. The waviness profile (w profile) is the low-frequency part of the surface profile that results within the measured length after the elimination of the gradient of the compensation lines and after the roughness has been filtered out. The maximum wave depth W_T is the vertical distance between the highest and the lowest point in the waviness profile. The graphs shown in Fig. 3.90 indicate the so-called P profile, which represents the surface profile with roughness and waviness but after the elimination of the gradient of the compensation line.

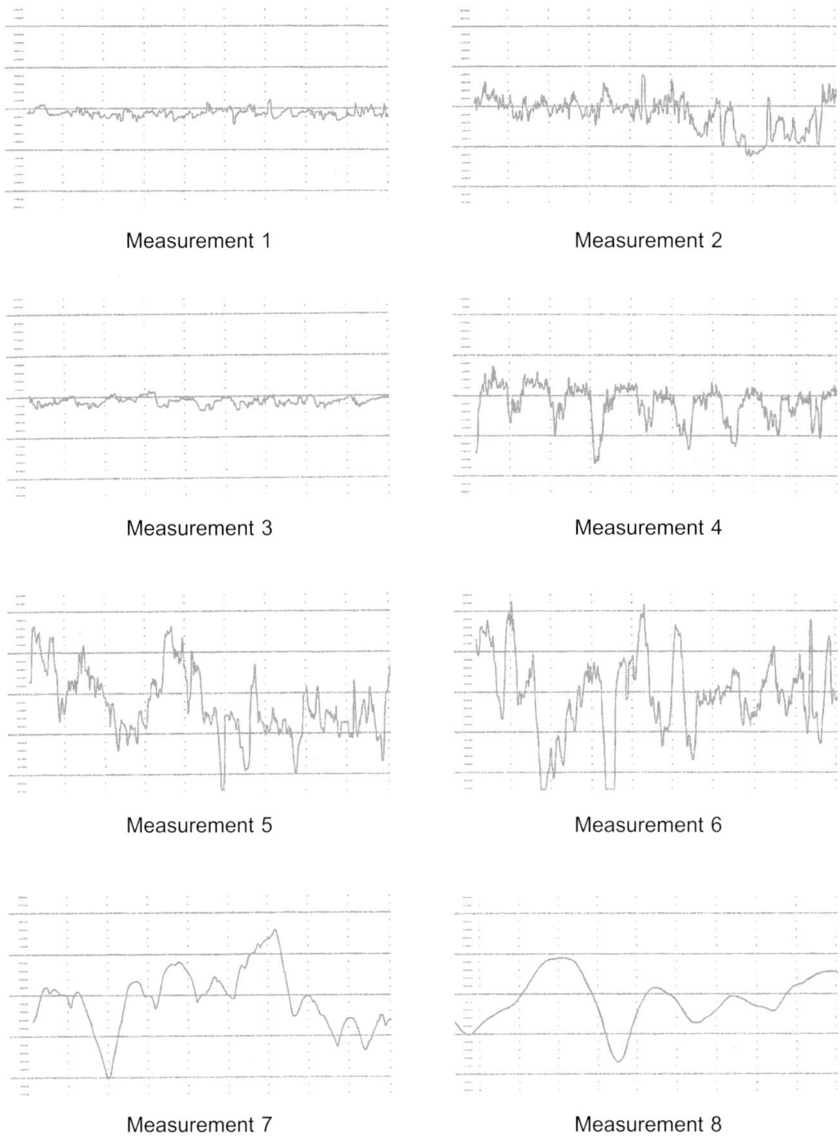

Figure 3.90 Development of roughness (R_A) and waviness (W_T) of SL and SLS surfaces of identical parts

In Fig. 3.90 and Table 3.3, eight graphs are shown. Roughness values were measured, and also the waviness of horizontal and vertical surfaces of stereolithography models with layer thicknesses of 0.125 and 0.250 mm and of selective laser sintering models with a layer thickness of 0.127 mm were measured.

The influence of layer thickness in stereolithography is very small and can be interpreted only as a slightly excessive scan structure caused by an increased curing depth if this should not just be taken for scattering. This cause must also be attributed to the differing roughness and waviness of the horizontal plane compared with the vertical plane in stereolithography: the STAR-Weave technique puts layer upon layer of netted structures on top of one another whereby the mesh is not cured during the scanning (option: no fill).

Table 3.3 Surface Roughness (R_A) and Waviness (W_T) of Two Identical Parts Executed in Stereolithography (3D Systems, Acrylate) and Selective SLS (3D Systems, PA Unfilled)

	R_A [µm]	W_T [µm]	Measurement (see Fig. 3.90)
Horizontal surface			
SL (0.125 mm)	6.44	93.2	2
SL (0.250 mm)	7.60	87.0	4
SLS (0.127 mm)	16.70	147.5	5
Vertical surface			
SL (0.125 mm)	3.13	10.4	1
SL (0.250 mm)	3.49	11.4	3
SLS (0.127 mm)	43.70	140.0	6
SLS glass-blasted	24.50	111.0	7
SLS manually smoothed	5.70	107.0	8

These can be found in the measurements as a result of their loosening during extended construction processes or during cleaning. The vertical planes do not show such structures and are therefore better by a factor of two regarding the roughness and by a factor of eight to nine regarding the waviness. New measurements on the inside area of a cube that is opened on the bottom and the top with different wall thicknesses show values in the same dimension (Fig. 3.91).

With selective laser sintering, on the other hand, the horizontal areas are better by a factor of two than the vertical areas (Fig. 3.90). In addition, the waviness is overall significantly higher but is the same for horizontal and vertical areas. While this is attributable to the scan strategy (raster process), the greater degree of roughness on vertical areas is explained by the fact that these end in the powder bed and neighboring particles can be melted on by heat conduction; the significantly smoother horizontal areas end in the nitrogen atmosphere.

The last two measurement drafts show the influence that finishing procedures have on the surface of sintered models. Draft 7 shows how the surface (roughness) can be improved by half again by glass blasting, while waviness is not so much influenced thereby (significantly $<20\%$). Intensive manual polishing achieves surfaces similar to those of stereolithography processes (roughness). Waviness is hardly influenced positively by polishing, probably because the process is manual.

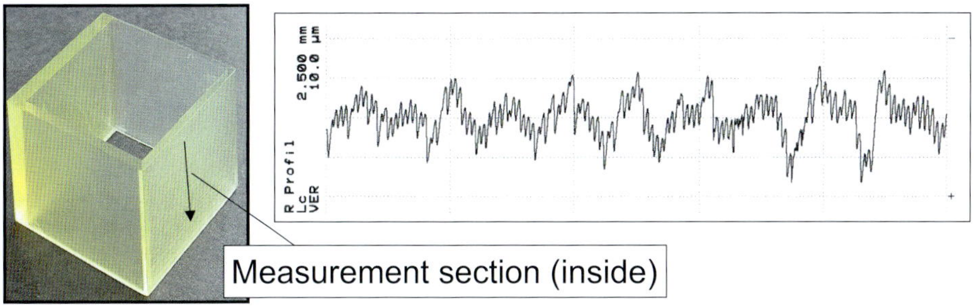

Place of measurement: cube inner wall / Wall thickness	Measurement distance Ra	Arithmetic mean roughness Ra	Roughness depth Rz	Max. roughness depth Rmax	Max. profile peak height Rp	Core roughness depth Rk	Roughness depth Rt
	[mm]	[µm]					
1mm	17.5	3.917	24.80	29.60	13.70	13.10	29.60
2mm	17.5	5.622	33.70	43.50	19.20	18.80	43.70
3mm	17.5	5.111	28.00	34.10	14.60	17.30	37.30
4mm	17.5	5.808	33.30	40.60	18.10	19.90	40.60

Figure 3.91 Roughness measurements in the inside of a cube at different wall thicknesses (Source: FH Aachen/Schwarze)

3.8.4 Benchmark Tests and User Parts

Benchmark tests, or so-called user parts, provide sufficiently exact information about the achievable machine accuracies of a system. Here, it is certainly legitimate not to question the results regarding the multiple calibrations of a machine or the multiple constructions of one specific part. On the other hand, the user needs to be aware that the achievable accuracies often require an effort that cannot be economically fulfilled in the daily routine. These processes are, however, well suited for the characterization of machine performance.

Benchmark Parts

When the first prototypers were introduced, models were constructed and built simultaneously, usually by more or less independent third parties, for the sole purpose of proving the abilities of a prototyper in the reproduction of details. Various so-called benchmark parts were developed in quick succession that were used primarily for comparing different systems. The common disadvantage of all benchmark parts is that, even when the details are selected reliably, they are still not as universally usable as they are presented to be. Trade-specific features, the experience of the designer, and the knowledge of any weaknesses and strengths of particular machines often find their way into the construction of those parts. In the end, most benchmark tests do not provide a solid base for comparisons because important circumstances are not taken into account. Among other factors are all of the influences on process accuracies and, for example, the questions: Who did the building? How often has he built before? How regularly was the machine calibrated? Were there transport influences? How soon after completion of the construction process were measurements taken? Who took the measurements?—and many more.

A closer look at so-called *user parts* shows that they are much more informative. It can be assumed that these are parts that were specially constructed to demonstrate the advantages of *one specific machine*. It can also be assumed that the machine was precisely calibrated and operated by qualified personnel to achieve the accuracies published by the producer. Of course, the storage and handling of the material will have been carried out with the utmost care. Taking for granted that this applies to all producers, user parts are quite a good basis for comparison. Apart from the accuracies achieved, the constructive designs of user parts especially tell us something about the strengths and weaknesses of a machine.

Measurement of Accuracies

Complicated measurement and interpretation processes followed in the wake of the first comparative examinations, where only a few main measurements were taken. Today these examine a huge number of points by means of automatic coordinate-measuring devices pinpointing the accumulated faults and the frequency of their occurrence. Graphs show (see Fig. 3.92) what percentage of the measured points are within a given tolerance range. Such graphs show both the error distribution and, above all, the largest single error. Figure 3.92 shows that when using vinyl ether in an EOS STEREOS MAX, an epsilon-95 value of <66 µm results [Serbin, 95].[27] In a company brochure, 3D Systems also published an epsilon-90 value of 65 µm [Jacobs, 94], so it can be assumed that, under optimal conditions, the machines achieve approximately the same good results. These reports should be viewed against the background that in 1989 comparable models still had epsilon-90 values of 400 µm.

Because the speed is an essential characteristic feature of prototypers, it is also important to ask about the achievable accuracies and how reliably these accuracies can be achieved. It has to be questioned, therefore, whether accuracies are achieved, so to speak, in the daily run of the mill with occasional calibrations in the course of routine maintenance, or whether it is necessary to prepare the machine with an extensive calibration process for each construction process to achieve a defined accuracy. Such specifications cannot be given in general and should therefore be checked by the user personally.

[27] The older measurements serve to illustrate the principle.

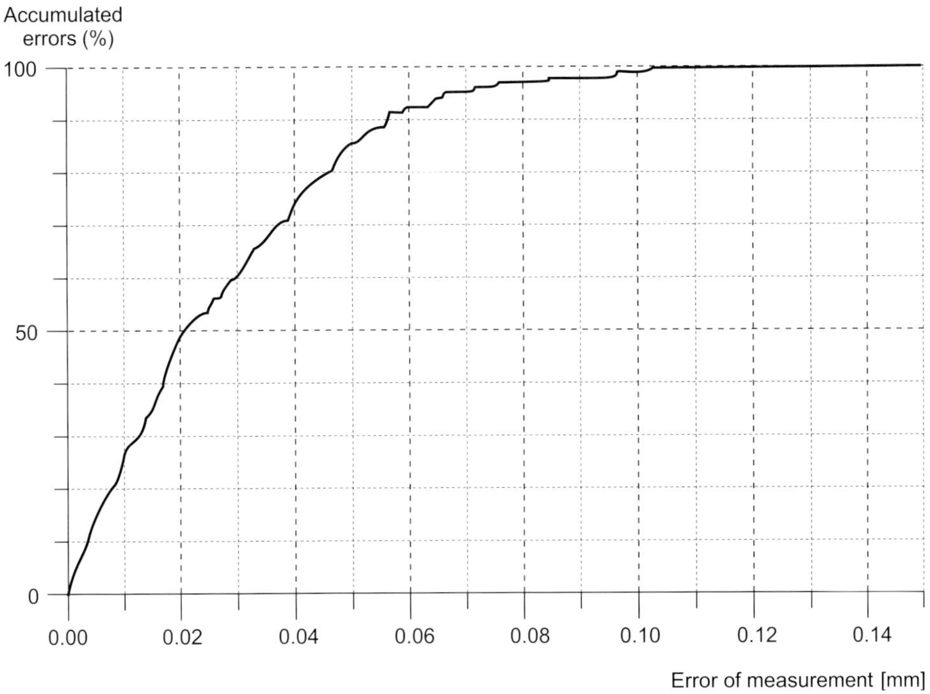

Figure 3.92 Accumulated errors as a function of error of measurement (108 measuring points[27])

As a proof of accuracy of parts, today an initial sample test report is created. Therefore, a huge amount of defined points is controlled by measurements, and the result is documented. Fig. 3.93 shows an example of the measurement of a plastic molded pulp. The measurements are dedicated to defined points in the CAD and labeled in terms of color for better clarity. The left-hand part of the image serves only for clarity. Green values are within and red values are beyond the tolerance. The measuring took place with a coordinate measuring arm. For that purpose the part is placed on a gauge.

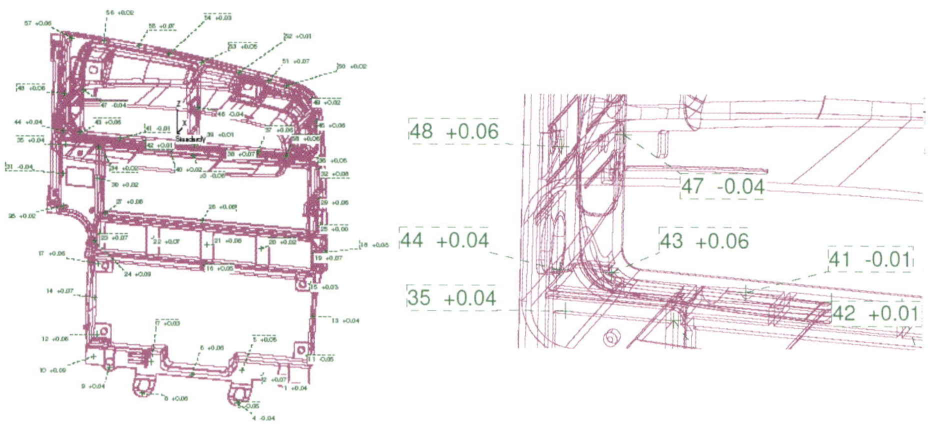

Figure 3.93 Detail of an initial sample test report: overview (left), detail (right) (Source: FARO, FH Aachen)

3.9 Planning Targets

The weak points of today's 3D printers can be summed up in three groups:

- 3D printed or additive manufactured parts show a comparably poor surface quality with visible stair steps. Additionally, depending on the process, the surface is rough, for example because of the particle size of the powder used for sintering and melting. Manual postprocessing is long, especially if many parts need to be treated. Additionally, this bears the risk of manually caused inaccuracies that may differ from part to part.
- 3D printed parts are made from process-dependent materials. The mechanical and physical properties of the parts often are very different from that obtained by parts made from the same material but using traditional high-volume manufacturing processes like plastic injection molding. This is also true for materials that are nominally identical. As an example, parts printed from standard polyamide sintering material show properties that are very close to high-volume PA 11 or PA 12. This is valid for some quasi-stationary values, but one can doubt whether it is true for dynamic loads as well. Even if only a few investigations in this field have been published, obviously a bigger void fraction with many microflaws reduces the fatigue durability.
- 3D printing processes normally deliver one or only a few parts. For more support of the product development process, usually 5 to 50 parts are needed, and for the simulation of automated assembly lines one needs more that 100 parts. Making parts by 3D printing often is not economically successful because there is, depending on the process and the part's dimensions, almost no economy of scale. If

a higher number of the same part is made, the time-saving effect of 3D printing is reduced significantly, which is a very important fact if direct manufacturing is addressed.

These points directly lead to the future requirements of 3D printing and the planning targets as follows:

- Improvement of surface quality.
- Elimination of the unwanted effects of anisotropic behavior due to the layered process, mainly in terms of accuracy and surface quality.
- Improvements in the reproduction of details and reduction of the stair-stepping effect by reduced layer thickness and continuous contouring in the *z* direction.
- Improvement of the material properties and of the variety of available materials.
- Reduction of the build times and the overall process time of the building process.
- Reduction or elimination of the manual postprocessing, preferably by automated postprocessing.
- Concentration on the development of one-step processes by redesign, simplification, or elimination of intermediate process steps.

3.10 Follow-up Processes

If, in order to make a defined part, 3D printing processes are followed by nonadditive process steps, the latter are called *follow-up processes*, and the entire process is called an *indirect process*. In most cases, the geometrically exact part is made additively and regarded as a master model. The desired number of parts and the desired material, color, and so on are obtained by follow-up or copying processes.

There are different processes and process chains in use, depending on whether the desired part (final part) is made from plastics or from metal.

3.10.1 Target Material: Plastics

Small series of plastic parts are obtained from tools made by casting or countercasting of the AM masters.

Tools can be soft or hard. The processes belong to the application "indirect tooling" as described in Section 5.3.2, "Indirect Processes for the Manufacture of Tools for Plastic Components."

Soft tools mainly are made from silicone rubber, and hard indirect tools are obtained from countercasting of hard plastics or low-melting-point metals.

This must be distinguished from hard tools made by direct tooling. The tooling and its application are described in Section 5.4, “Direct Rapid Tooling Processes.”

3.10.2 Target Material: Metal

If target parts made from metal are desired, the follow-up process is casting.[28] The most important ones are investment or precision casting, also called the lost-wax process (Section 5.3.3.1, “Investment Casting with AM Process Steps”). Applications for the casting industry are presented in Section 6.5.2.2, “Casting.”

This must be distinguished from the direct additive manufacturing processes for making metal parts (Section 5.4, “Direct Rapid Tooling Processes”).

[28] Except for low-melting-point metals or cold-cast materials that, from the process point of view, are treated as plastics.

4 Rapid Prototyping

Rapid prototyping[1] describes the direct additive manufacturing production of samples and prototypes by layer manufacturing technology. The products are models and the machines are prototypers.

This chapter is self-explanatory, provided that a basic knowledge exists, thereby enabling a quick approach to the subject. Repetitions are unavoidable because of the nature of this concept, but these are kept to a minimum.

4.1 Classification and Definition

Prototyping is known as the technology to create a first physical design of a product. It is also known as experimental modeling or design modeling.

4.1.1 Properties of Prototypes

The broad public combines the definition prototype with a single production in the automotive industry. In general, the word prototype describes every physical sample, model, or mock-up that is used to improve and shorten the development of an idea into a product by realizing the virtual model as a touchable part.

Prototypes are not products.

Prototypes are used and produced to test the product characteristics. They are highly abstracted. Ideally, they are produced so that they only contain the characteristics that are absolutely necessary for the evaluation, so they should be as simple as possible.

---

1 In general, rapid prototyping also means special methods that are used for fast concept development, to present a product, or to evaluate proof of a concept. Thus, rapid prototyping could also be used in the field of information technology in system development to design complex system architectures quickly.

For example, a prototype for the evaluation of the haptic behavior of a cup handle only needs a cutout of the geometry. The exact design is only required for the surface of the handle, the surface of the cup around the handle (and only there), and the weight of the cup. It is not important that the cup is hollow or that it consists of the series material, which are essential features of the final cup.

In dependence of the evaluation, prototypes made of the same dataset could be different. Glass panels covering the light sources of cars must have excellent geometric and optical qualities if they are shown on automotive exhibitions. Features like impact strength, UV-resistance, dismantlement or recycling are not of concern. This is a similar protoype process for a test vehicle for evaluating the drive chain required; the surface quailty and the UV-resistance take no account, but the impact-strength and the optical characteristics are fundamental.

Depending on the industrial sector and the application, prototypes are also called models, mock-ups, solid images, or concepts. Usually prototypes are models and not final parts or products.

If the prototypes used have the character of three-dimensional pictures, they are called solid images. If they show some relevant characteristics for the evaluation of the final product, they are called functional prototypes.

There is no need for prototypes to consist of the series material. Often they are made out of model materials that are easier or faster to handle, if they represent the final characteristics closely enough.

Completely defined 3D data sets are necessary for the production of prototypes. The data sets could be generated in the construction or from measurements. Construction or design rules are not necessary for prototypes. That is also why they do not have the product characteristics for industrial mass production.

Prototypes must be made available quickly and simply. Because of their short lifetime, they do not have to last a long time, but the recycling should be easy and safe. As a temporary object of the product development, they should be cheap.

4.1.2 Characteristics of Rapid Prototyping

From the end of the 1980s, the term rapid prototyping was influenced by stereolithography. Since the 1990s, rapid prototyping was used to describe additive manufacturing production technology in general. Today, rapid prototyping describes a subset of the additive manufacturing production technique that is used to produce models and prototypes.

One of the most important properties of production by additive techniques is that the production is based on the 3D CAD model. Rapid prototyping delivers 3D facsimiles and can be seen as a 3D version of a 2D screenshot. It can be used in every phase of product development to provide different prototypes that best fit the requirements of each stage.

As mentioned, the additive manufacturing machines used for the production of prototypes are called prototypers. Today they are often referred to as 3D printers. They are characterized by

- simple usability,
- fast disposability,
- low installation cost,
- low machine budget,
- acceptance of manual production steps,
- use of cheap and simple-to-clean materials, and
- the secondary relevance of the productivity (not of the build time!).

If the production of the prototype is not finished by the additive manufacturing process, but is built up on an additive produced model, the whole process is described as indirect rapid prototyping or indirect prototyping. This is not a separate additive process, but rather a sequence of nonadditive processing steps that supplement the additive process (see Fig. 1.7).

4.2 Strategic Aspects for the Use of Prototypes

Independent from their practical or operative meaning, the use of prototypes has a strategic dimension. Because prototypes, as for all other elements of the product design process, are seen mainly as cost factors in relation to a special product, all of the costs of the operative sector must be evaluated, especially by first-line management.

4.2.1 Product Development Steps

Prototypes are mainly used for industrial product development. Independent from industrial sectors or products, the German VDI 2221 shows a systematic process, which is also used for new approaches and company guidelines. Figure 4.1 shows the classic stages of product development by these guidelines. It is the basis for the next discussion.

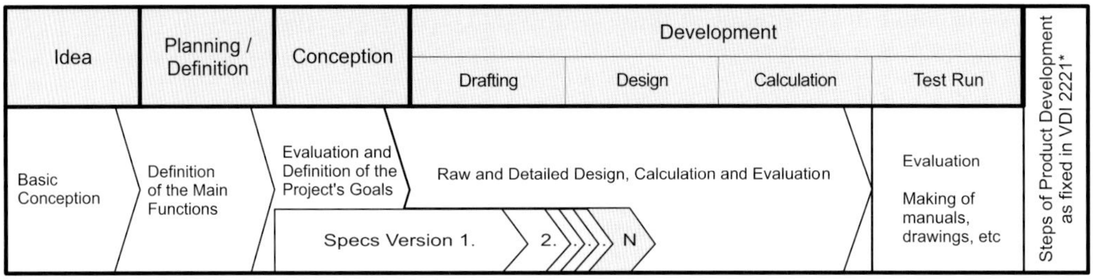

Figure 4.1 Definition of the product development steps by VDI 2221

4.2.2 Time to Market

This section demonstrates that all critical factors in the success of a product, but especially time and cost, can be condensed into one key element: the "time to market." This is the time that elapses between the decision to develop and produce a certain product and its introduction into the market.

Analysis shows that cash savings are greatest when the time to market is minimized. Figure 4.2 is based on a 1983 survey by McKinsey & Co. (US) [Reinertsen, 83]. It deals with the influence of various success factors on company profit against the background of various strategies. The disproportional influence of "time to market" on company profits is clearly apparent. For a product with less growth, a smaller fall in prices, and a longer life span, "time to market" loses its significance, while the importance of production costs increases. The business strategy therefore plays a central role when judging influences. The survey was conducted in 1983. Today, consumer goods with a life span of five years are considered durable products. As we have seen, product life spans have fallen dramatically, and the trend is continuing. This makes it clear that, for this reason alone, the time to market is becoming increasingly important for very nearly all products.

This example shows that if the budgeted development costs are exceeded by 50%, the budgeted profit is reduced by only 3.5%, whereas an overshoot of 9% in the budgeted production costs results in a reduction of 22% in profit. Exceeding the planned development time by six months with a resulting late entry into the market, however, will reduce the planned profit by a third.[2]

[2] The relationships are valid in analog form for other products with different ratios of product lifetime to product development time, which lead to different results. There are no detailed surveys as yet.

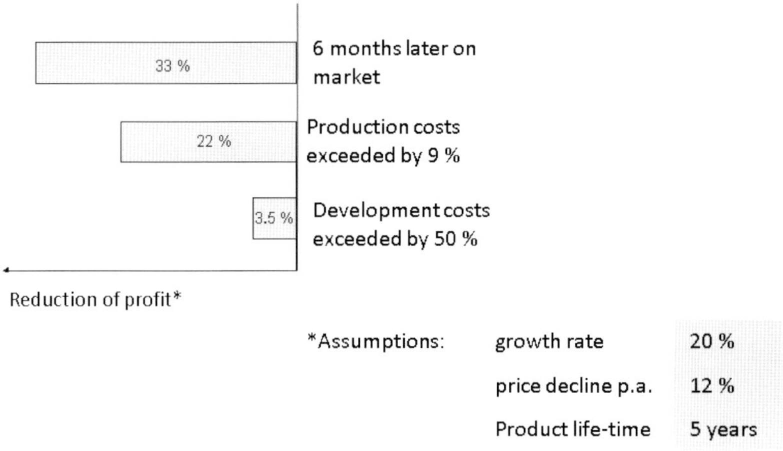

Figure 4.2 On the significance of "time to market": how missing the aim for various reasons influences the realizable profit

These results are confirmed by a survey made by Karjalainen and Tuomi [Karjalainen, 98] concerning the development of car tires. With a product life span of six years and also accounting for turnover decline caused by competition and declining appeal with growing product age, they calculate a 31 % reduction of profit, which they attribute solely to a six-month delay in entering the market.

This shows clearly that time to market is the key factor for the success of a product and becomes the most important criterion for management.

4.2.3 Front Loading

This dominance of time over money, typical for today's products, has not only an absolute but also a relative dimension. It is not solely a question of making the right decisions within a short period of product development; of equal importance is making those decisions as early as possible. In addition, it should be realized that although the accumulated expenditures for product development early in the process are still low, a large percentage of future costs is already determined over the course of the development. With the aid of the diagram published by Altmann, it can be shown that after the completion of the draft phase, only approximately 5 % of the total costs have been incurred, 75 % of the total costs have already been predetermined by the development (Fig. 4.3) [Altmann, 94].

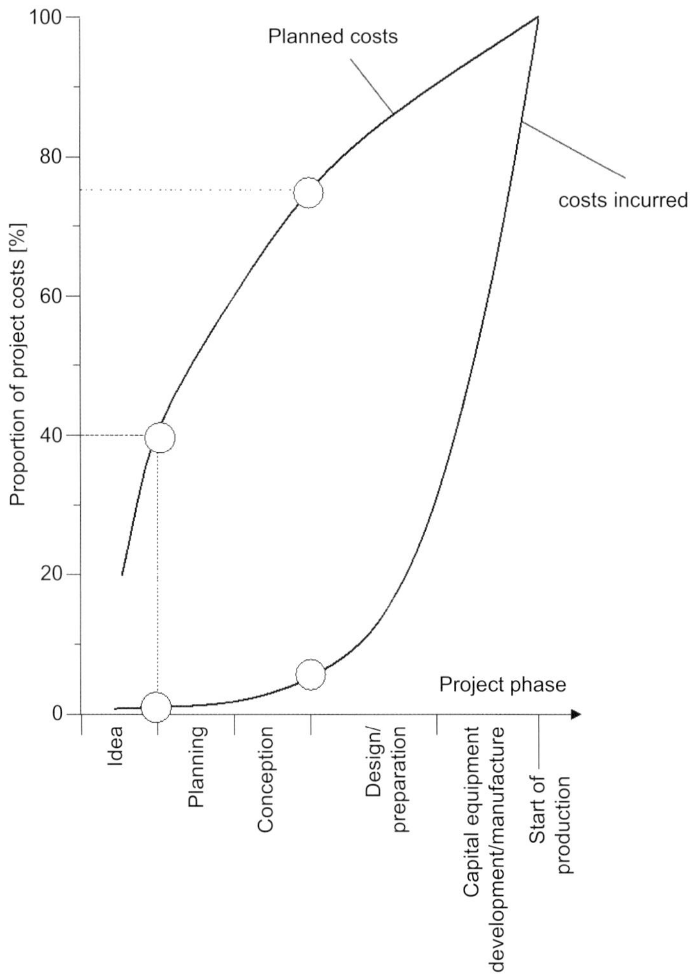

Figure 4.3 Comparison of total costs determined over the course of product development with the expenditures

Figure 4.3 also shows clearly that the relationships are even more dramatic at the commencement of product development: 40% of the total costs are already definitely fixed after the idea and draft phase, although at this point only negligible costs of 2% to 3% of the total costs have been incurred.

Engineers are often surprised to discover that it is important not only to make the right decision as early as possible, but also to make that decision final. The later changes are made, the more expensive they will be. Figure 4.4 shows that the costs for a certain late change to a product grow exponentially with the progress of product development. In a logarithmic scale this appears as a straight line. Here the same change is shown but at different stages of product development.

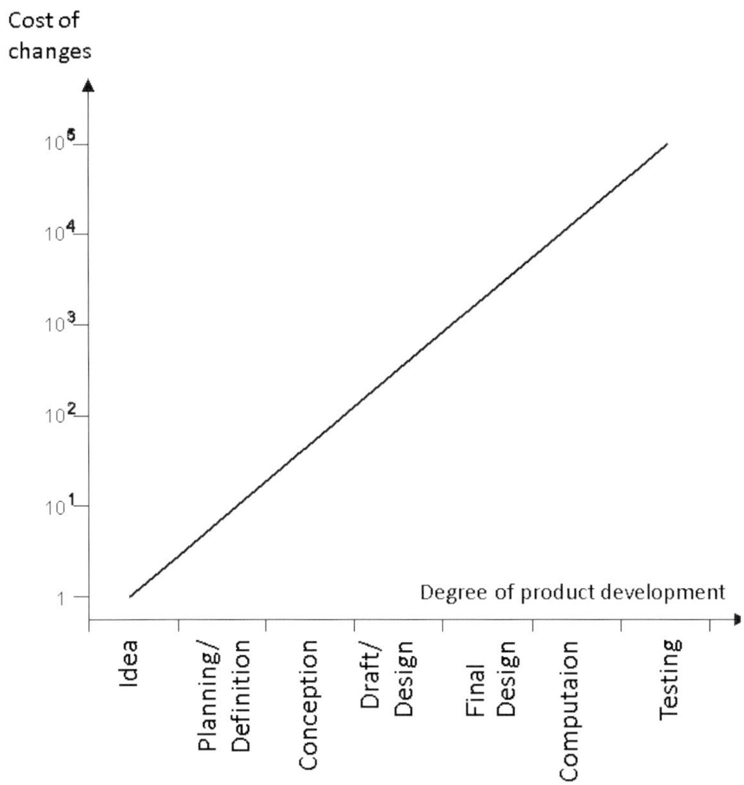

Figure 4.4 The costs for identical changes of a product at various phases of product development

It follows from this not only that changes to the design become more expensive and time-consuming the further the product development has progressed, but also that product deficiencies recognized too late can result in costs that threaten the viability of the entire project. Whoever doubts these findings should bear in mind that the cost for an alteration to a large machine tool can speedily reach several hundreds of thousands of dollars, and a recall of automobiles can easily top billions even though the defective part may be valued at mere cents.

To summarize, it follows that the minimizing of product development time is the key management objective, thereby enabling optimal total profit to be achieved and expenditure to be considered a time-dependent variable. When there is a focus on cost reduction, this is an important point, see Fig. 4.5.

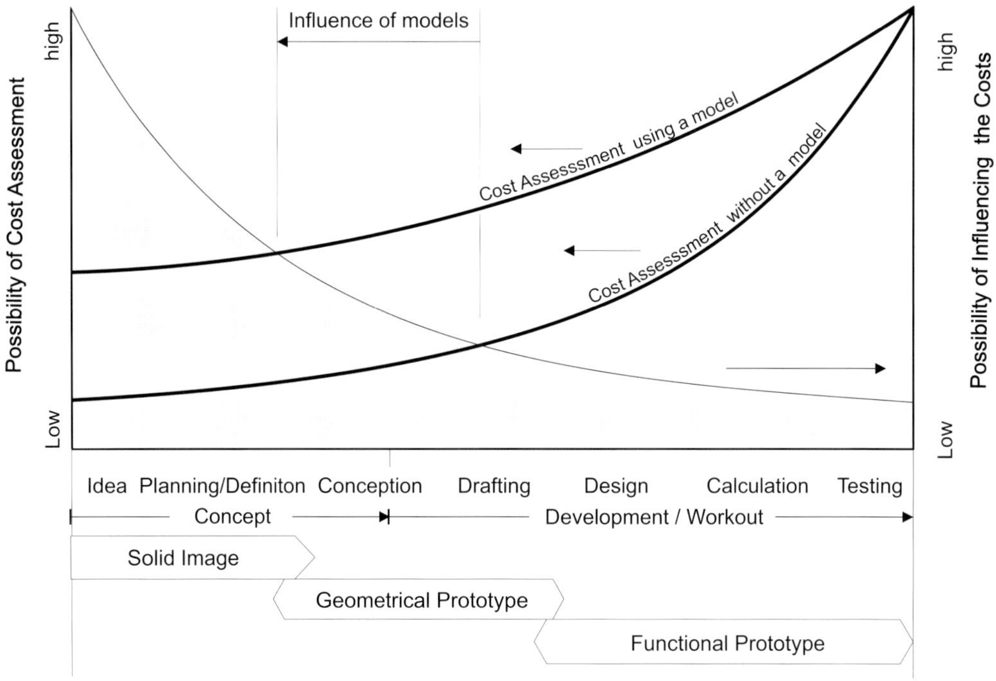

Figure 4.5 Evaluation of costs and the influence on them depending on the product development progress with and without the use of models

4.2.4 Digital Product Model

The discussed effects could be achieved with every kind of prototype. Additive manufacturing produced prototypes are even more appropriate because of the direct manufacturing from a digital data set, without the need for product-dependent tools. This leads to an always reachable, complete data chain (digital product model). With classic, partly manual model production steps, this chain was broken. Additive manufacturing production techniques have full access to the available data set at all times so that they can support the product development process.

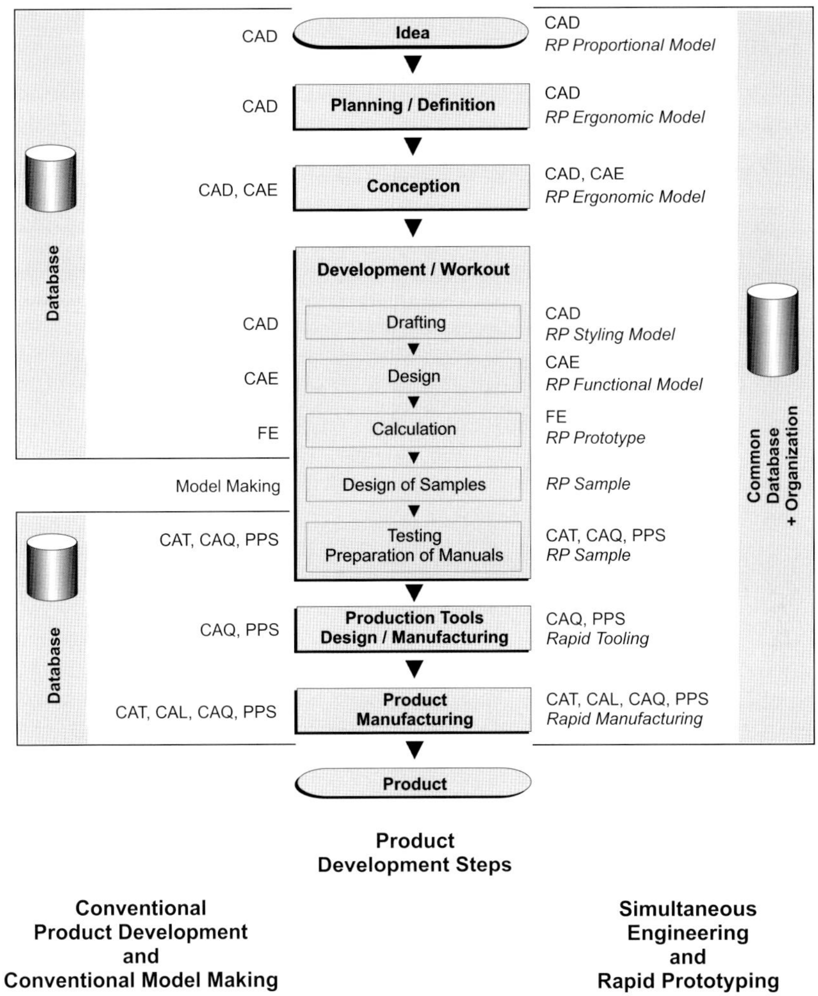

Figure 4.6 Partly closed classic CIM (computer integrated manufacturing) chain (left) and fully closed CIM chain by the use of rapid prototyping methods (right)

4.2.5 The Limits of Physical Modeling

With the continuing shortening of product development times, the need for shorter model build times also grows. Figure 4.7 shows an idealized trend of the readiness for the start of production for an arbitrary product. The start point is shown by curve 1. The end of product development is at point P1. Curve 2 describes a faster product development, which is already finished at point P2. If a prototype is produced in both cases at a readiness for production of 30 %, a model build time of t_M is necessary. At the same time, the development of the product goes on, but the construction could not be verified by testing results of the prototype, so the ongoing development during the manufacturing of the prototype is "blind" (t_B). If the model build times are the same, the blind part t_{B2} is much longer than t_{B1}. Thus it is unavoidable to cut off the build time directly proportional with the product development time. That is where fast and automatic model building techniques come into play.

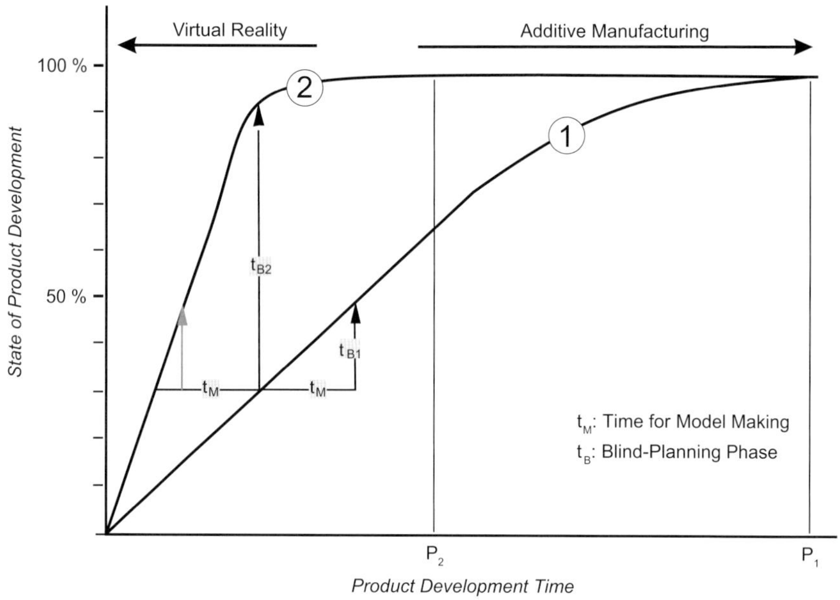

Figure 4.7 Influence of a shortened product development time on the blind planning times

4.2.6 Communication and Motivation

Independent from the specific strategy for the improvement of product development, the communication in the team that is developing complex products is getting more important, especially for decentralized working teams. Across cultural, language, and country borders, a picture truly says more than thousand words—and a model shows endless pictures.

If a circumstantial and long translation of a detailed product description can be replaced by sending a digital 3D model over continents by e-mail and this model can be realized as a physical 3D object on site, this is a useful contribution to fast and comprehensive communication. Additionally it is a contribution to product safety and cost savings.

Finally, the meaning of motivation is an often-underrated factor. Especially for interdisciplinary and international development teams, it is important that every member of the team has the product in the truest sense of the word in front of their eyes; this is also why everyone is assured of the successful conclusion of the project.

■ 4.3 Operational Aspects in the Use of Prototypes

The use of prototypes is an application at the operative level. It is necessary to produce the best prototypes by respecting technical and economic aspects, using the manufacture techniques for their production, and use prototypes independent from strategic considerations (not costs) to improve the product development process.

There are two main application fields for the use of rapid prototyping, the boundaries of which are blurred: the use of models to develop products and their use as 3D visualizations of objects, and respectively their data sets.

4.3.1 Rapid Prototyping as a Tool for Fast Product Development

In the pursuit of fast product development, the meaning of rapid prototyping is to produce different models by using additive manufacturing production techniques and use only the digital product model to ensure the actual development phase and make a decision about the ongoing project or to fix design decisions.

Figure 4.1 is based on the VDI Guideline 2221 [VDI 2221], which defines the product development steps: idea, planning/definition, conception, and development/realization. For every step of the product development cycle, models with different characteristics are used and named. For example, models used in the conception are concept models.

For rapid prototyping, it would be very useful to use authoritative names for different models to show how the requirements from the product development and the requirements of the additive manufacturing production could be best connected.

The key is to give the developer some data from which he can see which prototype would be the best fit for the actual product development step.

4.3.1.1 Models

In general language, physical parts are called either prototypes or models. Technically speaking, in the first part of the product development cycle, parts are mostly called models. As soon as they show mechanical-technological properties, they are referred to as prototypes.

If they are identical with the end product, they are called components or end products, even if they are produced in very small series or on an individual basis (see Chapter 6 on direct manufacturing).

4.3.1.2 Model Classes

In the relevant literature, a large number of terms and suggestions for model definitions can be found. They are often characterized by their intended use and by the specialties typical of certain branches.

It would be very useful to apply a very detailed definition as formulated by the German Association of Industrial Designers and Stylists (VDID) and the German Design Council (Rat für Formgebung). It differentiates six model classes: proportional model, ergonomic model, styling model, functional model, prototype, and sample, as shown in Table 4.1.

The definition covers all steps of the product development, including the manufacture of close-to-production prototypes. These are applications that are not included in the rapid prototyping discussed here. Prototypes and models in the case of the VDID are not prototypes in the sense of additive production techniques.

Table 4.1 Model Definition According to VDID

Proportional model	Shows the outer shape and the most important proportions. Facilitates communication and motivation, supports fast exchange of communication about the intended product properties, and enables a fast consensus on the product idea. The production process must be fast, simple, and cheap. Disposal and recycling are very important. Proportional models are often called “concept models” or “show-and-tell models.” Degree of abstraction: high; degree of detailed specification: low; functionalities: none
Ergonomic model	Supports fast decisions about feasibility (is it possible to develop this product and should it be done?). Shows important details for operation and use and also, if applicable, important partial functions. Degree of abstraction: medium; degree of detailed specification: medium; functionalities: some
Styling model	Shows the outer appearance as close as possible to the (series) sample. Surface finish needs to have “showroom” quality. Supports fast decisions on construction and manufacturing methods. Enables third parties (customers, sales, press, suppliers) to pass their judgments at an early stage, and enables public relations work. Degree of abstraction: low; degree of detailed specification: partially high; functionalities: some

Table 4.1 Model Definition According to VDID

Functional model	Enables the proof of the numerical simulation calculations and the early testing of certain functions (how it could be assembled, easy maintenance, kinematics). Shows some or all important functions if necessary, without showing the outer shape. Forms the basis for inquiries by customers and suppliers. Gives relevant information for tool and mold manufacturing, and for the construction and installation of the means of production. Degree of abstraction: low; degree of detailed specification: high; functionalities: several
Prototype	Resembles the (series) sample closely or, if necessary, exactly. Is produced according to production documents. The only difference from the series product lies in the production process. Enables the testing of a single or several product properties (how it can be assembled, ejectablility, start of special approval processes). Enables the production of tools (rapid tooling). Enables the preparation for market introduction by press campaigns. Degree of abstraction: none; degree of detailed specification: high; functionalities: all
Sample	Already produced in series, possibly a pilot batch, production batch, preproduction, or principal batch. Enables the entire testing of all product properties. Supports the training of production and maintenance personnel, supports the start of mass production, and enables the adjustment of the production and assembly sequence. Supports the detailed planning of customers and suppliers. Degree of abstraction: none; degree of detailed specification: high; functionalities: all

To complete the list, the prototype and sample classes were added, even if they are not model classes in the sense of rapid prototyping. To add a connection to additive manufacturing, the criteria degree of abstraction, degree of detailed specification, and functionalities were added.

The allocation by the VDID is very good for getting to know which kind of prototype is usable for a special application, which characteristics it needs to have, and which characteristics it does not.

In practical use, only three of the four model classes that are referred to in rapid prototyping are used. These are the classes for concept models, proportional models, and functional models. Technical prototypes are products in the sense of rapid prototyping.

There is also a definition in the German VDI 3405 guideline, additive manufacturing process, which is shown in Table 4.2.

Table 4.2 Model Definition According to VDI 3405

Concept model	The concept model is the first possible physical realization of a product design or product concept. Material, function, and dimensions are not as required by the product specification. Most important is the appearance of the product. Therefore scaled, additive manufactured parts (proportional models) could be used. Application: Verification of the aesthetic impression in the field of application
Geometric prototype	Geometric prototypes are additive manufactured parts that are used to evaluate the dimensions, form, and position. The material characteristics are not important. Application: Evaluation of the geometry (e. g., installation study)
Functional prototype	Functional prototypes are additive manufactured parts that should fulfill functions that are a component of the repetition part. An analysis of a single function or all functions are possible. The shape or figure could differ from the repetition part. Application: Evaluation of (part) functions
Technical prototype	Technical prototypes do not differ considerably in required functions from the repetition part, but they could be produced by a different manufacturing method. Application: Evaluation of parts in testing and preproduction
Product	Parts for a designated use, which are ready to ship. Application: small series, rapid manufacturing, individual products

The model definitions from Table 4.1 and Table 4.2 can be linked to the classification of additive manufacturing techniques given by Fig. 1.6. This is shown in Table 4.3. The definitions by VDI are presented in italics and are completed by short descriptions, to explain the assignment of the three model classes from VDID to only the two classes given by VDI 3405. An important, added term, which is not part of the definitions, is data control model. These are models that are used to show and control very complex 3D data sets.

Table 4.3 Link between the Model Definitions Given by VDID (Table 4.1) and VDI 3405 (Table 4.2) for Structuring the Additive Manufacturing Methods in Fig. 1.6

Model classes		Classification of the additive manufacturing processes	
VDID	**VDI 3405**		
Proportional model	Concept model	Solid imaging Concept modeling	Rapid prototyping
Ergonomic model	Design prototype		
Design model	Geometric prototype Data control model		
Functional model	Functional prototype	Functional prototyping Prototype tooling	
Prototype Sample	Technical prototype	Direct manufacturing Direct tooling	Rapid manufacturing
Product	Product		

4.3.1.3 Model Classes and Additive Processes

In Table 4.2, the definitions given by VDI 3405 were assigned to the different fields of application, which are useful in choosing the correct technique. To characterize the fields of application in the following articles, the requirements are discussed and weighted. These are summarized later on in Table 4.4. In Section 4.3.1.4, the fields of application are assigned to the additive manufacturing techniques.

Based on Table 4.2, there are three model classes: concept models, geometric models, and functional models.

Concept models are mostly used in the field of aesthetics or to give the impression of the outside appearance, the optic and haptic. Concept models are associated with the imagination of a solid image. They are used for communication (show and tell). They allow the visualization of the proportions and the basic outer appearance. Details, insofar as they do not represent proportions or the quality of the outer surface, have only minor meaning. The goal is the production of models that are simple, mechanically or technically not very resilient, but in the outer form and appearance, near the end product. Often model materials are used if they are cheap and fast to process. The degree of abstraction is high. Normally there are no functions. To produce the models as often as they are needed, a cheap price is important. Fast availability is necessary to reduce the blind planning time as shown in Fig. 4.7. Both mean that the priorities for the models are low investment and material costs and not the resilience of the models. Because of the fast availability of the models, but additionally because of the confidentiality of the parts, manufacturing in place is useful. Therefore there is a need for cheap and easy-to-use machines.

The sustainability is not important because of the short lifetimes. However, the safe disposal of waste is important, because of the rapidly growing number of prototypes. The reproducibility in the sense of a quality management system is not a focus.

Examples for concept models are shown in Fig. 4.8.

(a)

(c)

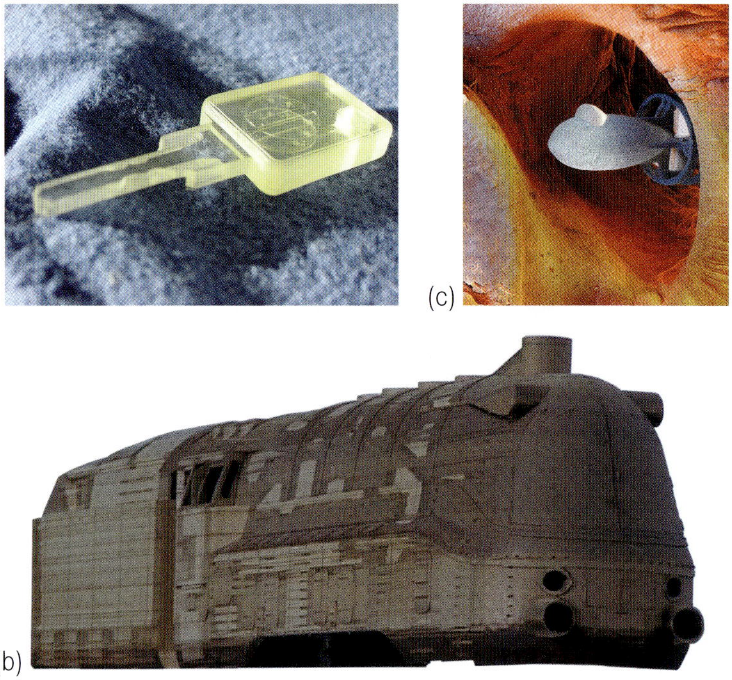

(b)

Figure 4.8 Examples for concept models[3]: (a) key model, layer laminate method (Solidimension/3D Systems); (b) locomotive model, layer laminate method (Kinergy); (c) submarine in blood vessel, microstereolithography (microTec)

Geometrical prototypes are used to control the geometric parameters from the data set to investigations of the fitting of parts.

Details are only important if they influence the parameters. The same applies to proportions, which are realized automatically by an exact geometry. The quality of surfaces is mostly not important. If possible there will be an abstraction in the meaning of effective model building. The mechanical-technical characteristics of the models are not important. Model materials are used if they are cheap and fast to process. The degree of abstraction is high, and functionality is normally not integrated.

Because geometry prototypes are not used as often as concept models, a cheap price is not as important. Fast availability is quite important to reduce the blind planning times. This makes a production in place useful. To realize this, cheap and simple-to-use machines are necessary.

The sustainability is not important because of the short lifetimes. Safe disposal of waste is also not as important because there are only a small number of prototypes.

[3] The choice was made only from a didactic point of view and does not imply any kind of valuation. It would be also possible to make another choice with a multiplicity of other geometries and methods.

The reproducibility in the sense of a quality management system is necessary for statements of assurance.

Examples of geometric prototypes are shown in Fig. 4.9.

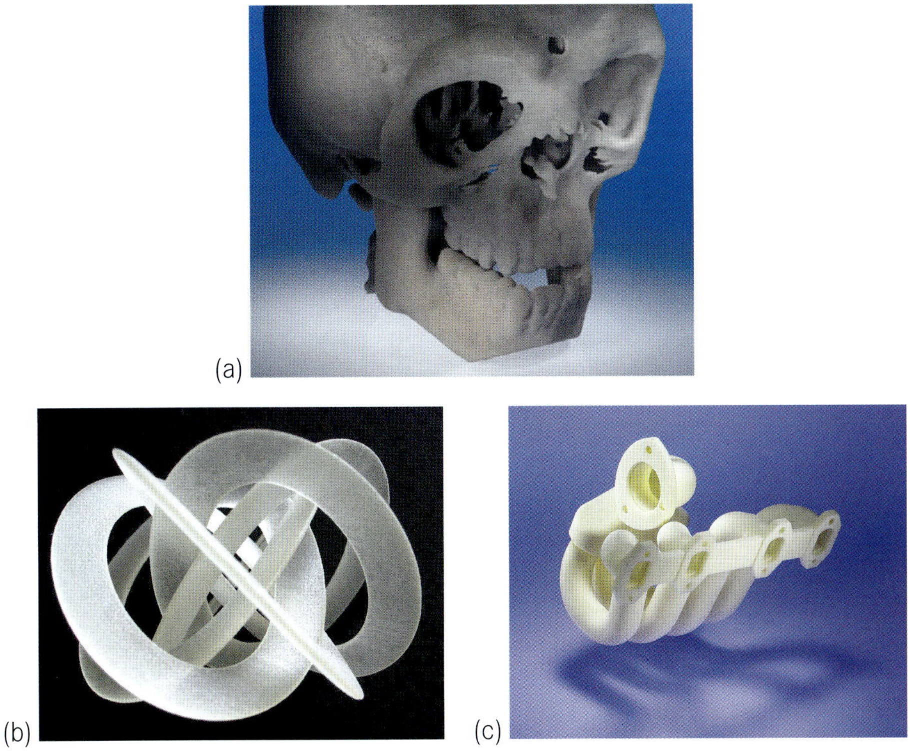

Figure 4.9 Examples for geometry prototypes: (a) Skull model for the planning of operations, 3D printing (CP-GmbH); (b) Moebius strip, polymer printing (Objet, Zedax, FH-Aachen/Schmidt); (c) Exhaust manifold, laser sintering (CP-GmbH)

Functional prototypes are used to evaluate one or more functions. Functions are defined as physical properties and could contain mechanical-technical characteristics (elongation, temperature resistance) such as color, texture, or geometric proportions. For aspects of the outer appearance, the optic and the haptic, this also means that the proportions, the details, and the surfaces are irrelevant if they do not influence the function.

Achieving models that have the same mechanical-technical properties as the end products is the goal of the production of functional prototypes. Model materials cannot be used, so construction materials are necessary.

Low price is not very important because the avoidance of costs for the production of tools and the small number of models in comparison with the concept models are the main reasons to use geometry prototypes. The rapid availability is in line with the production times for conventional tools manufacture of prototypes.

Because of the different machines that are necessary to process the different materials for all kinds of prototypes, in-house production is generally not recommended. Production is usually outsourced to specialists.

Sustainability and good disposal of waste are not key factors.

Examples of functional prototypes are shown in Fig. 4.10. The cola bottle neck is a functional prototype because it was used to develop a bottle collar that is compatible with most people's anatomy.

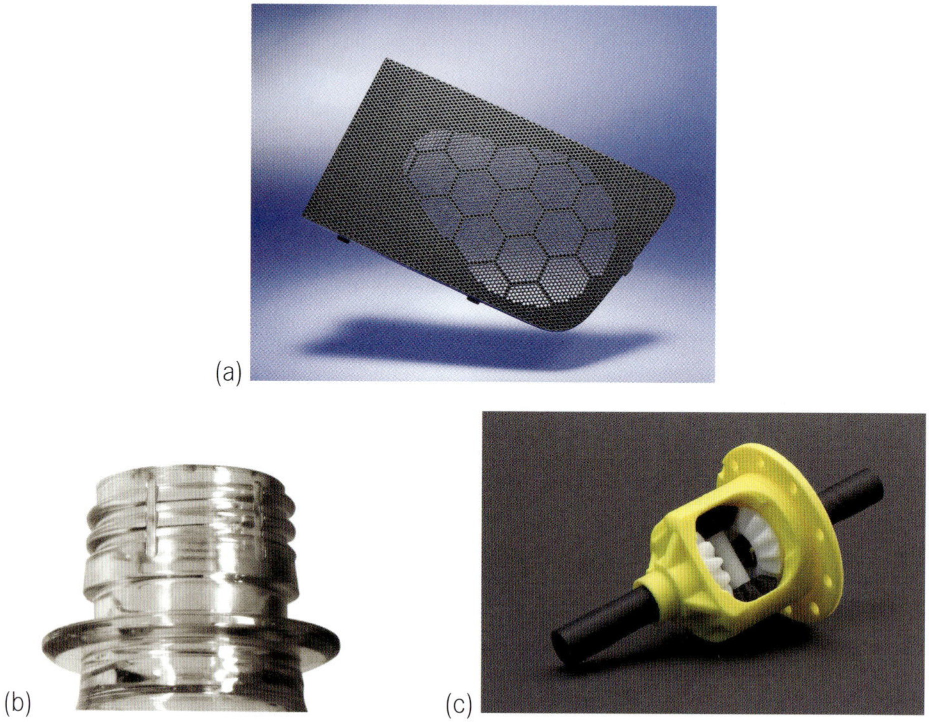

Figure 4.10 Examples of functional prototypes: (a) loudspeaker grid, stereolithography, painted (CP GmbH); (b) neck of a cola bottle, laser stereolithography (3D Systems); (c) gearbox, extrusion method, FDM (protozone-at)

Table 4.4 summarizes the given information and weights them.

Table 4.4 Requirements of the Model Classes by VDI 3405

Kind of model as defined by VDI 3405	Application: Evaluation of:	Proportions	Details	Surface	Mech.-technol. properties	Materials: ▪ Model material ▪ Construction material	Build time	Fast availability	Low cost	Sustainability	Recyclability	Reproduction
Concept model	outer appearance (3D image)	++	-	-	- -	M	+	++	++	- -	+	o
Geometry prototype	dimension, form, and position (data control, assembling)	+	++	-	- -	M	o	+	+	- -	o	++
Functional prototype	single or all functions	-	-	-	++	C	o	o	+	+	o	++

Legend: ++: very important +: important o: neutral -: less important - -: unimportant

4.3.1.4 Assignment of Model Classes and Model Properties to the Families of Additive Production Processes

For practical uses, the connection of model classes and model properties to additive manufacturing machines is very helpful.

The production of prototypes needs to respect the aspect that different models could be built with the same machine and that superior-grade models could be used for simpler investigations. But for economy reasons this should be avoided. Furthermore, the requirements for a model do not definitely always lead to a model class. Therefore, there must be a compromise between the connections of model classes to special machines on the one hand and a very rough connection of model classes to prototypers or fabricators on the other.

Such a compromise is given by a connection to the additive manufacturing families described in Chapter 3. The family of coating that includes melting by powder nozzle could not be used for prototyping because it works only with metals. For rapid prototyping these are relevant:

- Polymerization,
- Sintering/selective sintering: melting in a powder bed,
- Layer laminated manufacturing,
- Extrusion processes: fused layer modeling (FLM),
- Three-dimensional printing (3DP): powder-binder process.

As a summary of Chapter 3, the characteristic properties of the manufacturing families are shown in Table 4.5.

Table 4.5 Characteristics of the Five Usable Rapid Prototyping Additive Manufacturing Techniques

Application family	Details	Surfaces	Mech.-technol. properties	Materials ▪ Model ▪ Construction	Build time	Price	Sustainability	Recyclability	Reproduction
Index	1	1	1	–	2	2	1	1	2
Polymerization	++	++	+	C	+	++	–	–	++
Sintering	+	+	++	C	+	+	o	o	+
Layer laminate manufacturing	– –	o	–	M	+	o	–	–	–
Extrusion	–	–	+(+)	C	+	+	+	o	+
3D printing	o	o	– –	M	– –	– –	–	++	–

Legend:	++	+	o	–	– –
Index:					
1	very good	good	neutral	bad	very bad
2	very high	high	neutral	low	very low

From a comparison of Tables 4.4 and 4.5, the fields of application for the manufacturing families can be filtered. These refer only to the application level of rapid prototyping and models made from plastics or model materials. This further discussion is graphically summarized in Fig. 4.11.

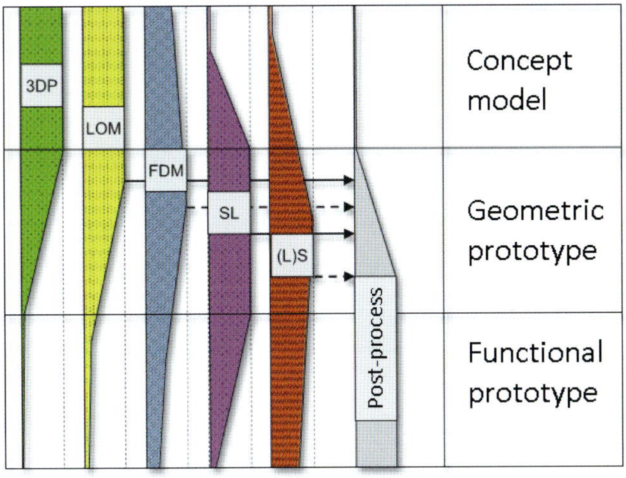

Figure 4.11 Assignment of the additive manufacturing families by Table 4.5, defined by the model requirements of Table 4.4, to the model definitions of Table 4.1 (VDID) and Table 4.2 (VDI 3405). This graphic can only be used for the application of rapid prototyping and for models made of synthetic or model materials

Figure 4.11 shows that the layer laminate manufacturing (LLM) process is very good for the production of concept prototypes. The 3D printing process is fast and cheap, but because of the necessary infiltration of the models it has a very high postprocessing cost, like the LLM process, because of the unpacking and fixation. This could be changed by using partly linking powders in the future. Extrusion processes (FDM, fused deposition modeling) require less effort, even if they do need to be washed. But they are more expensive than 3DP or LLM models. By getting more complex, they are more superior. Acceptable geometric details, especially functions, could only be achieved by using extrusion, polymerization, and sintering processes. The polymerization processes have the best surfaces, followed by the sintering processes. Because of the visible extrusion structure, which depends on the orientation of the model, the extrusion processes are not as good as the others. Good mechanical-technical properties are shown by the sintering processes, directly followed by the extrusion processes and, depending on the material used, the polymerization processes.

All discussions only relate to the model as it is produced by the additive process. In most cases, the requirements for the model are only reached by finishing it. This can be done by grinding, painting, coating, or something similar. Often a usable model is only reached by a postprocessing, which uses the additive model as a masterform (Section 3.10). This is why the postprocessing step is integrated into Fig. 4.11, at the point where a better result could not be expected by direct additive manufacturing. For layer laminate manufacturing and sintering processes, postprocesses are possible, but the manual cost to prepare the surface makes them economically useless for the most part.

4.3.2 Applications of Rapid Prototyping in Industrial Product Development

The use of prototypes for the optimization of the product development process will be shown by examples in the following chapters. The examples concentrate on the advantages resulting from the use of rapid prototyping. The various procedures are in the background. The techniques used or the amount they are used is not a criterion for a better or worse eligibility of a process.

The building times and costs are not shown because the basis of information and data on this topic were not enough to make serious specifications in this book.

4.3.2.1 Example: Housing of a Pump

To facilitate the feeding of two separate heat-circulation systems by one central boiler, a double-pump housing is necessary, which can be adapted with little effort to various boilers on the one hand, and on the other allows the accommodation of mass-production circulation pumps.

This part was constructed according to the general producer instructions with a 3D CAD model (IDEAS).

The 3D visualizing software enables all concerned to observe the first construction step and to give their approval.

The STL data were generated from 3D CAD data, and a stereolithographic model (Fig. 4.12, right) was produced. With the help of this model, the construction data were substantiated. Geometric and also forming technical assessments were necessary. Owing to higher mechanical and thermal strain, a nylon model was sintered for the flow studies (Fig. 4.12, left).

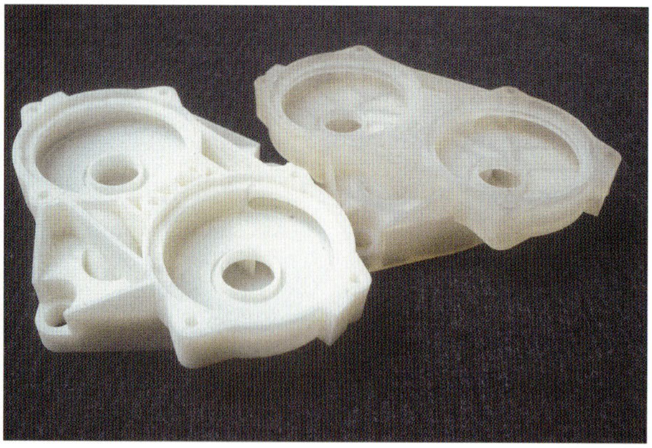

Figure 4.12 Double-pump housing. Stereolithography model (right), sinter model (nylon, left) (Source: M-Tec)

Because of the quick and reliable assessment by a 3D visualization and the stereolithography model, the production of prototype tools was not necessary. Therefore, the series tool was produced immediately. A special feature of the injection-molding technique is the use of melting cores. This technology enables the production of a one-piece pump housing and avoids potential leakages as well as “seams” in technically critical flow areas due to a mold-parting line.

Figure 4.13 shows the injection-molded part before the melting out of the cores and a separate core such as the one inserted into the injection-molding tool.

The cores themselves, which are made of low-melting-point alloys, have to be produced in a tool. Here too, rapid prototyping processes offer an alternative to traditional eroding and milling processes. The melting cores can be produced, for example, by a tool molded of high-temperature silicone based on a rapid prototyping master pattern.

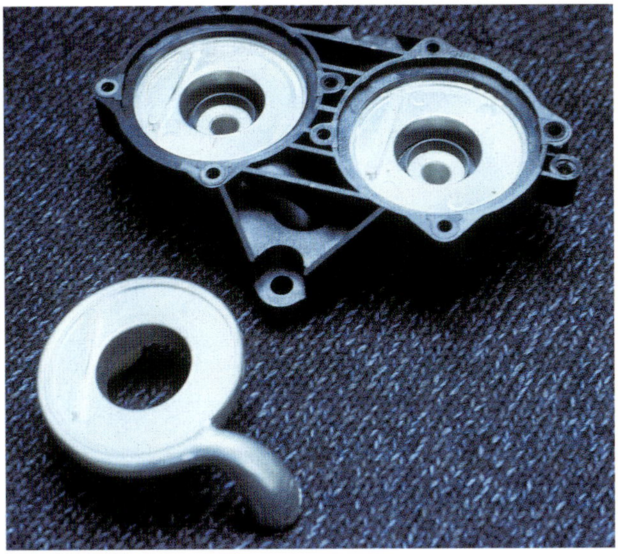

Figure 4.13 Injection-molded part with melting cores, melting core (front) (Source: IKM)

4.3.2.2 Example: Office Lamp

A producer of lamps for offices and shops asked his construction and production department to develop a new line of lamps. Prior to this, painstaking market research had been carried out. The decision for the development of a new line of lamps therefore came from the marketing department.

The development target was, apart from the customary light-technical, electrotechnical, and production-technical requirements, to produce a distinctive but still rather conservative design that would integrate itself into business surroundings.

In addition to various pendant versions that could be combined, a wall version and a floor version were required. A further requirement was the technically optimized production of this line of lamps.

In view of the restrictive pricing requirements for the final product, a feasible solution was quickly found. It was based on a simple basic concept using an economical production technique in which variations could be realized by individualizing the basic concept.

Based on this idea by the industrial designer concept (TRENO, Finland), which stems from the renowned German Rail Zeppelin concept (Fig. 4.14), a four-part basic concept was developed that consists of three different elements. An elongated lamp body, which was already in the prototype phase and constructed from a continuous casting aluminum profile, forms the actual lamp together with two cap elements at the head ends. A suitable mounting element allows the alternative designs for a floor lamp or a pendant lamp. The subsequent drafts assist in forming a consensus concerning the basic design of the central part and the head ends of the basic lamp. Based on the design drawing, a support element for the floor version was favored (Fig. 4.15, left).

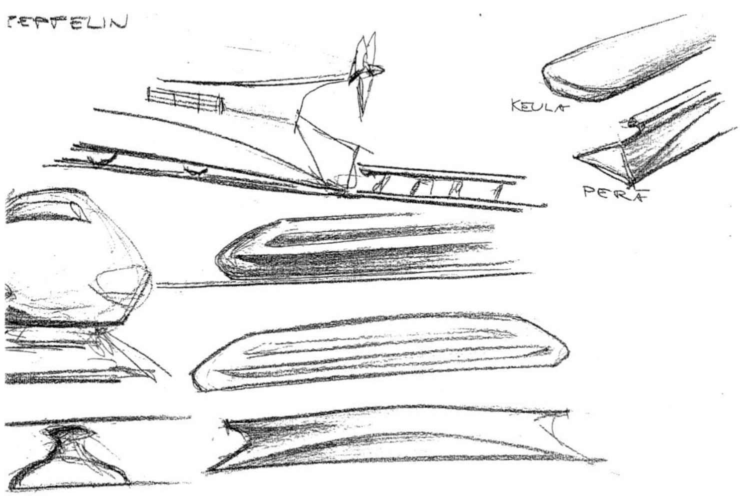

Figure 4.14 Design draft "Rail Zeppelin" (Sketch: TRENO)

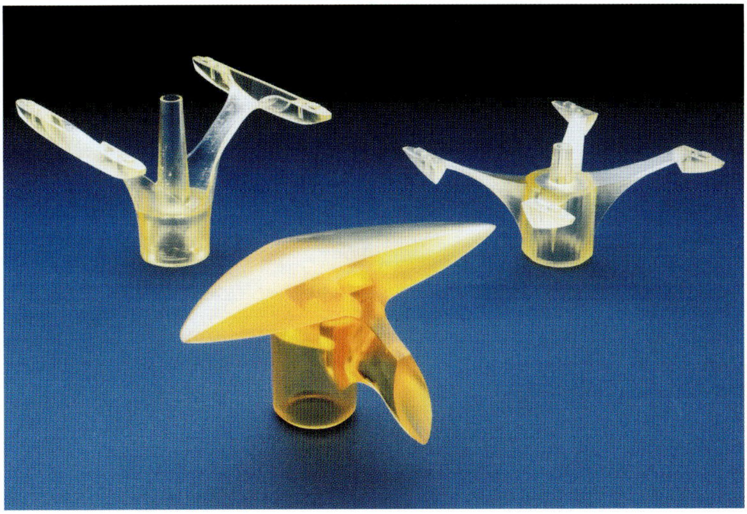

Figure 4.15 Support elements, floor version (Source: THORN)

To promote further product development, it was decided to produce prototypes for the whole lamp, and especially of the head ends and the support elements. As the company had its own conventional model-making department, it would have seemed understandable to employ a manual or semiautomatic conventional production. For time reasons, however, it was decided to use the then-new technology of rapid prototyping, specifically laser stereolithography. The fact that the extremely intricate freeformed area of the draft would have been very difficult to produce by conventional means made the decision for rapid prototyping technology easier. Here, laser stereolithography in particular promised a fast and geometrically exact solution. The costs of making the models was taken into consideration only insofar as it was not allowed to be much higher than that of a conventional production. The costs had to include the 3D CAD representation, which was not necessary for conventional production. The crucial factors for the decision were, however, the complexity of geometry, the speed of model making, and the possibility to test this promising new technology extensively while bearing the company's own product requirements in mind.

During the first project phase, it was already decided not to produce the entire lamp as a stereolithography model but to produce each of the main elements by the specially suited process and to assemble the lamp from these parts. The central element of the lamp was made of a continuous casting aluminum profile, as already mentioned, while the two head elements and the support element were made as stereolithography parts. The stereolithography parts were meant to be used as a sample. For further product development it was decided to use them as master models for second vacuum casts. The vacuum casting became necessary owing to the number of pieces required but mainly because the stereolithography material showed a different translucency than the series material.

The head pieces immediately met with general approval and were altered only insignificantly later. The stereolithography model and the corresponding vacuum cast are shown in Fig. 4.16.

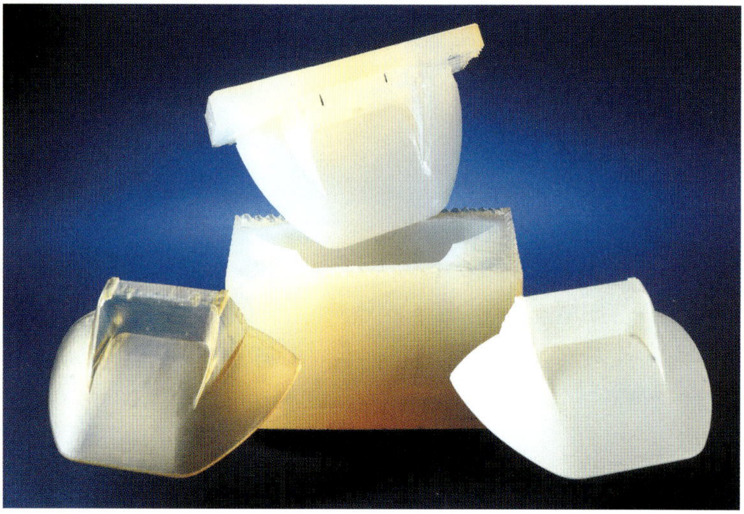

Figure 4.16 Head pieces. Stereolithography model (left), vacuum casting mold (middle), vacuum second casting (right) (Source: THORN)

In contrast, the authentic scale production of the mounting elements (Fig. 4.15, left) showed that this element was too delicate in comparison to the entire construction and optically too unbalanced. Although the second version of the mounting element met with approval when presented as an alternative draft (Fig. 4.15, right), it was later discarded after having been produced and duly assessed as a 3D stereolithography model.

Only the third variant (Fig. 4.15, middle) met with general approval, and was taken into series production. A final judgment was first possible after this element, which was difficult to represent in a drawing, was actually presented as a physical model. In such a complex construction dominated by freeform areas, which also had to accommodate the internal cable, the advantages of rapid prototyping processes became obvious, including, in this specific case, the ability of stereolithography to depict freeform areas very accurately.

All in all, the implementation of stereolithography enabled an earlier and more reliable appraisal of the product and its basic elements and thereby a reduction in product development time. The multinational dialogue, which in international companies is at times inhibited, was also positively influenced. The product could also be presented in its final form at important trade fairs in the form of stereolithography models and their subsequent vacuum casts. Figure 4.17 shows the series product (head piece). The pendant version especially is clearly reminiscent of the German "Rail Zeppelin" (Fig. 4.18).

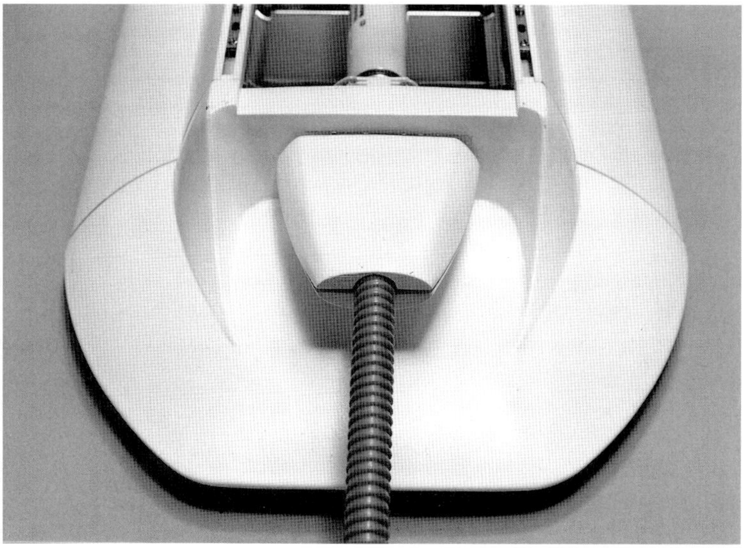

Figure 4.17 Series lamp, head piece (Source: THORN)

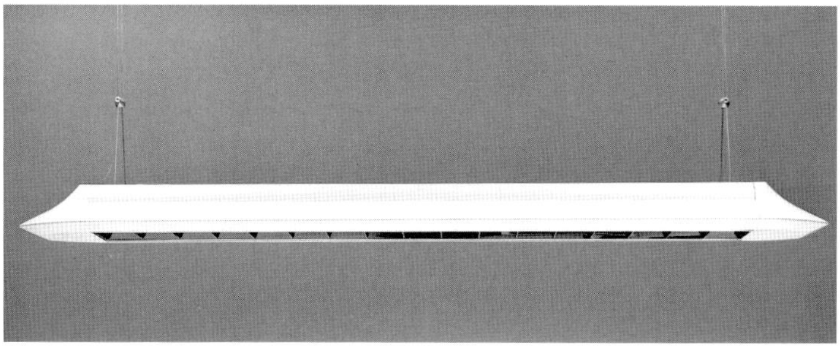

Figure 4.18 Series lamp (Source: THORN)

4.3.2.3 Example: Recessed Lighting Socket

While developing a three-piece casing for a plug-in halogen luminous element, a functional prototype was required that proved the accuracy of the fit of the multi-panel box socket, adaption ring, and socket case to be monitored. As required of the series product, the functional model had to be relatively stable under high temperature to enable the functioning of protruding brackets and the fine guidances especially to be tested. The actual surface finish was of secondary importance as long as it did not impair the function. Therefore it was decided to sinter this part of nylon. Figure 4.19 shows the sintered piece and a series piece that was checked against the sintered piece and the production technically derived from the sintered piece. The final part is produced in series as an injection-molded piece. The rapid prototyping model allowed the piece to be optimized by using the original luminous elements before an injection-molding die was made.

Figure 4.19 Lamp socket prototype (three-piece, left), injection mold (right) (Source: CP)

4.3.2.4 Example: Model Digger Arm

To clean and empty hot-metal ladles, pouring ladles, and converters of heavy slag conglomerates ("skulls"), special digger arms fitted with a ripper tooth are mounted onto standard digger undercarriages. The basic construction and the functioning of the special digger arm are shown in Fig. 4.20.

The producer of these special digger arms required a standing model to a scale of 1:50 for advertising purposes and also a fully functional model with remote control to a scale of 1:20 for trade fairs and exhibitions.

Because authentic scale models of diggers are available in shops, the development focused on the production of matching digger arm models. The smaller arm was generated with the aid of stereolithography because this demanded detailed accuracy, and the large arm was sintered because this demanded mechanical resilience.

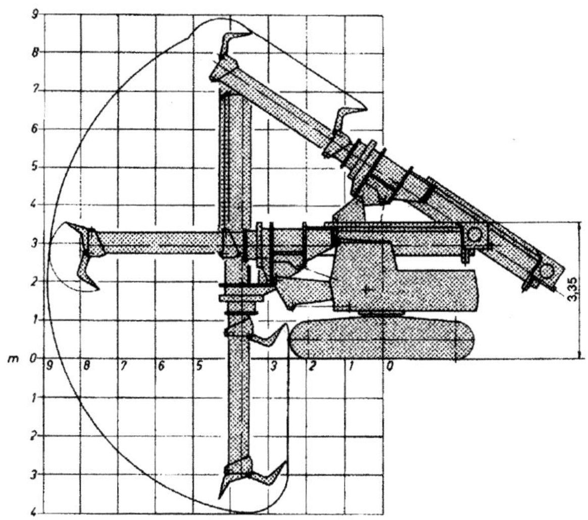

Figure 4.20 Special digger arm, kinematics, and work area

The main challenge lay in reducing the scale, which only superficially resembles the reduction of all geometrical measurements by a certain factor. In transferring the characteristics of the original to the model, details such as screwed connections need to be significantly larger in scale than in a linear reduction. The aim of using a model for design is to discern a high degree of detail and to be able to perform all important kinematic functions of the original digger arm. This demands minimum requirements concerning functional areas and wall thicknesses that differ from the geometrical reduction and need to be correlated with the above-mentioned optical requirements. The requirements on the model grow to the extent to which assembly and production requirements for a subsequent production of small series–from several hundred up to 1000 pieces–in metal or plastic are further integrated into the model.

Because the original construction was not based on a 3D drawing, the model making of the 3D CAD model was the first step. Figure 4.21 shows the 3D CAD model in the assembled form and an exploded view of the two main elements of the digger arm: the external structural components and the internal structural components of the digger arm. The two exploded views together show every single part that is later generated by rapid prototyping processes.

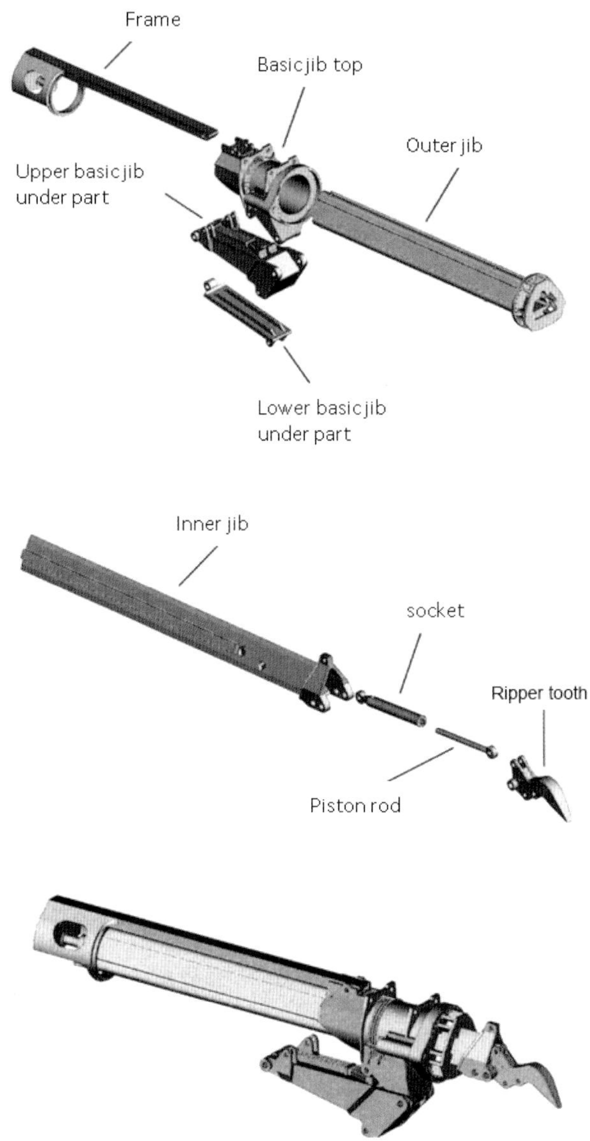

Figure 4.21 Special digger arm, external and internal structural components, assembly drawing (seen from above) (Source: van Crüchten/FH Aachen, Germany)

Figure 4.22 shows the ripper tooth as a stereolithography model, Fig. 4.23 as a sinter model.

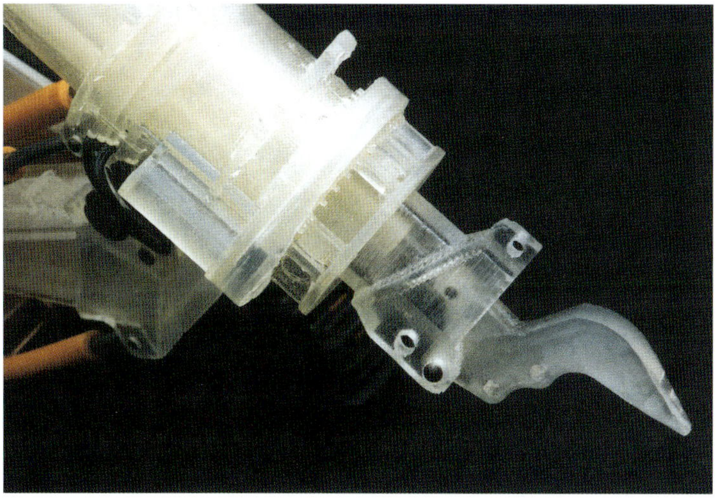

Figure 4.22 Ripper tooth, stereolithography model
(Source: van Crüchten/FH Aachen, Germany)

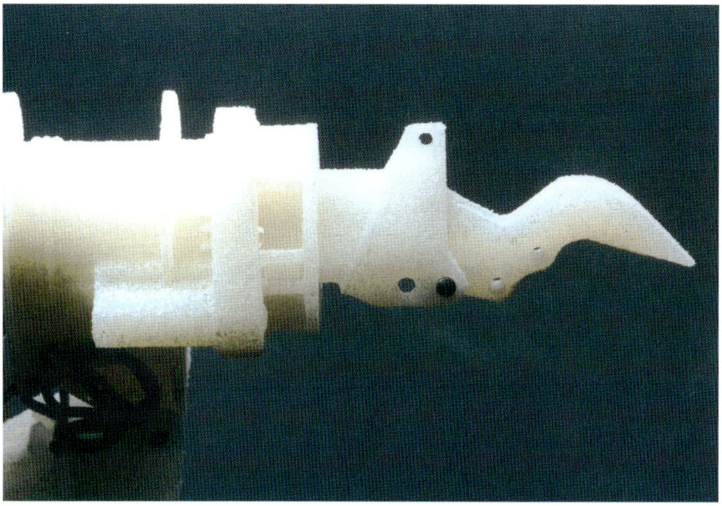

Figure 4.23 Ripper tooth, sinter model (nylon)
(Source: van Crüchten/FH Aachen, Germany)

Figure 4.24 shows the finished stereolithography model in the assembled form and after having been mounted onto a typical toy digger undercarriage to scales 1:20 and 1:50.

Rapid prototyping processes allow locally different scales to be optimized and thus achieve the overall optical impression of series quality. The desired miniseries is not economically feasible owing to tooling costs. This perspective is opened up by combining rapid prototyping processes with molding processes.

Figure 4.24 Digger model to scales 1:20 and 1:50
(Source: van Crüchten/FH Aachen, Germany)

4.3.2.5 Example: LCD Projector

LCD projectors must become more and more compact while their light-transmitting capacity must increase. In addition, the demand for a functional, up-to-date, and representative design is growing. These development targets sometimes work in opposite directions, and an intensive coordination between designers and technical engineers is necessary. The model shown (Fig. 4.25) has a very compact construction inasmuch as it can be folded for transport.

This reveals two clear demands on the model maker:

- The design, the geometry, the kinematics, the fitting conditions, and the principal function must be monitored.
- The thermal resilience of the casing must be monitored.

The visualization of the design, its external form, the kinematics, and thereby the principal function make great demands on the quality of the surface, the reproduction of details, and the fitting qualities of the model parts. Thermal and mechanical strains are quite slight in this phase. For this reason, a multipiece stereolithography model was built initially (Fig. 4.25).

It is not possible to monitor temperature changes, which also involves checking the necessary cooling systems for continuous running, with a stereolithography model because continuous running temperatures of over 60 °C already pose a problem. The geometrical form, therefore, was molded by vacuum casting (Fig. 4.25). The required temperature stability of 130 °C, however, creates difficulties with most pourable resins. For this reason, in this case, a suitable temperature-stable pourable resin was used.

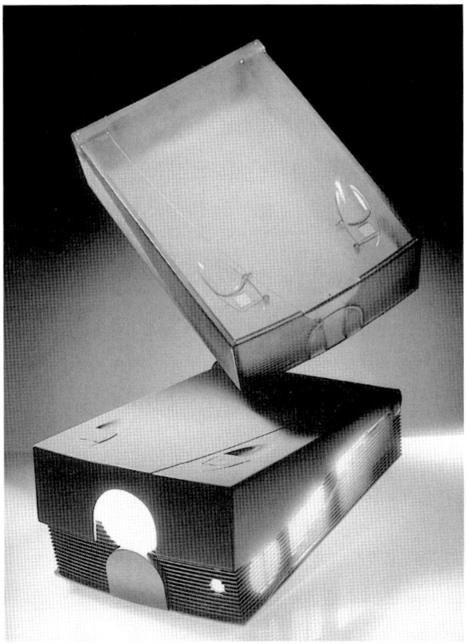

Figure 4.25
LCD projector casing, stereolithography model (above) and vacuum casting (below) (Source: CP)

A further advantage of vacuum casting is that in this way several casings can be produced in spite of the large and complicated contour, owing to many fine cooling ribs. At the same time, the castings aided the production of series injection molds that was taking place parallel to the prototyping.

The extremely compact construction of the device together with the mechanism that puts the lens into the projection position when the casing is opened is a great construction challenge both mechanically and—because of the high temperature—thermally. For such complex flow patterns, no reliable theoretical calculation methods exist as yet. The necessary experimental monitoring of the construction was made possible in a fast and reliable way with the aid of a series-similar prototype generated by stereolithography and vacuum casting long before a series tool was produced.

4.3.2.6 Example: Capillary Bottom for Flower Pots

Helped by a prototype, a plastic tray is to be tested that waters flower pots automatically by using capillary forces. The plastic tray used for larger flower pots is nearly square and has an integrated folding reservoir that is connected to the main part by a film hinge.

It was decided to produce prototypes by means of selective laser sintering and to test the functioning of the film hinge and other mechanical-technical properties; the functioning of the parts produced by the injection molds under authentic usage conditions was simulated (Fig. 4.26). The fact that the polyamide used is hygroscopic and that the part would therefore change to a small extent when used was not considered as a disadvantage for its intended purpose in these tests because the functionality was not impeded by such tiny tolerances.

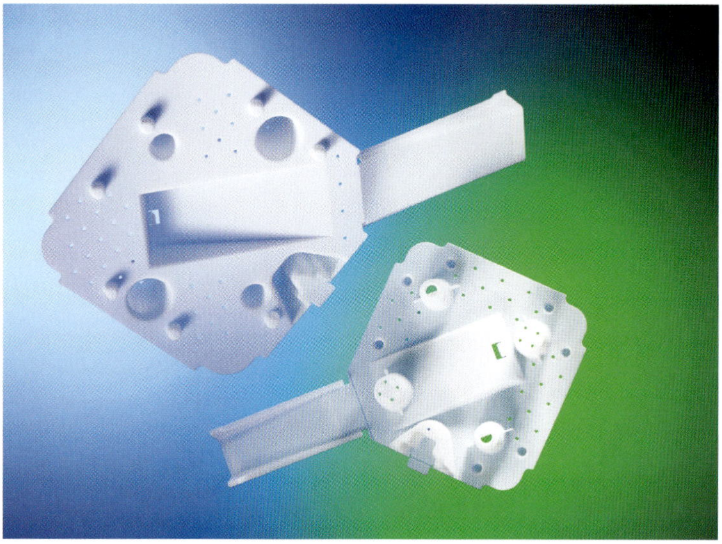

Figure 4.26 Capillary bottom for flower pots produced by selective laser sintering (Source: CP)

Directly after completion of the 3D construction, rapid prototyping made it possible to test all relevant characteristics of a functional model under the same conditions as those under which the series parts would be used. This was the basis on which the construction was optimized before tools were produced.

4.3.2.7 Example: Casing for a Coffeemaker

For a coffee maker, the models had to be produced in various colors, the piece parts had to fit perfectly, and high demands were made on the quality of the surfaces. Therefore, stereolithography was chosen as a rapid prototyping process. To realize various colors and to be able to produce several prototypes, the stereolithography master models were molded by vacuum casting.

With rapid prototyping it was possible to cast casings in various colors, some of them even transparent, without having to produce tools at this very early stage of product generation. The use of series parts such as glass jugs, turning knobs, lamps, and so forth gave the impression of this model being a series product long before the decision to proceed with series production was made.

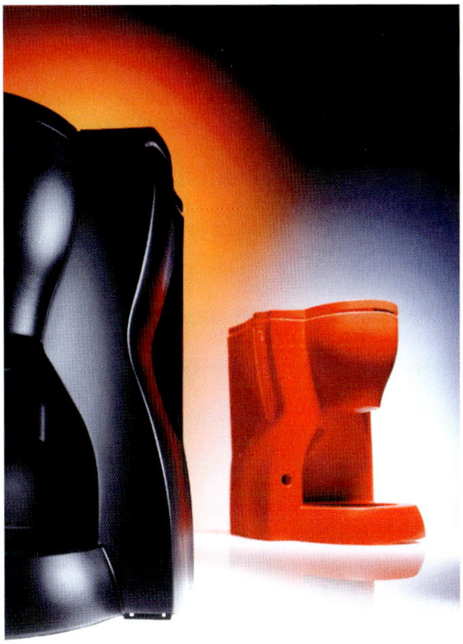

Figure 4.27
Casing of a coffeemaker produced by vacuum casting (Source: CP)

4.3.2.8 Example: Intake Manifold of a Four-Cylinder Engine

Induction pipes for internal combustion engines are becoming increasingly more complex because adjustable elements are being built in to ensure the continuous adjustment of the flow conditions to the engine characteristics. In addition, a number of sensors and actuators are built into the induction pipe. Aerodynamic tests are inadequate even with numerical 3D calculating methods primarily because of the unsteady flow. After a theoretical preoptimization, therefore, it is still indispensable to run a test on the engine test bench.

Mounted on the vibrating engine, the demands made on the mechanical and to a lesser degree on the thermal resilience of the prototype are very high; the active baffles must function, and the sensors and actuators must be introduced. For these reasons, therefore, a sinter component made of polyamide is usually chosen, especially in cases where the later series component is produced by plastic injection molding. For the suction pipe shown (Fig. 4.28), an FDM component made of ABS plastic would also have been a good alternative.

With rapid prototyping, a usable functional model is available very shortly after the 3D construction is finished. Geometrical alterations, which may become necessary after the test run on the engine test bench, can be effected in an optimized prototype comparatively quickly and economically. Another advantage is that alterations to the theoretical calculation program and the resulting practical consequences can be correlated directly on the engine test bench and are thus the basis for an optimization strategy.

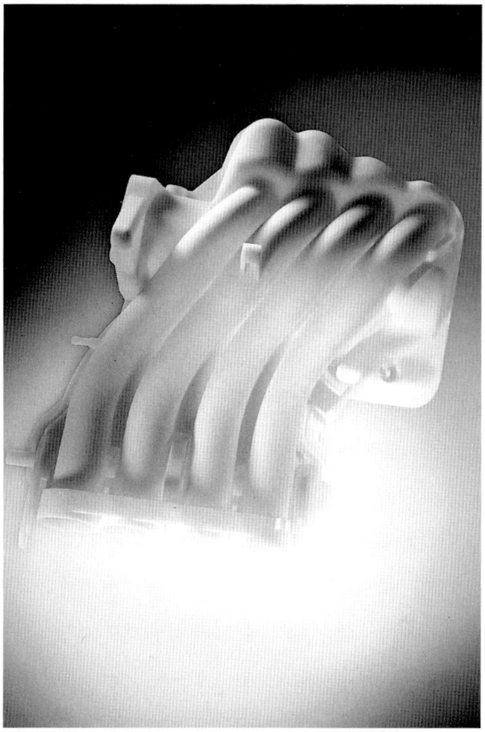

Figure 4.28 Induction pipe of a four-cylinder engine made by selective laser sintering (Source: CP)

4.3.2.9 Example: Cocktail Glass

The design of a new kind of cocktail glass, from which multiple persons can drink their cocktail at the same time by the use of straws, should inspire a new kind of lifestyle. The glass is defined by freeform structures and large transparent components, as well as integrated lighting that is used to stimulate with an animated color-changing play of light. To cover the optical impression of the transparent and partly transparent components, as well as the lighting in the bottom surface, a prototype was required that corresponds to the series in the appearance. The structured walls of the container and the freeform base were best described by a stereolithography model. The components were produced by highly transparent PU (polyurethane) vacuum casting, the base was metallized, the lighting installed, and the bowl assembled.

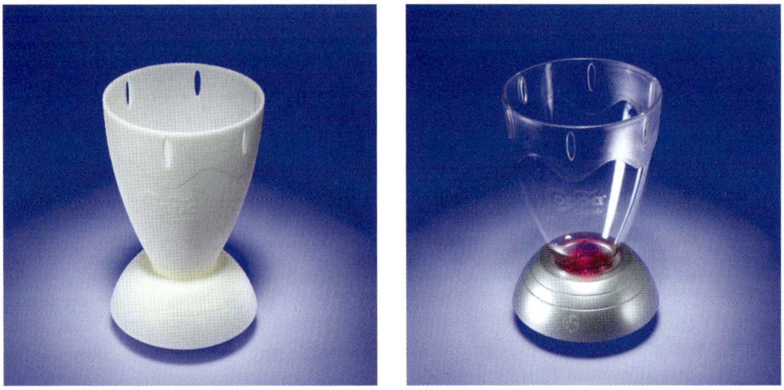

Figure 4.29 Cocktail glass; master model, laser stereolithography (left); prototype, transparent vacuum cast (right) (Source: Pfefferkorn/Toorank/CP-GmbH)

4.3.2.10 Example: Mirror Triangle

The growing integration of functions into car doors increases the requirements for the so-called the side mirror triangle, see Fig. 4.30. This rubbery-elastic triangle is a complex, three-dimensional assembly basis, lead-through for electrical wiring, and must seal. The flexible materials used today for additive manufacturing do not meet the required tests. This is the case for the Shore hardness, especially in dependence on the work temperature, but also for the reset forces or hysteresis in the case of pressure release. Because of that, a stereolithography model was produced and then cast.

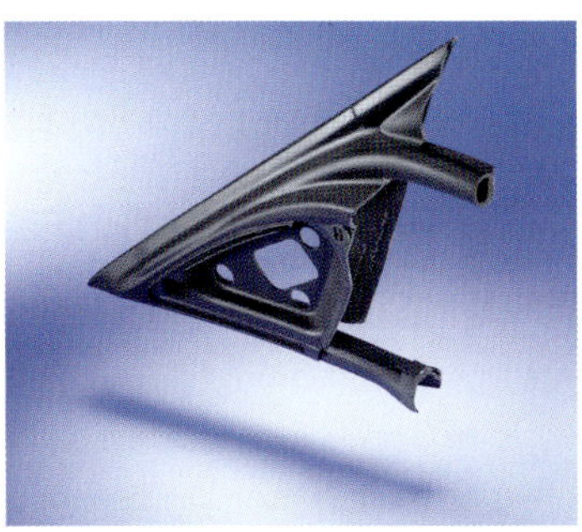

Figure 4.30
Mirror triangle, vacuum cast of the master form (rubbery elastic PU)

4.3.2.11 Example: Convertible Top

The demands on the convertible tops of cars are constantly increasing.

The user wants a good-looking convertible in the closed and opened positions with a top that is easy to handle and fast to open and close. It should use only a small amount of storage room in the closed position. The arrangement of levers should be invisible, except as a part of the design.

For the manufacturer, this results in more and more complex and intricate constructions, which need to be electrically driven and secured by a mechanically activated mechanism.

The construction is complicated by the elasto-mechanical characteristics of the convertible cloth, the elastic behavior of the preferably thin rod construction, the homogeneous force transmission of the drive-unit into the rod construction, and the higher number of parts, which are connected by joints, which lead to a higher bearing clearance and friction.

Additionally, the convertible top is in an intense interdependency with the car body, especially at the frame of the windshield and the freestanding side windows. The necessary closing forces need to be differentiated from the requirement of easy handling.

To fulfill this combination of requirements with coherent goals, the designers still use experiments to evaluate their constructions.

It would be too late in the process if the experiments began with a close-to-series prototype. That is why the experiments have to begin at an earlier time. To realize this, concepts of the design are necessary. These concepts need to be stronger the earlier the experiments start.

In practice, the following steps are proven:

1. The coordination of the conceptual approach and the design-technical solution by a concept model.
2. The evaluation of the data set by a data control model.
3. The validation of the kinematics by a functional model.
4. Optionally the detailed evaluation by additional defined models.
5. The production and test of a close-to-series prototype and the approval of production.

In the eyes of a prototype manufacturer, points 1–3 and sometimes point 4 are relevant. To build these prototypes, additive technologies are used. By the use of additive techniques, it is possible to build concept and functional models at a very early stage of the product design process, which helps to accelerate the product development process. They also improve the quality, because they help to identify risks early and eliminate the source.

For point 5, the prototypes should be of a quality that is close to series. The connected operations take place at a late step of the product development and were completely at the manufacturer's liability.

Concept and data control model. The first step (point 1) is the coordination of the creative and construction approach and the validation of the data basis.

For optimal communication of the developer team between the designers, who are responsible for the outside appearance, and the suppliers, as well as the manufacturer, a three-dimensional, physical model is better than a description or a screenshot. The model shows all elements and the results of their interaction, in addition, it gives an impression of the whole construction, even so it is scaled, as shown by Fig. 4.31.

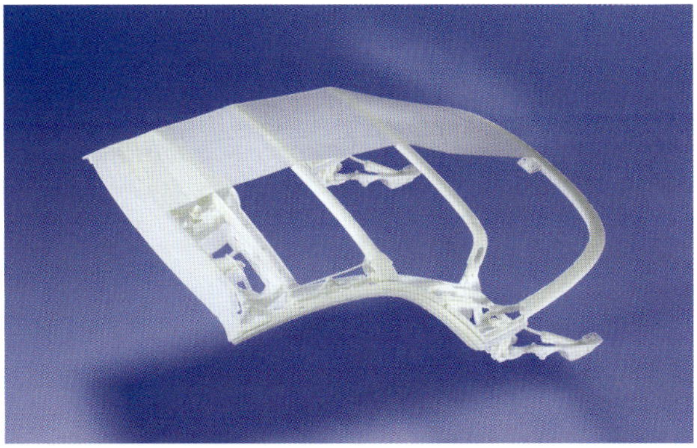

Figure 4.31 Data control model of a convertible top. One-half of the model is covered and the other not covered; scale 1 : 50 (Source: Karmann/CP-GmbH)

These models consist of two half models, as shown by Fig. 4.31. One shows the closed top stringed and the other half shows the not-stringed rod construction, so the whole construction could be quickly recognized and discussed.

The sintering process was preferred over the stereolithography process because of the stronger models. Except for the intricate structures of the convertible top, it is possible to pass it from hand to hand without damaging it.

By stereolithography it would be possible to show finer details, but the model would be more sensitive. Additionally, it would be quite difficult to remove the necessary support structures from the fine rods of the construction.

To validate the concept, 3D CAD models were used to develop simulations.

Functional model. The kinematics of the very intricate and complex construction of the rods could not be shown by the concept model. Even the simulations of the top could take the real elastics, clearances, and frictions only partly into account, so an exact simulation is not possible. In particular, they transfer no sense for the opening and closing process of the construction, which is needed to improve trust in the construction.

This is why the decision to build a 1 : 1 model by additive manufacturing, assemble it, and investigate it manually was made. All team members kept in mind that the polyamides used have a different stiffness and weight than the later-used product materials. But the main reason to build this model was to get a feeling for the construction and to "sit in the construction" (Figs. 4.32 and 4.33). This further development gave the assurance that the concept is correct.

Also for the functional model, to produce all of its parts, the laser sintering process was used.

Figure 4.32 shows the closed top rods (scale 1 : 1), Fig. 4.33 the half-open top in a side view.

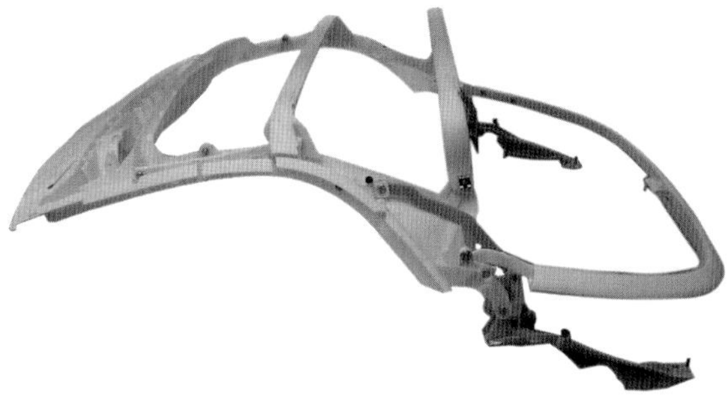

Figure 4.32 Top rods, closed, scale 1:1. Sintered elements made of PA (Source: Karmann/CP-GmbH)

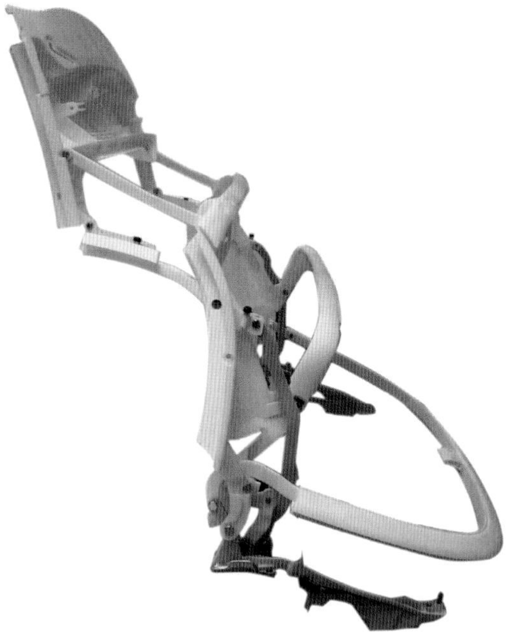

Figure 4.33 Top rods, half opened, scale 1 : 1. Sintered elements made of PA (Source: Karmann/CP-GmbH)

The parts made of polyamide (PA) could be mechanically processed, for example, grinded and completed with metal studs, washers, and screws for linking. Figure 4.34 shows a detail of the rear, left-side hinge point.

Figure 4.34
Top rods, detail, scale 1 : 1.
Sintered elements made of PA
(Source: Karmann/CP-GmbH)

With the help of these models, CAD simulations, and their experience, the manufacturers could complete the detail construction quickly and effectively and produce and test the prototype for a release of series production.

4.3.3 Rapid Prototyping Models for the Visualization of 3D Data

Applications that use the additive manufacturing model only as a three-dimensional picture must be kept separate from the tools for product development. The additive manufactured parts are visualizations of the 3D data sets, but the production of three-dimensional pictures is a process not intended to be product development. It is a discrete process that leads to something like products, a three-dimensional picture. In the sense of production technology, this is not production because the characteristics of a sellable product are mostly missing.

Rapid prototyping models for the visualization of 3D data sets are investigated in conjunction with their application. The main applications are in the fields of medicine, design, art history, and architecture, or they are used for the visualization of simulation results. Furthermore, applications, for example, in advertising, are possible but would not lead to new methodological problems.

4.3.4 Rapid Prototyping in Medicine

The application of rapid prototyping for the production of medical models is used to achieve two main goals: the visualization of nonvisual structures (Fig. 6.16) and the production of individual implants (Fig. 6.11) or splints.

4.3.4.1 Characteristics of Medical Models

Medical models were characterized by a set of criteria that have no or little meaning for the product development or in mechanical engineering.

4.3.4.1.1 Big Data Sets

Because of the huge amount of basis data (for example a CT of a whole skull with 1 mm layers and 512 × 512 dots per layer) and the complex structures of the models, the data sets of medical models are quite huge. A factor of around 10 larger, in comparison to models from mechanical engineering, is common.

4.3.4.1.2 Model Dimensions Not Precisely Defined

Models for medical use often encompass more than the specific area of interest. For the medical treatment of all kinds of deformities, traumata, and large-area resections following a tumor disease especially, models are used not only for the limited local judgment of the defect and for introducing the optimal therapy, including the production of possibly necessary implants, but they also need to fulfill aesthetic functions. With these models the physician can judge the results of his or her work on the general aesthetic. Therefore, it is usually necessary to have a model of the complete cranium, for example, and not just one of the immediate surroundings of the defect.

4.3.4.1.3 Multiple Models

If models are needed for planning operations, and trial operations are carried out with them, a second model that documents the initial situation will usually be necessary. In the future, considerations of quality control will be taken into account.

4.3.4.1.4 Transparency

Anatomic structures often show closed cavities that do not exist in technical models. For example, the frontal sinus, the maxillary air sinus, and the mandibular canal in the lower jaw have such cavities. Often it is important to evaluate such cavities. In this cases stereolithography models are used. Stereolithography models are transparent or at least translucent. Internal structures and hollow spaces (e.g., the mandibular nerve canal) are visible.

If transparency is required, the use of models made by other means is limited because they are not transparent, and internal structures cannot be judged.

4.3.4.1.5 Sterilization

The planning of operational strategies, trial operations, and the molding and fitting of implants make sense only if the model can be taken into the operating theater. This can be done with stereolithography models of acrylates. Sintered models of nylon, polycarbonate, or polystyrene, however, are less suited because powder still contained in the internal hollows could escape when the model is used because the models lose material constantly through abrasion, as is the case with milling models of polyurethane foam. Such abrasion can be reduced by impregnating the model, as several options and materials are available. The same applies to 3D Printing-models. Therefore, these models cannot be sterilized in the strict sense. For the FDM process (see Section 3.3.4.1), there is a material that can be sterilized.

4.3.4.1.6 Biocompatibility

Fully polymerized acrylate derived from stereolithography processes is considered the only biocompatible material suitable for rapid prototyping. Epoxy resins are most certainly not biocompatible. As the material properties are insufficiently known at this point, it is not advisable to use rapid prototyping models as direct implants. After all, toxic fumes are produced during the sintering of polystyrene. Sintered polystyrene, however, is not toxic. Polyacetate and bioresorbable materials, such as hydroxylapatite or tricalcium phosphate, are becoming more and more available and are used for laser sintering and 3D printing.

4.3.4.1.7 Support Structures

The calculation and the optimal installation of support structures required in stereolithography, along with their expert removal, especially with complex structures such as the cranial base, demand certain medical knowledge of the model makers. Support structures that can be dissolved by a washing process have advantages in this case.

4.3.4.1.8 Unconnected Model Parts

When bone structures are segmented, soft tissues such as connective tissue, cartilage, and muscles disappear from the data record, resulting in unconnected model parts. For medical therapy it is often very important to know the exact position of these unconnected model parts in relation to each other. With the aid of CAD programs, such parts are fitted with connecting clamps. In this way it is, for example, possible to display the exact position of the mandible joint in relation to the socket in the maxilla independently of the segmentation in the computer tomography image.

4.3.4.2 Anatomic Facsimile Models

Three-dimensional physical models are preferably used for diagnostics to supplement the two-dimensional imaging techniques. Because the models are three-dimensional copies of the originals, the term “anatomic facsimile models” is used.

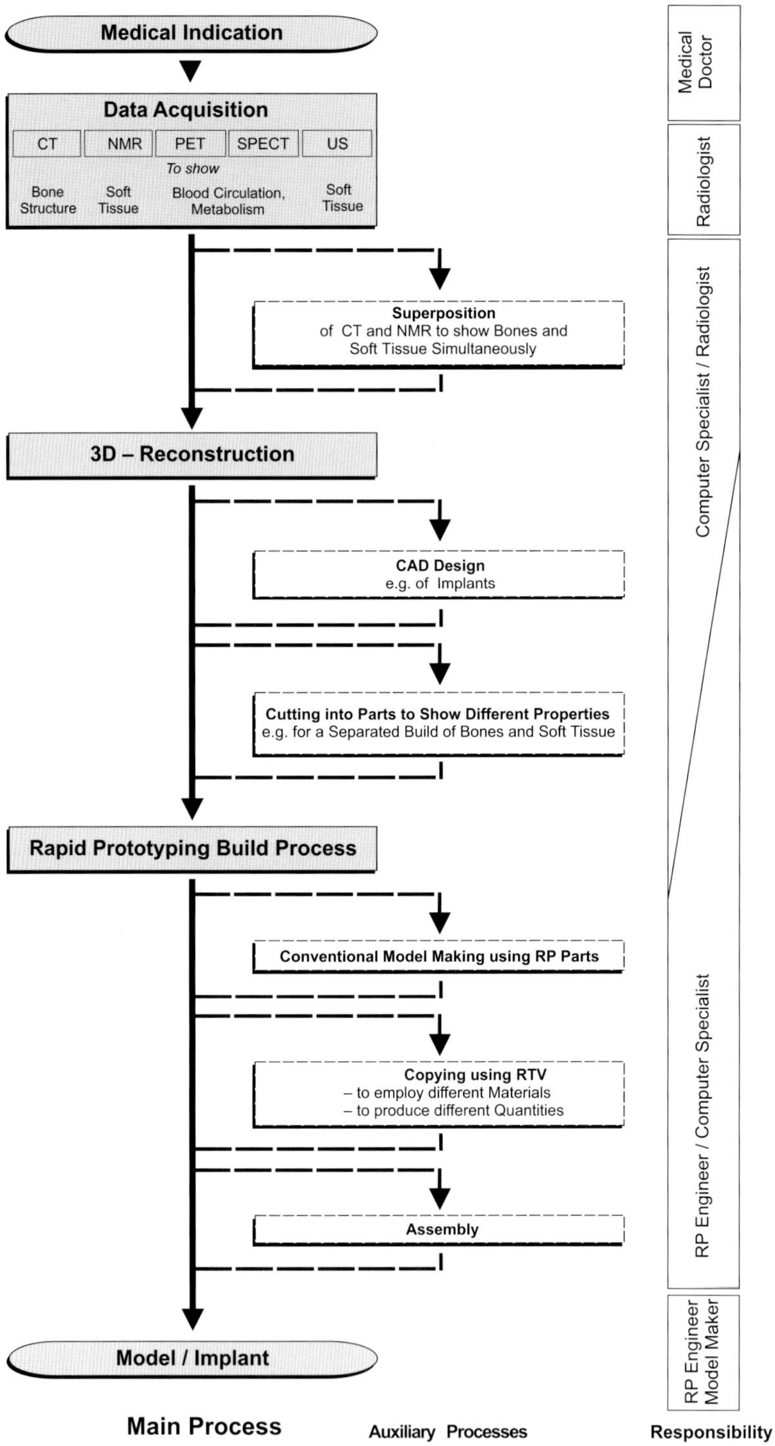

Figure 4.35 Process chain to produce an anatomic facsimile model

The process chain starts with the medical indication that requires a model for diagnosis or therapy (Fig. 4.35). The data are generated with the help of classical techniques. Computer tomography (CT) is used to visualize bone structures, magnetic resonance imaging (MRI) and medical ultrasonics for the visualization of soft tissue. PET (positron emission tomography) and SPECT (single photon emission tomography) are used for the visualization of blood circulation and metabolism processes. Even though the CT images contain layer information, it is not possible to use it directly for rapid prototyping because the layer height is between 1 and 2 mm (minimum 0.5 mm), which is too big for rapid prototyping layers (these are around 0.1 mm).

With the help of special image analysis techniques, it is possible to generate 3D models from the measurement values. The single layers are segmented in the structures, which should be visualized. If bones and soft tissue should be visualized at the same time, as is necessary for the larynx, part pictures could be overlaid. From the segmented layer information, a virtual picture on the computer is generated with the help of 3D reconstruction techniques. In addition to the data formats that are used for medical routines, modern CT scanners and image analysis devices are able to produce STL data, which could be directly used for the control of prototypers. Where this is not possible, special software like MIMICS is used. They also have workarounds for the typical problems of rapid prototyping, like the positioning and orientation in the build chamber, support structures, and so on. With this technology there is a raw data basis that is equivalent to the CAD models in mechanical engineering.

The rapid prototyping model can be used directly or as a master form. With the help of casting techniques, for example, soft tissue or details that should be colored could be better reproduced. Depending on the assignment of tasks, it could be necessary to produce multiple models by casting.

4.3.4.3 Example: Anatomic Facsimiles for a Reconstructive Osteotomy

As a result of a congenital deformity, a two-year-old girl needed surgical rearrangement of the entire cranium geometry, known as a craniosynostosis. To prepare and plan the complicated operation, the deformity was at first reproduced as a 3D stereolithography model (Fig. 4.36). In the course of the preoperative planning, the osteotomy lines were drawn in (Fig. 4.37). Because the stereolithographic material can be worked on with tools that are normally used by physicians, the most important steps of a trial operation could be carried out. The condition after the osteotomy is illustrated in Fig. 4.38. By using osteosynthesis plates the cranial bones were newly configured (Fig. 4.39).

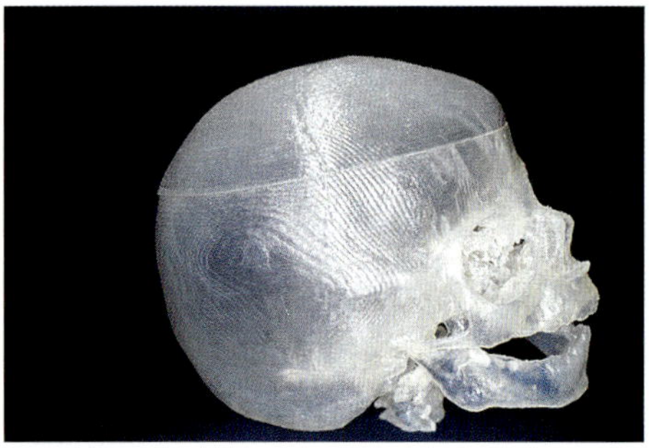

Figure 4.36 Craniosynostosis: original state, stereolithography model

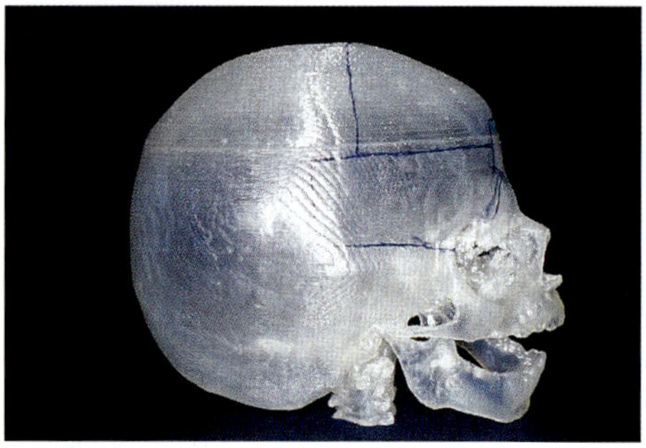

Figure 4.37 Craniosynostosis: osteotomy lines drawn in

Figure 4.38 Craniosynostosis: condition after the osteotomy

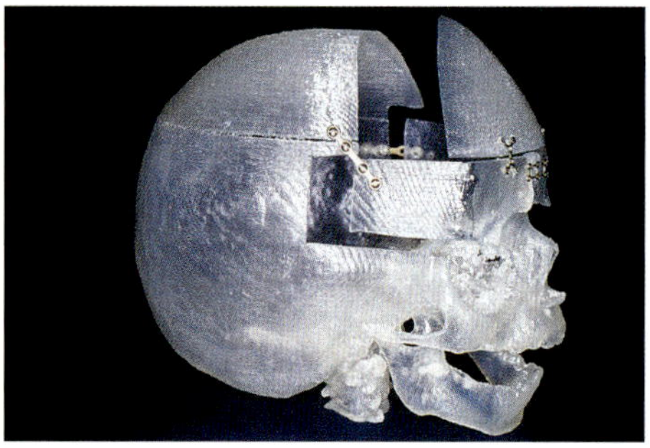

Figure 4.39 Craniosynostosis: newly configured cranial bones

(Source: Bier, Charité, Campus Virchow-Klinikum, Berlin, Germany)

With the aid of such stereolithography models and their ability to make trial operations possible, complicated surgical treatments can be planned effectively; the necessary information can be conveyed to the team, thereby enabling a faster and better-quality operation and causing less stress for the patient.

4.3.5 Rapid Prototyping in Art, Archaeology, and Architecture

4.3.5.1 Model Making in Art and Design, General

CAD systems that are made for designers (e.g., Rhinoceros) are so far developed today that creatively influenced, nearly playful work with the computer is possible. In this way surfaces and bodies can be designed as imagined and not controlled by mathematical rules.

For designers, the appearance of parts, the visualization of surfaces, and haptic points of view are most important. In this case, questions of kinematics or material science aspects have to be defined before the rapid prototyping method.

One limitation has to be accepted today and in the near future: designers, who are used to evaluate the design of a car body by water surface profiles, would not be satisfied by the rapid prototyping surfaces.

4.3.5.2 Example of Art: Computer Sculpture, Georg Glückman

Since 1987, the artist Georg Glückman, Berlin (www.glueckman.de), has been working with a CAD technique. He looks for methods and processes to produce his "computer sculptures" as three-dimensional physical objects. The first objects were produced by 2 ½D milling of Stryrodur. These objects were presented in Berlin at the Blumenautomat-Galerie, in Vienna at the Pinx-Gallery, and in Hannover at the Artware.

In 1995, the first additive manufactured plastic was produced and shown in exhibitions at Antwerp and Stuttgart. It was followed by "Plastx," a three-dimensional structure in which the elements seem to glide into one another and steadily interpenetrate; see Fig. 4.40. It was produced by laser sintering and presented at the international triennial for small sculptures in Stuttgart and Vienna.

Figure 4.40 "Plastx," a computer sculpture by Georg Glückman, 1995, laser sintering (Source: CP-GmbH)

4.3.5.3 Example of Design: Bottle Opener

A bottle opener was designed as a gift for good customers. It needed to be easy to handle, have a pleasant and unassuming design, be attachable to a key ring, give the impression of precision, and be produced by a process used by the rapid prototyping service company CP (Centrum für Prototypenbau GmbH, Erkelenz/Düsseldorf, Germany). The company colors and the logo had to be integrated into the design as discreetly as possible.

The first designs were based on the letters C and P, the initials of the company. Figure 4.41 shows some of the first drafts. In the course of further development, the design shown in Fig. 4.42 crystallized. It consists of a laser-cut metal plate into which a cutout was incorporated in such a way that it could open bottles. In addition, the plate contains the company's initials and a loop for the key ring. A number of holes are drilled around the edge, which on the one hand enable a secure and firm plastic coating to be made, and on the other prevent the hole for the key ring from being ripped. The model was verified as a stereolithography model. As mentioned above, the cutouts were made by a laser.

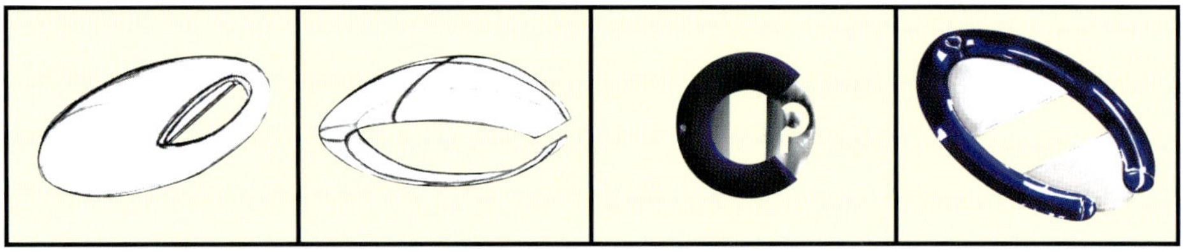

Figure 4.41 Bottle opener, first drafts (Source: C. Schoenen/CP)

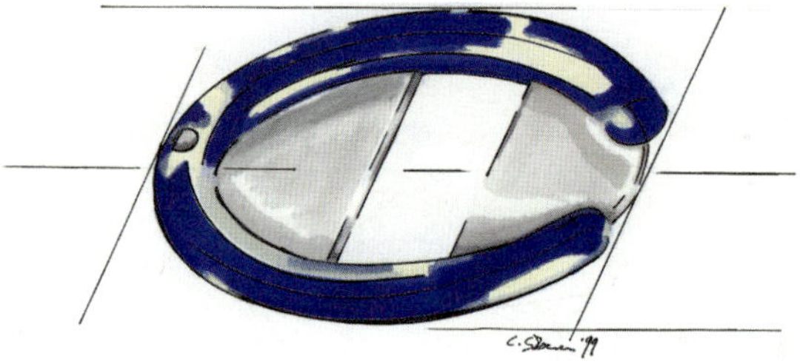

Figure 4.42 Bottle opener, final design (Source: C. Schoenen/CP)

After several pre-series had been produced by vacuum casting techniques, an aluminum mold was produced by high-speed milling, and from this, the required number of injection-molded parts were produced. Figure 4.43 shows such a series part.

Figure 4.43 Bottle opener, series part (Source: C. Schoenen/CP)

By using vacuum casting techniques, it was possible to test critical design elements, such as the exact application of the relatively fine plastic rim, the secure function of the key ring hole, and how far the device was bent when used, without having to produce a series injection mold. The decision for a design and thereby the decision for production was made solely on the basis of the vacuum-cast prototype.

4.3.5.4 Applied Art: Statuary and Sculpture

By using additive manufacturing, new ways for sculptors are opening, especially in virtual modeling. They also allow the production of anatomic facsimiles of living people and even of their inner structures. For example, in 2000 the sculptor Adrian Schoormans, Düsseldorf, exhibited his own skeleton at the "Glaskasten Marl" [Schoormans, 01]. "The rapid prototyping has the same meaning to sculpting, as the photography had to painting," he says.

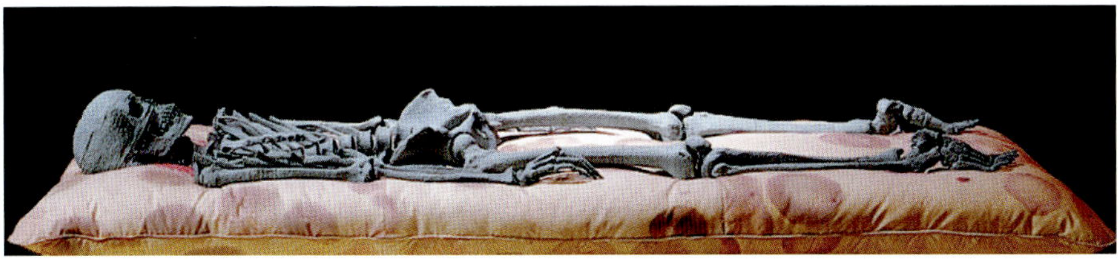

Figure 4.44 His own skeleton, Adrian Schoormans, Magnetic resonance imaging and 3D printing (Source: Schoormans/CP-GmbH)

The Berlin artist Karin Sander is also working with facsimiles of living people. In her exhibition "Konzeptkunst, Karin Sander" in the Staatsgalerie, Stuttgart, she made the visitors themselves into exhibition pieces. All of them had to enter a 3D scanner at the entrance. Their data were directly and openly prepared for everyone and then sent to two 3D printers (Z-Corp), which printed the models in the scale of 1:7. The numerous figures can be seen today at the permanent collection at the Staatsgalerie.

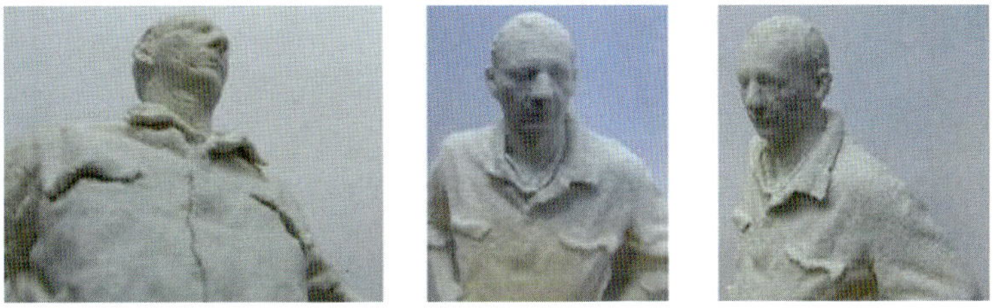

Figure 4.45 Figures, Karin Sander, body scan and 3D printing (Source: Karin Sander/Staatsgalerie Stuttgart/RP-Labor FH-Aachen)

The sculptor Alisa Minyukova, New York, is mainly creating plastic models of famous artists. She devoted several of her works to the Russian poet Alexander Puschkin (Pushkin). One larger-than-real bust is located at the Puschkin home in St. Petersburg. A small, 14-cm-tall sculpture with top hat and cape was scanned as a draft, laser sintered, and postprocessed by manual handwork. To produce a small line of nine single pieces, wax blanks were produced via silicone mold and vacuum casting. After this they were individually reworked and investment cast by hand by R. Ellerbrok, Aachen.

Figure 4.46 Alexander Puschkin, Alisa Minyukova, laser sintering, vacuum casting, bronze casting (Source: Minyukova, CP-GmbH, Ellerbrock)

4.3.5.5 Example of Archaeology: Bust of Queen Teje

The Egyptian Museum in Berlin, Germany, not only houses the world-famous bust of Nefertiti but also a rather unassuming bust of her mother-in-law, Queen Teje (aka Tiye), 9.5 cm high. Her status changed from queen to goddess presumably when her husband, King Anophenes III, died. At the same time her royal headdress was covered by a hood made from paper maché set with blue pearls. After the X-ray technique had been discovered, it was verified during the 1920s for the first time that the few golden elements that were visible through some slight damage in the outer hood really were part of an extensive gold headdress fitted onto a silver hood. Ever since, it has been discussed whether the scientifically and very interesting assessment of the headdress and the silver hood should be made accessible by removing the also scientifically important outer hood that would be damaged by the procedure. When 3D modeling on the basis of a computer tomogram became better known, it was decided to use this technology and to produce a 3D model of the bust beneath the hood with all of its details.

A computer tomogram of the bust was taken in cooperation with the BAM (Bundesanstalt für Materialprüfung, Berlin) (Federal Institute of Material Research, Berlin, Germany), and the data were processed for construction in a stereolithography machine. In this case, it was helpful that stereolithography enables the bust to be built in double size without difficulty so that some details could be better discerned. The data taken were separated mathematically so that the viscerocranium, the silver hood, and the jewelry could be built as individual parts and the complete model could be built as an assembly of those parts.

The results were received among the experts with such enthusiasm, especially with regard to their art historical and museum educational value, that the four stereolithography models are firmly established in their place right next to the original in the Egyptian museum in Berlin, Germany (Fig. 4.47) [Wildung, 95].

Figure 4.47 Stereolithography model of Queen Teje's bust with the bonnet removed, original (Source: Wildung, Egyptian Museum, Berlin, Germany)

4.3.5.6 Model Building in Architecture, General

In architecture and town planning, models serve two main purposes. First, they help when solutions for special spatial-artistic designs are required. Here the model construction is a kind of experimental medium for finding solutions. Second, at the end of a development process, they help to represent the architect's final design and give a clear picture of his or her ideas. In both cases, the new methods of rapid prototyping are additions to the design repertoire. The ability to combine CAD with computer-supported model construction directly gives valuable support to the work of architects and town planners. The design process is not merely a matter of computer graphics in visual abstraction but also a produced design and perceptible object as known by architects and town planners from time immemorial.

The methods of rapid prototyping are used especially for complex design models in which the internal geometric and topological structures of buildings are important–for example, the layout of rooms inside a building–and cannot be realized with other computer-aided model-making methods such as milling processes.

The use of rapid prototyping models and the necessary technical effort connected with it are worth the trouble only when complicated freeform areas or other architectural details must be reproduced true to life. Therefore, in architecture especially, it is preferred to use so-called "hybrid models," which are a combination of classical elements of architectural model construction and rapid prototyping elements.

4.3.5.7 Example of Architecture: German Pavilion at Expo '92

Figure 4.48 shows the design of the German Pavilion for the World Exhibition Expo '92 in Seville, Spain, which won first prize in the competition for the design of the main building, though was not constructed for cost reasons (design: Auer & Weber, Stuttgart/Munich, Germany).

The Expo pavilion on the scale 1:200 places high demands on the model-making process with regard to the geometry and the detailed content owing to its extremely complex support structures with their long slim supports and small profiled girders.

Figure 4.48 Design of the German Pavilion for the Expo '92 in Seville, scale: 1:200, stereolithography (Source: B. Streich, University of Kaiserslautern, Germany)

4.3.5.8 Example of Architecture: Ground Zero

The redesign of the former World Trade Center site in New York is a very complex task, because not only must the architectural draft convince, but it must also show how the new building will be integrated into the urban surroundings. Architectural models show proportions, but they do not show the absolute dimensions.

Virtual 3D architectural models demonstrate important views, but not the integration into the cityscape. For this reason, most architectural studios generally use at least three different views. A two-dimensional rendering visualizes the old and the planned buildings in a fixed view. A screenshot of the three-dimensional simulation is important from the architectural perspective. In the presentation the simulation can show any view. Finally, the rapid prototyping model can show the whole building and, separated by color and material, the buildings directly surrounding it. For the print media, this model is also shown from a fixed view, but in the presentation everyone can watch it from infinite views. These three components, the two-dimensional rendering, the three-dimensional simulation, and the rapid prototyping model, show the strengths and weaknesses of the different possibilities for renderings.

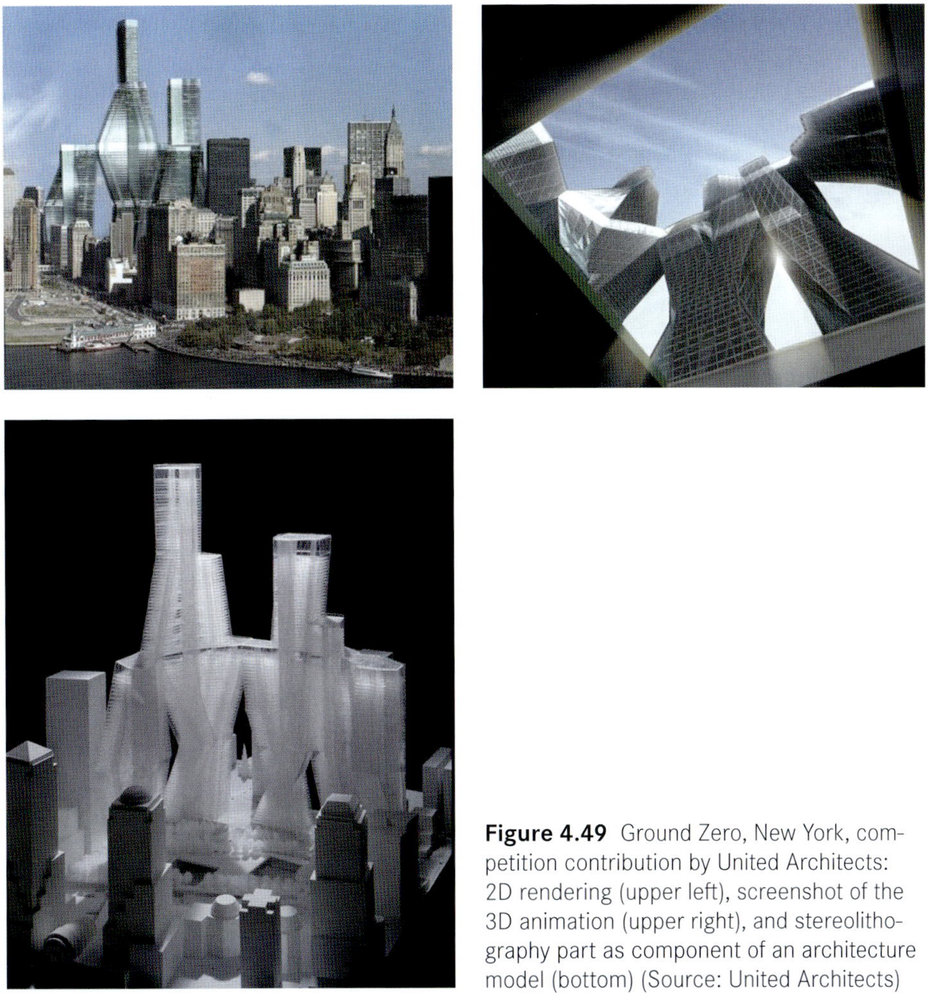

Figure 4.49 Ground Zero, New York, competition contribution by United Architects: 2D rendering (upper left), screenshot of the 3D animation (upper right), and stereolithography part as component of an architecture model (bottom) (Source: United Architects)

4.3.5.9 Example of Architectural Monuments: Documentation of Buildings Relevant to Architectural History

The Aachen cathedral is one of the most important sacred buildings north of the Alps. It was built towards the end of the 8th century and for over 200 years, was the building with the highest arches and the biggest width of a dome. The central building is the Carolingian octagon, in which most German kings were crowned and that contains the marble throne of Charlemagne, on which the kings not crowned in Aachen, without exception, also took their seat.

In 1978, the Aachen cathedral was the first German building to be listed as a UNESCO world cultural heritage site. The 35th anniversary in 2013 was the reason for the University of Applied Science and the Einhard Gymnasium in Aachen to start a school project. The task was to construct a 3D model of the cathedral and build it with the use of additive manufacturing techniques. By using images and footprints of the dome, the proportions were fixed and transformed into a 3D model by an extra Autodesk Inventor course. In addition to the about 25-cm-long full model, the most important single buildings in the building history were manufactured separately. They could be assembled into a whole model and were the basis for a study of the historic building and expansion process of the dome, which took place over centuries.

As an additive manufacturing process, an extrusion process was chosen, and because of the partly necessary support structures, the FDM process was used. The visible stair-stepping, which generally presents a disadvantage, in particular for small models, was deliberately used to produce the characteristic construction detail on the roofs of the cathedral, which could not be achieved easily by a conventional CAD system.

Figure 4.50 shows (a) the whole complex, (b) a detail of the roof, and (c) the single parts, on which taps are attached to assemble them. The black material highlights the quality rating of the model.

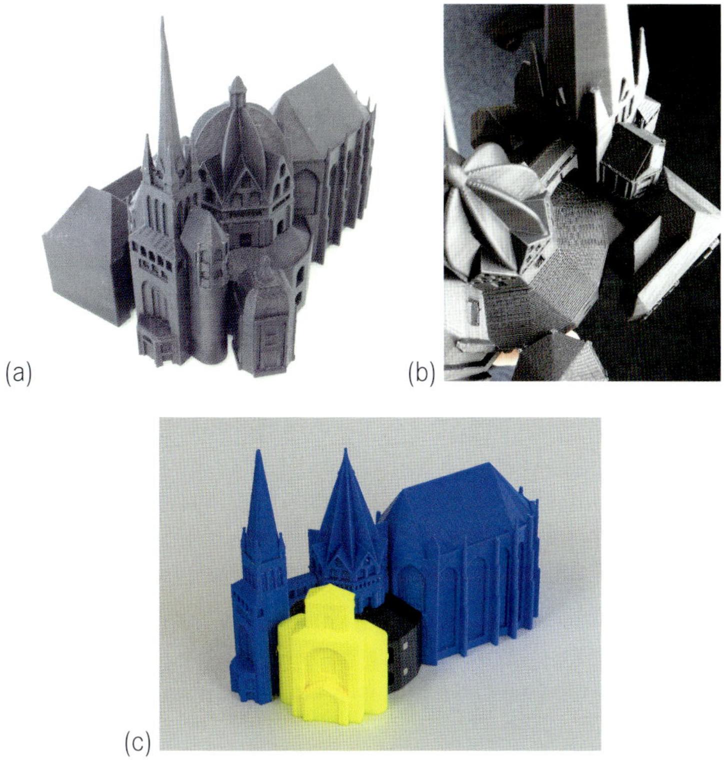

Figure 4.50 Model of the Aachen cathedral, fused deposition modeling. The CAD data were reconstructed from images and processed for model building: (a) full model, scale around 1 : 468, (b) detail of the roof, with deliberate extrusion structure, (c) important parts of the building history produced as single parts (Source: FH Aachen/Einhard Gymnasium Aachen)

4.3.6 Rapid Prototyping for the Evaluation of Calculation Methods

4.3.6.1 Photoelastic and Thermoelastic Stress Analysis

Mathematical-physical calculation models allow the simulation of product properties, like stability, vibration behavior, and temperature resistance, and of manufacturing technologies, for example, casting simulations that are based only on theoretical approaches directly at the computer. The use of such methods, which are also the basis for virtual reality, is the fastest route to new products.

One disadvantage of the calculation models is that they are more or less simplified, so their results cannot be fully applied on the repetition part and there are more or less uncertainties. Therefore, the variables of the models have to be changed or corrected constantly by the results of evaluations, which were carried out on repetition parts. A fast and efficient way to get such correction values is to carry out the experimental evaluations on rapid prototyping parts instead of repetition parts.

A longer-used method is the so-called photoelastic stress analysis on specially developed acrylic glass parts. Polished rapid prototyping models could also be used here and can enhance the geometric possibilities.

A newer, but also significantly more complex, approach is the detection of the stress level by a systematic analysis of analogies between the optical and the thermal properties of the part. This method is called thermoelastic stress analysis.

4.3.6.1.1 Photoelastic Stress Analysis

The approach is to use the refraction properties of stereolithography models for photoelastic analyses and thereby verify the predictions of computer calculations by testing at an early stage of product development.

Epoxy resins, especially those used for stereolithography, have the property of displaying optical double refractions, which is the basis for a stress analysis aided by photoelasticity. An interference pattern in the transparent stereolithography component is obtained that is proportional to the stress within the component and that can be interpreted manually or automatically.

This process adopts the concept of photoelasticity, which has been known since the beginning of this century. With the aid of stereolithography models, today's greatest problem can now also be solved—that of making 3D models from conventional photoelastic materials.

Epoxy resins are exceedingly well suited for the generation of models for photoelasticity because of their high transparency. They hardly differ from the standard materials (Araldit B, Ciba Geigy) used for photoelasticity. Acrylates are less well suited because of their lower degree of transparency. If used for stereolithography models, care should be taken that the model is cured as evenly as possible, so as few air bubbles as possible are created, and that the surface is flawless. Some kinds of construction and exposure methods, for example, those generating hollows intentionally to increase the quality and dimensional stability of SL components, are therefore unsuitable for photoelastic checks. It also has to be taken into account that semipolymerizing processes (whereby complete polymerization occurs later in a postcuring oven) thus create optically effective parting planes. From the aspect of photoelasticity, components made of epoxy resins SL5170 and SL5180 that use the build style ACES (3D Systems) are especially suitable.

The basic setup for photoelastic checks (Fig. 4.51) consists of a pole filter, a light source, and two $\lambda/4$-wave plates, which facilitate the separation of the isochromatics (lines of the same principal stress difference) from the isoclines (lines of the same principal stress) so that then only the isochromatics and the transparent model, which is placed between the two $\lambda/4$-plates, can be viewed.

Regardless of the type of resin used, the necessary material parameters must first be defined before the photoelastic test can be carried out and evaluated. The most important value is the photoelastic constant, which is determined from a calibration test and which denotes the relationship between stress and isochromatic order. The photoelastic material parameters of the most important stereolithography resins are listed in Appendix A3.15, “RP Materials and Casting Resins” [Steinchen, 94]. Should other resins or other kinds of constructions (scan strategies) be used, these calibration tests ought to be repeated.

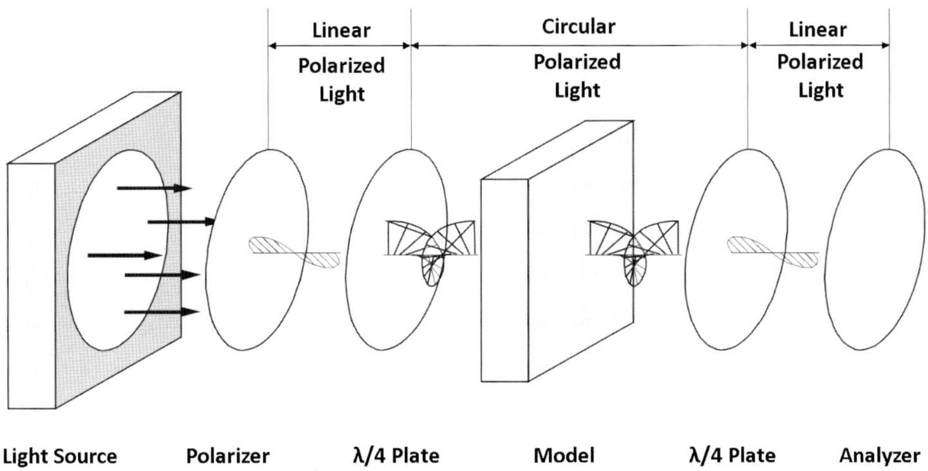

Figure 4.51 Setup for a photoelastic stress analysis (Drawing: Steinchen)

4.3.6.1.2 Thermoelastic Stress Analysis

The development of safety-relevant devices such as, for example, steering assembly parts in automobile manufacturing, still relies on tests made on series identical models. Rapid prototyping processes change little here because the plastic parts and also modern metal parts can either not be tested under series conditions or, if tested, they do not allow any applicable conclusions to be drawn.

Thermoelastic stress analysis (THESA) is successfully employed in testing such components without extensive field trials. It allows single parts to be simulated on the test bench instead of monitoring the module or doing a driving test. THESA is based on the fact that metal components under stress show temperature changes that are proportional to the given stress, provided the load stays within the elastic sector. These temperature fields and their fluctuations can be recorded using appropriate high-resolution thermal cameras; tensions can be related to the temperature fields and correlated with the strains. This process was modified in cooperation with an automobile manufacturer, so it is now possible to use this process, initially meant for metal components, for plastic components also [Gartzen, 98]. In principle, this has paved the way for the optimization of highly stressed components with the aid of plastic rapid prototyping models.

4.3.6.2 Example: Photoelastic Stress Analysis for a Cam Rod in the Engine of a Truck

The starting point is a 3D CAD model of the component to be tested. This is produced as a stereolithography component. Care should be taken that the value of the photoelastic constant is kept as low as possible. The most important parameter here is the aftercure under UV light [Steinchen, 94].

Further action depends on whether a 2D test at room temperature is to be run, or a 3D photoelastic test. If a 3D test is required, the stereolithography component must first be "frozen." Using an oven with a programmable cooling curve, the model is heated up to approximately the glass transition temperature, loaded, and then cooled down to room temperature at 2 to 3 °C per hour. Afterward, the elongations, and thereby also the tensions, are "frozen."

The photoelastic test itself is now run on 2D models. In the case of a 3D photoelastic test, this involves the cutting out of a cross section at the test-relevant area for use as a 2D model. A normal band saw can be used for this purpose.

When placing such a 2D (Fig. 4.52, top) or 3D (Fig. 4.52, middle) model of the valve lifter (Fig. 4.52, bottom) between two polarizers or quarter-wave plates (Fig. 4.51), a usable isochromatic pattern is recognizable in the 2D model. For the 3D model of the valve lifter, the isochromatic images of the single layers are superimposed so that a clear definition is not readily possible. In general, it can be observed that the stress concentration is higher the more isochromatics appear at a particular point and the closer they are together. If the quarter-wave plates are removed, the isoclines are recognizable. Whether an interpretation of the isocline picture is necessary depends on the method of analysis used. An automatic analysis using an electronic image processing system is also possible. The aim of the analysis is to define the main extensions and the main tensions to allow a conclusion to be drawn concerning the number of tensions.

After the interpretation of the isochromatic picture is complete, the results of the test can be used for dimensioning prototypes with the aid of similarity laws. The similarity law for the static case is, for example:

$$\frac{\sigma_P}{\sigma_M} = \frac{F_P}{F_M} \cdot \left(\frac{l_M}{l_P}\right)^2 \tag{4.1}$$

As Equation 4.1 shows, the results are transferable without great effort. To summarize, it can be stated that photoelasticity together with stereolithography is a reliable and economical method of implementing an experimental optimization of the subsequent product already in the design phase [Jacobs, 95], [Kramer, 92], [Susebach, 93], [Steinchen, 94].

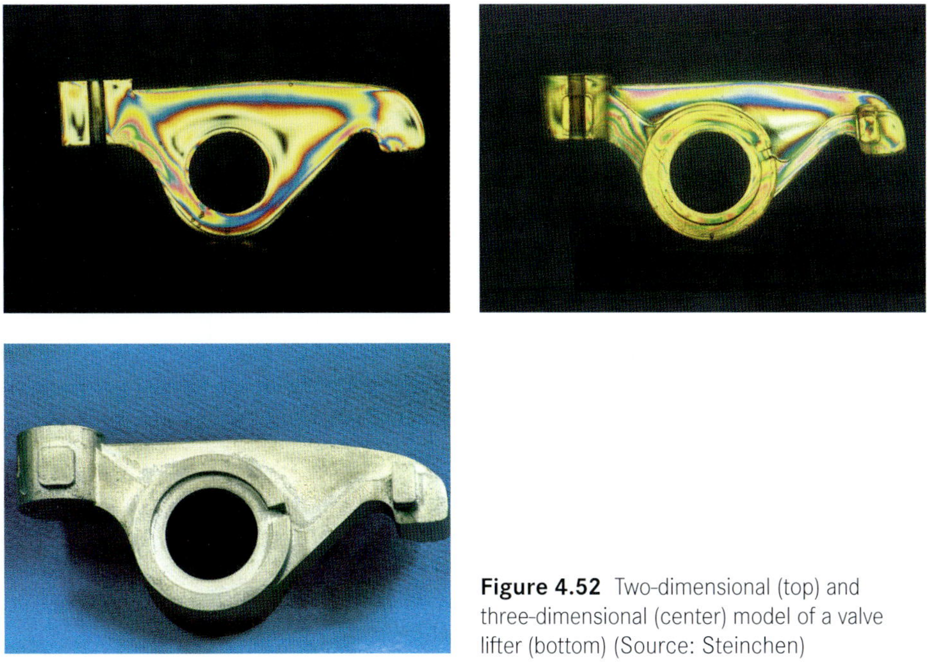

Figure 4.52 Two-dimensional (top) and three-dimensional (center) model of a valve lifter (bottom) (Source: Steinchen)

4.3.6.3 Example: Thermoelastic Stress Analysis for Verifying the Stability of a Car Wheel Rim

The wheel rims of sports cars are highly stressed safety components that are increasingly required to be lightweight and attractive while the dynamic stress on them increases continuously. A hollow rim promises to fulfill these requirements. The development is difficult, however, because the prototypes of hollow rim segments cannot be produced by milling as in solid constructions. For optimization purposes, therefore, the components must be cast as in series processes. One set of molds is necessary for each casting and has to be made as custom-built models from wood in a model workshop.

The development until now has been correspondingly complicated (Fig. 4.53, left string). From the CAD data, which are checked and optimized by FEM processes (for reasons of clarity, the optimization loop is not shown in the graph), a second cast in aluminum is obtained by conventional mold making. This cast is finished, mounted, and tested. Depending on the results, this loop is performed several times and the results iteratively improved.

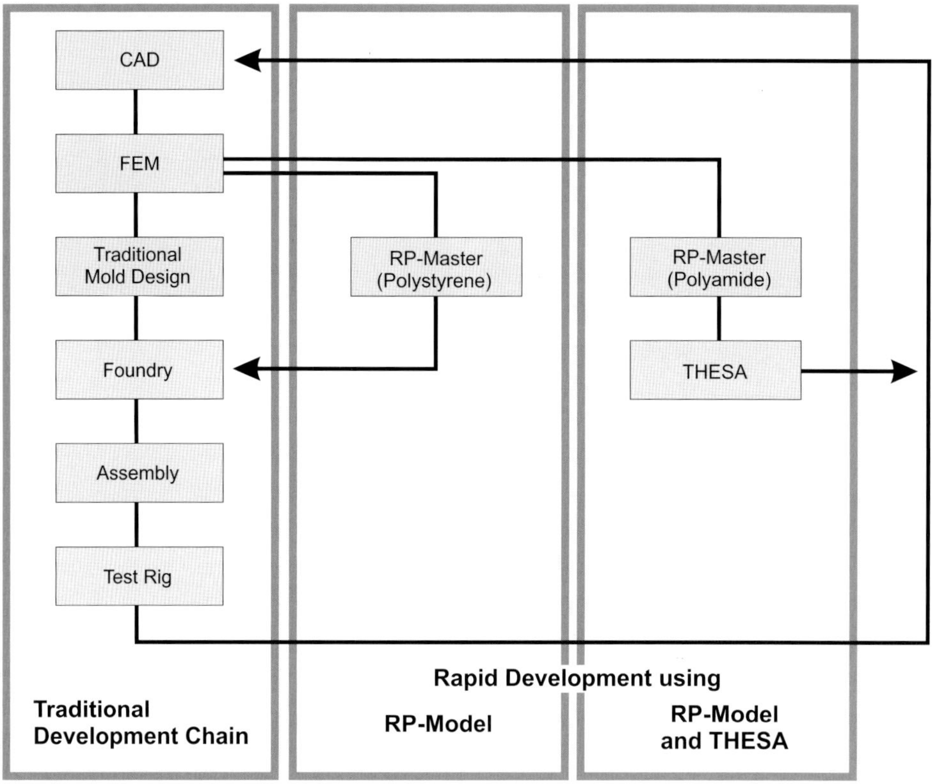

Figure 4.53 THESA, schematic steps in the process chain and the shortening thereof by the use of rapid prototyping (Source: Schwarz/CP)

When THESA is employed, this process is shortened because assembly and driving tests are omitted.

When using rapid prototyping processes, the process can be further shortened insofar as the mold making is substituted by a rapid prototyping model that is cast directly in a precision casting process.

The decisively shorter development process is achieved by being able to test the sintered polyamide rapid prototyping component directly by means of THESA, thereby eliminating the necessity of the entire casting process. Figure 4.54 shows an exemplary thermographic image with a defined stress and, in comparison, the same situation with a molded aluminum wheel rim. The conformities are excellent, as documented by the totals of the main tensions given in the example over the radius at the center of the component (Fig. 4.55).

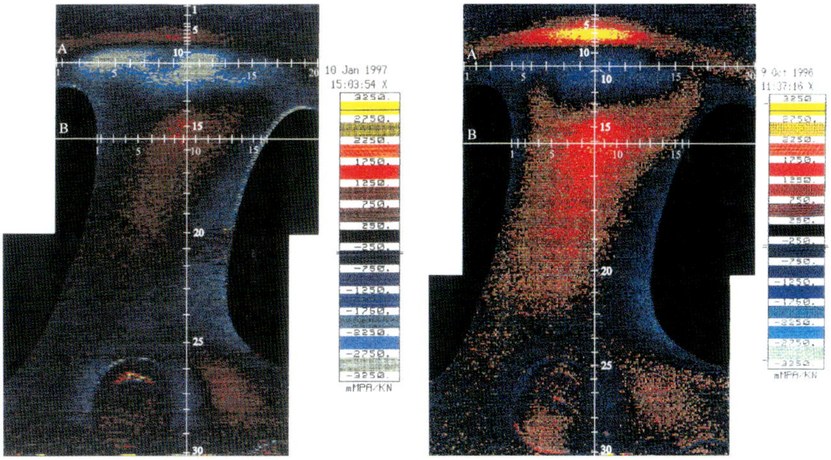

Figure 4.54 THESA, thermographic reproduction of a rapid prototyping component (polyamide) (left) and of an aluminum series part (right) (Source: Schwarz/CP)

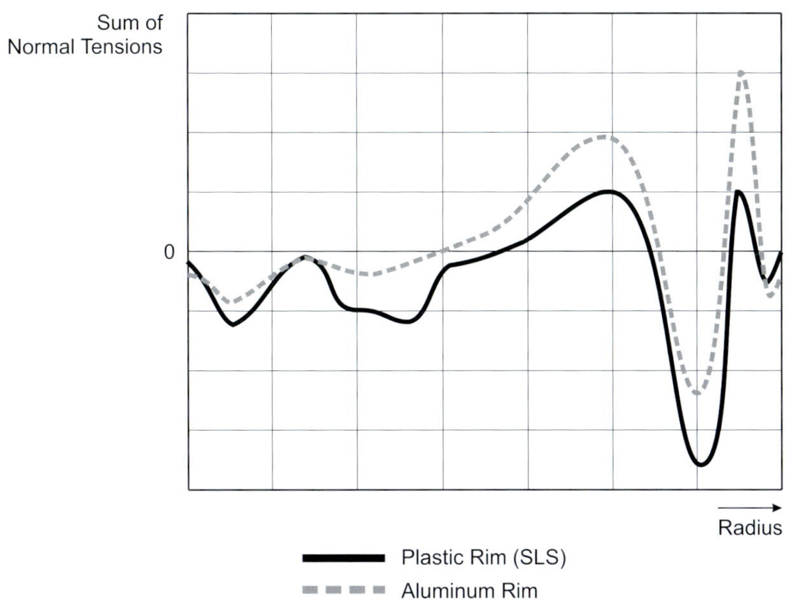

Figure 4.55 THESA, total of the main tensions over the radius in the center of the component (Source: Schwarz/CP)

When THESA is used for sintered plastic components generated by rapid prototyping, the time needed for the iteration loop is reduced to 20% of the time previously needed. The greatest single effect is the reduction of production time from 18 days for a cast component to three days for a polyamide component. By applying this modified THESA process, the entire development process can be reduced dramatically, although not down to 20 % of the previous time because the final test for a safety component still needs to be run with a component made of the original material.

4.4 Outlook

For solid images and concept models, the trend is toward cheap and easy-to-use personal printers that can be used at home or in the office. The production of models requires less or no data preparation and also no or less manual postprocessing. The price limit for the mass production is today at around 5,000 € (just over $5,000), and for self-made systems and kits the price is between 1,000 € and 4,000 €. The ProJet 1000/1500, which was presented as V-Flash in 2007, shows the way.

The development of machines for the production of functional prototypes goes in another direction. The coupling of machines and materials today allows the given assignment of model requirements to the additive processes. In the future, development will concentrate on the materials, and from this there will be a bigger material range for every machine available. The models and prototypes will become more and more powerful, but the choice of a process is getting complicated. From today's optimization of two groups of process parameters, the model definition and the additive process, three groups of parameters will grow: the process, the model definition, and the material.

The development of machines will lead to a reduction of the necessary manual postprocessing and partially reduce costly finishing. Models and prototypes will then be easier to reproduce, and also be available quicker. Office-usable processes that produce hazardous waste and require the use of solvents, alkalis, and the like will be less accepted.

5 Rapid Tooling

Rapid tooling describes the additive manufacturing of tools, tool inserts, gauges, and molds. The parts are termed *tools*. Depending on the application and process, they can be prototypes or serial tools or their elements. The additive manufacturing machines are *professional printers or production printers*. They are described in Chapter 3, "Machines for Rapid Prototyping, Direct Tooling and Direct Manufacturing." Tools that are used for the production of plastic moldings are preferably considered in the following section. This concentration does not represent a process-specific restriction.

Applications for blow molding, deep drawing, embossing, forging, extruding, glass molds, and also for gauges and devices have been published, and the manufacture of such tools follows the same rules in terms of manufacturing technologies.

Premised on the basic fundamentals, this chapter on rapid tooling can be understood independently and thus enable a succinct introduction to the topic.

■ 5.1 Classification and Definition of Terms

To give the manufacturing of negatives, or tools, its own name, "rapid tooling" has historical roots. The additive manufacturing of tool inserts is older than those of final products (see rapid manufacturing, Chapter 6, "Direct Manufacturing: Rapid Manufacturing") and was introduced in the early 1990s, primarily from a marketing point of view as a new application of rapid prototyping. Therefore, classification of *additive manufacturing* processes with rapid tooling as its own horizontal application level can be found in the literature. But from the perspective of the technology of additive manufacturing, there is no reason to distinguish between the production of positives, or parts, and negatives, or tools. Rapid tooling thus summarizes different applications of additive manufacturing for the production of tools and molds together vertically (see Fig. 1.6).

5.1.1 Direct and Indirect Methods

If the tools are created directly in the AM machine, this is called the direct method. If the tools are created in downstream nonadditive process steps through molding of directly generated master patterns, this is called the indirect method. Indirect methods are predominantly nonadditive and can only be treated in the context of additive manufacturing techniques if they have at least one AM process step. In most cases it is the first step and leads to the master pattern.

Direct Methods: Prototype Tooling and Direct Tooling

If for the direct manufacturing of tools and tool inserts, rapid prototyping methods are used, in the sense of functional prototyping (Fig. 1.6), they lead, depending on the application, to tools with a prototype character, preferably made of plastic. This is called *prototype tooling* (Fig. 5.1). Prototype tooling is a subset of functional prototyping.

If rapid manufacturing processes are used for the manufacturing of tools, then tools or tool inserts with series character (direct tooling) are created, which preferably are made of metal. This is called *direct tooling* (Fig. 5.1). Direct tooling is a rapid manufacturing process.

To underline the additive character and to differentiate them from the indirect methods, the prefix *direct* should be added in the direct methods, as in *direct prototype tooling* and *direct rapid tooling*, but this doesn't usually happen in practice.

Indirect Methods

Indirect methods or *indirect tooling* are also called *follow-up procedures* or molding processes. They are nonadditive and don't belong to the AM processes. Nevertheless, they are briefly discussed for two reasons. First, they are closely related to the AM processes, and with the production of the master pattern always have a first AM process step. Second, they should be clearly differentiated from (direct) AM processes. Linguistically, a differentiation usually does not take place because indirect methods are often not called indirect tooling but instead *rapid tooling*. This often happens consciously and mainly for marketing reasons. The name of a process therefore usually doesn't allow a conclusion as to whether it is an AM process or not.

Bridge Tooling

In practice, there are applications of prototype tooling that do not lead to prototype moldings, but instead to small quantities of moldings with series character. Because prototype tooling bridges the gap to the direct tooling processes suitable for series production, this subset of the prototype tooling processes is also called *bridge tooling*. Bridge tools can be manufactured directly additively manufacturing, but they can also be the result of a molding process. The name bridge tool is, however, also selected for non-AM processes if they have the same function.

The correlations are shown in Fig. 5.1 (a detailed section from Fig. 1.4). This again makes it clear that rapid tooling does not refer to a horizontal technological structure of the AM processes but is summarized thematically in terms of a vertical layout for different applications for the manufacturing of tools and tool inserts. Figure 5.1 also shows that the indirect tooling on the basis of the functional prototyping depends on the application, and further processing either results in pure prototype tools or in tools with the character of bridge tools. To be correct, the names should have the prefix "indirect" added. This was done in Fig. 5.1, but it is usually omitted in general usage.

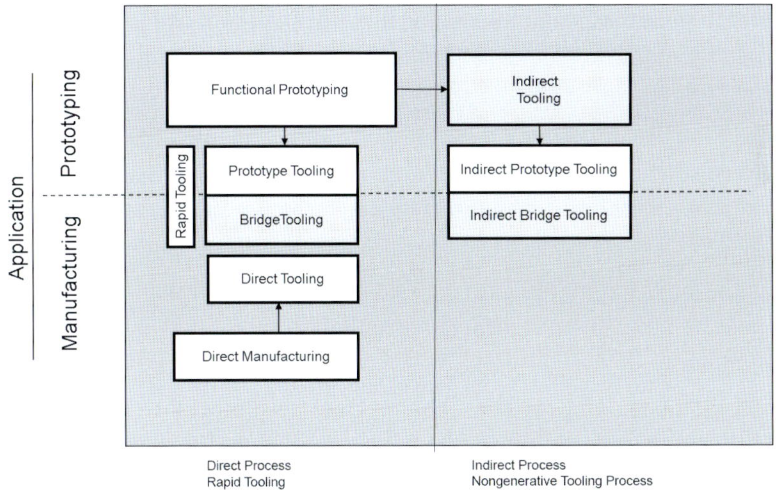

Figure 5.1 Structure of rapid tooling in the context of additive manufacturing, functional prototyping, and direct manufacturing

■ 5.2 Properties of Additive Manufactured Tools

Additive manufactured tools differ both in the strategic approach as well as in the concrete fabrication technique of nonadditive manufacturing, so-called conventional, tools.

Despite the name, in practice, tool inserts are preferably made and completed with standard elements and removal processes. Rapid tooling leads only in exceptional cases to complete tools.

5.2.1 Strategic Aspects for the Use of Additive Manufactured Tools

The additive manufacturing of tools and tool inserts in particular is targeted toward speeding up the manufacturing process and realizing new technical concepts.

The criterion speed usually leads to prototype tooling and thus to nonmetallic tools, which are preferably produced by the molding of additive manufactured master patterns or directly from AM plastic processes.

The implementation of new technical concepts happens along with the process of direct tooling and thus exclusively with additive metal or ceramic processes.

5.2.1.1 Speed

The decisive progress that has given the prefix *rapid* to the AM processes consisted first in obtaining complex plastic moldings without the use of any tools and thus faster by many weeks, in addition to being cheaper than previously (at the end of the 1980s).

But already in the early 1990s, new methods for the manufacture of tools and tool inserts were starting to be used, although manufacturers actually wanted to avoid the use of these methods in the future through the use of AM processes.

This was mainly due to the inadequate material properties of the AM parts and the fact that the additive manufacturing of larger quantities was quickly becoming very expensive because no significant economies of scale occur, as is the case with injection molding.

Therefore, methods and procedure chains were developed for the direct and indirect manufacturing of additive manufactured tools and tool inserts. In order to quickly obtain moldings and very quickly manufacture tools for that, other requirements were subordinated, such as the quality of the finishing of the parts; for example, the tolerance of bare spots and sink marks. Besides a lower application rate, significantly longer cycle times, and if necessary, also manually handling loose parts instead of sliders were accepted.

To ensure the primary goal–speed–AM processes for the manufacture of plastic components are chosen as the starting point. To acquire the tool directly leads either to prototype tools or bridge tools or to the corresponding indirectly produced tools or to mold inserts through follow-up procedures (Fig. 5.1).

5.2.1.2 Implementation of New Technical Concepts

In the search for unique features, two sets of applications were defined at an early stage that illuminate new technical aspects: conformal cooling and the variation of the material above the component cross section. Both attempts support an effective tempering of injection molds that target quality improvement through a more even temperature distribution as well as cost advantages through shorter cycle times. They are both directly linked to metal tools in production quality.

The geometrical freedom allows the manufacture of internal cavities, which can be placed as spatially curved channels or networks following the contour of the cavity just below the surface, and therefore in particular, they allow for a much faster tempering (Fig. 5.2). Uneven temperature distributions and local overheating, so-called hot spots, can be selectively avoided as well. Such internal cavities can also be used for the realization of air ejectors that need to be taken into account because of the cavities, which are traversed with cooling structures, as an alternative to the drilled ejector channels with straight ejectors.

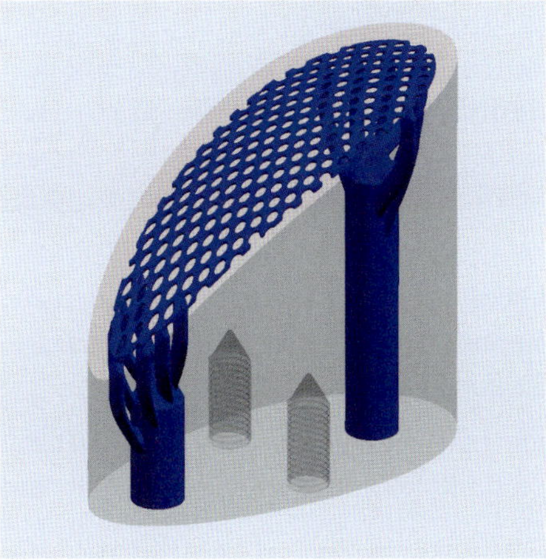

Figure 5.2
Conformal cooling, CAD design of a cooling network (Source: Concept/Hofmann)

The AM processes, especially layer laminate manufacturing (metal foils, Section 3.4), coating with a powder nozzle (Section 3.3), or if one imagines the variation of the powder quality during the coating process (metal sintering or melting, Section 3.2), open up the opportunity to process materials with different thermal conductivities in order to achieve a defined, improved tempering.

The variation of the material properties over the component cross section is implemented in the approach, with different materials, to the so-called metal foil LOM or the metal laminated tooling (MELATO, Section 3.4.10.1). Figure 5.3 shows the CAD model and the finished form for the photocut tool. It is an internally cooled steel mold insert that also has a copper input in the thermally critical zone.

Figure 5.3 MELATO, Tool inserts made of steel and copper, CAD design (far left, second from left), finished tool (far right), and sliced insert with a view of the cooling zone (second from right) (Source: Techel, IWS/rtejournal)

5.2.2 Design Properties of Additive Manufactured Tools

Strategic considerations are directly reflected in the design of additive manufactured tools and tool inserts.

Fast methods are based on the casting of quickly available additive-manufactured master models. The soft or hard molding compounds used for this purpose lead to prototype tools with corresponding characteristic structural properties.

New technological approaches such as conformal cooling are combined with directly manufactured tools and tool inserts suitable for series production. They are used in series production.

5.2.2.1 Prototype Tools

Prototype tools are based on the principle of the separation of properties. Starting points are quickly available as are geometrically exact rapid prototyping master models. They are molded with different molding methods. After the demolding, the molding compounds with the cavity contained therein form the tool.

The molding or subsequent processes used are not all new. They themselves are not additive and not exclusively applicable to additive manufactured parts.

5.2.2.1.1 Soft Molded Tools

Molding processes often indicate soft tools of silicone or rubber. A CAD-based tool design is not required for soft molded tools and also not easily possible, so the tool design is done manually. The parting lines are defined by means of transparent adhesive tape on the model, casting sprues and gates are attached in form of sticks and wires, and the model prepared thus is positioned in a casting frame and cast around with the molding compound. After the solidification, the hardened molding compound is manually sliced with a scalpel at the parting line and the model is taken out. The mold is sprayed with a release agent, closed, secured against opening, and then poured out in a vacuum chamber. The casting process is semimanual. Castable thermosets, known as cast resins, and preferably polyurethanes (PU) are used.

Because the molds are soft, no drafts are necessary. Even slight undercuts can be forcibly demolded. But the elasticity brings about a loss in dimensional stability and dimensional accuracy with it. Large molds deform under their own weight and are therefore reinforced by depositors. The soft mold material receives a fairly high amount of abrasion, whereby the output quantity is limited to about 15 to 20 castings per mold. Fine details, for example, if they are characterized by free-standing domes, can easily tear off, which is why the aforementioned output quantity strongly depends on the geometry. Depending on the cast resin and the mold material, chemical interactions lead to surface defects or to shape wear through the hardening of the material and material fractures. Heavy undercuts can be made by loose parts or multidivided molds, but that increases the complexity of the shape. Inserts made of metal, for example, can be placed in the molds and cast around. The cycle times are quite long, about 1 to 2 hours.

The molds are preheated for the cast. This preheating process can be at least used for the fine-scaling of a main dimension, because silicones have an almost linear expansion behavior in the temperature range around 70 °C.

5.2.2.1.2 Hard Molded Tools

Hard tools results when the molding compound hardens completely, which is the case for example with aluminum-filled epoxy resin. For hard tools, a complete CAD tool design is required. It includes mainly determination of the parting lines, the contribution of the draft angles, the dimensioning of the gating system and of the ejector, and the definition of any necessary loose parts. Slides are usually omitted to facilitate a simple tool assembly. If they can't be avoided, they are milled as loose parts, molded to the original model, and included in the casting process. The tools are preferably uncooled. The embedding of cooling pipes in the casting process is easily possible. For hard tools, each mold half must be individually mounted on forming plates and cast separately.

Hard cast tools are preferably mounted as mold halves into a master frame and the moldings produced on injection-molding machines. This much more complex tool manufacture compared with soft tools comes with the advantage that the moldings can be manufactured from thermoplastics suitable for series production. The application rates are furthermore higher and can achieve, depending on the geometry and material, several 10,000s of pieces. The cycle times are in the range of minutes and are longer when several loose parts must be demolded and inserted manually.

5.2.2.1.3 Hard, Direct Manufactured Tools and Tool Inserts

The direct manufacture of ready-to-use tools and tool inserts is based on an additive processing of tool-suitable metals or ceramics directly in the machine. A complete CAD tool design is an essential requirement for the construction of hard tools. The parts should contain at least the accuracies and surface qualities that non-AM processes have in tool manufacture as well. So of course the parameters for serial tools apply. Currently, these goals are not yet reached directly in the process. However, the accuracies and surface quality apply to conventional tooling as well. Elements produced by chip-forming machines are postprocessed or supplemented by erosion and automatic and manual polishing.

Current developments in the AM processes continuously contribute to improving the surfaces. During the process, this is primarily happening through optimized exposure strategies and closed-loop control algorithms based on measurements of the temperature field in the construction layer and on the detection of the actual geometry through cameras in the process chamber. Additive manufacturing is still only useful for components that cannot be manufactured conventionally well.

In principle, all tool elements that can be manufactured by machining should only be manufactured by machining. Even components that can be purchased as standard elements or as semifinished products with standard element character should never be additive manufactured. Figure 5.4 shows the basic setup of a typical steel standard mold unit [Lucke, 95] for plastic injection molding. The tool itself consists mainly of standard elements, in which the cavity (not shown) is mounted.

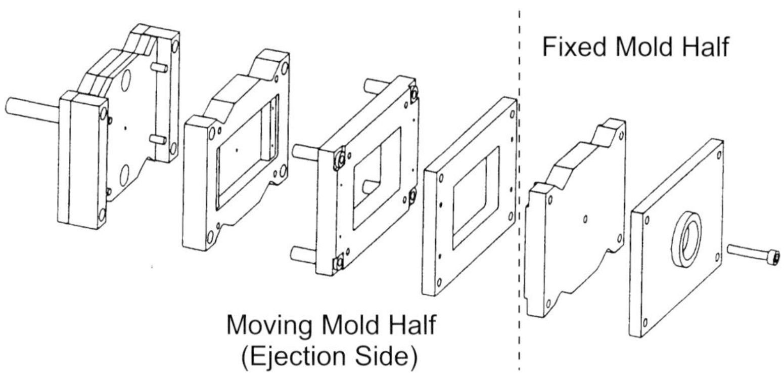

Figure 5.4 Basic setup of a steel standard mold unit for plastic injection molding, exploded view

In Fig. 5.5, a modularly constructed demonstration tool is shown. To illustrate the construction, the panels consist of Plexiglas. The purchased guides, springs, ejector, and the nozzle holder as standard elements are incorporated as originals and easy to see. In the center one can see both halves of the additive manufactured mold insert (cavity). In front of the tool the molding tool can be seen from the front and rear views.

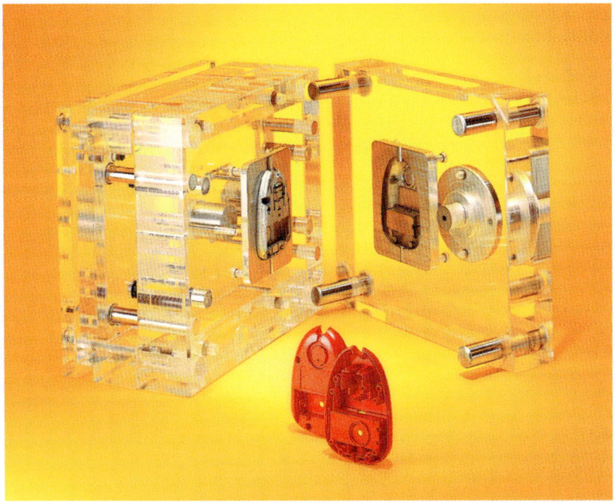

Figure 5.5 Basic setup of an injection mold with plates made of Plexiglas for demonstration purposes (Source: 3D Systems)

In practice, it is useful to combine elements of additive and conventional manufacturing by additive building on a purchased standard element. Some AM metal machines support this, and at Concept Laser (see Section 3.2.8), it is part of the strategy.

It is obvious that the direct AM production of tools requires a peripheral tool design, as is usual in series tooling. Preferably cavities or parts of them are additive manufactured, but never whole mold halves, as is the case with cast tools.

Today's AM tools must have properties that cannot be achieved with non-AM processes. These include the design implementation of the principle of conformal cooling (see Section 5.2.1.2 "Implementation of New Technical Concepts" and Figs. 5.2 and 5.3) and the integration of different materials in a tool. Tools equipped with tempering channels or networks are much less massive than traditionally designed tools. A stress analysis by finite element method simulation and of course a mold filling simulation are therefore part of the tool layout and construction.

5.2.2.2 Supply of Data

Soft cast tools only require an additive manufactured master model. The data path for its preparation corresponds to that of Fig. 5.6.

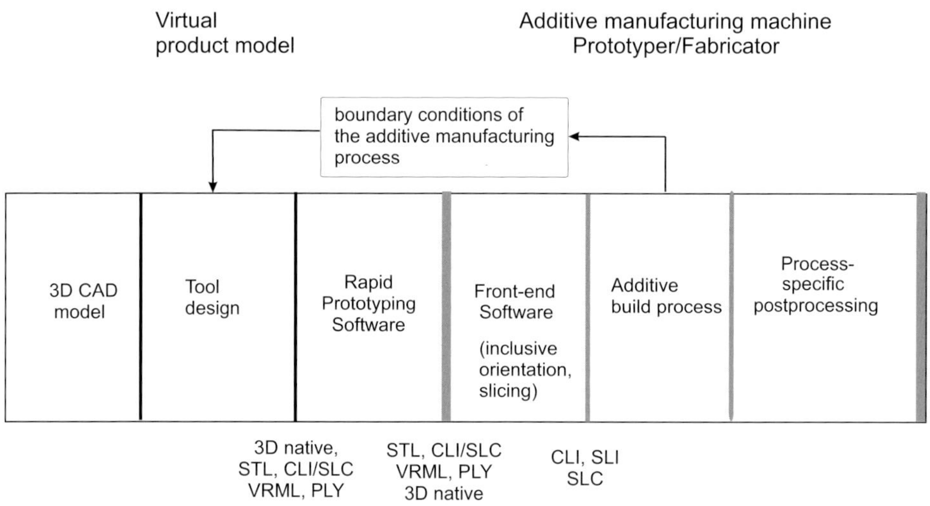

Figure 5.6 Modified data path for rapid tooling

Hard tools require a tool design. Therefore, the data path for the rapid tooling of hard cast or directly manufactured tools and tool inserts must be modified as compared to the basic data path shown in Fig. 5.6.

In rapid prototyping, the AM part could be regarded as a facsimile of the 3D data set because the AM part and the CAD data set are ideally identical; the geometry of the tool and the molding are basically and in some cases significantly different. A tool design must be done that follows the actual CAD design of the target part but is highly cross-linked with this content. Neither, however, can be done without consideration of the later AM process. The orientation in the installation space is essential, such as when different materials are used or steps affect the circularity of the channels or their roughness and thus influence the flow conditions. Also, the effort involved in the metal cutting processes, for example for the manufacture of system holes for ejectors or nozzles, may depend on it. In principle, to manufacture the part perpendicular to the direction of demolding has proven itself. In the mold design, therefore, the later-used AM process must already be selected, and thus the constraints must be incorporated into the tool design.

5.3 Indirect Rapid Tooling Processes: Molding Processes and Follow-up Processes

Indirect methods are not necessarily linked to AM processes, but are implemented at the same level as non-AM master models. They only offer an advantage if the AM master models make the process in total either faster, of higher quality, or even make the process possible at all. Therefore, the consideration of indirect rapid tooling processes must be preceded by a discussion of the suitability of different rapid prototyping processes for the manufacture of master patterns.

5.3.1 Suitability of AM Processes for the Manufacture of Master Patterns for Subsequent Processes

Master models need to have all of the details that are to be seen in the later molding. The same applies to the structure of the surface. Because of the demolding characteristics and the mold wear, usually the highest demands are made on the surface. To enable demold, master models need to have strengths that withstand the demolding forces safely. Therefore the details should not be too fine. No undercuts can occur with hard tools and only the smallest ones with soft tools. Steps that are opposite to the demolding direction must be avoided.

To fulfill these requirements in all processes, a careful finishing of the surfaces, usually through grinding, is necessary. Porous models, such as sintering models or 3D printing models, are infiltrated prior to the demold, and usually are also lacquered. For specific surface effects, textured coatings are also applied. It is important to ensure that these operations do not change the geometry.

For all processes, the compatibility of the materials of the AM master models with those of the mold materials has to be checked. This applies particularly in regard to fillers and lacquers.

The different AM processes are thus differently suited for these follow-up procedures:

Polymerization Polymerization processes are generally well suited because of the high detailing and the good surfaces that come directly from them. Polymerization processes are the AM processes that are mainly used for molding. Because of the soft surfaces, the finishing is relatively easy. Processes without supports or those in which the supports are washed out are advantageous in internal cavities such as flow channels.

In the application of reactive resins, the production of heat has to be taken into account, due to which polymeric materials with low glass transition temperatures can be softened or melted.

Sintering (of plastics) The surface quality has to be improved by fillers. The infiltration of porous components is recommended. The defined porosity of the mold, which is necessary for the vacuum forming of plastic foils, can be set on the process side. Sintering is explicitly supported by some manufacturers, for example, Speed Part (S), see Section 3.2.11.

3D printing To overcome the demolding forces, an increase in strength through infiltration is necessary. In addition, an extensive finishing to achieve the required surface quality is unavoidable. In view of this effort, 3D printing processes are less suitable for molding processes. This does not include lost molds for metal investment casting, for which there are special building strategies, such as Z-Cast from 3D Systems, Section 3.6.2, or DSPC of Soligen, Section 3.6.4.

Models for the manufacture of parts with internal cavities, which can be completely washed out, occur at suitable building parameters and if the infiltration is omitted intentionally. But in this case, the handling is very complicated.

Layer laminate manufacturing Infiltration against the delamination of the layers and paint finish are necessary. Freestanding walls with wall thicknesses less than 0.5 mm and aspect ratios over 10 are barely able to be realized considering the demolding forces. The finishing is quite complicated.

Fused layer modeling or extrusion process The typical surface structure generally requires very complex fillers and grinding. FLM processes therefore are of little industrial importance as master models for follow-up procedures. Extrusion processes that directly manufacture wax models are, on the other hand, very suitable, especially for the investment casting process.

The sequential or molding processes are divided into those that are aimed at plastic components (Section 5.3.2, “Indirect Methods for the Manufacture of Tools for Plastic Components”) and those that are used for the production of metal parts (Section 5.3.3 “Indirect Methods for the Manufacture of Metal Components”); and today, of lesser importance, of ceramic parts. As part of the prototype tooling, mainly indirect procedures for the manufacturing of plastic moldings parts are used.

Metal tools especially occur directly with direct tooling processes.

Molding processes for the manufacture of metal components are preferably associated with casting processes, especially with the investment casting process, and used directly or produced via wax or melt-out models as intermediate processes (see Section 5.3.3 “Indirect Methods for the Manufacture of Metal Components”).

5.3.2 Indirect Methods for the Manufacture of Tools for Plastic Components

The most common nonadditive follow-up procedures for the manufacture of plastic parts based on AM master models are subdivided according to the properties of the tools. Soft and hard tools are molded. The corresponding methods for soft tools or molds are

- vacuum casting,
- nylon casting,
- silicone casting,
- photocasting, and
- spin casting.

For hard tools or molds, the methods are:

- cast resin tools, and
- mask tools (polyurethane casting).

All of these methods are non-AM and are therefore only briefly described next to put them in the context of AM master models.

Soft tools have the greatest technical importance for vacuum casting (room temperature vulcanization, RTV, or silicone transfer molding). Prototypes and very small series are manufactured. For hard tools, the most important application is the injection molding (plastic injection molding) of thermoplastic moldings.

Whereas vacuum casting is used for the manufacture of prototypes, but also often for the manufacture of end products, cast hard injection molds are mostly only used for the manufacture of prototypes and very small series with the character of end products.

5.3.2.1 Casting in Soft Tools or Molds

Soft tools made of silicone or rubber compounds are used for vacuum casting, nylon casting, silicone casting, photocasting, and spin casting. The methods have in common that a master model is molded in a tool, which is usually made of two mold halves. For large undercuts, even more complicated tool designs with loose parts may be necessary.

5.3.2.1.1 Vacuum Casting

Vacuum casting is the most important molding process in prototype and specialized small series production. The parts are created by molding of two-component polyurethane (PU) resins in a silicone tool under vacuum. The advantage of the molding under vacuum consists in the uniform filling of the mold and the prevention of defects caused by bubbles. Also very small parts are produced (for example in the jewelry industry) and very large parts (for example car front spoilers).

Figure 5.7
Front spoiler, vacuum casting, and finishing (Source: CP-GmbH)

The special advantage of vacuum casting is that no CAD tool design is necessary for the tool and that the silicone molds expand at temperatures around 70 °C, nearly linearly with the temperature, and therefore are scalable, which can also be used to compensate for the shrinkage. Therefore the parts, at least in one main direction, can be adjusted by tempering.

The manufacture of the mold happens as discussed in "Prototype Tools," Section 5.2.2.1. The AM master model often gets damaged during the molding. If it remains intact, other molds can be produced for the production of a larger series.

The cycle time–the production time for one part–is around a half an hour and up to more than one hour, depending on the geometry and the size of the part. Here, the actual casting only takes about 4 to 8 minutes; the hardening in the warm (around 60 °C) mold is done in a heating oven and lasts up to one hour. The part can then be removed from the mold. Depending on the vacuum casting material and complexity of the mold, 10 to 30 parts, and in exceptional cases up to 50 parts can be produced out of one mold. For complex geometries, 15 casts per silicone mold are assumed on average. The absolute amount of casts therefore is limited mainly by economic criteria.

Manual and automatic shot sleeves in different sizes are available for production. Industrial models (for example, Fig. 5.8, top) automatically mix the components of the cast resin and fill them through a funnel and a gate hose in the underlying mold (Fig. 5.8, bottom). During the casting process, the silicone mold is located in the lower part of the vacuum chamber.

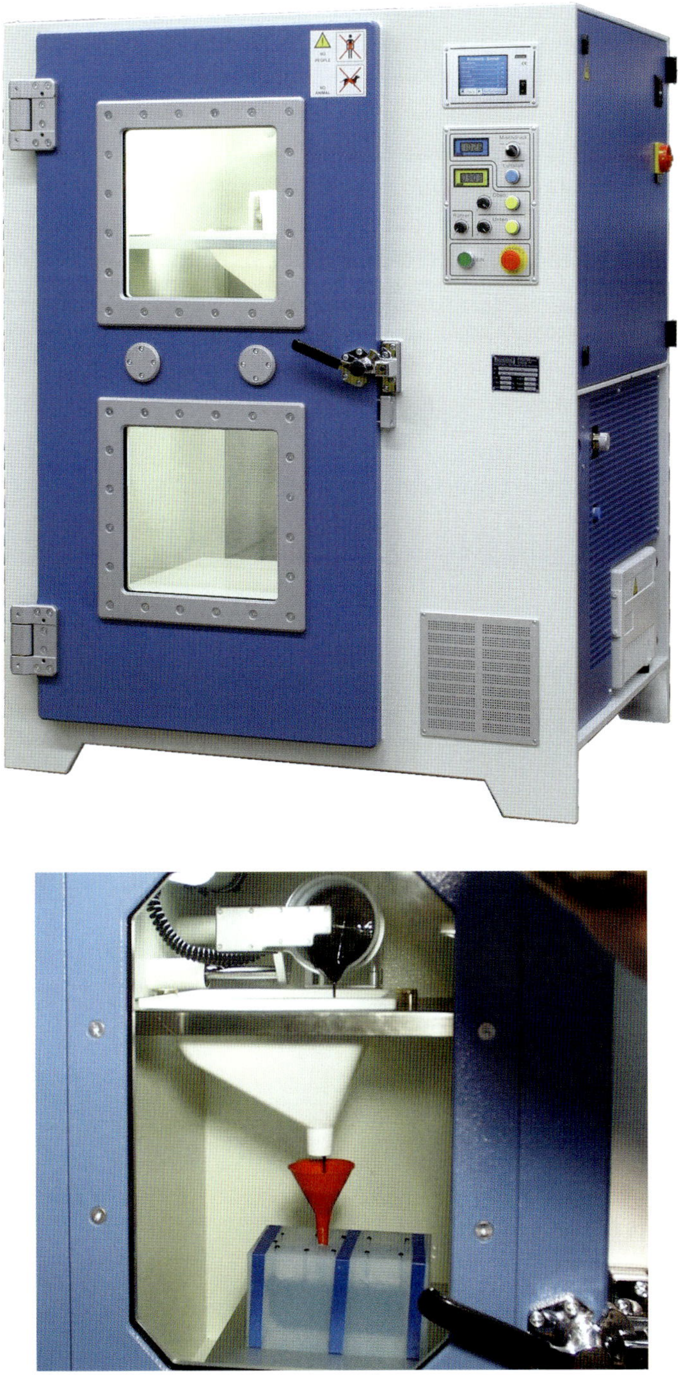

Figure 5.8 Vacuum casting machine (above), process chamber (below) (Source: Schüchl)

Regarding their physical-technological properties, vacuum casting materials cover a wide range of thermoplastics and lead to parts that can be compared, at least for selected properties, with later injection-molded large series parts (see Appendix, Table A3-12, casting resins). They are also available as filled resins containing mineral and metallic fillers or glass fibers to increase the thermal and mechanical strength. There are vacuum castings that are crystal clear (see Fig. 5.9), translucent, or in defined colors, for example, according to Pantone Matching System, colored throughout. The parts can be soft (from approximately Shore hardness 35) or hard, have surface structures according to VDI 3400 (eroded structure), and include inserts of other materials (e. g., metal conductors) or with other colors (simulation of the two-component technology). Depending on the material, they are heat resistant up to about 170 °C.

Figure 4.29 depicts the stereolithography master model of a cocktail cup and the vacuum cast in a highly transparent material. In Fig. 5.9 on the left is shown an opened silicone mold with a cast. To the right are entire system plugs made of different translucent materials. They consist of an upper and a lower part (one mold is needed for each) and the electrically live metal elements.

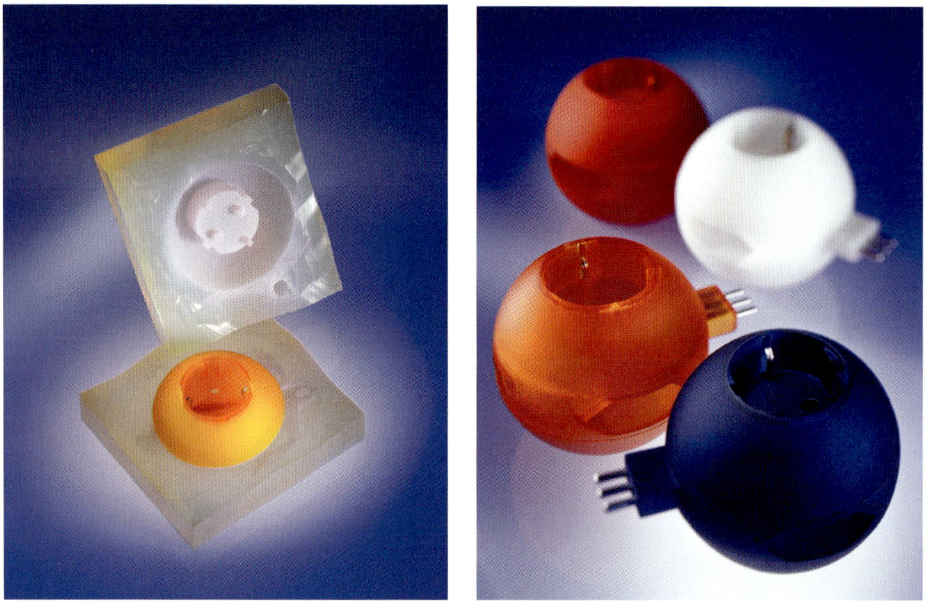

Figure 5.9 Silicone mold with cast (left), mounted system plug made of translucent material (right) (Source: meis van schoen DESIGN, CP-GmbH)

5.3.2.1.2 Nylon Casting

For the casting of polyamides (nylon, PA 6), the company MCP-HEK developed nylon casting as a variant of vacuum casting. For this purpose, a special nylon module is integrated into the machine so that the outlet is located approximately at the location of the mixing bowl, and the material can go from there directly into the silicone mold. Due to the higher temperature, relative to the vacuum casting process of about 140 to 160 °C, temperature-resistant silicone must be used, which is not usually transparent and therefore requires an accurate preparation of the mold separation.

For this process, the upper and lower vacuum chambers must be pressure-tight and shielded against each other, which by default on many machines is the case or can be upgraded.

The module is basically a special RIM (reaction injection molding) device. It mixes after appropriate tempering a reactive polyamide that goes directly into the mold. The process is automatically controlled by PLC (Programmable Logic Controller). The cycle times are specified for 6 minutes; the pure casting process only takes about 2 minutes. According to the manufacturer, the parts have the same properties as injection-molded parts of PA 6. Small ridges, down to 0.3 mm wall thickness, can be cast.

5.3.2.1.3 Silicone Casting

Silicone casting is a very simple variant of vacuum casting for casting in ambient conditions, thus without the use of a vacuum chamber. The process is, based on the manufacturing of the mold, identical with vacuum casting. Several ten cast (sets) can be produced with one mold. Although the process is quite simple, the manufacturable models have unsatisfying mechanical and technological properties. They are able to drive only very small loads and are preferably used as demonstration models.

5.3.2.1.4 Photocasting

Photocasting differs from silicone casting only by the highly transparent mold material used. A photosensitive resin is filled into the silicone mold and cross-linked with UV radiation through the walls of the mold, thus similarly hardened as in the stereolithography process through photopolymerization.

5.3.2.1.5 Spin Casting

Spin casting is a variant of the centrifugal casting process. The rapid prototype positive is put into a rubber compound. This is hardened through vulcanization and then separated along a predetermined separation layer perpendicular to the later axis of rotation. Several of such forms, of the same kind or different, can be disposed, concentrically around an axis of rotation, into a rotating tool. The feed of the casting material is carried out from the inside so that the material is pressed into the molds due to the centrifugal force. In addition to duroplastics, low-melting Zamak[1] alloys are also suitable as cast materials, whereby the spin-casting method represents the connection to the metal casting processes.

5.3.2.2 Casting into Hard Tools

Hard tools are produced through metal spraying, recasting of cast resins, polyurethanes, or proprietary metal alloys (3D Keltool). AM master models are molded. Hard tools require a tool design. Cavities or parts of cavities are produced as inserts, but not usually entire mold halves. The rest of the tool assembly is done as usual. The tools are usually used on plastic injection-molding machines in which they are mounted in steel frames. For the most part, those molding parameters (pressure, temperature) can be achieved that also apply to steel tools in the series.

5.3.2.2.1 Metal Spraying

In the (low-temperature) metal spraying process, plastic rapid prototyping parts are provided with a several millimeter thick layer that is usually made of aluminum. The metal layer is removed as a shell and reinforced with fillers on the back (backup). The resulting cavity is inserted into a tool frame, postprocessed by machining, and completed with standard parts. A part constructed as described and inserted into a steel mold master pattern is shown in Fig. 5.10 [Gebhardt, 97].

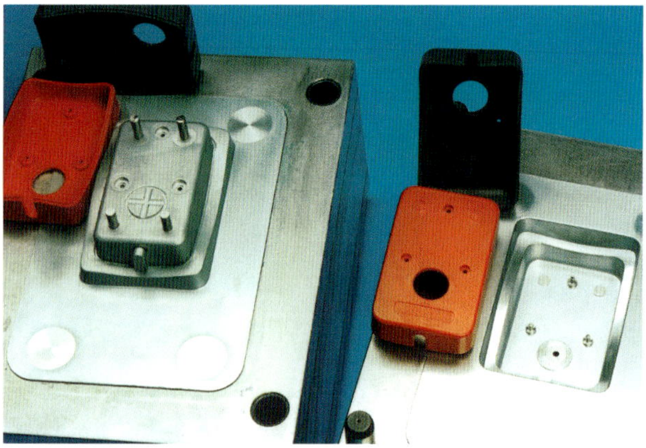

Figure 5.10 Aluminum cavity derived by metal spraying from an SL master model, with epoxy backing, milled insert, and steel master frame (Source: CP-GmbH)

[1] ZAMAK is a low-melting metal alloy. The name is an acronym of the first letters of the ingredients zinc, aluminum, magnesium, and copper.

Metal spraying tools tend to have breakthroughs or cracks in the layer. They arise as a consequence of the softening of the thermoplastic backfill when the heat accumulates locally. The causes are inclusions of air between the model and the metal layer or a nonuniform temperature dissipation at internal corners.

Galvanizing is an alternative to metal spraying in small cavities.

5.3.2.2.2 Cast Resin Tools

Cast resin tools are generated by the casting of scaled AM master models through hardenable molding compositions, preferably by means of aluminum-filled synthetic resins.

The mold casting is separately generated in accordance for each half, by the use of mold plates and frames. The thus-produced tool halves are completed with standard parts, mounted in steel master tools, and operated on industrial injection-molding machines with serial parameters. Relatively simple, preferably two-piece, and mostly uncooled tools result from this. They usually do not have pushers. Undercuts are realize by loose parts.

Depending on the part, several hundred to a thousand parts can be injected in such molds. The cycle time is variable, depending on the amount of loose parts. The application rate is therefore primarily limited by economic criteria.

To increase the lifetime of loaded edges or to locally reach higher accuracies that cannot be reached just by casting, the affected areas are milled out of aluminum, modeled to the master model, and cast. Loose parts can be used in the same way to mold undercuts. They are then molded together with the molding at every cycle, removed from it, and manually again put into the tool afterward. See Fig. 5.11.

Figure 5.11 Cast tool of aluminum-filled synthetic resin cast through countercasting from an SL master model, injection-molded part (material: PA 6 GF) (Source: Elprotec)

Many undercuts increase the complexity of the tool and make it expensive and complicated to use. The advantage of a quick and easy tooling then shrinks continuously.

Cast resin tools can fail because of pores inside or edge chipping. The repair is expensive and requires a generous cutting out of the affected areas and the insertion of contoured inserts that commonly have to be milled out of aluminum.

5.3.2.2.3 Mask Tools, Polyurethane Casting

Polyurethane casting is defined as the manufacture of so-called mask tools of unfilled plastic (polyurethane). Preferably, stereolithography master models are cast. The production is quite similar to those of cast resin tools. The tools are mostly used for the manufacture of lost-wax models for the investment casting, but also plastic moldings can be manufactured by vacuum or nylon casting.

The usually two-part polyurethane mold is cast separately in one prepared aluminum mold counter to the AM master model, whereas parting lines should be created. The targets are stable molds that are dimensionally stable even during the manufacture of large, delicate, or thin-walled wax models. To a large extent, such molds substitute for much more complex milled and assembled aluminum molds for wax spraying. One needs to do a (partially) manual tool design for this process. See Fig. 5.12.

Figure 5.12
Mask process; mold halves in an aluminum frame (Source: BeNe-Gusstechnik)

5.3.2.2.4 Low-Pressure Injection Molding, Reaction Injection Molding

In low-pressure injection molding or reaction injection molding (RIM), a two-part closed mold is filled with a two-component reaction resin (PUR), which is hardened in the mold. Essential for this is a machine that mixes the two components and presses it into the mold under 4 to 6 bar. The mostly undercut-free mold is created by the milling of model materials (e. g., Ureol) or aluminum, but it can also be produced directly by sintering methods or indirectly produced by casting processes. The process is advantageous for molds of great length because the thin fluid material can fill and cast even thin wall thicknesses. RIM is also used as a series production process.

5.3.2.2.5 3D Keltool, Course4 Technology

The Keltool Course4 Technology process is based on the long-term, low-temperature sintering of a special metal-plastic mix powder.

From an AM, preferably stereolithographic master model within an intermediate casting, a high-temperature silicone cast is produced (see Fig. 5.13). This is cast together with the molding compound of polymer-bound tungsten carbide and tool steel, which changes into a hard mold then. Depending on whether the positive or the negative is used as the master model, a single or a double casting process is necessary.

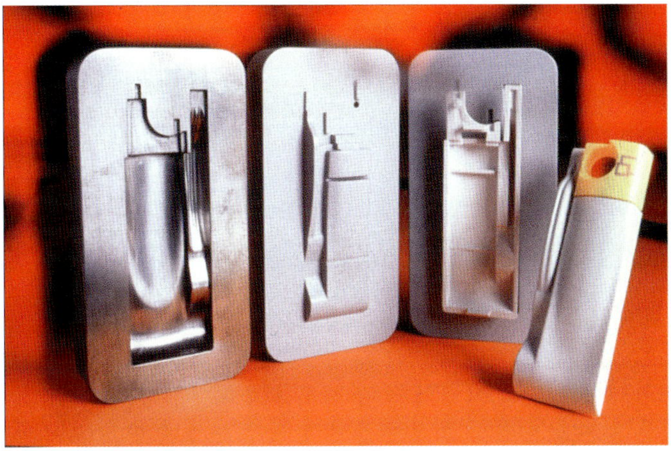

Figure 5.13 Keltool: both tool cavities (left), left cavity with enclosed injection molded part (second from right), and injection-molded part (right) (Source: 3D Systems)

The advantage of this method is the very detailed cast that results.

Figure 5.13 is an example the two mold halves of a Keltool mold insert, one injection molded part still in the mold, and the finished product.

A variant of this process allows the use of copper-tungsten carbide alloys for the manufacture of electrodes for the die sink. Surface qualities up to RA (surface roughness) = 0.4 μm are possible.

The Keltool/Course4 Technology process is closely related to the *low melting point metal* (LMPM) method, at which the resulting mold consists of a low-melting metal alloy (see Fig. 5.13).

This process currently has no great industrial importance.

5.3.2.3 Other Molding Techniques for Hard Tools

Closely related to metal spraying are methods that apply ceramic layers onto the master models. Because of the relation to the manufacturing of lost ceramic molds for investment casting, they are also called *Shaw methods*. The developmental aims are thin layers that can be applied quickly and at the lowest possible temperature in the required layer thickness. Ceramic intermediate layers are also common. The material for the backfilling of the thus-obtained mold shell must combine mechanical strength, good heat condition properties, a good adhesion behavior on the rapid prototyping model, and a good processability. If necessary, spatially curved preformed cooling channels can even be inserted and cast into the backfilling.

To these non-AM processes, though mostly described by the suppliers as rapid tooling processes, also belong the Ford Sprayform method and rapid solidification process (RSP) tooling method.

5.3.2.3.1 Ford Sprayform Method

Ford Sprayform is the current description of the method from Sprayform Holding Ltd., which was taken over by Ford. Via intermediate ceramic shells, steel molds with series character are produced, which are suitable for plastic injection molding, metal die-casting, blow molding, and deep drawing of sheet metal. This process is used for production internally as well as offered as a service to external customers.

5.3.2.3.2 Rapid Solidification Process

The rapid solidification process, also called RSP tooling, is closely related to the Sprayform method. A steel layer is sprayed onto substrates of ceramic that are cast from rapid prototype master models. After cooling, the mold is fired, bound, and backfilled. This method is offered by special services providers.

Further casting methods for the manufacture of tools, for example, CEMCOM (nickel coating), dynamic tooling, express tooling, Kirksite[2], and others are not represented in Germany, or are so only via specialist service providers. Like all molding processes, they are above all not AM processes. Therefore, they are not discussed further at this point.

---

[2] Low-melting zinc aluminum alloy (94% Zn/6% Al).

5.3.3 Indirect Methods for the Manufacture of Metal Components

One of the most important indirect methods for the manufacturing of metal parts is *investment casting*. Investment casting is non-AM and is independent of the type of manufacturing method of the mold. Therefore, series identical parts are created if suitable models and molding compounds are provided. Parts made of aluminum, titanium, or magnesium are manufactured. The additional procedure of investment casting thus can be seen as a multistage rapid manufacturing process.[3]

5.3.3.1 Investment Casting with AM Process Steps

Investment casting (wax melting process) and its traditional and alternative AM process steps are displayed in Fig. 5.14.

The starting points are lost-wax models (grape cast) that are usually mounted to a grape for casting, with risers also made of wax.

Traditionally, lost-wax tree molding are produced by wax injection in a milled aluminum wax injection mold (left section). Alternatively, scaled plastic master models are additive manufactured. With these master models a silicone mold is created as in vacuum casting. By injecting or casting the wax, the number of wax patterns needed for the cluster mold is obtained out of the silicone mold.

The wax is less abrasive than the polyurethane in vacuum casting and therefore permits the manufacture of a much larger number of wax models from one mold (usually around 50 or more). Because wax reacts like a thermoplastic, a special heated wax container should be used for the casting, which should take place in a vacuum chamber.

In some vacuum chambers, the upper and lower chamber halves can relatively easily be separated into two chambers with different pressure levels. That allows, by means of an appropriate module to increase the pressure in the upper chamber, an increase in the quality of the wax models and extension of the range of usable waxes to highly filled variants.

One variant that completely avoids the tool is the direct additive manufacturing of the wax model. This is particularly true for large components such as cylinder blocks or unique items, which are then cast as single pieces and not as lost-wax tree molding. Often, no wax is taken, but a lost sintered amorphous plastic model that is waxed or manufactured by extrusion or by 3D printing.

[3] Casting applications in terms of a business solution are discussed in Section 6.5.2.2.

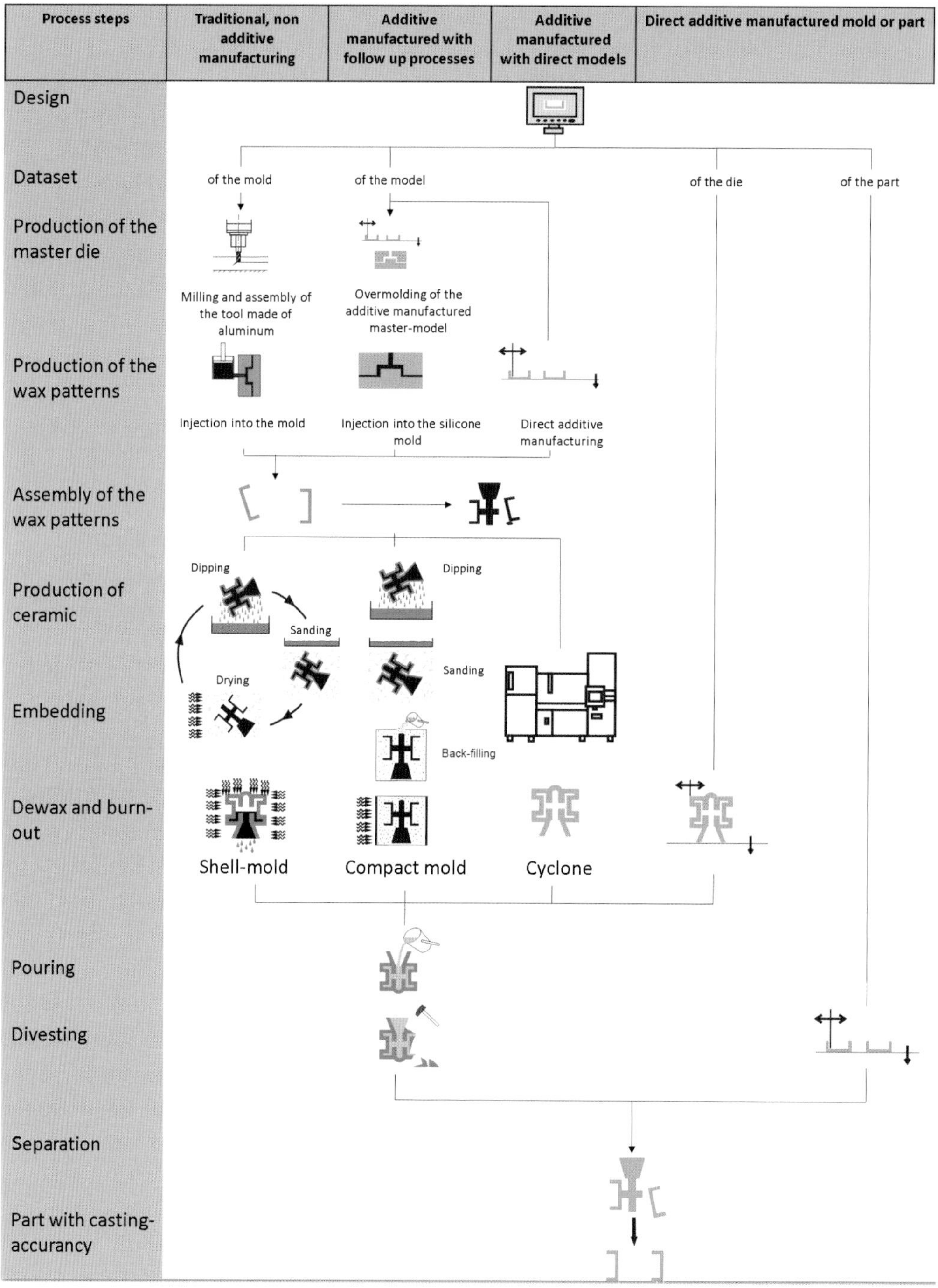

Figure 5.14 Investment casting (wax melting process); substitution of non-AM process steps by AM process steps

In order to use stereolithographic parts directly in investment casting without risk of breaking the shell (shell cracking) during firing of the master model, nearly all manufacturers have developed special hollow construction methods. At 3D Systems it is called *Quick Cast* (Fig. 3.13), at EOS, *Skin and Core,* and at Fockele & Schwarze, it was called *Fast Sculp*. Figure 5.15 shows such a lost stereolithographic model and the molded part.

Figure 5.15 Hollow constructed lost stereolithographic model (Quick Cast) and associated investment casting component (Source: 3D Systems)

Classically, in several working steps, the cast grapes (or wax trees) are immersed into a ceramic slurry, sanded, dried, and then immersed again. This process is prolonged and takes several days until a 4–8 mm thick layer forms, which is smooth on the inside. On the outside, it becomes increasingly rougher as the sand grain gets coarser. This is independent of how the wax tree was manufactured. Common attributes are the shell shape, in which the self-supporting shell arises over several cycles, and the compact form, in which only the sanding of the surface is done, and the backfilling, as well as usually also a cuvette to ensure stability (Fig. 5.14).

An automated process (Cyclone, MK Technology, 2004) enables the method to be reduced to a few hours [Weihnacht, 05]. The automatic process is based on an optimized molding compound and particularly on an accelerated drying process. The machine integrates an automatic silting and sanding as well as infrared drying and manufactures shells up to a part size of 500 × 500 mm. The finished ceramic mold is fired in the machine. In conventional processes, this happens in an external oven. The machine, which was once developed for prototype and small series casting, is meanwhile offered as interlinked modules for series production.

In prototype construction, instead of ceramic molds, one will also find plaster molds as a standard, which are cast in cuvettes.

Alternatively, the entire mold can be additive manufactured, for example, with a variant of 3D printing, the DSPC method (Section 3.6.4), the Z-Cast process (Section 3.6.2), or the ExOne machine (Section 3.6.3.2), or directly manufactured by sintering, for example, Phenix (Section 3.2.10), out of the casting compound.

The entire mold is molded with liquid metal and destroyed after hardening and the molded model released of the ceramic adhesive. After that follows the cleaning, the removal of risings, and a postprocess that is mostly cutting.

Because the required metal part can be directly additive manufactured, there is a further alternative method. For example, metal sintering and melting processes that are also suitable for the direct manufacturing of metal tools (Section 5.4.3 "Direct Tooling: Tools Based on Metal Rapid Prototype Processes" or Section 3.2), are used. The direct manufacture of metal parts includes the complete substitution of investment casting by an AM process.

Because the availability of aluminum in the AM sinter process (concept, Section 3.2.6, 2006 and MCP-HEK (today SLM Solutions), Section 3.2.5, 2007), a method for simulating and evaluating aluminum investment casting parts is also available. Of course, especially for direct additive manufacturing (rapid manufacturing, Chapter 6), it has a great potential.

Literature on this topic includes [Ambos, 95], [Shellabear, 94], and [Weihnacht, 05].

5.3.3.2 Tools by Investment Casting of Rapid Prototyping Master Models

All rapid prototyping models (positives) can be cast in moldable tools, and out of these, metal parts can be produced by investment casting.

The classic investment casting and AM alternatives are described in Section 5.3.3.1, "Investment Casting with AM Process Steps." Of course, it is also suitable for the manufacturing of tools and tool inserts, preferably of aluminum, by casting.

In practice, investment casting, especially for the production of positives, has proved its worth. For the manufacturing of tools and tool inserts, this process is fairly uncommon.

5.4 Direct Rapid Tooling Processes

Tools and tool inserts that are directly manufactured in the AM machine can be made of plastic. They are then allocated to the category prototype tooling and possibly also to bridge tooling. But in focus here are tool components of metal, possibly of ceramics, that belong to the category direct tooling. This can be entire cavities, but preferably exchangeable inserts as slides or the like. They are a part of series tools and must have appropriate properties.

The method is referred to as having one or more stages, depending on whether the finished part comes out of the machine, in the sense of AM, or if process-related steps like an infiltration are necessary.

The manufactured parts must be postprocessed with non-AM, manual, or automatic methods and integrated into a tool.

5.4.1 Prototype Tooling: Tools Based on Plastic Rapid Prototyping Models and Methods

Prototype tooling describes processes for the fast and cheap manufacturing of a tool for a small number of molds. Therefore, compromises are generally taken into account concerning the dimensional accuracy, the cycle time, the material, or the unit price. For injection-molding applications, it is important to note that the temperatures and pressures commonly used in conventional tools and large series and, consequently, the flow conditions in the melt are usually only approximately accessible with prototype tools. Therefore, the tools must be designed for lower pressures and temperatures, larger flow cross sections, a greater number of gates, and the flow behavior of the plastic. In any case, one has to check whether and to what extent these abstractions limit the informational value of the tests.

In the following, methods that are based on plastic rapid prototyping methods are presented for the manufacture of prototype tools.

5.4.1.1 ACES Injection Molding

Polymerization processes are the most exact AM processes, in terms of the tolerances and surface quality. A direct application of stereolithographic tools therefore makes sense. Because of the glass transition temperature of the stereolithographic material of 55 to 60 °C, it was long considered to be impractical. Nevertheless, systematic developments led to the ACES injection molding process (AIM). Using a special stereolithographic building style, ACES (Accurate Clear Epoxy Solid, 3D Systems), cavities are manufactured and directly used for a very small series up to 200 pieces as inserts in injection-molding tools (Fig. 5.16). These are usually very simple on-off tools. For this purpose, the AIM mold halves are inserted into default master frames and backfilled with epoxy resin.

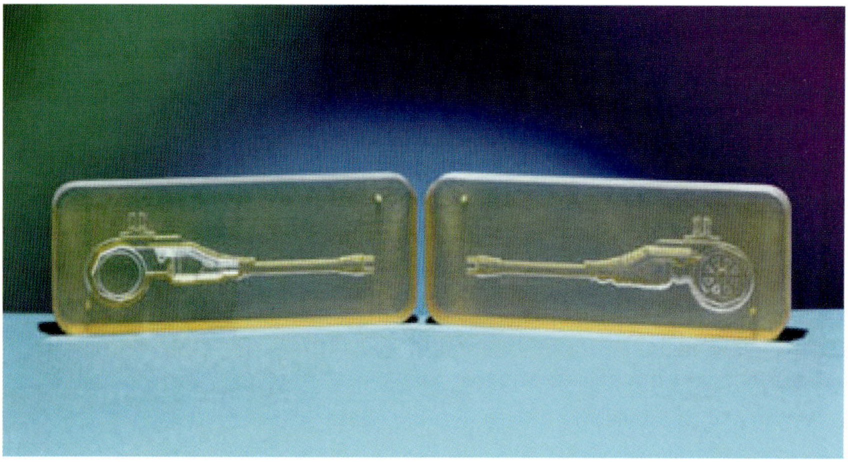

Figure 5.16 Tool inserts of stereolithographic material for use in plastic injection molds (AIM) (Source: 3D Systems)

Compromises are made primarily during the cycle time, which usually takes several minutes. In spite of several successful applications, this method is not of great importance.

That could change with the application of higher-duty, filled materials or composite materials. This is supported by the trend toward smaller quantities and smaller part weights.

This method aims to bridge the economic gap between vacuum casting and injection molding. It is therefore attributed to bridge tooling. At least one comparison published by the manufacturer between an AIM and a conventional tool with respect to the manufacturing time, 4 hours versus 90 hours, is impressively in favor of the AIM process [Jacobs, 97].

A transfer to other AM processes has not yet been published.

The process is not to be confused with the proprietary method of the service provider AIM Plastics, Clinton Township, Michigan.

5.4.1.2 Deep Drawing or Thermoforming

The deep drawing of plastics is also called vacuum deep drawing or thermoforming. The contour of a permanent model is transferred to films or sheets of plastic that are mostly of polystyrene, polyethylene, or ABS. Preferably, films starting at a thickness of about 0.1 mm, but also sheets with a thickness up to several millimeters, are processed.

The film is stretched in a frame and heated by infrared lamps so that it is deformable but remains safely within the thermoelastic temperature range. The frame is positioned above the model and sealed against it. Between the heated film and the model a vacuum is drawn so that the deformable film abuts against the contour of the model. After the cooling the film retains its shape. The method is mainly used for large tools with simple topology.

To draw the vacuum, holes and slices must be located in the model, but a special porous semifinished product is also usable. In AM, sintering processes are used. The required porosity can be adjusted via the process parameters. The current material of selective mask sintering, In Section 3.2.9, a glass-filled polyamide is expressly designed for the production of porous structures for vacuum forming.

Except for sintering and possibly 3D printing, which has not been done so far, other AM processes aren't applicable, as the required porosity cannot be assured in the process.

5.4.1.3 Casting of Rapid Prototyping Models

In principle, all rapid prototyping parts can be manufactured as negatives and then used as molds for casting. This requires that all details can be reproduced according to the requirements; thus, the surface quality and the casting materials are matched to each other so that demolding is possible and a postprocessing of the cavities is either not necessary or at least possible without a loss of accuracy. EOS promotes the process as *direct pattern*. The example in Fig. 5.17 shows the sintered mold of PA for a shoe sole, which is cast to optimize the geometry of soles for rain boots with elastic material. The mold is used to vary the product in the shortest time.

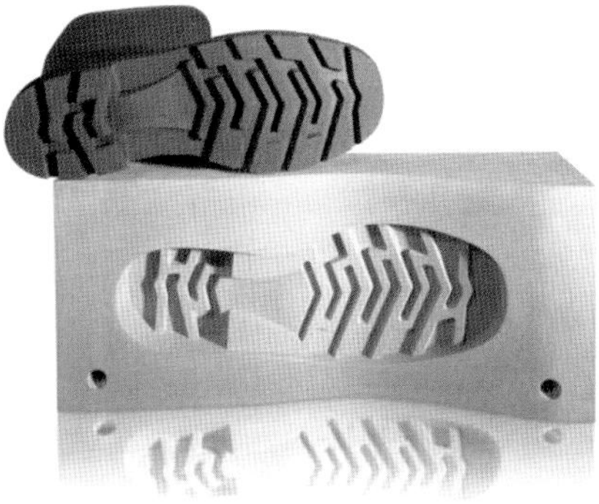

Figure 5.17
Direct pattern; sintered pattern of PA for a shoe sole product (Source: EOS)

5.4.1.4 Manufacture of Cores and Molds for Metal Casting

Molds and cores are tools or tool components for metal casting. They are directly manufactured with different methods.

5.4.1.4.1 Sand Casting

Sand cores and sometimes also molds are manufactured by laser sintering or 3D printing.

Based on the plastic process, EOS offers its own series of sintering machine for the processing of foundry sands: EOS S (Section 3.2.4). With this, in direct casting, the direct manufacturing of cast parts is possible whose cores cannot be manufactured conventionally or whose quantities do not make sense for an economic manufacture.

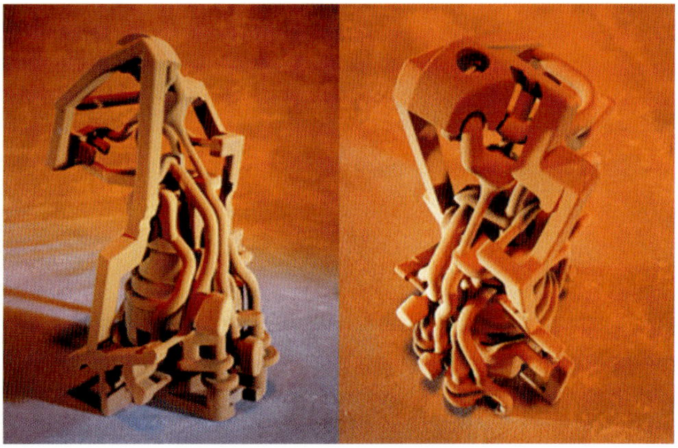

Figure 5.18 Direct cast sand cores (Source: EOS)

The materials are quartz sands or silicates, for example, quartz (90%) and zircon (10%) mixtures. The EOS materials, Quartz 4.2 and Quartz 5.7, are phenolic resin coated quartz sands that are seen as standard materials.

Ceramics 5.2 is a phenolic resin coated aluminum silicate sand (synthetic mullite). Because of its high heat capacity and low thermal expansion, it is especially suitable for high-temperature casting.

The ExOne GmbH distributes the S-Max, and the smaller S-Print derived from it, for the manufacture of cores, core packets, and entire molds for sand casting (Section 3.6.3.2); their system was previously also known as the Direct Core and Mold Making System. In particular, the big machine can be developed for a partially automated manufacturing machine. It is a casting machine.

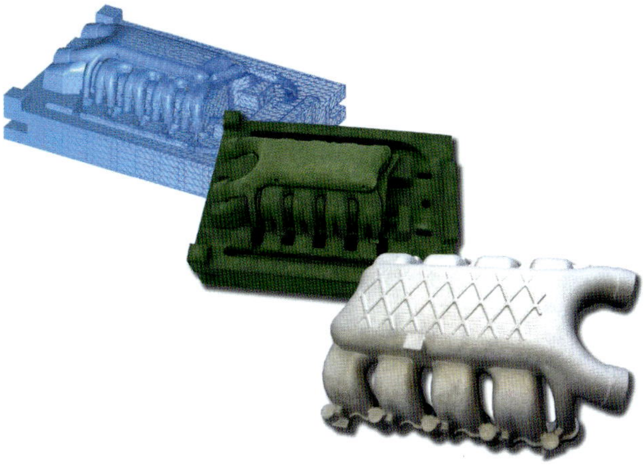

Figure 5.19 3D prints of sand (Direct Core and Mold Making System) CAD (left), sand mold (middle) with inserted core packet, and casting part intake connection (right) (Source: ExOne)

5.4.1.4.2 Die-Casting

Permanent molds for metal die-casting can be manufactured with the same direct methods as the permanent molds for a plastic injection mold (Section 5.4.3.2, "Single-Component Metal Powder Methods: Sintering and Additive Manufacturing"). There are applications mainly in magnesium die-casting [Reinhart, 98].

5.4.2 Metal Tools Based on Multilevel AM Processes

By 1993 there were the first approaches to the sintering of metal powders. They were made on the basis of plastic-coated metal particles in unmodified plastic machines.

One variation of coating is the embedding of the metal particles in a paste-like polymer; another is the 3D printing of metal powders with polymer binders.

All variants lead to green parts, which must be relieved and sintered to obtain dense and robust tool components.

5.4.2.1 Selective Laser Sintering of Metals: IMLS by 3D Systems

The first metal process by 3D Systems works according to the indirect method. It was formerly known as the IMLS (indirect metal laser sintering) process. Later this designation from 3D Systems was no longer used, and now it has been replaced by the SLM process. Because many systems have been sold, it is discussed briefly here.

The process has two stages. In the AM machine, a metal powder mixture is sintered by using a polymer binder contained therein to form a green part. The process and the associated material LaserForm A6 are described in Section 3.2.3.

With this method, tool inserts are manufactured for preproduction but basically also manufactured for the start of series production. Figure 5.20 shows a component that is inserted as a core slider in a multipart injection-molding tool [Breitinger, 95].

Figure 5.20 Core slider after the selective laser sintering of metals (IMLS) (Source: Breitinger)

Functional prototypes can also be manufactured that have the final product character. One example is the turbine casing of an exhaust turbocharger for hot gas tests (Fig. 6.26, Section 6.5.1.1).

5.4.2.2 Paste Polymerization: OptoForm

The stereolithography process of the paste polymerization OptoForm is described in Section 3.1.13.1 and can be used with the appropriate filling of the resins for the manufacture of ceramic and metallic molds and tools. OptoForm has published appropriate examples of ceramic forms.

5.4.2.3 3D Printing of Metals: ExOne

The 3D printing of metal parts with the M-Print (so far also known as a direct metal printer) and with the M-Lab is described in Section 3.6.3.1. This AM process does not need any supports or bases. The process is cold and therefore causes hardly any warps in the construction. After the building process follows a sintering and infiltration step that codetermines the manufacturing time and the material properties. For the construction of internal cavities, as in a conformal cooling, it should be noticed that they must be cleaned of supporting powder. Examples of tool inserts are given in Fig. 5.21.

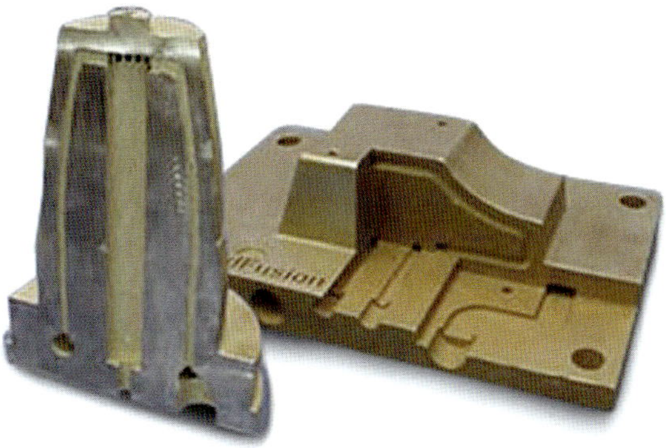

Figure 5.21
Prometal direct metal printing tool insert with internal conformal cooling channels (Source: Ex-One)

5.4.3 Direct Tooling: Tools Based on Metal Rapid Prototype Processes

The aim of AM tool making is the manufacture of metallic tools and mold inserts by direct processing of single-component metal powders. Here, the materials should be based solely on the demands on the tools. Compromises that are forced by some of the processes described previously will no longer be necessary.

The demand for standard materials is widespread and reasonable, especially from an economic point of view. However, in terms of the technical aspects, it is only superficially compelling, because the material properties not only depend on the powder, but arise in the AM process, or at least are largely influenced by it (see Section 2.3.2.1), and therefore must be qualified in the context of the process.

Direct AM processes are also suitable for the manufacturing of cores and molds for metal casting. Those applications of the casting technology are discussed within the context of the single processes in this chapter. They are also discussed together with the machines in Chapter 3.

5.4.3.1 Multicomponent Metal Powder Laser Sintering

An intermediate step on the way to one-component powders is the processing of multicomponent metal powders. They consist of a mixture of high- and low-melting metal powders. In the process in a modified plastic sintering machine, the low-melting component is melted and acts as a binder for the solidification of the high-melting component. The binders are thus components of the part and will not be removed. The methods are single-stage. Infiltrations are useful or necessary depending on the porosity.

The development was driven forward in the mid-1990s, particularly in the wake of the EOS Electrolux process. The construction material is a multicomponent powder (direct metal) that compensates for the resulting shrinkage through components with different expansion coefficients.

Figure 5.22 shows a tool with four pushers (1998), whose single parts are all manufactured by the direct metal process. The powder used was a Direct Steel 50V2 from the Electrolux powder. The components of the tool were fitted by wire EDM.

Figure 5.22
Direct metal (1998) with EOS Electrolux multicomponent-material tool with four pushers (Source: EOS)

The process was further developed by improved laser sources, scanning strategies, and scanners, a protective gas atmosphere, and especially by new materials into a unique one-component process.

The current process is described in Section 5.4.3.2.1, next.

This process provides a good base for the development of materials of pseudo-alloys, those compositions that cannot be produced by melting metallurgy.

5.4.3.2 Single-Component Metal Powder Methods: Sintering and Additive Manufacturing

To overcome the limitations of the plastic-based or multicomponent processes with regard to process and material, direct one-component sintering processes for the processing of tool steel, but also of CoCr steel, titanium, and aluminum, and the corresponding machines have been developed worldwide. Variants are also suitable for the processing of ceramics. They are described as beam melting or selective laser melting (SLM).

The problems of development included not only a higher local sintering temperature in the machine, due to the higher melting points, but also some physical effects that lead to greater porosity than in the comparable plastic sintering process. Due to the higher surface tension of liquid metals during melting, large droplets and, after solidification, a very porous structure arise. In addition, an oxide film that worsens the wetting properties is formed. Beyond this, the thermally induced stresses produce distortions. In view of the high melting temperatures of metals, most of the technical solutions do not provide the heating of the entire powder bed to a few degrees below the melting temperature, in analogy to the plastic heat sintering process. However, heating of the building platform and panel heaters is quite common.

Finally, advanced lasers with high intensities and very good beam quality, such as disk and fiber lasers, optimized scanning strategies, and appropriate protection and process gas atmospheres led to a breakthrough. Currently, at least five one-component metal processes are available worldwide that manufacture the part in the powder bed; four of these come from Germany. Beyond this, the material range is constantly being expanded by the machine manufacturers and increasingly by independent material manufacturers and distributors.

5.4.3.2.1 Direct Tool, EOS

The application of direct metal laser sintering (DMLS) by the company EOS, Section 3.2.4, is described as the DirectTool process. Steel, stainless steel, CoCr steel, martensitic hardenable steel, and titanium are available. The grain size of the powders goes down to 20 µm, so details with characteristic dimensions down to 0.5 mm are possible.

In the literature, time reductions of 50–80% are given in comparison to the traditional tool making. These values only apply to the special case and must be checked for individual applications. Figure 5.23 shows a two-cavity mold with sintered mold cavities that is manufactured with the DMLS process.

Figure 5.23 DMSL process, DirectTool, two-cavity mold with sintered mold cavities (Source: EOS)

5.4.3.2.2 Laser Melting, SLM Solutions

The laser melting process that is based in the machine family SLM Solutions GmbH is described in Section 3.2.5. An example for a mold insert is shown in Fig. 5.24.

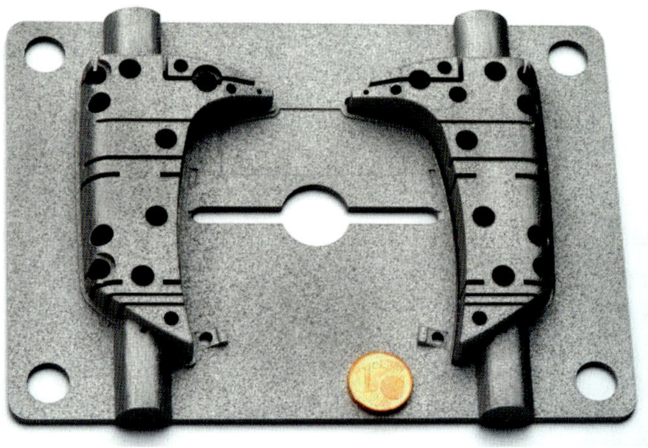

Figure 5.24 Laser melting, mold insert (Source: MCP-HEK)

An application for the manufacturing of a forming tool is shown in Fig. 5.25. The constructed die that was additive manufactured on a substrate is clamped onto a bending machine and used to transform the sheet metal parts shown.

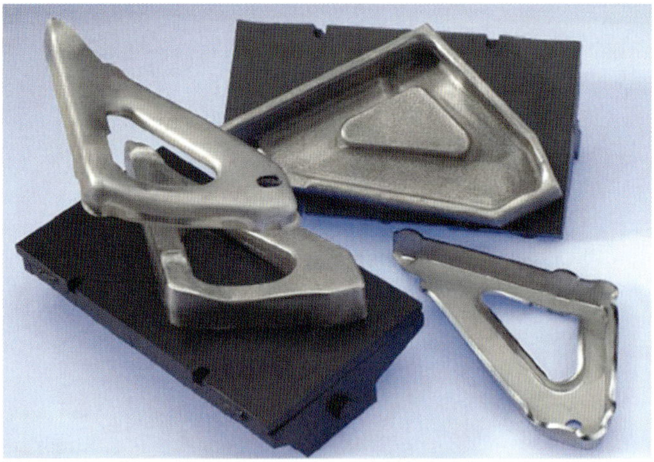

Figure 5.25 Laser melting, die for sheet metal processing, transformed parts (Source: MCP-HEK)

5.4.3.2.3 LaserCUSING, Concept Laser

"LaserCUSING" creates dense parts and is based, like laser melting, on the entire melting of the material. It is described in Section 3.2.6. Tool making is the focus of the development. This is reflected in the materials, which also include the hot work tool steel CL50WS (hardness 54 HRC), but especially in the design of the machines. The construction of the tool components is preferably happening on system holders (e.g., Erowa 3 R), so postprocesses such as milling, drilling, and EDM are possible without any complicated alignment. Exchangeable plug-ins for 3D ablation and for labeling are also based on the needs of tool making. See Fig. 5.26.

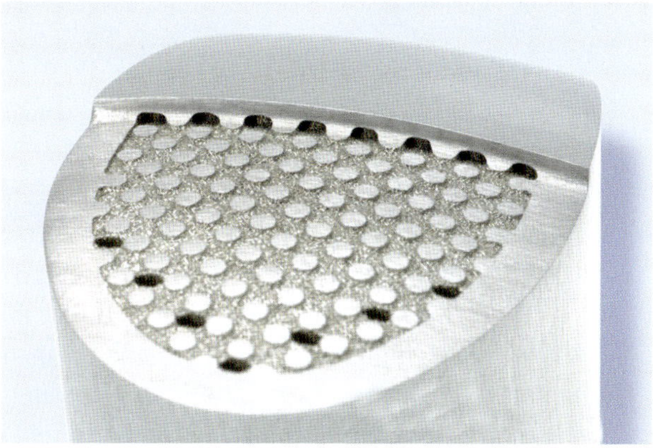

Figure 5.26 LaserCUSING, mold insert with conformal cooling, sliced part (Source: Concept Laser)

Published application examples concentrate on the further development and application of the conformal cooling. This applies to both complex structures for conformal cooling, which may include the entire cavity as loops or meshes, and cooling pins (Fig. 6.30), which are sold as standard parts (Section 6.5.2.1).

5.4.3.2.4 Direct Laser Forming, TRUMPF

The laser melting process of TRUMPF, Laser Forming, FL, is described in Section 3.2.9. The two-construction place concept emphasizes high productivity. The machine is designed as a universal machine for the production of different metal components. This includes applications for tool making. Figure 5.27 shows a mold insert of 1.2343 (53 HRC) for a large series tool for the manufacture of a toothbrush head after additive manufacturing and in the postprocessed, ready-for-use condition. The generated part also contains start drillings for wire EDM. The machine is no longer manufactured, but the machine and its parts underline the potential of tool making through AM.

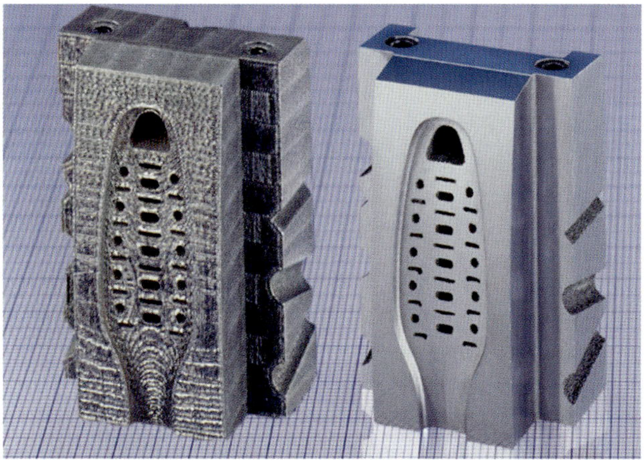

Figure 5.27 Laser forming; mold insert with internal cooling channels after the AM process (left) and postprocessed, ready to use (right) (Source: TRUMPF)

5.4.3.2.5 Electron Beam Sintering, Arcam

Arcam calls its metal melting process electron beam melting (EBM). It is described in Section 3.2.8. The published applications focus on the direct manufacturing of components for the air and space industry (Arcam A2) and for medical technology (ARCAM Q10). The material developments therefore concentrate on Ti and CoCr alloys. Of course these applications are also have potential in tool making. Figure 5.28 shows a tool insert of 1.2344 (H13) steel.

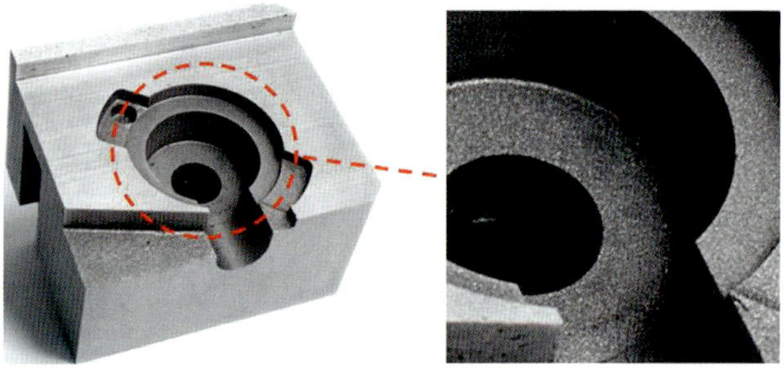

Figure 5.28 Electron beam melting, cavity, postprocessed (Source: tomskcad, Arcam)

5.4.3.2.6 Laser Sintering, Phenix

The Phenix method for high-temperature sintering of metals and ceramics is described in Section 3.2.12. The unique selling proposition of Phenix is the sintering of ceramics into intricate end products. This processing of metals concentrates more on materials for the air and space industry and medical technology and also contains, for example, nickel-free CoCr powder for the manufacturing of dental implants. Applications for tool making are likely to remain in the background. An example cavity is shown in Fig. 5.29.

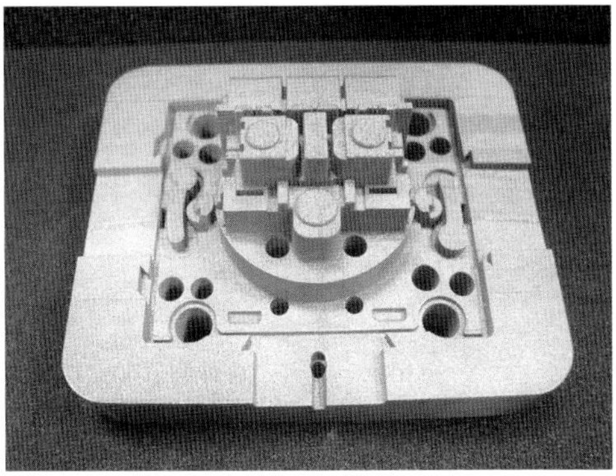

Figure 5.29
Phenix, laser sintering, cavity
(Source: Phenix)

5.4.3.3 Laser Generating with Powder and Wire

Both coating with a powder nozzle, see Section 3.3, and deposition welding with wire, see Section 3.7.1, are more of a repair than a manufacturing process. As processes of laser material processing, they work with power lasers in the kilowatt range, providing high melting performances. This results in layers that have to be finished by machining.

Through optimized-protection gas pipes, preheating, and building strategies, the classic problems like lack of surface quality, warp, and insufficient density are increasingly better controlled. Powder-based nozzle concepts have the advantage of the diversity of the manufacturable powder mixtures over melting with wire, and they have the advantage of a simple machine setup over sintering or melting in the powder bed. Powder nozzles can generally be moved freely in space and are not bound to the stepwise-construction horizontal layers.

5.4.3.3.1 Laser-Engineered Net Shaping, Optomec

The laser-engineered net shaping (LENS) process (Section 3.3.2) is based on the principle of laser coating with the powder nozzle in a protective gas atmosphere. The focus is on the material development and thereby on the manufacturing and repair of parts for the air and space industry and on applications in medical technology. Tool concepts with conformal cooling and inserted cooling zones are rarely published. See Fig. 5.30.

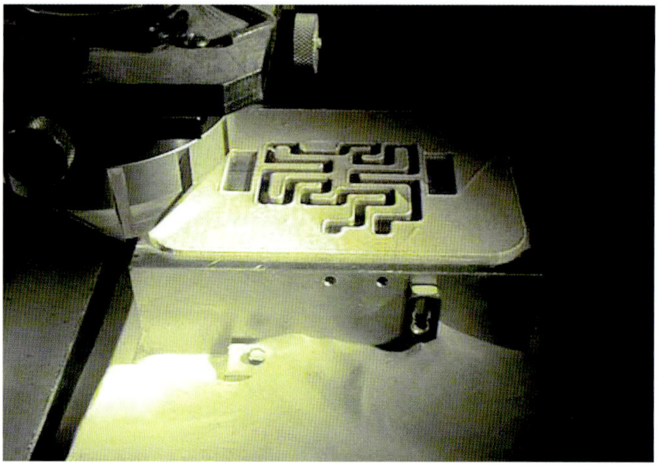

Figure 5.30 LENS, cavity during the construction process (Source: Optomec)

5.4.3.3.2 Direct Metal Deposition

The direct metal deposition (DMD) method from TRUMPF is described in Section 3.3.3. Besides repair strategies and partial hard coatings for forming tools, a so-called bimetal tool has also been released. This integrates a functional surface made of tool steel (1.2344) and a heat sink made of a wear-resistant copper alloy[4]. Both materials have been additive manufactured in the same process. This material combination enables a reduction of the cycle time of up to 25 %. See Fig. 5.31.

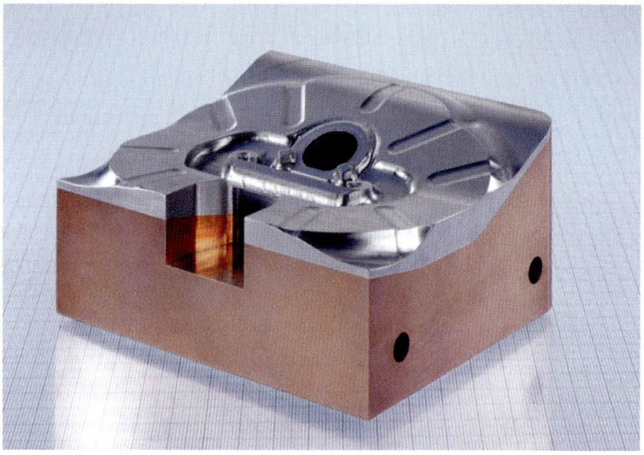

Figure 5.31 Direct metal deposition, bimetal tool of 1.2344 and CuCr (Source: TRUMPF)

[4] AMPCOLOY 940 (CuNiSiCr) was used, which is a beryllium-free copper alloy that is mainly used for resistance welding and mold designing, for example, for thermally highly stressed parts such as die-cast pistons.

5.4.3.4 Layer Laminate Process, Metal Blade Tools, Laminated Metal Tooling

The majority of different methods that have arisen from the layer laminate approach that use metal blades, Section 3.4.10.1, are not commercialized. Most of the approaches are not automated and therefore are not AM processes in the strict sense. The quality of the tools is largely determined by the joining process used.

5.4.3.4.1 Ultrasonic Consolidation, Solidica/Fabrisonic

The ultrasonic consolidation method is discussed in Section 3.7.2. The company is aiming the application at the manufacturing of fully capsuled sensors. Applications in tool making have been occasionally published. Ultrasonic consolidation is, up to now, the only method that delivers completely dense aluminum tools for injection molding. This company is delivering parts as a service, currently only in the USA.

5.4.3.4.2 Lamella Tools, Weihbrecht

The method of manufacturing of lamella tools after Weihbrecht is described in Section 3.4.10.1.1. Lamella tools only consist of the die. A stamp is necessary for the molding. Either a hydromechanical method is used, during which water is pressed into the mold, or a plastic stamp is cast directly into the die, which is therefore covered with an offset film.

Metal sheets are molded with maximum dimensions of 800 mm × 900 mm and a tolerance of ± 0.5 mm. The limitation of this application is given by the dimensions and the power of the press used (currently 280 t, 1000 × 1200 mm) and by the strength of the stamp and of the lamella tool material. With a thickness of the lamella of 1 mm and the use of tension film, parts with a radius down to 6 to 8 mm are possible. One can create up to 500 parts per lamella shape. The output range can be increased by the use of lamellar sheets with greater strength. See Fig. 5.32.

Figure 5.32 Blade tools, surface of the die with a thickness of the blade of 1 mm (Source: Drechsel, 4 marcom + PR!/Weihbrecht)

Applications are wheel housings, door interiors and floor sheets, tank filling ports, inner parts of washing machines and dishwashers, and similar products.

5.5 Future Prospects

The application of AM processes for the manufacturing of tools and tool inserts will be more clearly divided into the three areas discussed above:

1. Indirect tooling: Molding processes based on (not only) additive manufactured master models
2. Prototype tooling and bridge tooling: Direct molding of additive manufactured prototypes of plastic and model materials
3. Direct tooling: Direct processing of additive manufactured tools of metal and ceramic materials

Despite their different approaches and intentions, the common core problems are the quality of the surfaces, the reproducibility, and especially the material properties. It is not only the physical-technological properties of the starting materials, but also the development of user-friendly systems that provide the materials in the context of process parameters and generate reliable features from it. Especially for the directly used metal components, today's scattered parameters are hard to tolerate.

Instead, machines regarded as all-rounders, preferably manually operated, will be used in the research and development laboratories.

The machines used in industry more and more will have the character of fully automatic special machines, and above all, regarding the organization of work and the production flow, they will be integrated into the specific manufacturing process, such as tool making.

Modern approaches such as conformal cooling and derived strategies such as cooling networks or the variation of material properties within the component (gradient tools) will set new standards in tool making. This applies in particular if we succeed in supplementing the tool making with AM elements and resist the desire to build entire tools with AM.

Standard parts with internal cooling (example pin, Fig. 6.30) open up new applications even for nonadditive manufactured tools.

6 Direct Manufacturing: Rapid Manufacturing

Direct manufacturing is a subset of rapid manufacturing and refers to the direct additive manufacturing of end products. The machines used for this purpose are fabricators or production printers.

This chapter can be understood in isolation and thus enables a rapid introduction to the subject given the essential basics provided. Repetitions cannot be avoided but were limited to a minimum.

■ 6.1 Classification and Definition of Terms

Rapid manufacturing refers to the application of additive manufacturing for the direct production of (end) products and series production tools (Fig. 1.6). If the focus is on components, the process is called direct manufacturing. The components have all the properties of marketable products. They are commonly but inaccurately called series products, generally as target parts.[1]

The method of direct manufacturing is largely identical to that of rapid manufacturing and is based on the same principles as rapid prototyping. The machines currently are largely identical with those of rapid prototyping. The initial approaches show that they will be clearly different from those in the future.

Direct manufacturing is currently developing and is very dynamic. It is therefore difficult to properly and fully present the situation and foreseeable developments reasonably in a textbook. In order to enable the reader to follow the contemporary development, sources in the form of Internet addresses are indicated, despite the associated problems, and deviating from the first chapters.

[1] A target part is a component that has exactly all properties assigned to it while designing (properties the designer has targeted on). Target parts are often but inaccurately called series parts, to distinguish them from prototypes and patterns and because they usually start with tool-based production methods in large numbers, so-called series.

6.1.1 Terms

In industrial environments, the definitions above are often not followed exactly, and the term rapid manufacturing is used instead of direct manufacturing as well in the context of additive manufactured end products. Direct manufacturing is often an alternative to the standardized preamble additive manufacturing and is, in common parlance, increasingly also designated as 3D printing.

There are also a wide variety of terms used, including creative neologisms. Some terms are increasingly used by manufacturers and some are even trademarked. A few examples:

- Digital manufacturing, DM
- Direct digital manufacturing, DDM
- Advanced digital manufacturing, ADM
- Free-form fabrication, FFF
- Solid freeform manufacturing, SFM
- Laser additive manufacturing, LAM
- Integrated computer-aided manufacturing, ICAM
- Desktop manufacturing, DTM
- On-demand manufacturing, ODM

Generalizing and therefore not as specific, but especially in the sense of the definitions from Fig. 1.6, incorrect, the terms are:

- Rapid prototyping, RP
- Rapid tooling, RT

In publications, terms such as *e-manufacturing* and *digital warehousing* are also being increasingly used. These are also used to denote new electronically assisted Internet-based marketing and sales strategies and are thus an indication that the technological aspect of the new manufacturing technology and logistically oriented marketing are approaching one another continuously. In this context, the term cyberproduction is increasingly used as a continuation of the term cybercommunication.

The term "manufacturing" is also exchanged for "production" and more rarely with "fabrication."

Machines that manufacture the products are called fabricators or fabbers (see footnotes 5 and 6, [Burns, 93], and [Gershenfeld, 99]). The term 3D printer has been established with high dissemination speed for all types of machines that practically use the layer manufacturing technologies.

6.1.2 From Rapid Prototyping to Rapid Manufacturing

Additive processes are not distinguished by the type of components they produce. The use of independent designations for rapid prototyping and rapid manufacturing is only understandable from a historical perspective.

In 1987, through stereolithography, it was suddenly possible to generate a physical three-dimensional model from any virtual three-dimensional data structure. The records were optimized for later (series) manufacturing processes that were based on the current design, but these were no longer important for building an additive model. The model was a facsimile of the data and not of the product. Thus, it was a prototype.

From the beginning, it was the dream of the inventors to transfer the special properties of additive production to the production of end products. The production should be decentralized, without tools, and unit-independent. Designs that cannot be manufactured with nonadditive methods should be realized. Above all, parts that cannot be clamped and thus cannot be machined were of concern, as well as internal structures. Examples are the Mobius loop and the model of the skull in Fig. 4.9. The University Duisburg-Essen (Witt, Sehrt) and the designer Abdurrahman Tüylüce show the application of a product in a project with surfboards. A concept for manufacturing surfboards with internal irregular honeycomb structures was developed instead of foam or an industrially produced regular honeycomb (Fig. 6.1). The board is more lightweight and stiffer and can be manufactured automatically.[2] Examples for internal cavities are inserts with integrated conformal cooling for injection molds, as seen in Fig. 5.2, or medical models like maxilla and sinuses, see Fig. 6.2.

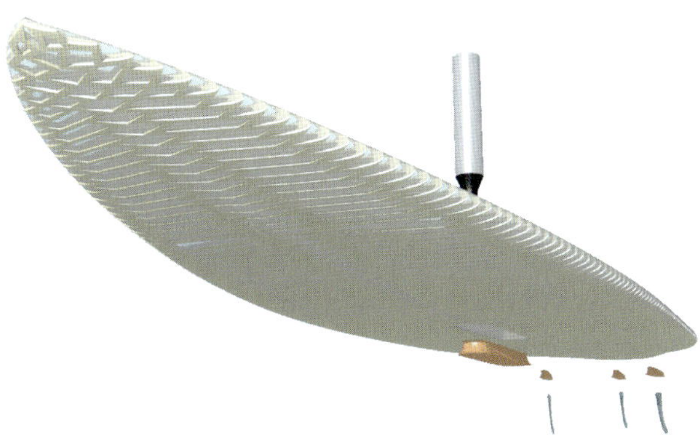

Figure 6.1 Surfboard with internal irregular honeycomb structures, laser sintering (Source: Tüylüce/Witt/Sehrt)

[2] The product-specific aspects were intensively examined from the designer Abdurrahman Tüylüce (at-dezigns@hotmail.de), but in the context of this book, they are rather placed in the background.

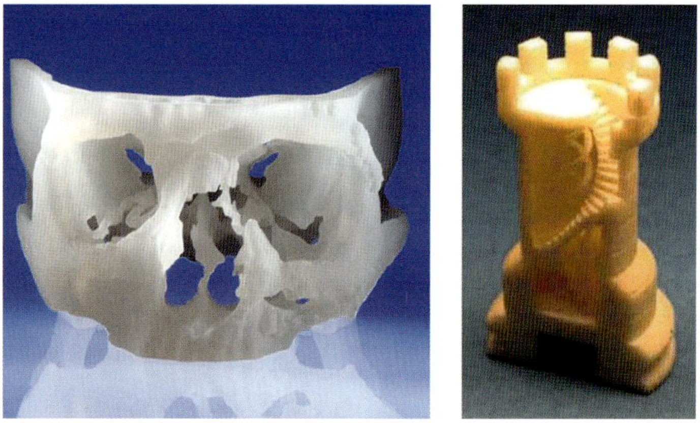

Figure 6.2 Internal cavities, skull model (left), and cut tower from Fig. 1.2 (right)

6.1.3 Workflow for Direct Manufacturing

As already discussed, whether an additive manufactured component is a prototype or a product depends less on the machine used than on the associated process. It is crucial that the component has all of the physical-technological properties that were provided by the design.

The result is a workflow for the direct production, which also includes the responsibilities of design and manufacturing.

6.1.4 Requirements for Direct Manufacturing

Direct additive manufactured parts have to show the mechanical-technological properties of the target parts. Therefore, safe processes, suitable materials, and particularly a design system for additive manufactured components have to be available. In the very detailed book of Breuninger, Rommel, et al. [Breu, 13], various approaches for that are in relation to plastics.

It is desirable, but in contrast to the common opinion, not necessary, that the material properties from additive manufactured methods correspond exactly to the properties of conventional methods. It is crucial that the components have the properties defined through the design, so that they are target parts. Hence, it is increasingly important to develop a design system that is adjusted to additive manufacturing.

In order to emphasize that the methods of additive manufacturing of end products are at least equivalent to traditional manufacturing, but entail new possibilities beyond that, a new term was coined: direct manufacturing.

6.2 Potential for Additive Manufacturing of End Products

What are the perceptions of additive manufacturing? What is potentially considered to be strong enough to revolutionize today's manufacturing technology? What needs to happen for rapid prototyping to turn into rapid manufacturing?

The following approaches arising from today's perspective.

6.2.1 Increased Design Freedom

Progress in design and manufacturing cannot be considered independently. Limitations in the design are mostly analyzed by the limits of production, and vice versa. The boundaries of the following influences are therefore fluid, and individual aspects often apply to several stages.

6.2.1.1 Advanced Design and Structural Opportunities

Additive manufactured products are not affected by geometrical and technical manufacturing limitations of today's mass production. In plastic injection parts, wall thickness differences no longer necessarily lead to sink marks, and weld lines are no longer a result of the choice of injection points. Drafts are not necessary anymore, separating planes no longer dominate the tool design that is necessary for production, and undercuts require no slide.

Deliberately introduced internal cavities for temperature control, labeling, and the installation of sensors and actuators are easy to implement.

In addition to the technical construction, it is mainly because of the design possibilities that the additive manufactured products are increasingly coming into the spotlight. They allow increasingly sophisticated designs that are detached from manufacturing chains to be realized. The difference between the distinctive finish and the mass production facing industrial design is fluid.

Examples are the lamps project from Freedom of Creation, FOC-Twister (Fig. 6.3) or the sintering chair project (Fig. 6.4). Both use the geometrical and optical possibilities of additive processed materials effectively. Both are sold commercially.

Figure 6.3 Twister lamp (Source: Freedom of Creation)

Figure 6.4 Laser-sintered chair (Source: Voigt-Weizenegger)

6.2.1.2 Functional Integration

Additive manufactured components can be geometrically complex. This reduces the number of items, the assembly costs, and the associated problems with accuracy and adjustment, for example, as compared to nonadditive methods. Further, the integration of functions into products often leads to designs that cannot, or only with very great effort, be realized with nonadditive methods.

Examples are grippers (see Fig. 6.5) that have been developed for use with robots at the Fraunhofer Institute IPA. Figure 6.5 shows a sketch of different published principles and a sintered gripper. Details and further examples can be found in [Breu, 13].

MEMS also belong to functional integrated components; see Section 6.2.2, "Production of Traditionally Not Producible Products."

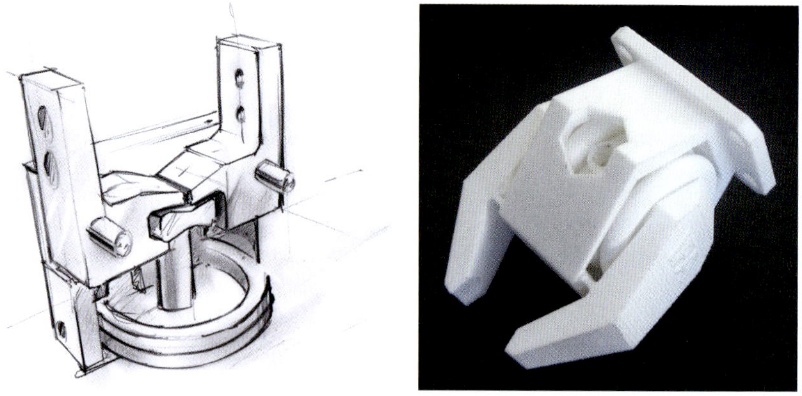

Figure 6.5 Functional integration in the example of a gripper schematic diagram (left), part laser sintering (right) (Source: Becker/Fraunhofer IPA)

6.2.1.3 Novel Design Elements

The internal implementation details (Fig. 6.2) and the functional integration in individual cases also can be realized in terms of additive internal structures within the concept of three-dimensional trusses that have the character of design elements. They are related to the hollow construction strategies for melt-out, preferably stereolithographic components. These were developed as QuickCast (3D Systems), Skin and Core (EOS), and Fast Sculp (F&S) to prevent shell cracking when burning lost molds for investment casting. As a design element it can take on arbitrary geometries and can be spatially nested arbitrarily (Fig. 6.6). Conceptually, this approach leads from crash boxes to cancellous (bone) structures.

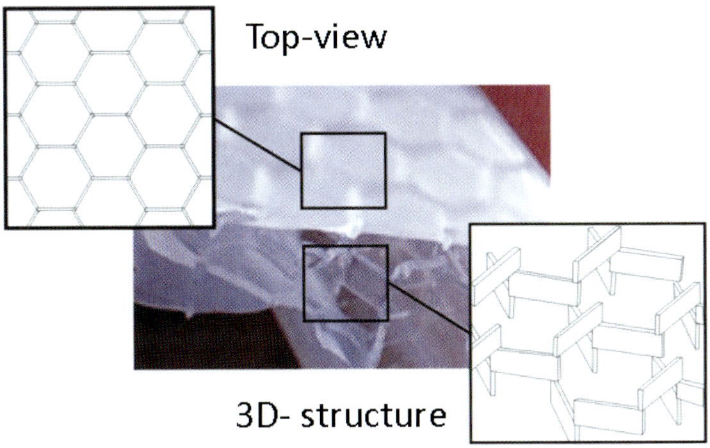

Figure 6.6 Internal space structures as a design element, plan view and sectional (Source: CP-GmbH)

6.2.2 Production of Traditionally Not Producible Products

In addition to the examples from Figs. 6.3 and 6.5, which both show structures that cannot be made traditionally with nonadditive methods, it is mainly microcomponents where the production reaches its limits because of the small dimensions. This has led to the development of complex processes (for example the LIGA process).

For this purpose, additive methods have already begun to conquer the market, but especially in the area of so-called mesocomponents with dimensions between macro- and microcomponents. Typical dimensions are a few 10 µm up to a few 100 µm. In addition to isotropic[3] components, mainly integrated functional components

[3] Isotropy is the independence of the direction of physical and chemical properties of a material.

(MEMS, microelectromechanical systems) and systems for the investigation of chemical processes with very small amounts ("lab on a chip") are produced outstandingly and mostly unrivaled with additive manufacturing processes. The production is often associated with the development or application of special materials, for example, to set the specific hydrophilic or hydrophobic surface properties (see Section 3.1.6, "Microstereolithography, microTEC").

Also, the battery concept of A123 Systems (www.a123systems.com) cannot be realized with traditional manufacturing methods. Only the layered structure of the batteries, produced using a modified 3D printing method, allows for a specific porosity, higher power density, smaller dimensions, and shorter load times. Application areas are small and portable (laptop, mobile phone) but also large, for example, vehicle batteries. The project is supported by a large battery manufacturer.

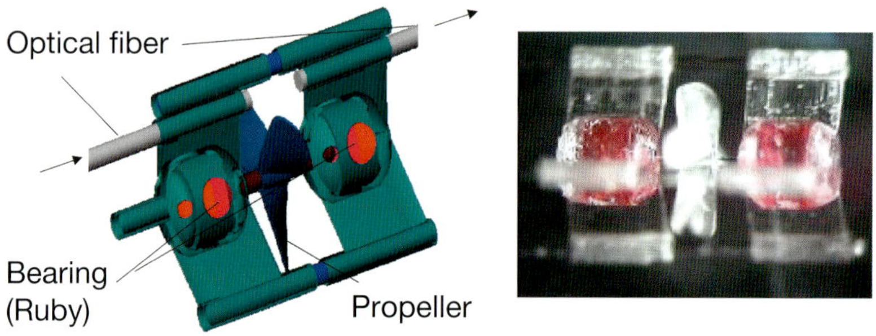

Figure 6.7 MEMS flow meter construction (left), view (right) (Source: MicroTEC)

6.2.3 Variation of Mass Products

The variation of bulk products means that their adaptation to customer groups or market segments is the manufacturer's responsibility. Mostly small batches or rare items are produced. Examples are individually different plugs, battery compartments, or a variable cable holder as in Fig. 6.8.

What is possible in tool-based production with additive manufactured production batches, for example, with some effort, through the exchange of inserts and for a sufficiently high number of items, in any number of pieces down to one can be made, without the economic disadvantages. The shoe on Fig. 6.9 can be easily scaled in any direction. The idea basically can be applied to many other marketing methods.

The shoe is also an example of passive personalization of mass products; see Section 6.2.4.1, "Passive Personalization: Manufacturer Personalization."

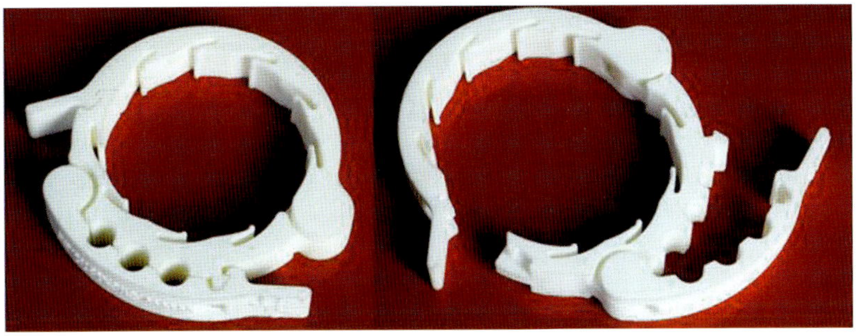

Figure 6.8 Cable holder, one-piece sintered. Closed (left) and open (right) (Source: FIT)

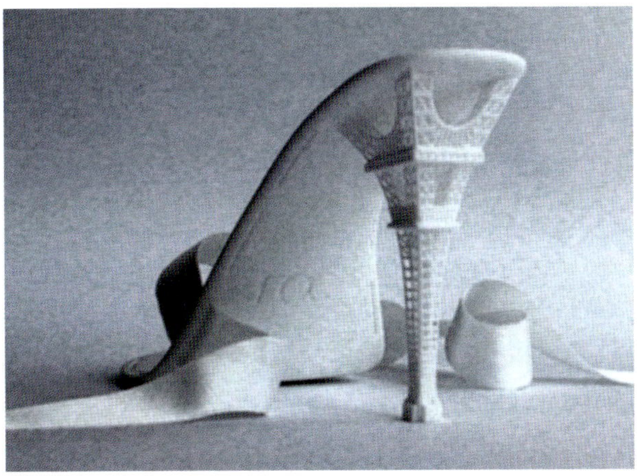

Figure 6.9 Paris sandals, laser sintering (Source: Freedom of Creation)

The continuous improvement of products, but also adaptation to the norms and rules of any other market, can be implemented in a timely manner at no additional cost. Series risks and barriers to exports can be reduced especially for SME (small- and medium-sized enterprises).

6.2.4 Personalization of Mass Products

Personalization means the adaptation of products to the needs of individual customers. Therefore the term individualization or *customization* is also common.

The additive manufactured product or its individual design is in each case initiated by the customer, even if the impetus originates from the manufacturer providing appropriate routes, for example using the Internet.

When a customer describes his individual needs and the manufacturer implements them, it is called passive personalization or passive customization, or manufacturer personalization or *manufacturer customization, MaCu*. When the client is involved actively in the design process, this is called active personalization or active customization, or customer personalization or *client customization, CliCu*.

6.2.4.1 Passive Personalization: Manufacturer Personalization

In passive personalization, the product responsibility, including the design solution and compliance with the relevant rules and regulations, rests with the manufacturer. The customer either brings his ideas, such as with variant catalogs, or his biometric data on the corresponding measuring devices (e.g., as computed tomography, CT).

Applications are in the area of individually enriched products, for example in the production of 3D models (Fig. 6.10), planets, and other celestial bodies, each with individual peaks.

Figure 6.10 Individual globe (Source: Relief Globe Company/RTejournal)

Particularly important are products of the human-machine interface and the field of medical technology. Examples are immachines (Fig. 6.11), braces (see Section 6.5.2.3.1, "Braces: Aligner, Invisalign"), and hearing aids (see Section 6.5.2.3.2, "Hearing Aid Shells, Ear Molds"), which are all based on three-dimensional medical imaging.

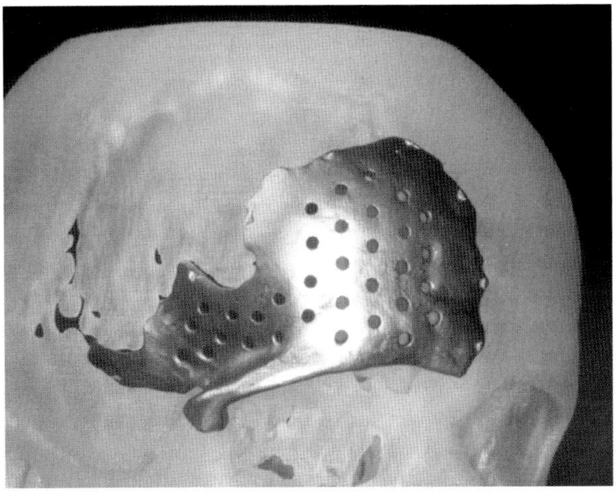

Figure 6.11 Implant for craniofacial surgery (Source: CP-GmbH)

In the operation of industrial products, such as in the driving of motor vehicles, in the entertainment industry, and in the household, customer-specific solutions with an adaptive character will complement the basic industrial design with an individual product in the future.

The biometric 3D data can also be used for the production of three-dimensional portraits and sculptures, opening a previously very exclusive market for wide layers.

Figure 6.12 shows examples of the conceptual art exhibition by sculptor Karin Sander, State Gallery, Stuttgart, 2002. The body scans of visitors were implemented on the ground in 3D printing sculptures (see Fig. 4.45).

Figure 6.12 Group of figures, concept art by Karin Sander (Source: Sander/Eder/Schmidt, FH-Aachen)

6.2.4.2 Active Personalization: Customer Personalization

In active personalization, the customer implements his ideas of the product. He or she reaches directly into the compilation of data, for example, by the operation of a (preferably simplified) CAD program. As an example, DigiSmugg (www.ifi.unizh.ch) shows how a piece of jewelry can be customized using fewer parameters by the customer on the screen.

In addition to the role of the consumer, the customer assumes the role of the designer and the producer (prosumer, a blend of producer and consumer).

Preferably consumer goods are recorded. The production can be done by the manufacturer and is controlled by the production planning of the order. However, it can also be triggered directly by the customer via Internet portals.

In each case, different requirements agreeing to the customer-modified products are summarized and manufactured in a construction process.

The question of product stewardship is increasingly getting more complicated with the intervention of the customer. The customer is responsible for the individual design, but product responsibility remains with the manufacturer or designer. This also applies if the customer sends the finished data via the Internet to a fabricator. A new area for the legal field is developing here due to these complications.

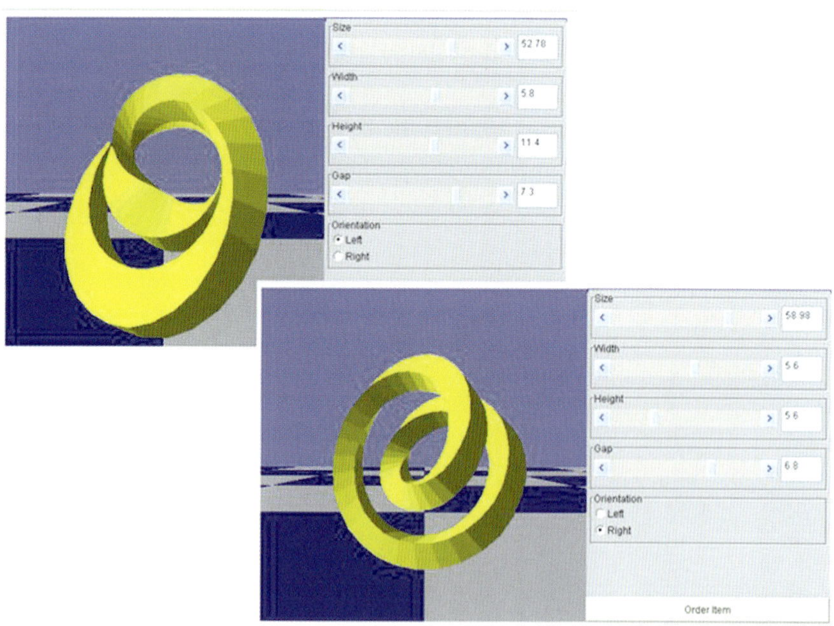

Figure 6.13 DigiSmugg, individualization of predetermined designs by the customer via the Internet (Source: IFI, Uni Zürich)

6.2.5 Realization of New Materials

Today's design and manufacturing technology are characterized by the use of isotropic materials. Although composites are becoming increasingly important, they are produced mainly from isotropic layers, especially in certain sectors such as aeronautics or yacht building. Even today, powder metallurgical components consist of homogeneous mixed powders and thus can be regarded as isotropic materials. With additive manufacturing methods it is possible to continuously vary material properties beyond the part and generate a defined anisotropic behavior. Examples are hard-soft compounds in plastics manufacturing. In the field of metals, any metal-powder mixtures are possible, both locally modified compositions and pseudoalloys, i.e., those that cannot be produced by melting. In addition to continuous changes in terms of *graded materials*, discontinuous variations are possible. Examples are the manufacture of medicine tablets, where a variety of drugs are embedded into the support matrix, and also detectable markers, which are introduced for nondestructive wear control or to generate emergency operation characteristics in tribological systems (medicines for machines) that are abrasive. Examples are continuously colored components (here the physical property of color is varied) and also the battery project mentioned in Section 6.2.2, "Production of Traditionally Not Producible Products."

Figure 6.14 Continuous change of the physical property color, 3D printing (Source: FH-Aachen)

6.2.6 Realization of New Manufacturing Strategies

Today's glorification of continuous tool manufacturing, and therefore the necessary manufacturing organization, storage, and distribution (logistics), can be arbitrarily discontinuous in the application of additive processes, and thus made much more flexible. This affects both the spatial and the temporal distribution. With the use of fabricators, products can be made anywhere on Earth in any quantity. In addition to the individual production of products, as mentioned previously, there are many advantages, especially for the supply of spare parts. A simple example shows the functional model of a camshaft of an internal combustion engine, as is usually used in automobiles. The damaged original was modeled in CAD and manufactured on an FDM system as a single piece. The camshaft could be adjusted to the shaft that was no longer dimensionally stable because of abrasive wear–a requirement that parts off the shelf cannot meet.

The spare parts problem that is currently in the manufacturing industry can be improved with additive manufacturing. In the future, the varieties of spare parts are loaded from databases and additive manufactured as *spare parts on demand*. Therefore, they are no longer produced on stock, so mainly no need for storage.

Additive manufacturing processes have the potential to add material to semifinished products. This allows the individual repair of components, as is already common today in the repair of turbine blades for aircraft engines by hardfacing. But it also provides for the manufacture of new devices based on the old. In the case of their design systems and product strategy, worn-out components can be upgraded without their destruction in newly functional form with additive manufacturing processes. Today's recycling, which takes place at the level of materials and against the background of the "small cutting" and "meltdown," actually represents a down-cycling that can become a resource-efficient, sustainable "upcycling" process instead.

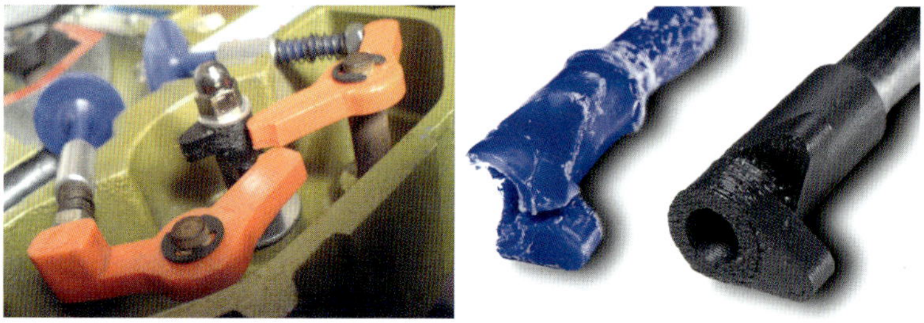

Figure 6.15 Functional model of an internal combustion engine with remanufactured camshaft (left), defective camshaft (center), and generated camshaft (right); fused deposition modeling (FDM) (Source: FH-Aachen)

6.2.7 Design of New Labor and Living Alternatives

Additive processes support the design of new alternatives for working and living.

Fabricators today have the potential to develop into individual manufacturing machines. Just as today's printers, in connection with the personal computer, have allowed private offices to do individual printing, the *personal fabricator* (PF) [Gershenfeld, 99] will enable the production of goods in the private sector. Conceivable are the self-made elements of system toys, spare parts, and decorations. Fabricators are, in addition to self-sufficient, a form of self-employment.

Approaches in terms of a cooperative process include the *imaginary manufactory*, DIM. It opens with an emphasis on design and currently still largely a conventional production technology, a competitive position for people with visual impairment beyond today's "compassion article" (www.blindenanstalt-berlin.de/).

Further, the philosopher Frithjof Bergmann recently presented his theory of the new work as a book [Bergmann, 04] and outlined work and life patterns that are suitable to solving today's problems of unemployment and isolation, and also solve the impoverishment of large sections of society [Irle, 04]. In particular, in Chapter IV, he shows decentralized production models that will include additive manufacturing with integrated fabricators that are preferably implemented in the form of centers for new work (www.newwork-newculture.net/).

■ 6.3 Requirements on Additive Manufacturing for Production

Even if much from today's perspective occasionally appears, the outlooks that have been discussed may not be described perfectly. Of course, the different ideals can be implemented only with individually optimized machines and often not exclusively with fabricators, but in conjunction with them. New strategies, processes, materials, and machines are required. What requirements must they comply with? How should today's additive machines and processes be developed in order to realize optimal future outlooks?

For a discussion of these issues, the characteristics of today's additive (prototype) methods are analyzed and compared with the requirements of future manufacturing processes.

The discussion is based on six selected criteria:

1. Process
2. Material(s)
3. Organization
4. Design
5. Quality assurance
6. Logistics

The explanations are based on current additive production of one or more (individual) parts (Section 6.3.1, "Requirements on Additive Manufacturing of a Part"). On this basis, the requirements on the production of series with today's methods are considered (Section 6.3.2, "Requirements for the Additive Mass Production with Current Methods"), and finally, the conditions for a future additive series production are derived and discussed (Section 6.3.3, "Future Efforts on Additive Series Production").

6.3.1 Requirements on Additive Manufacturing of a Part

6.3.1.1 Process

The starting point is an overall optimum construction process for the respective application. There are currently over one hundred commercially available additive machines, so this discussion requires an extensive knowledge of the individual processes. This applies even if you restrict yourself to the five most widespread industrial process families: 1) stereolithography and polymer printing, 2) sintering and melting, 3) 3D printing, 4) extrusion, and 5) sheet lamination. For example, the same data set leads to visually very different results even with two different methods (Fig. 6.16).

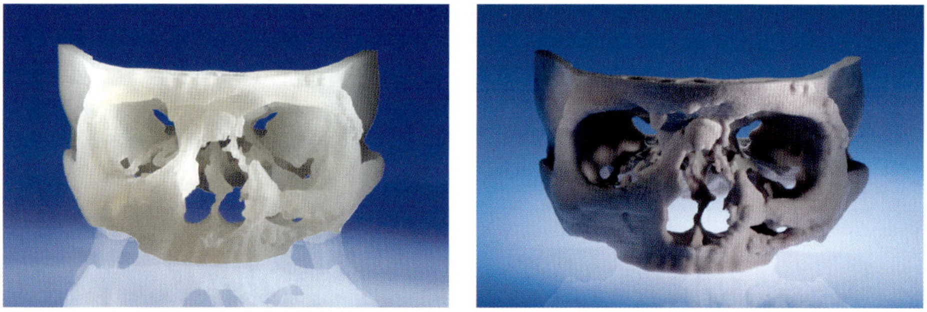

Figure 6.16 Model of a skull, stereolithography (left), 3D printing (right) (Source: CP-GmbH/FH-Aachen)

However, the choice of a suitable method is only the first step. Each method allows one to set many parameters that have an influence on both the accuracy and the speed of construction and thus on the economics. As this must be done depending on the geometry, this requires a detailed knowledge of the process. Figure 6.17 shows the influence of the parameters on the example of an FDM part, which, with the same geometric data set but with different parameters, was built with different orientations in this case.

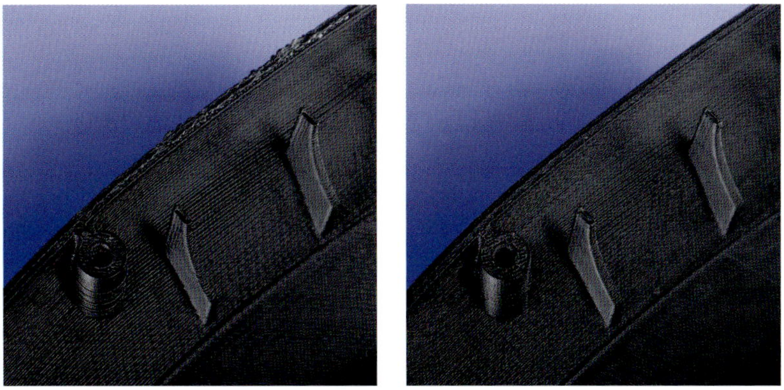

Figure 6.17 Influence of different process parameters (orientation) on the part's surface for the example of an FDM part (Source: FH-Aachen)

Today's additive processes are thus characterized by a complex parameter optimization that must be done individually depending on the complexity or features of the part.

In addition, these processes show an unsatisfactory reproducibility in construction results (see Quality Assurance, Section 6.3.1.5). This is reflected in a tendency for variations in quality. Possible causes may be found in the material but also in the process. For the process that usually means a lot of calibration.

In the manufacture of a part in terms of a prototype, these aspects also have positive effects because the parameter optimization is an additional optimization criterion, if you know your machine well. But the production of the final goods is rather hindered.

In any case, the postprocessing and finishing must be included in the analysis of the process.

6.3.1.2 Materials

The materials for additive processes also lead to different part properties due to the manufacturing process, if the part and the materials for the nonadditive series production are nominally identical (e.g., chemically). A polyamide (PA) melt that is placed on an injection-molding machine under high pressure and turned into a solid form, for example, leads to a part with different mechanical-technological properties than locally unpressured PA powder that is molten in layers. This basic problem is exacerbated because the additive manufacturing process usually requires a special processing of the material. For example, polyamides for processing in selective laser sintering not only need a much smaller grain size fraction than those for injection-molding machines, but they must also be modified with additives to ensure the flowability.

Although there has been tremendous progress in material research for additive processed materials, the level of construction materials is significantly higher. This can be clearly seen in the example of the property pyramid for plastics (Fig. 6.18).

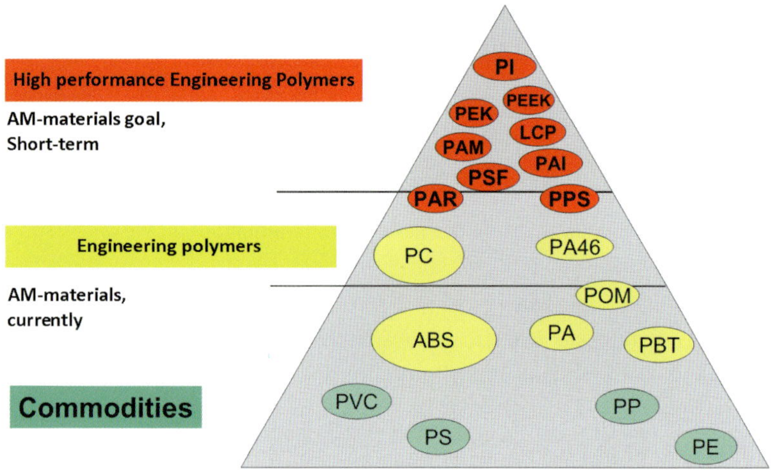

Figure 6.18 Property pyramid of plastics, niveau (levels) for AM materials

In addition to the additive processable materials that are chemically identical with the current construction or series materials, there are a number of model or AM materials for which there are no similar construction materials. This applies especially for methods in which binders are used, such as the 3D printing process (particle + binder) or the layer laminate method (layer + adhesive or bonding layer). For these methods, it must be checked in each individual case whether the produced parts will have the required properties.

Most materials are also proprietary products with a correspondingly narrow machine and process coupling. The user therefore is faced with problems that do not consider or have incompletely available material data and often involve an unfavorable pricing policy. With the growing importance of the additive process, this problem is becoming defused. On the one hand, independent material suppliers are increasingly becoming established, and on the other hand, the construction of a machine that works with marketable materials, especially the sintering machines, increasingly apply as a development goal. However, the validation of the component properties still has to be performed, in view of the fact that the mechanical-technological properties of the part are affected by both the material and by the process.

Additive processes enable the complete or partial re-use or recycling of the material that was used but not melted in the building process. This is the case for sintering and melting, for example. But depending on the level and number of heating operations that a material was in, its physical and technological properties can change. The problem can be avoided by the disposal of the old powder and the use of only new powder, but this is not economical. A corresponding powder management (general material management) therefore is extremely important from the perspective of the quality of the part and its cost.

6.3.1.3 Organization

The organization includes the manufacturing organization, from order to delivery of the part. It still corresponds, especially for additive manufacturing, to a craft-oriented single-part manufacturing. Those manufacturing jobs that exist in the form of 3D data sets are arranged in the construction space primarily by geometric criteria. The manufacturing process starts when the construction space is filled. This results in a kind of batch processing, which can run overnight unattended. Thus it can be coordinated with the daytime manual manufacturing process.

Manual tasks especially occur in the preparation of the machine for the construction process and in the finishing of the models. For preparation, the machine must be released of material remaining after each construction process and if necessary be cleaned. Construction materials as well as raw materials and supplies (if necessary binders, toner, and so on) must be checked and filled up. The proper function, including the calibration of the machine, must be checked. The filling of the machine is done mostly manually, which means a worker carries the material in appropriate containers to the machine and fills it up. Automatically indicated machines work with a cache that refills the machine automatically but must be filled manually.

After the additive manufactured parts have been taken out of the machine, depending on the construction process, a very extensive manual finishing is required. It consists of steps that are directly related to the process (postprocessing), such as cleaning, the removal of supports, and post-cross-linking, and of process-independent (finishing) steps, such as sandblasting, grinding, and manual deburring, filling, and painting. A rule of thumb illustrates the dimensions: a large stereolithography system employs up to five people for the postprocessing. The organization is therefore geared toward a largely manual operation for the production of parts with item character.

6.3.1.4 Design

The production data are derived directly from the 3D-CAD design. The parts thus represent facsimiles of the 3D data. These generally neither have the tolerances nor are they otherwise redesigned with respect to the manufacturing.

Design interventions in relation to the production technology are limited to the positioning in the construction space and optionally the attachment of supports. In additive manufacturing applications, the additive production is not taken in account from the design viewpoint.

6.3.1.5 Quality Assurance

Quality assurance is often limited to the verification of the geometry, which is often complicated by the fact that the data sets contain no tolerances. Mechanical-technological properties such as tensile strength or temperature stability are not verified in general. At this point, one is mostly satisfied with the information provided by the material supplier. Occasionally, in addition, test bodies are built. Questions regarding the durability and the sustainability normally play no role in the discussion. The latter is because prototypes are only needed for a short time. In fact, many components after some time lose important properties such as dimensional stability, elasticity, or color. But as part of the material development these disadvantages will be systematically eliminated.

Dynamic characteristics are almost completely missing for plastics and therefore are consequently not checked. For metals they exist sporadically. Appropriate work is ongoing at many institutes, as well as in industry, which usually retains the results for it.

The quality of the part is substantially influenced by the reproducibility of the process. This was a major problem mainly in sintering processes because of the high vulnerability to fluctuations in temperature and the temperature distribution, but this is more and more fading into the background because of new machine concepts and optimized controls.

The problem of the mixture of new and used equipment in some processes has already been discussed in Section 6.3.1.2, "Materials."

6.3.1.6 Logistics

The additive manufacturing of prototypes and single parts is a pure single part production. The clock is set by the incoming orders and only slightly affected by the synchronizing effect of the machine capacities. As in prototype construction, basically the shortest possible delivery times are wanted; manufacturing and logistics that are structured after other criteria is almost impossible to obtain.

6.3.2 Requirements for Additive Mass Production with Current Methods

Today's additive series production means the production of comparably low amounts of the same parts in the sense of the traditional terms of a series. The quantities are mostly between 20 and 500, although there are no technical restrictions on that.

The production is either done by repeating the single construction process or by so-called follow-up processes (Section 5.3, "Indirect Rapid Tooling Methods: Molding Processes and Follow-up Processes").

In molding, through follow-up processes, the rapid prototyping part is seen as the master model. From this master model preferably simple and fast-to-implement tools are derived (indirect rapid tooling), and with those, the required number of parts are constructed in the desired material. Whereas the great advantage of additive manufacturing is the tool-less manufacturing, molding processes are tool-bound. Furthermore, they are nonadditive and are not further considered here. In the repeated construction of the same part by AM, all criteria apply that apply for single parts as well, but there are some additional technical manufacturing characteristics.

6.3.2.1 Process

The additive manufacturing processes are too slow compared to series production. Whereas the direct manufacturing of a single complex part without using tools is considered to be extremely fast (rapid) when it is finished after several hours, the series production of AM parts must be measured in terms of series conditions.

The benchmark is no longer the manufacturing time of the tool and the associated potential for savings in time and money, but instead it is the cycle time in the manufacturing process, and thus the productivity. For a plastic injection-molding process, this means a time regime in the range of seconds. Regarding the additive construction time of several hours (as a comparison time for each part), cycle times of many minutes result if many parts (often more than 100) can be manufactured in one additive construction process.

All additive processes that are illuminated by a mask method, meaning they can apply the entire contour information for a layer at a time, are fundamentally superior to the methods favored today that work with the vector or raster method, the stereolithography, polymer printing, sintering and melting, the extrusion process, 3D printing, and the layer laminate methods. However, the vector methods are generally more accurate.

New illumination strategies are in great demand. The RMPD (Rapid Micro Product Development) process from microTec that was already designed and marketed as a manufacturing process (Fig. 6.19), for example, works with a microlens system that allows the simultaneous illumination of several (but equal) components.

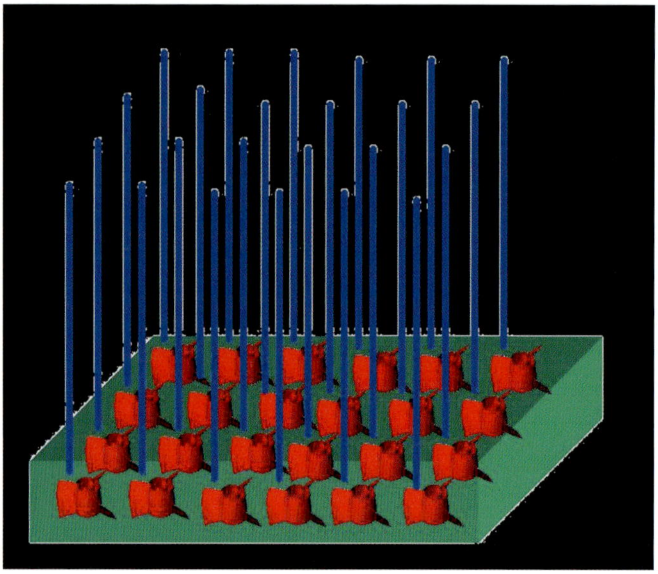

Figure 6.19 Mass production of microcomponents by means of laser beams, which are positioned via a microlens system (Source: microTec)

The direct additive manufacturing of today is economical only under special conditions. For example, for small quantities, when the tooling costs of the competing nonadditive production that are allocated to the part drive their unit costs above the price of additive manufacturing. This particularly applies for complex parts that require time-consuming and therefore expensive tools and, if necessary, additional assembly works for nonadditive manufacturing, however the grade of complexity hardly affects the AM process.

Today, additive manufacturing has the character of a so-called *blind production*. This means that for hours, between the initiation of the manufacturing process and the completion of the part, the production can only be observed but not influenced. On the machine side, a fairly accurate control of construction time is supported in most processes; moreover, often information on the stock of material, and when appropriate, the construction space temperature, is available. A visual inspection is usually possible, but it provides only basic knowledge about whether catastrophic events have occurred during the production, which suggest a termination of the construction process. In some processes, this is supported by the software.

In the manufacturing of prototypes, it is frustrating if the manufacturing process could not be finished properly and the part must be constructed again. Regarding the production of series, the fact that a first quality check is only possible after the end of the construction process represents an economic problem.

The completion of a construction process does not always mean that a similar subsequent process in a few days or weeks will be completed successfully because, as mentioned, the occasional poor reproducibility with the manufacture of large quantities can be problematic in the production technology.

The process is dependent on calibration, which further limits the ability for mass production because of the partially complex calibration procedures, though at least it reduces the availability due to large *downtime* ratios.

6.3.2.2 Materials

When producing a single part, the materials do not often correspond to the construction materials commonly used today. On the other hand, one has to keep in mind that due to the fixed design and the actual part properties there is a difference from the later series manufacturing method and the already-used materials as these do not occur in the additive processes.

In some processes, the mostly welcomed opportunity to use material that was not used for the part in the construction process leads to quality variations. These variations can be avoided by the expensive method of not using the recycling materials or by avoiding consuming material management processes. Series are more affected by this than single parts.

But the additive manufacturing principle also opens up new perspectives. With the aluminum-filled polyamide powders (EOS: AluMide; 3D Systems: DuraForm AF), materials are available that have excellent aluminum-like surfaces that are easy to polish and reach better strengths compared with the unfilled powder. Since 2004, EOS has already used this material for the manufacturing of parts for its own products, for example for the direct production of the handles of the EOS sinter machines. A comparable injection-molding material is currently not known.

6.3.2.3 Organization

The organization does not change fundamentally for the production of prototypes. It will tend to be simpler because it can be assumed that, given the manufacturing of series, the number of customers decrease per construction process and thus lessens the organizational effort. However, in an additive series production, the productivity is of significantly greater importance than in prototype construction. In considering the time advantage, underutilized construction spaces can not be tolerated, also because the customer is no longer willing, given a (possibly piecewise) continous supply, to pay extra for faster delivery.

6.3.2.4 Design

The design of the parts is still done in regard to a later nonadditive series to be produced. An adaptation to the special requirements for additive manufacturing occurs only insofar as the parts are oriented prior to the construction process and are scaled, if necessary. If several additive processes are possible, there is potential to select the most suitable method. It is also possible to assemble the part from single parts that come from different additive (and nonadditive) construction processes. Particularly with respect to the conceivable complexity of a part, further opportunities of additive methods are not being fully used constructively.

6.3.2.5 Quality Assurance

As mentioned, the problem of insufficient reproducibility is becoming more important in the production of series than in the manufacturing of individual prototypes. Additive parts not only have aberrations from the target geometry, but also often from part to part. This can even occur if they are made within one construction process on the same construction platform. Therefore a single part control is necessary as needed. In the production of several identical parts, the common practice of today is to transform the received 3D data into a physical part in a loose manner, though as much as possible, and then tolerances should be defined and controlled.

6.3.2.6 Logistics

The logistics still has the character of single part management. The difference is that there are differently sized groups of mostly the same single parts. Therefore the implementation tends to be easier compared to the single part production of prototypes. But more often, part deliveries must be managed. However, the pressure on price will be greater because a customer being delivered continuously is accustomed to series conditions, for example, the injection mold.

6.3.3 Future Efforts in Additive Series Production

In the future, additive manufacturing methods will be an equivalent alternative to classic manufacturing methods. They will be used for the manufacturing of individualized (end) products. The design will be orientated on additive manufacturing. Additive manufacturing will develop into a continuous manufacturing, even of completely different parts. The charges[4] can have different quantities and decrease down to only one piece.

---

4 The traditional definition of charge as a subset of the same products for which a manufacturing machine must be set up no longer applies. Additive production is independent of the number and sizes of the charges.

The ideal plan consists of manufacturing the highest possible number of different pieces per construction process. In contrast to the production of prototypes, therefore, the productivity is becoming more important. Any kind of manual activity therefore should be eliminated if possible. At the same time, the production of different parts in one construction process means that many of the methods and procedures that ensure quality and productivity in today's manufacturing processes can no longer be used. This applies in particular for non-destructive tests, for processes based on statistics, and also for manual setup.

Additive series production therefore must simultaneously satisfy the requirements of today's series production and prototype production.

The performance of such individualized production processes and the quality of the produced products are no longer measured on prototype criteria but on series criteria.

6.3.3.1 Process

Everything that applies to the production of prototypes is valid for the actual production process. But construction speed, process reliability, repeatability, sustainability, and productivity are of a much greater importance.

The main requirements of the process are the development of discontinuously to continuously operating processes and from manual to automated machines.

The first approaches have existed for some time in the form of semiautomated powder management (EOS, 3D Systems). It includes an automatic powder supply and the ability to take the still-warm powder cake as a whole from the sintering machine, so the manufacturing process is not interrupted by cooling phases (EOS, Fig. 3.36).

With the sintering Station Pro, 3D Systems launched a machine concept in early 2005 that contains important elements for additive production (see Section 3.2.3). But from the view of manufacturing technology, it still require high manual effort. Figure 6.20 shows the additive machine, the unpacking station, and the material handling and storage system. On the far left in the picture, connected to the tempering station (OTS, Offline Thermal station), there is the Quick Release System (RCM, Rapid Change module) that contains the actual construction space. It is manually connected with the OTS for preheating and cooling, pushed into the machine for construction, and docked on the cleaning station to unpack the parts.

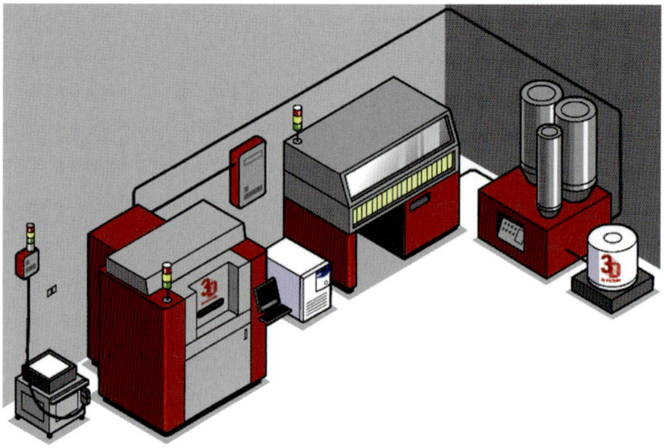

Figure 6.20 Sinter station PRO Rapid Manufacturing System with RCM, OTS, cleaning, and stocking module (Source: 3D Systems)

For the current sPro there is a monitoring system, the Real Monitor & Graphic Viewer Advanced software. It detects and exports the important events as .log data, for example, events of the laser, the heat-up module, and the sensors.

The EOS system shows similar approaches and beyond with the One-Part Property System at the level of individual construction data.

For the additive manufacturing of microparts (microTec), for years there have been production technological approaches.

For a virtually continuous manufacturing, processes and machines need to be specially developed where the part is ejected after the construction process is complete and new material is automatically injected. One approach in this direction has existed for some years in which parts are still inserted in the ongoing construction process during the laser sintering as far as the available volume and the remaining time permit. As an intermediate step to the virtually continuous process sketched above, discontinuous parallel working methods are also conceivable. Conceptually, key issues are the automation of replaceable modules that are currently moved manually.

The world's first continuous 3D printer (VXC800) was presented at EuroMold 2012 by the company Voxeljet. It shows out a potential path for continuous additive manufacturing.

The focus is on the following tasks: To eliminate the manual rework or at least minimize it, and a much better surface quality must be achieved directly in the additive manufacturing process. Besides parameter optimization, equipment integrated in the process for finishing are conceivable.

The parts must exit the machine properly after the additive manufacturing step. A later sorting out because of individual tests should be avoided. Controlling the process on the basis of experience is not sufficient any longer. A process control, in the sense of a multivariable control of nonlinear systems with state estimators and state observers, must be established. Therefore, the process must be simulated by numerical process models with sufficient accuracy and at an accelerated rate. The expected results must be detected (preferably camera-based) and correlated during the construction process with the corresponding online measurements (observer status). Depending on the result, the process parameters have to be adjusted in the process or it must be decided to stop the production of the component and restart with new parameters. Theoretically this approach agrees in some aspects with the "look-ahead" functions that are already implemented in today's machine tools.

6.3.3.2 Materials

Materials should be available that can reproduce the designed and defined product properties. Therefore the data of the current materials for series production has to be also available for the materials used for additive manufacturing. But these are still very incomplete. This is true for many fixed characteristics. The measured values are often recorded but not the norms that are beyond the measurement. Another problem is repeatedly converting to other units and other numeric characteristics.

Dynamic material characteristics are largely missing; for example, optical properties are published only selectively.

The characteristics of each material are, strictly speaking, only valid for one machine and one set of parameters. Therefore the corresponding parameterized values have to be available for different processes.

Note that the construction parameters may be varied during a construction process by the person responsible for the manufacturing. There must be comprehensible rules for this process.

Because it is a matter of the production of competitive products, high-performance materials must also be available. For the plastics it is not sufficient to achieve the properties of today's mass-produced plastics, so-called *commodities*. Currently, high-performance materials such as PEEK are available, but major construction materials such as PA 6 are still under development.

The material palette was expanded considerably in recent times, especially for metals. This is partly because independent material suppliers have come to the market. The use of commercially available materials does not relieve one of the responsibility for the product characteristics and does not exempt one from evaluating the material for one's own product (see Section 2.3.2.1).

6.3.3.3 Organization

In addition to the process that is already used for prototypes, it should be remembered that the manufacturing operation likely will need direct consumer contact if one wants to understand the direct transmission of the modified product data from the Internet as the form of customer contact. In addition, there are changes in requirements for the deliverable product.

The additive manufacturing company no longer delivers a prototype according to individual specifications, but an “additive manufactured part” or an “additive manufactured product.” Parallel to the production, for example, an instruction manual and a package must be created, and one must ensure that both are immediately conveyed together to the finished product.

The legal conditions are unlikely to change to the detriment of consumers for individualized products compared to today. So product liability for individualized single products must be considered and how the resulting requirements are implemented organizationally. Also, insurance constraints also have to be reconsidered.

Lastly, the purchasers of individualized products are also entitled to repairs and the provision of spare parts. This is where further challenging tasks in the field of additive manufacturing and its organization arise.

6.3.3.4 Design

In additive manufacturing, the data set of the design goes directly into the machine. There is no more additional production-related design, as is the case with many nonadditive processes. Design details and material selection with regard to the selected method for additive production are therefore design parameters and need to be optimized as part of the design.

Therefore, a design scheme for additive parts to be manufactured must be developed, a “design for additive manufacturing,” DfAM (or perhaps because of the introduced concept of a “design for rapid manufacturing,” DfRM).

These design schemes must also consider that the designs for many additive manufactured components are no longer in accord with today’s principles. Overall, particularly the ability of additive processes must be exploited to produce extremely complex parts and significantly reduce assembly costs in this way. In plastic molded parts, for example, the avoidance of undercuts to prevent slides or on which location “diving cores” should be placed must no longer be prevented. Changes in wall thickness are less critical, and draft angles are omitted just like the parting line and injection points. Run lengths and weld lines are not critical points anymore.

Different than in today’s additive prototype methods, the later-used additive method must be already defined in the design, also in view of the specific process parameters. A production on any machine determined later, which is a general advantage of additive manufacturing, is no longer useful and above all does not ensure a good result.

So the designer must specify, not only the additive manufacturing process, but also the key manufacturing parameters. For example, in the FDM process, a few parameters are crucial in how the part geometry is built (see Fig. 6.17) and what its mechanical-technological properties are. This shifts the responsibilities of the operator onto the designer.

The first design aspect brings the principle of customization. Active personalization will mean that in the future, the consumer completely or at least partially creates his or her product. For this purpose, the consumer logs into the Internet. The consumer will be guided, even as a layperson, by a simple CAD system and thus be enabled to modify the 3D model of his or her product as desired. Theoretically this is a variant design.

Laypeople will successfully shape their own personal product via the Internet and using CAD only if all feasible changes were determined and saved by in a parameter field in advance. So theoretically the goal of the designer is no longer an optimal product, but several, if you like an infinitely optimized or nearly the optimum number of products that are within a predetermined process window. In contrast to that, the classical variant design leads to a finite number of related but individually optimized products. In terms of a photography analogy, the single part design equates to a camera with a fixed focal length, the variant design equates to a set of interchangeable lenses with different but fixed focal lengths, and the parameter field structure equates to a zoom lens.

It is clear that, with the involvement of customers in the design, the design process is not getting easier, but much more complicated. The expense for the documentation associated with the design is also much larger. There will also be portals for the end users that provide configurators for modifying a predetermined product or to develop their own part or product.

6.3.3.5 Quality Assurance

The manufacturing of individual parts creates new challenges for quality assurance. Given the batch size of one, all statistical methods and destructive test methods are limited to general preliminary tests for machine optimization. During production, the output is definitive.

One approach is to integrate an X-ray or CT unit in the additive manufacturing facility to provide a real-time analysis of the results.

The parts must be individually, permanently, and automatically marked in the process. Each part must be individually assigned at any time, so it can be individually packaged and shipped after its manufacture. These markings should also allow traceability of each part for product liability. Current approaches such as radio frequency identification (RFID) should be checked for their utility. The documentation must permit the production of a replacement part at any time.

6.3.3.6 Logistics

Logistically, the additive manufacturing process is therefore a "manufacturing on demand," or a *just-in-time* manufacturing of single parts. The difference from the logistics of current big series is not irrelevant, because due to the single part character, the parts are not delivered in containers or other bulk packaging for further assembly, but must be placed as individual items in the usage site.

Variant production by the automobile manufacturers is already an indication of this integrated logistics and production engineering task. In the Audi A6 (built in 2005), for example, 18,000 different door panels can be installed. A journal on automobile production headlines it accordingly: "Insanity with method" [Automobil Produktion, 05]. With these numbers, it is still a matter of parts that are in series, albeit small, and produced with current production methods.

Therefore, series products are combined with other series products to form individualized mass products with a unique image. But all part products relied on for individualization are part of series and logistically can be treated as anonymous series products charge-wise.

Additive manufactured individualized products are unique in manufacturing technology as well. In the future, "single part-uniques" are completed with other "single part-uniques" to complete "product-uniques." Logistically, consequently, all elements of the whole supply chain are to be treated as individual parts.

Logistically, the task also is to implement a suitable merger between the product with all of its parts and all necessary accessories, also individualized, such as manual, warranty certificate, and packing material. Finally, everything must be archived so that the individual access is ensured at all times and over a long period of time.

6.4 Implementation of Rapid Manufacturing

In the application of the methods of additive manufacturing, different scenes are possible. AM machines are integrated as elements of a mostly nonadditive multicomponent fabricator in the manufacture process as well as decentralized for complete additive manufacturing. They are used in industry or privately as personal printers. Depending on the application case, different machine concepts will emerge. The developments conceivable today are introduced next with examples. The transition from current additive manufacturing in the concept of rapid prototyping to rapid manufacturing takes place smoothly. It had already begun some time ago.

Classifications are not reflected accurately in these following examples because, in practice, mixed forms tend to occur.

6.4.1 Additive Manufacturing Machines as Elements of a Process Chain

In industry, rapid manufacturing that has already been introduced by additive methods was integrated into a production network with nonadditive methods. Parts are manufactured that are additionally machined, finished, and assembled with other products in the course of the production chain. An example of a simple combined process is the production of a titanium structure for aviation (Titanium 6Al-4V); see Fig. 6.21. The structure is additively manufactured on a milled plate, heat treated, machined to size, separated from the stabilizing base plate by machining, and measured. The structures have characteristic dimensions up to 2500 mm. The additive method consists of an adapted LENS process (see Section 3.3.2) and machine finishing. It is marked by a high percentage of manual operations.

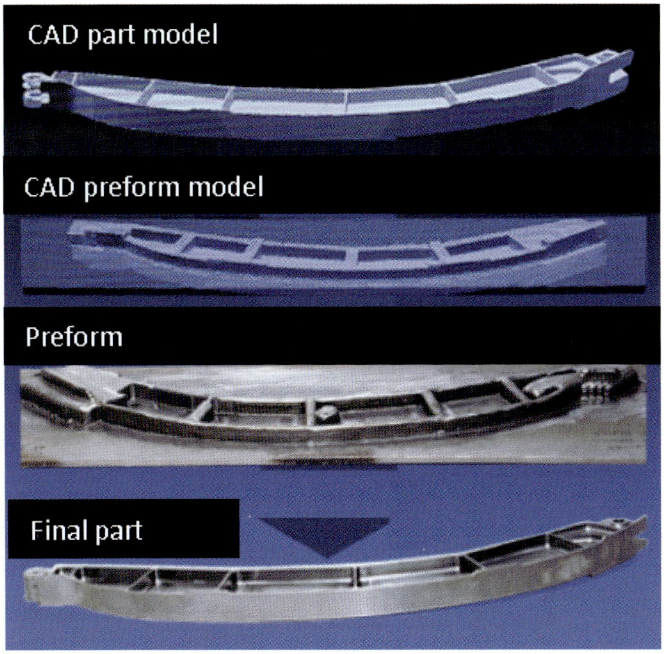

Figure 6.21 Combined additive and conventional manufacturing process for aviation components made of titanium (Source: Aeromet)

In the field of plastic molded parts, a combined conventional and additive manufacturing is also possible if plastic injection-molding machines do large quantities, while parallel to that, the additive machines provide the single parts.

For that, the additive machines must be free of manual work parts (according to Section 6.3.3, "Future Efforts in Additive Series Production") and be integrated into production lines.

6.4.2 Additive Machines for Complete Production of Products

Additive manufacturing methods have the potential to directly manufacture not only components, like plastic moldings, but also complex products (see Section 6.2, "Potential for Additive Manufacturing End Products"). Such approaches represent a paradigm shift from today's multipart single part and assembly-oriented production technology to an additive one-step manufacturing of complex products.

6.4.2.1 Industrial Complete Production

In the form of complete manufacturing machines, fabricators will enable the manufacturing of selected parts or complete products. One example is the 3D printing machine of Therics Theriform that can produce up to 40,000 pills per hour with complex inner structures and distributed active components. The pills leave the machine already packed (see Fig. 6.22).

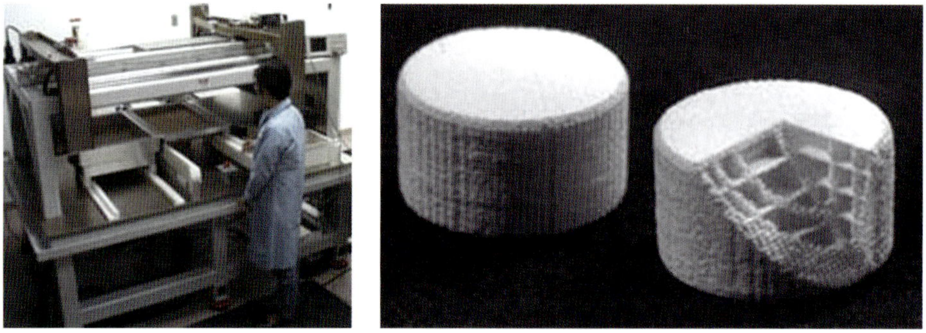

Figure 6.22 TheriForm 3100: manufacturing of pills by 3D printing machine (left), pill outer view and in section (right) (Source: Therics)

Because producing additive manufacturing machines do not have to be integrated into production lines, but constitute the lines themselves, and because any spatial separation of design and production is possible, their operation and installation can be decentralized. A central design with a decentralized manufacturing can be realized, and vice versa, as well as all conceivable combinations. Especially attractive would be the equipping of customer service centers for the direct production of the required accessories and spare parts or the adaptation of hull products with country specific, locally additive manufactured attachments.

The industrial fabricators for those applications must include all of the properties just discussed in future AM machines and in particular, to ensure a high productivity.

Examples of products are a one-piece additive manufactured cable clamp with two integrated hinges and cable guides with integrated clamps (Fig. 6.8) or the gripper in Fig. 6.5.

The nozzle plate produced by microTec (Fig. 6.23) is an example of the micromanufacturing of components. As part of a catheter, it drives the turbine wheel of the mill. Its diameter is 4 mm, and the outlet cross section of each nozzle is 20 µm. The plate consists of biocompatible material. It is an example of microparts that cannot or can only with much greater effort be manufactured with conventional methods.

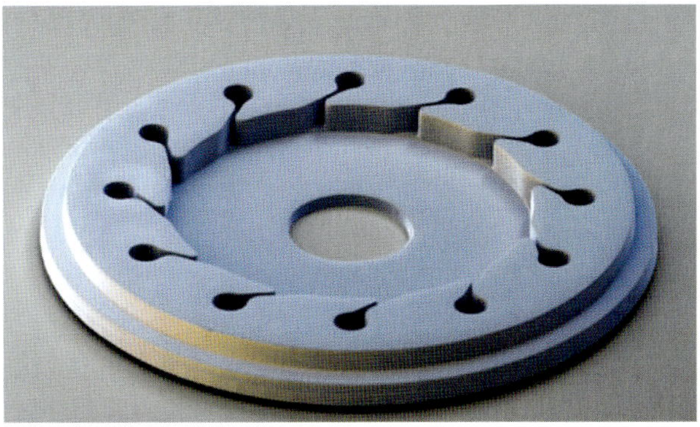

Figure 6.23 Nozzle plate (Source: microTec)

An example of an integrated product that, from the point of view of the manufacturer of mobile phones, represents one component, is a tapered helical mobile antenna manufactured with the M^{3}D Technology of Optomec. See Fig. 6.24.

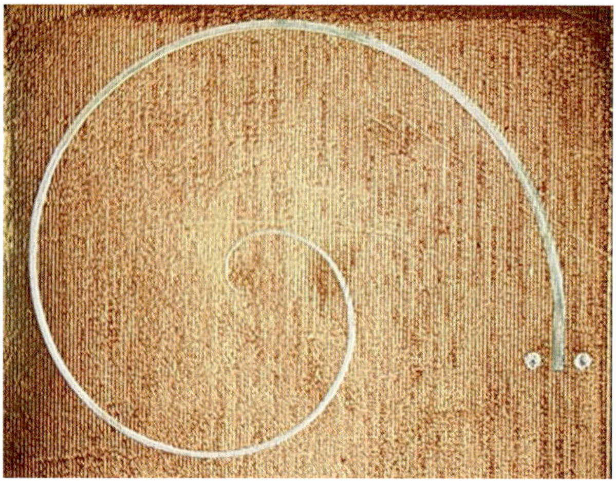

Figure 6.24 Mobile antenna, aerosol printing (Source: Optomec)

6.4.2.2 Individual Complete Production (Personal Fabrication)

With complete and in one step automated additive manufacturing machines, non-manufacturing technicians can also produce products. This is possible through personal printers or fabbers (Section 3.5.2). They are available as complete machines with good part properties (FDM Mojo, Polymerisation ProJet 1000/1500) or as fabbers with reduced component quality, but at very affordable prices, either as complete machines or as an assembly kit. The bottom price for a DIY fabber of the type Prusa Mendel is today (2013) at about 600 € (about $ 675).

In combination with a personal computer, those kinds of fabbers or personal 3D printers represent complete production systems for individual decentralized production. The implementation, especially in a private setting, has already begun.

The data come from a personal 3D design that can be created by anyone with programs like Google SketchUp or as a complete data set from Internet portals like 3D Warehouse or Shapeways.

With this, the over 10-year-old scenario is becoming realistic in which children design system construction kits and other toys on the computer on their own and directly manufacture them. The 3D printing blogs provide assistance and promote the exchange of experiences. Self-made spare and additional parts are therefore no longer ideal. The step from cybercommunication to cyberproduction has been taken.

The productivity for these applications mostly plays a subordinate role, and manual work parts are accepted.

An example of personal fabrication is the model of a camshaft wheel (Fig. 6.25). It was manufactured with the (up to 2006) easiest and cheapest machine, the LD3 Printer (Section 3.4.10). It is easy to imagine that in this way, system assembly kits, components of model toys, or spare parts can also be manufactured.

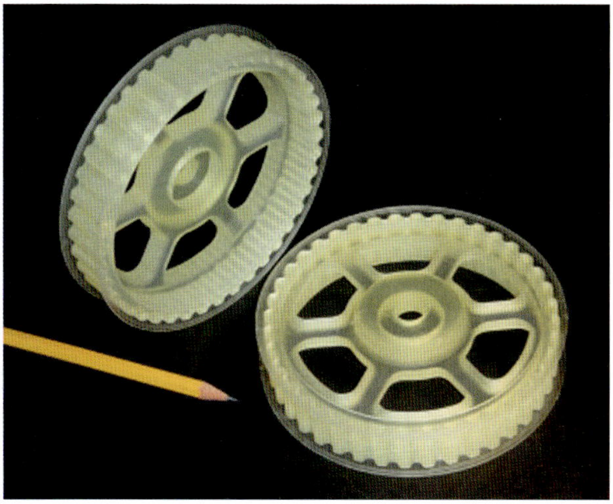

Figure 6.25 Camshaft wheel as example for system toys (Source: Solidimension)

6.5 Application Fields

During the discussion of additive manufacturing machines, it has already become clear that development is running in two directions. One of them leads to universal machines, the other one to specialized applications, also in the sense of branch solutions. Both trends will grow further in future.

More specifically, for the so-called universal machines there are also two lines of development. One leads to simple office-suitable or also privately usable machines that are summarily described as personal 3D printers.

The other one leads to complex additive manufacturing machines that require a manufacturing infrastructure and are suitable for developers, for in-house or independent service providers, or for research laboratories. With them, depending on the application, end products (production 3D printer) or also prototypes (professional 3D printer) can be manufactured.

The specialized machines aim at the processing of special, often also branch-specific materials or material families. But they contain also more often packaged solutions, consisting of the additive machine, a branch-specific software solution, and materials matched to the process. Both the total packages as well as the pure service are offered. In each case only finished products or their components are manufactured.

Therefore, from the point of view of additive manufacturing technology, application fields have been developed that it can be usefully structured by materials or by branch. Of course the areas have strong interactions with each other.

6.5.1 Application Fields for Materials

This chapter gives an overview of selected, particularly heavily developed areas of application. It does not claim to be complete.

The processes and the mechanical-technological properties of the materials will not be discussed in detail. These are described in Chapter 3 and compiled in Appendix A3.1/A3.12.

6.5.1.1 Metallic Materials and Alloys

The development of metallic materials is oriented preferably in niches that contain attractive boundary conditions for additive manufacturing. Niches with respect to the application are flow-conducting, thermally loaded parts, such as the turbine housing of a turbocharger in Fig. 6.26.

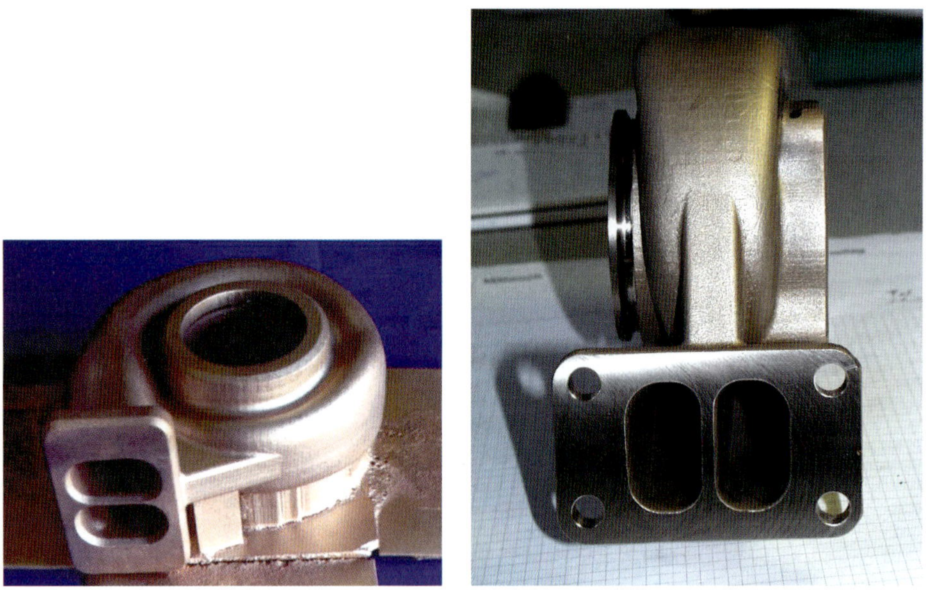

Figure 6.26 Casing of the turbine of an exhaust turbocharger. Selective laser sintering, after the building process (to the left) and after finishing (right) (Source: 3D-Systems)

Industry niches include aerospace and medical technology. The applications are characterized by small numbers or single parts. These industries are used to materials that are specially prepared and sensitively monitored, and to extensive testing of each part. Also, one is more willing to tolerate an appropriate price level for this effort.

Titanium and CoCr alloys are the keys to these branches. All suppliers of metal processes (Sections 3.2 and 3.3) have multiple relevant approved materials. The parts are all end products.

The niches also include the tooling, although altogether it is a very large industry. The basic requirement to offer tool steels was first fulfilled a few years ago. The tendency, compared to nonadditive processed steels, to higher strength and lower elasticities is increasingly overcome. The range is still small and often includes only a few materials per supplier, but it is growing very dynamically.

6.5.1.2 High-Performance Ceramics

Unnoticed by many and hidden by the efforts in the processing of metallic materials, applications have been developed in ceramic materials for all additive process families [Gebhardt, 07]. Used are 3D printing and extrusion processes, polymerization and layer laminate processes with subsequent sintering processes, and also direct applications of the additive sintering technique. Figure 6.27 shows a filter element made with the Ceraprint 3D printing process, which has been modified by Specific Surface based on the MIT license.

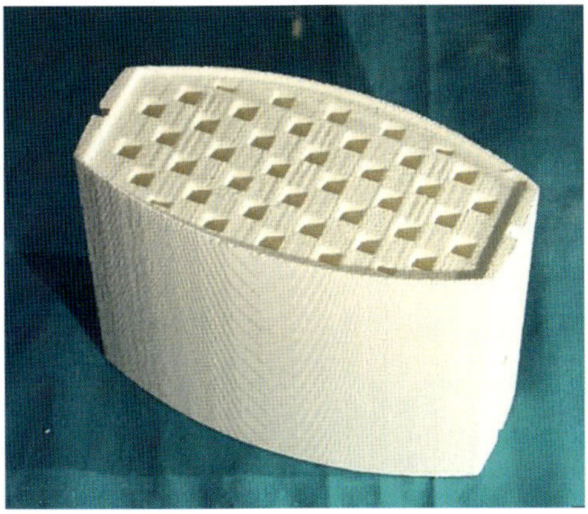

Figure 6.27
Ceramic filter element
(Source: Specific Surface)

Direct sintering in the high-temperature chamber was developed by Phenix and implemented industrially. The process is described in Section 3.2.12. Figure 6.28 shows a ceramic heat exchanger with an intricate internal structure.

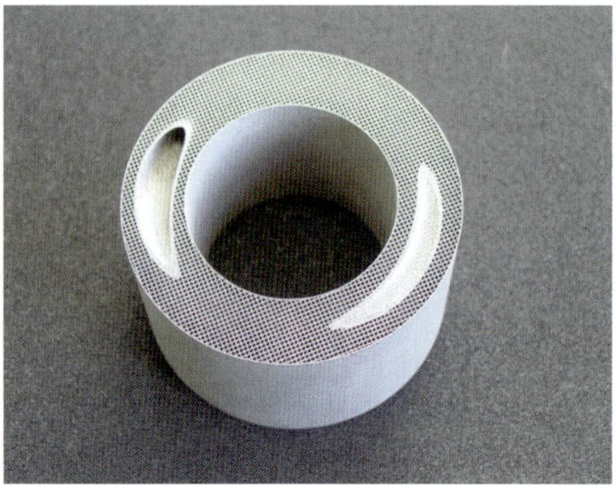

Figure 6.28
Ceramic heat exchanger
(Source: Phenix)

The range of ceramic materials covers almost the entire spectrum: shapes and parts made of Al_2O_3, SiO_2, ZrO_2, and SiC, fully sintered Si_3N_4, and so-called *graded materials*, such as zirconia-reinforced aluminum (ZTA) formed by coating ZrO_2 on an Al_2O_3 layer or a corresponding substrate.

Monolithic ceramics are manufactured, but flowed mainly through or with high temperatures impinged structures. Defined macroporosities are the basis for implants of resorbable bioceramics. Microporosities are used in reactors, but especially in tribological systems.

6.5.1.3 Plastics

Many critics focus primarily on devaluing additive processable plastics when parts do not have properties equal to those of nonadditive plastic parts, especially when processed in a thermoplastic injection mold. This is mainly a question of the design (compare Section 6.3.3.4, "Design"), but consequently this criticism causes intensive work on the improvement of existing materials systems, and the qualification of new ones.

Certain specific additively processed materials, such as polyamides, can come quite close to the properties of their nonadditively processed counterparts, but overall this is still some way away from a technically important range, especially for the high-performance materials (see Fig. 6.18). The qualification of PEEK has already been achieved, and the important PA 6 is scheduled to launch.

An important application of plastics is in the casings (shells) for in-ear hearing aids. If the relevant data were available, hearing aid shells could have been additive manufactured since 1991. In addition to the development of appropriate software packages for the design of the shells, the development of materials provided the breakthrough that means that worldwide an estimated 85% of all hearing aid shells are manufactured with (different) additive processes today (see Section 6.5.2.3, "Medical Equipment and Aids, Medical Technology"). The breakthrough came with the development of a material that can directly and permanently be worn in the ear (see the next Section 6.5.1.4, “New Materials”).

Alternative plastic materials are pigmented polyamide for the sintering process and colored resins based on (meth)acrylate for polymerization processes.

6.5.1.4 New Materials

According to the previous section, the development of hearing aid shells required new materials. Radiation-curable resins from the SLA technology are not usually biocompatible. The reasons are often the use of a combination of radical and cationic curing (hybrid system), and the selection of highly reactive unsaturated epoxy compounds. Therefore, new biocompatible materials have been developed for the otoplastics that are permanently in contact with the patient. By means of an anaerobic inhibitor, 2,2,6,6-tetramethylpiperidin-1-yloxy (TEMPO, free radical), the resins were stabilized, and the photosensitivity was adjusted so that a low radiation energy results in the greatest possible depth of cure with a high degree of polymerization, good green strength, and sufficient stability (material: Fototec, [Klare, 05]). The final products obtained by complete curing not only have good mechanical characteristics, but also exhibit excellent biocompatibility, are hard-elastic, and show a very low water absorption. The material is opaque or transparent.

In microtechnology, the material is often the deciding factor for a successful application, and frequently this material must be developed or modified. In this context, the use of a service provider is often the better choice. This is the business model of the company microTec (Section 3.1.6), which also undertakes individual material development.

New materials created through the composition of different layers of, among others, plastic, metal, ceramic, or even living cells, are produced by the M^{3}D method of Optomec (Section 3.6.6).

6.5.2 Application Fields by Industry

This section provides an overview of selected applications that have been strongly developed. It does not claim to be complete.

The processes and the mechanical-technological properties of the materials will not be discussed in detail. These are described in Chapter 3 and listed in the appendix.

6.5.2.1 Tooling

In addition to the basic requirement to offer tool steels that match the properties of nonadditive processed materials, the additive systems must be integrated in particular into the production flow in the tooling. Some systems are tailored to the tooling in this sense. They build up for example on a system holder or integrate pallet or clamping systems, which can be aligned with little effort for the subsequent steps in nonadditive machines. Examples were discussed in Section 5.4.3.

In tooling, it is essential to optimally match additive and nonadditive elements together. One example is the ecoMold project [Hennings, 04] that touches upon the DMLS processes (Section 3.2.4) but, in principle, is possible to implement with each sintered metal or fusion process. The starting point is a modularization in the sense of dividing additive and nonadditive manufacturable elements.

The nonadditive finishing work is enabled by clamping elements. Figure 6.29 shows an element after additive processing, a compilation of all of the elements and the clamping plate and the fully assembled mold half.

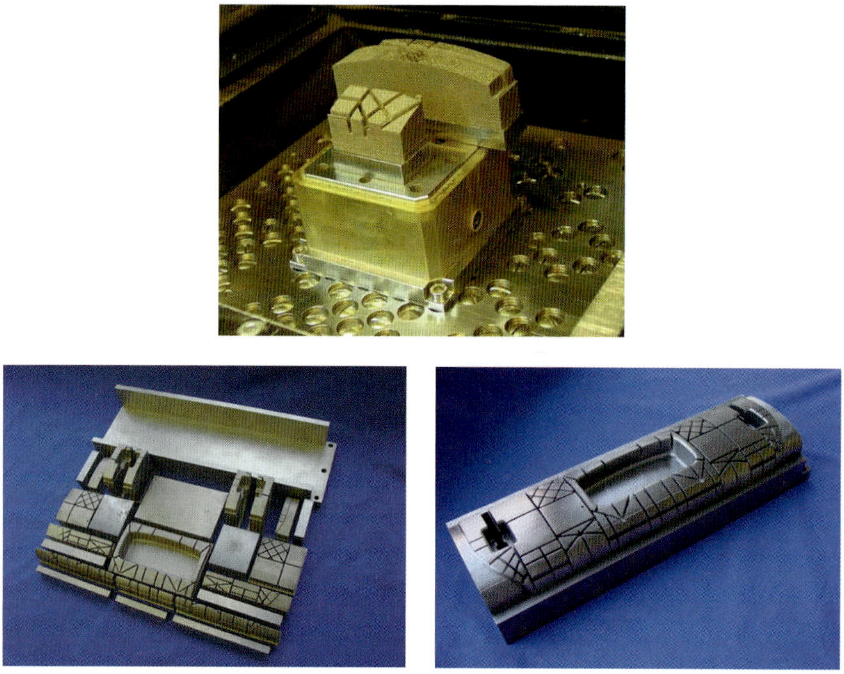

Figure 6.29 Ecomold: additive construction process (above), additive and nonadditive manufactured tool components (bottom left), and assembled tool half (bottom right) (Source: IFAM)

The application tooling is particularly discernible by design enhancements that are only additively implementable. Of special significance is conformal cooling. It was presented and discussed in Chapter 5.

The advantages of additive manufacturing will succeed insofar that on a broad line in which it is possible to overcome current limitations, especially the still-insufficient surface qualities.

Direct additive tooling, Section 5.4, exclusively leads to tools with serial character.

While discussions have been held almost exclusively on the additive construction of tool inserts, and thus individual, single part production, the company Hofmann has introduced a series product. They offer fasson pins with internal cooling that can be used in screw bosses to reduce sink marks. The cooling is done with air by an internal double spiral, which was produced by LaserCusing (Section 3.2.8). The pin is completed with nonadditive partial processes by turning and grinding. Fasson pins are available as standard parts with diameters of 3.5–6 mm (in 0.5 mm increments) and a length of 500 mm. They represent an additive manufactured series part.

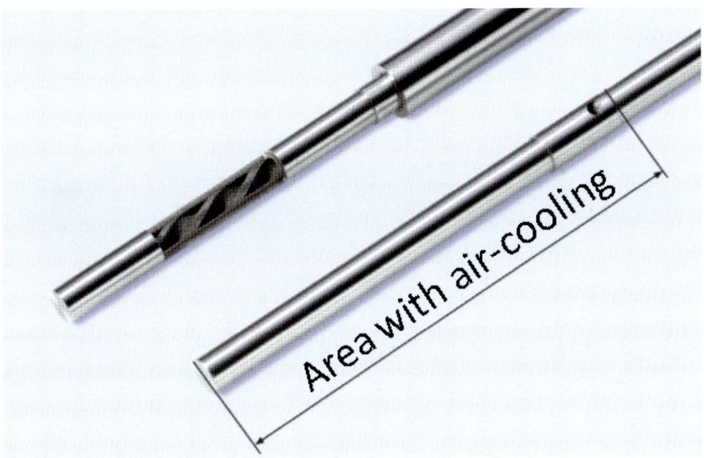

Figure 6.30 Fasson pin, LaserCusing (Source: Hofmann)

6.5.2.2 Casting

Additive manufacturing is used in casting for the direct[5] manufacture of models and moldings. The results are parts with series character because the casting process is conventional. However, depending on the application, a possible interaction between molding, binder, and casting material has to be considered, which can affect the outcome.

Additive methods are used in many casting processes. Of particular importance are the contributions treated in Section 5.3.1 for investment casting, the processes discussed in Section 5.4.1.4 for the manufacture of models and cores for

[5] Indirect methods can be assumed from any original model and therefore are not considered here.

sand casting, and the direct metal processes, Section 5.4.3.2, which are also used for the production of permanent molds for die-casting.

Applications for additive manufacturing technology for casting can be found in many industries. The focuses are mainly in dental technology, Section 6.5.2.2.1, and in the jewelry industry, Section 6.5.2.2.2, but also in industries that manufacture small and detailed parts, such as in precision engineering.

Investment casting applications are everywhere, where the parts are complex and their weight plays a big role, and thus especially in the aerospace industry.

The resulting parts have the character of end products.

6.5.2.2.1 Dental Technology

Caps, crowns, and multipart bridge frameworks for subsequent ceramic veneered tooth replacement are described as restorations. From the point of view of machine construction, it is a metal support structure. The nonadditive production is characterized by a molding and casting process characterized by grinding or milling operations, which is carried out in the laboratories of dental technicians with great manual effort.

Since 2009 when the first fully digital process chains were established, the standard process steps like aligning in the articulator, wax modulating, embedding, and casting through the digitization of the model, can be replaced by a virtual design of the restoration and direct additive manufacturing.

The direct manufacture was first possible after a suitable machine concept (TrumaForm 130, Section 3.2.7 of the third German edition of this book) was adapted to the required cobalt-chromium and gold-platinum alloys and was developed into the BEGO Medifacturing method (Fig. 6.31) (www.bego-medical.com) [Uckelmann, 05].

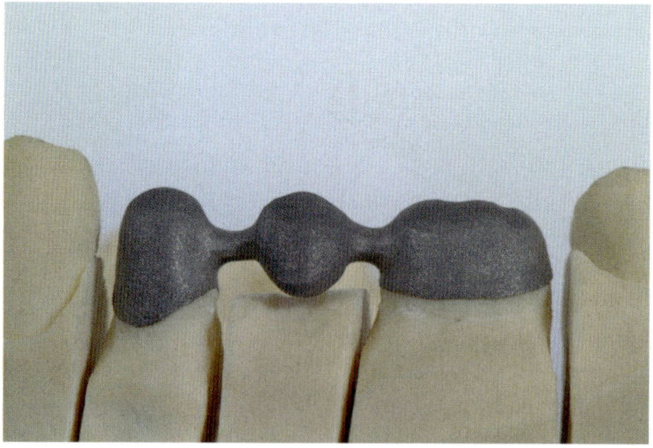

Figure 6.31 Cobalt chromium dental bridge, BEGO Medifacturing (Source: Bego Medical AG)

The model is typically taken from the patient's imprint, or now increasingly gained digitally by means of an intraoral scanner. The parts are manually worked out and reworked. The digital process chain therefore includes the traditional manual operation steps, but assigns modified tasks to them. The restorations are available sooner and are also cheaper in the medium term. In the course of quality assurance, the geometrical data and the minimum wall thicknesses are controlled, and samples are taken to determine the elongation at break, the yield strength, the surface roughness, and the material density. The dental technician outsources the production, reduces his fixed costs, and focuses on the ceramic veneer.

The market has increased tremendously in recent years. Since about 2009, there is hardly a manufacturer that does not offer a usable material in dental technology or even a modified machine for this industry. The direct production of metal parts is also offered by EOS, 3D Systems, SLM Solutions, Realizer, Optomec, and Phenix. Indirect processes tailored to cast-making are also offered by NextEngine, Sanders, and Envisiontec. An industry solution with scanner and software is the InVision DP (Dental Professional) System from 3D Systems.

For plastic processes, 3D Systems offers modified standard machines (Projet 3510 DP and MP). With this, the melt-out photopolymer VisiJet Dentcast can be processed for the production of wax-ups, VisiJet Pearlstone can be used as gypsum or with the transparent and colorless Stoneplast, and so all dental problems can be solved.

Alternatively, in polymer printing (iPro 8000MP) or by the laser stereolithography process (iPro™ 8000 MP), the slightly purple shimmering material Accura Amethyst, or Accura e-Stone, which appears to be classic dental material, can be used.

6.5.2.2.2 Jewelry Industry

Jewelry first became affordable once it could be manufactured in series. But its distinct character is diminished when the pieces are listed in catalogs. With additive manufacturing, a unique character can be created, although the individual pieces of jewelry are automated and thus produced comparatively inexpensively.

Mainly lost models for investment casting are made additively. Therefore, most of all processes and machines that are suitable for dental technology are also suitable for the jewelry industry.

Also in the jewelry industry, a breakthrough was connected with CAD programs to support the design process. JewelCAD (Hong Kong) is an example of such industry software (www.jcadcam.com). It features STL and SLC interfaces for the control of stereolithography machines (3D Systems and Stratasys-Objet) as well as wax, pressure, or extrusion (Sanders) methods, but it can also drive cutting plotter and milling machines.

For jewelry production, there are solutions from multiple manufacturers. For many years Sanders has focused on jewelry as a main application. The extremely delicate wax models can be directly rendered. A model, the cast, and the finished jewelry piece are shown in Fig. 6.32.

Figure 6.32 Solidscape meltable wax model, cast part, and finished jewelry piece (Source: Solidscape)

3D Systems also offers machinery in the Projet 3510 series for jewelry production. Envisiontec has, for delicate parts, the Perfactory in the variant "Mini" in the program. It supplies, depending on the focal length, a resolution of 32 microns in the *x*-*y* direction and layer thicknesses up to 23 microns. The then-available construction space with a base area of 41 mm × 31 mm (*x*, *y*) is sufficient for jewelry. Solid Cape (now Stratasys) provides the 3Z-STUDIO 3D printer for the jewelry industry, which specializes in melt-out models for wax casting.

The direct additive manufacturing of jewelry is one of the most attractive fields for the active personalization of products; see Section 6.2.4.2, "Active Personalization: Customer Personalization." Concepts like the there-outlined DigSmugg project quickly and easily find a very large customer base via the Internet. Production can be done centrally or decentrally.

6.5.2.3 Medical Equipment and Aids, Medical Technology

All products that are directly linked to biometric or anatomical characteristics of humans are especially suitable for additive manufacturing because they technically follow a common manufacturing principle, but nevertheless display unique pieces adjusted to the patient. The achievable accuracies are usually less of a problem than an exact determination of the data with the methods of medical resonance imaging and the development of biocompatible materials.

Products are formed when additive manufacturing is associated with a resonance imaging method and industry software.

The basic principle is the combination of virtual and physical process steps. Virtual process steps assume the task of creating geometries that cannot be taken or measured from the patient because they represent internal structures or because they describe the goal of the treatment but not represent the current state. Non-additive intermediate steps, for example replicas, are possible. Mostly, suitable materials have to be developed. The additive process steps are taken over by the production.

6.5.2.3.1 Braces: Aligner, Invisalign

Especially in the United States, but increasingly also in Europe, white teeth and straight teeth are important social indicators. Adults sometimes wear plastic braces called aligners. They are made of transparent plastic films and therefore can be worn during the day.[6] The production of the aligner is made by thermoforming a transparent plastic film on a tooth model. Because the aligners must have the future position of the teeth and thus cannot be detached from the existing teeth, a complicated process chain results.

The tooth position to be corrected is taken by a conventional dental imprint from the dentist, and the impression is sent to the company Align Technologies (www.invisalign.com). To achieve the required accuracy, the imprint is filled with a chalky white mass, creating a positive of the teeth in gypsum. Together with some 10 others it is embedded in a black contrast material. The embedded imprints are milled in thin layers and the resulting contours scanned individually, which brings very good results because of the high contrast. From the contour record a three-dimensional dental model is reconstructed and transferred into a computer program that enables all of the teeth to be moved virtually. With the help of this program, specialists design the new tooth position. Because with aligners only comparatively minor corrections can be done, the total correction is decomposed into separate corrections. The result is sent to the treating physician for evaluation via the Internet. From the data sets, all submodels that are necessary for treatment are manufactured by stereolithography, each as a complete tooth model. On a construction platform of an SLA-7000 (3D Systems), 90 tooth models are produced simultaneously.

[6] The dental aspects are illuminated in detail for example by Dr. Madsen (www.madsen.de).

Because the tooth models represent the respective tooth positions on the path to the desired teeth, they are used as a model for the deep drawing process. The contoured films are trimmed on the contour of the neck of tools (Fig. 6.33, left). The finished aligners are given to the patient. Under medical supervision, each pair is worn for about two weeks (Fig. 6.33, right).

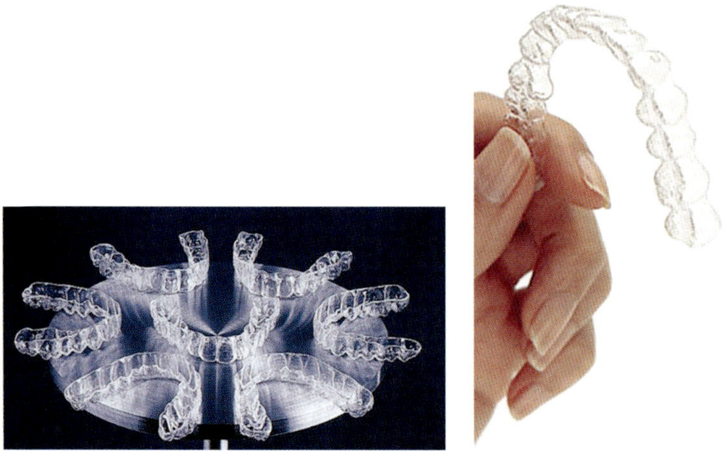

Figure 6.33 A set of objects formed and trimmed as aligners (left); finished aligner (right) (Source: Align Technology)

The company Align Technologies operates more than 50 SLA-7000 and iPro 9000 XL machines; about 8000 braces are made in parallel. However, this individualized mass production is not possible without the entire process chain. The additive process alone would only prove the feasibility of an interesting phenomenon. Only the combined virtual and physical processing makes the approach an economical system.

6.5.2.3.2 Hearing Aid Shells, Ear Molds

Shells for hearing aids, also ear molds or otoplastics, on the one hand, with respect to their outer contour, need to be optimally adapted to the anatomy of the patient and, on the other hand, also must contain a number of internal cavities. The traditional production takes place in a multiple molding process (positive-negative-positive, PNP process) with a multiple of single manual steps, a high technical effort, and a corresponding number of error sources.

In order to use the possibilities of additive manufacturing, a correspondingly modified process chain was developed. It begins with the traditional imprint at the ear, nose, and throat specialist. Further processing is carried out virtually with a special CAD program in the computer. For this, the imprint is scanned. External defects are removed, the ear canal is modeled, and an offset for optimum fit is applied. The conch (shell) form is defined, and the holes for the sound channel and the vent are placed. The virtual method, through the interaction with additive manufacturing, enables completely new geometries that cannot be produced with the traditional milling. The complete digital display also allows a comprehensive geometric quality control.

The data sets of the thus-generated ear pieces are arranged on a building platform for the actual construction process. Additive processes of different suppliers are used based on laser sintering (SLS, EOS, Section 3.2.4), stereolithography on the principle of digital light processing (DLP, Envisiontec, Section 3.1.8), and laser stereolithography (SLA, 3D system, Section 3.1.3).

The construction process and material properties are closely linked together.

The shells can be after-treated traditionally, for example, grinded and painted. This is very important for the integration of the new manufacturing process into existing structures.

The use of an additive manufacturing process, in combination with suitable materials and a special virtual design program, has revolutionized the production of ear pieces worldwide. Since 2007, several hundred systems from different manufacturers have been in use at many hearing aid manufacturers worldwide.

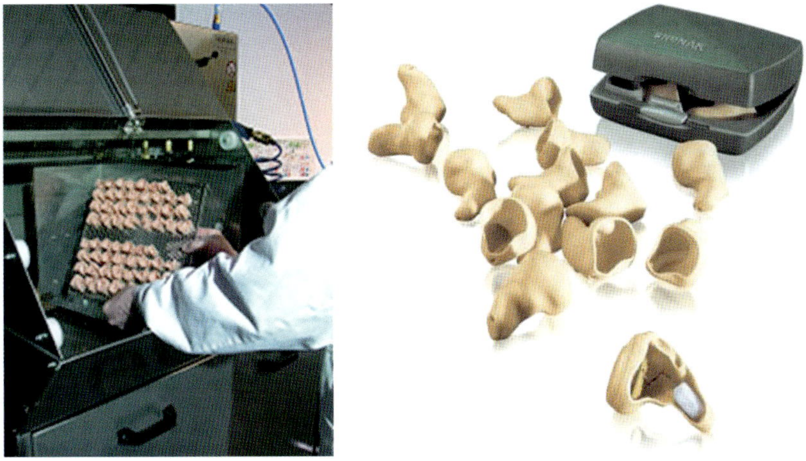

Figure 6.34 Hearing aid shells from Fototec material: approximately 40 pairs on a construction platform (left), finished shells (right) (Source: 3D Systems/Drewe Otoplastik/Phonak)

The market is estimated to be very favorable. 3D Systems has developed their SLA system Viper HA SLA especially for the production of hearing aid shells. It is optimized for the Fototec material (Section 6.5.1.3, "Plastics") and has a split resin container in which two different materials (or colors) can be processed simultaneously. On one platform, about 75 ear molds are constructed in seven hours. Thus, three construction processes per day are possible [Klare, 05]. A post-cross-linking cabinet optimized to these geometries is used.

6.5.2.3.3 Technical Medical Devices

Dental endoscopic camera: One example of the optimization and additive manufacturing of a medical device is an intraoral camera. The existing product was manufactured from sheet steel and welded with a laser. A worn tool set was the reason to reconsider engineering and manufacturing. The very small series and the desire to obtain greater design freedom, such as to adapt the grip part of the two-part housing to the anatomical conditions of the physician, were the reasons for the direct manufacturing with a metal laser sintering process (EOSINT M); see Fig. 6.35. The surfaces are either polished or, as in the traditional model, nickel plated.

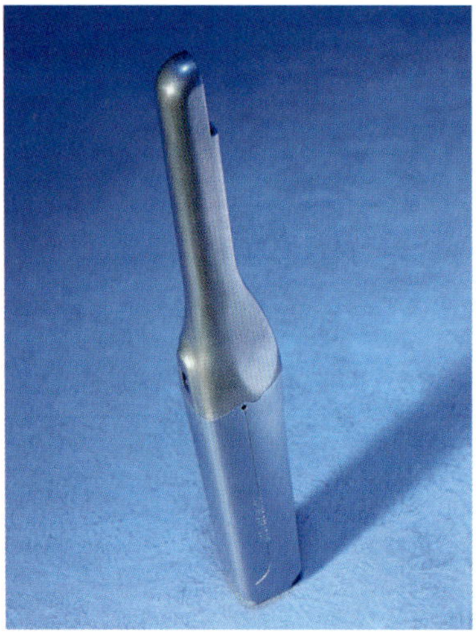

Figure 6.35
Endoscopic dental camera
(Source: EOS, Planmeca)

Container for a blood centrifuge: The company Andreas Hettich GmbH offers blood centrifuges, which enable the sedimentation and separation of blood within a single system. The core of the machine is a drum motor with six tanks and drip trays. The containers have a complex geometry and are made up of three parts: a drip tray, a chip board, and a housing connecting both. They are loaded at a rate of 2000 rpm and acceleration forces of 1200 g. The production numbers vary hugely between 10 and 1000 per year. Therefore, a batch-size optimized production was envisioned and then economically and technically evaluated, particularly with regard to the sterilization of the product [Lenz, 04].

The preoccupation with additive manufacturing soon led to a new geometry. Undercuts were now more a tool technical problem, and because different geometries could be produced at the same time, the adapter plates have been integrated into the part. An assembly was no longer necessary. Figure 6.36 shows the largest container of the old, three-piece design and the new, integrated, and one-piece additively manufactured one.

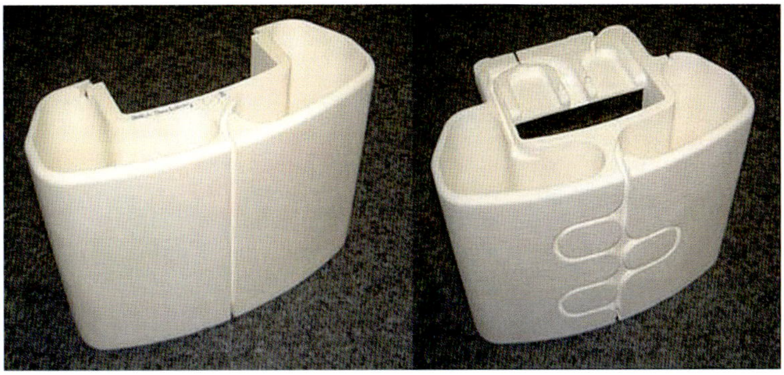

Figure 6.36 Conventional (left) and additively manufactured housing (right) of a blood centrifuge (Source: EOS/Hettich)

Further variants can be realized quickly and with minimal costs. Interesting, for example, are different versions of the housing for different blood bags.

Prospectively, the logistics can be simplified in favor of a request-controlled production (production on demand).

6.5.2.4 Design and Art

The direct manufacturing of articles that are attributable to design, to the arts in general, sculptural art in particular, and the history of art and the architecture, discussed in Section 4.3.5 in the context of rapid prototyping.

The argument is made that the parts used in design and art do not meet the requirements that are usually made on an industrially manufactured product. What applies to products that are attributed to industrial design does not have to apply to the same extent for art. Many nonadditive manufactured artifacts are not sustainable. If they have functions, they usually must not be used as intended. Overall, they need expensive protection from decay. Against this background, it is less the definition of a textbook but the manufacturer's and the user's assessment that determines whether an additively manufactured rapid prototyping component is seen as a product or not. In this respect, there is no clear division between the examples in Section 4.3.5 and the following.

EASYPUSH is an enlarged keyboard with a back shell in which commercially available mobile phones are inserted so that they can also be used by older or disabled people in workshops or during outdoor work. Additive manufacturing enables single part production for a variety of models. Tobias Fink was nominated with this design for the final round of the top five in the design competition in the Trade Fair and User Conference Rapid Tech 2007.

Figure 6.37 EASYPUSH: mechanics of the keyboard (right), opened shell with an inlaid mobile phone (center), and instrument ready for operation (left) (Source: Tobias Fink)

One-Shot Stool: The almost infinite geometrical possibilities of additive methods inspired the designers from the beginning. The company Freedom of Creation (FOC, www.freedomofcreation.com) became known early on for designs for additive manufacturing. Products such as lamps (see Fig. 6.3), watches, and fashion accessories were introduced. Sintered of one piece is the One-Shot Stool by Patrick Jouin (www.patrickjouin.com) (Fig. 6.38), the pioneer of designs for additive manufacturing. The picture sequence shows the structure from the folded object to the expanded chair.

Figure 6.38 Laser sintering in a single piece. Image sequence showing structure of the folding chair (Source: Patrick Jouin/MGX)

Eyeglasses: Eyeglass frames are a kind of wearable prototypes or samples that have been the subject of additive manufacturing for years (see also Fig. 6.39 in the third (German) edition of this book).

A step in the direction of mass production of high-quality consumer products was performed by the Berlin eyewear specialist MYKITA. It distributes its uniquely designed eyeglass collections in their shops in Berlin, Monterrey, Paris, Tokyo, Vienna, and Zurich.

The eyeglass frames have a visually and haptically comfortable surface structure reminiscent of wood. However, they are made of nylon by laser sintering and therefore are light, flexible, and comfortable to wear. MYKITA eyewear calls the material MYLON.

The surface is formed by a complex proprietary process that refines the surface of the material in six steps and includes grinding and dyeing operations. The temples and the frame front of the glasses are connected by novel hinges.

For their development process, MYKITA was awarded the iF Material Award 2011 (see Fig. 6.39).

Figure 6.39 Eyeglass frames, additive series production by laser sintering and postprocessing (Source: MYKITA)

For such individual mass production, the laser sintering process must be ensured to be reproducible around the clock. The company Kegelmann produces on several sintering machines annually a five-digit number of different eyeglass frames. Figure 6.40 shows a batch of glasses after sintering and a salable glasses.

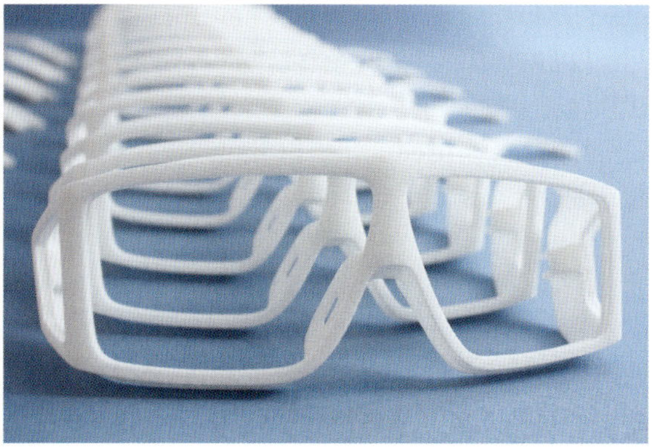

Figure 6.40 Eyeglass frames, additive series production by laser sintering. Frames after the additive process (Source: Kegelmann Technik)

Objects: Metal jewelry and art objects have been commercialized by the artist Bathsheba (www.bathsheba.com). He provides a large number of sculptures on the Internet that are produced by the Prometal Metal 3D printing process. The finished parts are passivated or artificially oxidized with sulfur, blasted, and then sealed with oil. Through variation of the process parameters, and according to the artist's specification, also the composition of the powder, the sculptures have different shades of yellow to steel gray.

Figure 6.41
Sculpture *Nexus*, Bathsheba, Prometal process (Source: Bathsheba)

Sketch Furniture: To be able to materialize thoughts is an old dream of mankind. Closest to this comes only 2D (freehand) sketching as one of the most effective methods of recording one's thoughts. In order to convert them into physical objects, 2D sketches must be translated into 3D drafts, then into complete 3D data for production and then by means of AM implemented into physical parts.

Even if we had the three-dimensional design in mind, indeed in our arms, how should we possess it differently? The Swedish design group FRONT (www.frontdesign.se) sketches directly into the air. The three-dimensional trajectory of the pen is tracked by a tracking system and converted into a 3D computer model. With additive processes, preferably with laser sintering, the 3D model of the sketches in the air can be converted directly into three-dimensional physical objects. This conversion is only possible with additive methods, because only they allow one to produce any three-dimensional objects without having to worry about limitations of the selected production process.

Figure 6.42 Sketched furniture: process of sketching with the visualized path of the pen (left); products: chair, lamp (right) (Source: Front Design)

3D Pen, 3Doodler: The 3Doodler, the first 3D printing pen, is one step closer to the aim of materializing thoughts. The pen, handheld in 3D space, allows the manufacture of a corresponding part by simultaneous extrusion of a thin plastic wire made of ABS or PLA (Fig. 6.43). The manufacturer is the toy company Wobble Works (www.wobbleworks.net).

Figure 6.43 Pen (left) and thus-made eyeglass frame (right); series product in the left background (Source: Wobble works)

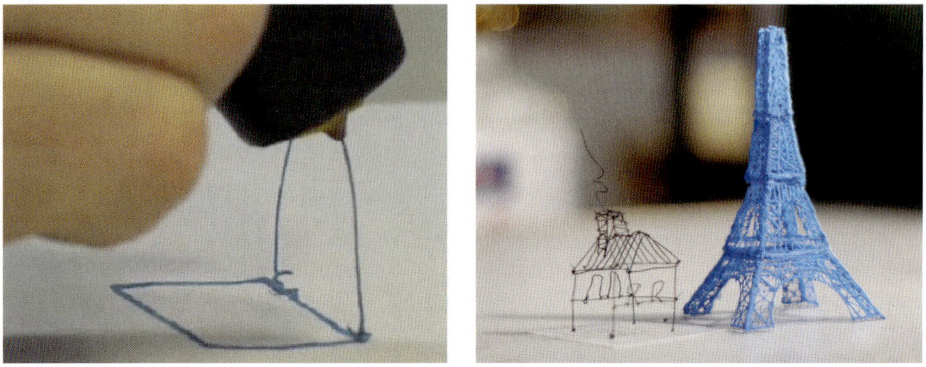

Figure 6.44 3D Printing 3Doodler: (left) structure during fabrication, (right) part Eiffel Tower (Source: Wobble works)

The direct extrusion in three dimensions does not only require spatial skill and drawing skill, but also compliance to a comparatively small but as constant as possible extrusion speed to ensure the defined cooling and solidification of the material. Figure 6.44 (left) shows a detail during construction; Fig. 6.44 (right) shows a section of a finished part.

Extrusion on paper is easier and more precise, also because the heat is dissipated safely. Here, 2D lattice-like structures arise, which can be regarded as structural elements and subsequently assembled into 3D parts. For this purpose, templates are sometimes used to improve the result but that expanded the idea of direct 3D manufacturing, so to speak, by drawing into the air.

Both approaches, Sketch Furniture and 3Doodler, do not belong to AM, according to the definition, because they are not additive layer manufactured and because they do not provide automatically reproducible parts.

■ 6.6 Summary

The components demonstrate that future fabricators will essentially be distinguished from current prototypers even if they both will increasingly be called 3D printers. They also demonstrate that the development does not result in a universal machine that can produce all kinds of products, but that there will be a greater number of specialized machines that are optimally set for industry-specific characteristics. Only a few will, according to the original idea of fabricators, produce completely finished products. Most will be integrated into automated production chains as effective links.

At the other end of the scale, in the field of private consumption, the personal fabricator or Fabber will find their way. The fab-at-home movement (www.fabathome.org) enables this with assembly kits, starting around $ 800 and sold on the Internet, and with service by company-close and independent organizations. The commercial aspects of private use will strengthen and accelerate the step to cyberproduction.

This development also forms a basis for new forms of work and life that prominently integrate into their philosophy individual production at the lowest possible investment. A pioneer in this direction is Prof. Frithjof Bergmann, Chairman of the New Work–New Culture movement (http://newworknewculture.com).

Processes: Industrial production will increasingly demand more powerful machines. Vector-oriented methods will be limited in favor of the productivity of processes that require this due to the necessary accuracy.

For a significant improvement in quality, it will be necessary to reconsider layer manufacturing as a basic principle of additive manufacturing. “Spatial” layer manufacturing in three-dimensional form of stored volume elements (voxels) stored in space are a possibility for reducing the number of steps and the defined roughness are possible. Different materials can be integrated into one part. Above all, extensions can be placed on partly finished parts.

Technically, this means the renaissance of the principle of ballistic particle manufacturing.

Materials: In material development, today’s manufacturing processes for colored prototypes will use the two-component printing method (Objet). The next development step will include varying material properties over the cross section (graded materials) for a wide range of plastics and also for metals. Finally, different materials, including plastics and metals, will be processed in one process (Optomec’s M^3D process shows the first approaches). The development of the layer to voxel orientation will set the way for complex products. The era of three-dimensional parts consisting of one material or of discrete layers of material would be overcome. The jump from the part to the functional material could be completed. The mobile phone out of the printer is no longer a fantasy.

Anyone who thinks that this is just a vision in the true sense of the word, and thus dreams, fantasies, and utopias, should take note that the technology of additive manufacturing for personal fabricators has already become a toy itself.

While parents still think about direct additive production, their sons and daughters have already designed and manufactured their own smartphone case.

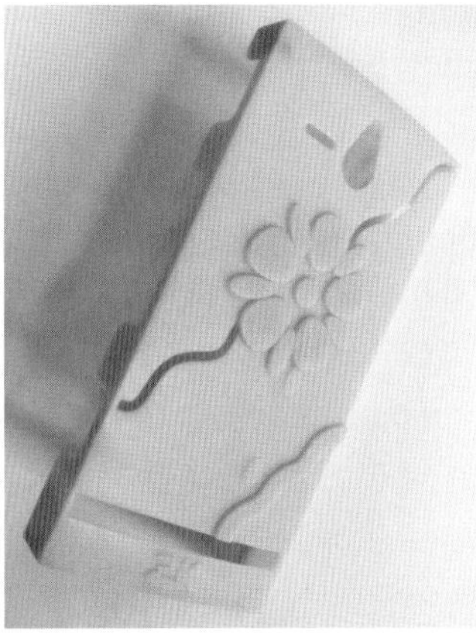

Figure 6.45
Individual smartphone shell, FDM, finished by grinding and painting (Source: FH Aachen, Jugendzentrum OT Talstrasse)

7 Safety and Environmental Protection

Aside from voluntary activities, safety and environmental protection are mainly based on knowledge and observation of the relevant fields of law.

Yet most of the additive manufacturing machines have been installed like normal printers or photocopiers without a discussion of the legal position and consequences. Normally, all of the requirements are fulfilled anyway when the setup, operation, and maintenance of the additive manufacturing machine follow the operating instructions and positioning advice given by the manufacturer.

Certainly, the operation of an additive manufacturing machine needs to fulfill all safety and environmental requirements required by law. But the necessary work was already done by the producer and distributor during manufacture, solutions were considered or changed into practical procedures.

In the future, the user can only rely on this kind of customer service as long as the mass market is affected. Office-friendly constructions and a closed material flow including the disposal of waste allow the user to pass the whole problem on to the distributor for good.

In general, this is not the case, and it is well known that a lack of knowledge does not protect one from consequences, legal or otherwise. Additive manufacturing machines are based on different building principles and use a great number of material variations. Depending on the process, a postprocessing may be needed that requires the use of solvent, hardener, and paint. Thus, general specifications cannot be made for regulations to ensure safety when additive manufacturing machines are installed or operated, or for which measures need to be implemented to protect the staff and environment.

In Germany, the legislation increasingly changes. Whereas in the past the rules had the character of very detailed regulations that one had to follow literally based on the specific case, they nowadays represent guidelines. With these legislative guidelines, a company has to prove that its actions conform to the law. It is expressed by phrases like "... handle it so that no risk occurs." For the company this means a preparation of a risk analysis and based on it a statement of itemizations for how according to the law the risks are dealt with. That is why small companies are often driven to hire a consultant. In the face of continuously changing laws and especially the ongoing process of Europeanization of laws and instructions, a review of current norms is essential. One of the most important characteristics of the current legislation is to consult technology guidelines as decision-making criterion. It means that requirements for the examination of that problem are essential. Access to that topic is given by the technical guideline. All affected laws, regulations, and instructions have the urgent aim of protecting the people involved and their environment. Relating to the construction, adjustment, factory, process, and materials, they define the boundary conditions. In general, the philosophy of factories growing dynamically with the state of technology is recognized. A permission given once has no continuous relevance without permanent improvement.

7.1 Labor Agreements for the Operation and Production of Additive Manufacturing Machines and the Handling of the Corresponding Material

The marketing and operation of additive manufacturing machines includes different arenas of law regarding the specification, construction, and method of operation. The rules are that every part of law is considered by itself separately, and have to be considered and fulfilled independently. In fact, there are several interferences that, with a closer consideration, could be used in a sense of synergy.

The following guidelines and laws are obligatory for the producer of the machine, the materials, and the operator:

- Building law
- Water law
- Emission-control law
- Property-protection law
- Waste-material law
- Chemical law

■ 7.2 Annotations to Materials for Additive Manufacturing

The user of additive manufacturing machines has to protect himself and his employees against injury when handling materials.

In most processes relevant to the industry, synthetic materials are manufactured. Either complex chemical reactions are carried out or synthetics are heated up to their melting point where, depending on the energy source used, partial vaporization cannot always be avoided. Polymerizations are made of monomers that should not contact skin. Powders and fluids need to be stored in containers and be protected against environmental influences like oxygen, moisture, and ultraviolet radiation.

When using a reactive powder to produce titanium and aluminum parts, attention is especially needed when storing the materials outside of the machine and when feeding the machine. Skin contact should be avoided. During the process, danger is limited because of the machine's construction, as long as the oxygen is continuously controlled in the building chamber.

The 3D printing of Z Corporation uses a harmless amylum (starch) powder. The powder, and also the water-like binders, contain additives to facilitate the workability and storability of the materials. Increasingly, new powder binder combinations are released that show partially cured characteristics and thus behave rather like synthetics.

Paper foils are mainly used for the laser-based LLM process. In the cutting of large models, a lot of paper and glue is evaporated because of sublimation. That is why good ventilation is necessary. Other powder binder processes, as well as the LLM process with polymer films, work with activators and glue.

Because the producers give their machines and materials names that make it difficult to derive the possible expected properties from the designation. Only the safety data sheet brings clarity.

■ 7.3 Annotations for Using Additive Manufactured Components

It can be assumed that the prescribed additively manufacturing built components do not constitute or contain any danger for the user.

Polymerization

When creating components with radiation cross-linking, and thus all types of stereolithography, the proper polymerization of the material is important. Liquid residual monomers may have extreme skin irritating or even hemolytic effects inside the body.

Even so, the hand soap and the compatibility of gloves should be checked first when skin irritations occur.

There are as yet no reliable findings of biocompatibility of rapid prototyping parts made from stereolithography. It is generally accepted that completely polymerized acrylates are biocompatible whereas epoxy resins are not. The fact that certain (not made by photopolymerization) acrylates and polyamides in the context of artificial implants are already being used shows their basic applicability. Statements on biocompatibility for materials cannot be made generally because it always depends on the specific case of application.

Considering the different systems, it can be assumed that today every kind of material, besides epoxides but also acrylates and vinyl ether, is on the market.

Sinter Materials

Polystyrene is an aromatic compound. The benzene is only bound slightly and can be dissociated easily. That is why traces of benzene can arise when decaying. Benzene is carcinogenic (the MAK value is zero!) because it can penetrate a cell's lipid membrane unhindered and cause irreversible nucleus damage (mutation).

Polyamide (nylon) is marked by a long unstable chain structure that decays into fragments when heated but is without aromatics.

Polycarbonate is quite stable. It is also aromatic, but the oxygen binding forces are very strong so there is only a very little chance of annulations of benzene.

An overall analysis of the trend shows that grain fractions are becoming ever smaller. Current 20 µm grains contain smaller particles as well. It can be said that below 10 µm the nasal and throat mucous membranes are strained. Grains under 6 µm are considered as lung accessible, and under 5 µm as explosive.

Extrusion Materials

Due to its combination of acrylonitrile, polybutadiene, polystyrene, and when heating, ABS (depending on the different types) can also have the unwanted buildup of benzene described before for polystyrene. Because the material is only liquefied in the intended operation, the danger of evaporation is less, although partial evaporations could occur.

Solvent

TPM (tripropylene glycol monomethylether) is used in stereolithography and is a combination of alcohol and ether, and it dissolves hydrophilic and hydrophobic materials. It is relatively harmless. Particularly for the solution of CIBATOOL SL, the following solutions can be used safely, each harmless when used separately: acetone, acetic acid ethyl ester, tertiary butyl acetate, and dimethyl sulfoxide.

Gloves should be worn in any case. To prevent against skin irritation, appropriate protection creams are recommended (for example, Silicoderm).

Biocompatibility and Sterilization

The uncertainty in biocompatibility has already been mentioned in the context of polymerization materials. It should be noted that materials that are biocompatible when they are not additively produced, are not necessarily biocompatible, even if the materials have the same chemical composition, if they are additive manufactured. Within medicine product law, evidence need to be collected individually.

Biocompatibility and sterilization are often not sharply separated from each other. All products that are intended for the operating room need to be sterilized. It is also important to know the type of sterilization in advance. There are materials that are not suitable for steam sterilization but are suitable for plasma sterilization. Individual evidence need to be collected here as well.

8 Economic Aspects

The question of economy in using rapid prototyping has two aspects: a strategic one and an operative one.

In considering the technology and its potential, the strategic aspect is of considerable importance, that is, the question of whether additive manufacturing should be employed. The assessment depends to a large degree on the competitive strategy chosen and is not generally applicable (regarding competitive strategies, see the appendix in Chapter 10, "Critical Success Factors and Competitive Strategies").

From the operative standpoint, one must decide which rapid prototyping process would be best and with which process chain it could be used with to the best advantage. The same considerations apply to service companies offering rapid prototyping and for rapid prototyping processes carried out in-house.

Quite apart from all other considerations, some technical criteria must be fulfilled simultaneously if AM processes are to be applied economically:

- The models must contain a complex geometry and in the future also contain a complex material mix.
- The shortest possible development time is of utmost importance for the product's market position (in entertainment electronics, this is a matter of months, while in heavy-duty engines, it is years).
- A large number of variations is advantageous for the market position of the product.
- A volume-oriented 3D CAD system must be used as a matter of routine.

Even if only one of these criteria is unfulfilled, it must be assumed that the use of additive manufacturing is then not technically nor economically viable.

- The personnel available for the construction of additive manufacturing models must be committed and skilled.
- Knowledge of the advantages and disadvantages of different additive processes must be present in the case of outsourcing.

8.1 Strategic Aspects

The discussion of the economic impact of the strategic use of additive manufacturing concentrates only on the product development. In additive manufacturing, a general consideration is no longer possible in view of the range of possible products. The aspects that speak to additive manufacturing and their general conditions were discussed in Chapter 6. The same is true for the use of additive manufacturing methods for the construction of tools and molds (Chapter 5).

8.1.1 Strategic Aspects of the Use of AM Methods in Product Development

A competitive strategy must be defined, for example, the pioneer strategy, in which time is the critical success factor. In the pioneer strategy, the economical optimization of the overall process is preferably reached by minimizing the product development time.

8.1.1.1 Qualitative Approaches

The key question must be:

(1) *To what extent will saving time by using models during product development really result in increased profitability of the product?*

The answer depends basically on the market and especially on the product's life span. The effect is certainly higher for consumer goods than for capital goods. The answer should be substantiated quantitatively. At least target specifications should be defined.

For the use of additive manufacturing, the question must be:

(2) *To what extent will the development time of the product be shortened as a result of using models?*

The quantitative relationships between the use of models and product development time are difficult and often not determined in the individual case. In this case, one's own estimate is better than an assumption of a hardly verifiable indication from the literature (80% savings is reported and 30% savings are believed to most often occur, often without knowing the valuation basis).

The further that question (2) is broken down into the individual phases of product development, the more likely one will find a satisfying answer. How much sooner the decision for the development of a product takes place, the question of when models are used instead of sketches can be answered thoroughly. A total estimation from the partial answers can then be developed.

This appropriately takes place in target specifications and will thus become mandatory for the development team as a whole.

Observations reveal that the economic influence of additive manufacturing processes can be judged only if interactions with the entire product development process and in the context of the competitive strategy of the company are included in the calculation. Those answers result from a number of softer factors. Such a situation is quite unpopular with the technician. It is improving, but only slowly, coming from the process side, in the fact that simulation models allow more and more reliable estimations and, coming from the market side, more and more model calculations are increasingly being developed and established.

8.1.1.2 Quantitative Approaches

For the quantitative evaluation of these effects, reference can be made to Siegwart and Singer's model modified by Steger for additive manufacturing [Steger, 95]. It enables the quantitative evaluation of the potential of a strategy, compared with alternative competitive strategies. The model is shown in detail in the appendix in Chapter 10.

The critical success factors of time, quality, and flexibility are condensed into an efficiency value, and from the factor expenses, a cost value is established. The ratio of efficiency value to cost value gives a characteristic value for profitability, WI, and defines the economic potential of a specific technology. This value may be weighted according to probability, thereby further evaluating the risk to be taken if this technology is used (the characteristic value of profitability WII).

The process is described in detail in Chapter 10, "Critical Success Factors and Competitive Strategies." Based on the assumptions made there, the following values for the cost-effectiveness ratios (shown in Table 8.1) result. The results are based on the conventional basis process "milling" and are listed accordingly as the quotient $WI_{additive}/WI_{milling}$ and respectively $WII_{additive}/WII_{milling}$.

The results show that competitive strategies based on the factors of time and cost are becoming approximately twice as effective through the use of the AM process (factor 1.75 to 2.02) than by the milling (factor 0.70).

Ultimately, it is difficult to obtain reliable absolute numbers with this method because the input variables are ultimately subjective assessments. But a variation of individual factors in terms of a sensitivity analysis gives a sense of confidence regarding the dependencies and therefore supports the necessary decisions.

Table 8.1 Calculated Efficiency of AM Processes for Different Competitive Strategies, According to the Model of Siegwart and Singer and as Compared to Milling Technology

Strategy	WI/WI conventional	WII/WII conventional
Cost leadership	1.84	1.75
Differentiating strategy	0.86	0.70
Pioneer strategy	2.14	2.02
Outpacing strategy	1.91	1.76

■ 8.2 Operative Aspects

Additive manufactured parts must be provided at market prices in competition with other rapid prototyping or conventional processes, regardless of strategic considerations.

To analyze the cost on the basis of individual approaches, in particular the prices of investment and consumption goods must be considered[1] (Section 8.2.2, "Establishing the Costs of Additive Manufacturing Processes"). Because different additive processes for the same component geometry sometimes result in very different process-dependent quantity structures, the consideration of prices for the optimization of the cost is not sufficient. The influences of different methods are therefore included in the discussion under economic aspects (Section 8.2.3, "Characteristics of Additive Manufacturing and Its Impacts on Economy").

For the basis of the cost analysis, first, suitable single processes or process chains must be selected due to the mechanical-technological requirements. This selection must be made by each user in response to the criteria shown and discussed in this book. A compilation of the most important chapter is facilitated with an overview (Section 8.2.1, "Establishing the Optimal Additive Manufacturing Process").

8.2.1 Establishing the Optimal Additive Manufacturing Process

The requirements for a part are the starting point, which can best be collected in the form of an individual request list. As the applications can range from baby pacifiers to the front spoiler on a car and from the model of a skull to a coffeemaker, it does not make sense to create a general list of requirements.

The fundamental influences concerning the selection of one specific process have already been discussed in previous chapters. The possibilities and limitations of

[1] A detailed calculation approach was developed in the IWB User Centre Augsburg [Willinger, 95].

single processes and their suitability for subsequent processes were discussed in Chapter 2 in general and in Chapter 3 with regard to specialized machines.

Selected parts properties are discussed in:

Accuracy	Section 3.8.2
Details	Section 3.8.2
Surfaces	Section 3.8.3
Material	Chapter 3 and Appendix A3.1 to A3.12
Dimensions	Chapter 3
Mechanical-technological properties	Chapter 10 (appendix) A3.1 to A3.12
Subsequent processes	Section 3.10 and cross references mentioned
Rapid tooling	Chapter 5

Further technical data and material properties are summarized in Chapter 10.

The analysis of this literature leads to one or more alternatively applicable methods, which then must be investigated in terms of their cost-effectiveness.

8.2.2 Establishing the Costs of Additive Manufacturing Processes

Each user must develop an individually tailored calculation system made for his or her purposes. The most important factors influencing costs are as follows:

Variable costs

- Material costs
- Operating costs
  - energy costs
  - repair costs
- Consumer Goods
- Recycling costs

Fixed costs

- Investment costs (Figs. 8.1 and 8.2)
- Additional investments
- Expenses for service and maintenance
- Accommodation expenses
- Personnel expenses

8.2.2.1 Variable Costs

The variable costs are usually based on the number of parts produced. In AM processes, they can usefully be related to a construction process and therefore to the number of parts produced in the construction space during this construction process.

8.2.2.1.1 Material Costs

Material costs include the cost of material procurement, as well as the costs for material handling and processing of the material.

Materials for additive manufacturing are expensive. A sintered powder of polyamide (PA) costs around 60 to 90 €/kg (about 67 to $ 101/kg) and therefore costs 20 to 30 times as much as PA powder for plastic injection molding. This is due to a more complex production and the relatively small amounts of production. In addition, the supplier takes the responsibility for the quality and often also for the material properties of the produced part.

Stereolithography resins cost around 200 €/kg (about $ 227/kg) and are significantly more expensive; 3D printing nonmetal powders cost between 50 and 70 € (about 57 to $ 79), depending on the quality and quantity, and LLM papers cost around 15 €/kg (about $ 17/kg) and less. These prices are intended for guidance only. The prices vary from manufacturer to manufacturer, and within a product group they can vary, especially because materials with very different properties are offered.

Materials available in large bundles require a more complex material handling, including storage and conditioning, and therefore also requires more staff. The workforce can be greatly reduced through automated material handling systems (for example, Fig. 3.35 and Fig. 6.20). The purchase of large bundles opens up the possibility of discount negotiations, but also requires financing, costing interest.

In some machine concepts, this is not just a business decision. The machines require a minimum filling that must constantly remain in the construction space, regardless of the number of parts produced. In a large stereolithography machine, therefore around €25,000 (about $ 28,300) is permanently tied up into the system. For sintering machines, this amount is less because of the lower price, depending on the volume of construction space.

Machines that do not require a process-related material stock, which significantly exceeds the volume of the manufactured parts, and if necessary, its supports, of course, must also store material. The amount depends on the size of the reservoirs, which are designed such that they enable a typical long-term operation of a process.

Especially for smaller machines, which are also operated in an office environment, the material is preferably supplied in refill containers or cartridges. This increases the material price significantly, but saves handling costs and storage requirements and integrates the disposal. Material cartridges for the FDM process cost around €380 (ABS) (about $431) and contain about 1 kg (920 cm^3).

When considering the material prices, not only the price per kilogram but the total price for one part and thus also the process-dependent material consumption must be considered. Influences were discussed in Section 8.2.2, "Establishing the Costs of Additive Manufacturing Processes."

8.2.2.1.2 Operating Costs: Energy Costs

The electrical connection values of the majority of additive machines range from 0.5 to 6 kW. The only exceptions are systems with power lasers from 1 to 5 kW (Optomec, LENS; TRUMPF/POM, DMD) with corresponding significantly higher power consumption.

In the energy debate, the high consumption of the laser is often discussed. This is not true across the board. A 200 W laser takes about 4 kW connection power and is still below the level of a 4 kW high-power lamp. A closer look is therefore worthwhile.

8.2.2.1.3 Consumer Goods

Consumer goods, in addition to materials, are binders, toners, print heads, nozzles, building platforms, filters, and similar, depending on the process. These costs at first often do not appear explicitly because they are included in the initial installation. Therefore, one should take a look at the price list of the manufacturer.

The lifetime of the laser remains low at 5000 to 8000 h/year, with smaller stereolithography systems only 2500 h/year. That of high-energy lamps (such as the Perfactory FC) is only 1000 to 1500 h. If one bears in mind that a year has 8760 h, in a three-shift operation, a laser must be replaced at least once a year, or the lamps replaced six to seven times a year.

Operating costs also affect costs for air conditioning, the process gases, and the blasting machines and processing equipment for the finishing work.

8.2.2.1.4 Repair Costs

Repair costs affect the machine parts that are not considered to be wearing parts. These mainly include scanners, sensors, heaters, and valves. Because many manufacturers do not have these parts in stock or only in small quantities, they often need to be ordered from suppliers. Actual repair costs also include downtimes. The repair is also often associated with high travel costs for the technician. This poorly calculable risk can be defused with a maintenance contract. Economically, this means, however, the transformation of variable costs into fixed costs and thus the opposite of what is usually sought.

8.2.2.2 Fixed Costs

8.2.2.2.1 Investments

The investment costs include the cost of the machine and the machine-oriented infrastructure. This concerns both the computer systems and their software and communication interfaces that are necessary for the preparation of the construction process, as well as for the physical infrastructure, such as for the air conditioning of the machine. Especially for large systems, investment costs that are close to the amount of the asset costs of the machine are defined.

For the machine itself, the focus starts on the asset costs, even if the depreciation is actually take into account in the calculation. The prices for prototypers and fabricators today cover an enormous range of less than 1000 € over some 10,000 € to one million euros. Roughly broken down, for a personal printer the range is from less than 5000 to 50,000 €, for a professional printer 30,000 to 120,000 €, and for a production printer or fabricator about 120,000 to 500,000 € and beyond. Moving downward to the fab-at-home movement (www.fabathome.org), a range of assembly kits that are available from about $ 600 is developing.

This information is mainly intended for orientation. The prices are of course a matter of negotiation and include, depending on the manufacturer, material and maintenance packages. Price information usually applies from the manufacturing machine. They may well rise by half, due to increased transport costs, customs, currency influences, and insurance to customers who are both local and worldwide. Again one has to look closely. Some manufacturers differentiate the costs into *local pricing* and *international pricing* and communicate thereby from the outset a realistic view.

More than 25 years after the launch of AM, there are also a growing list of used machines. Some manufacturers such as Voxeljet even offer used factory machinery.

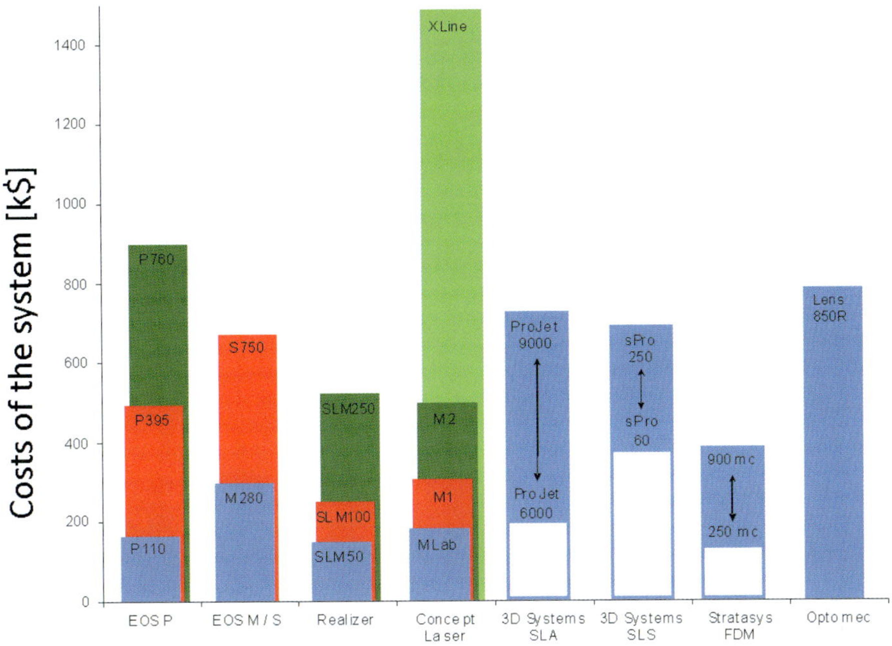

Figure 8.1 System costs, AM machines I

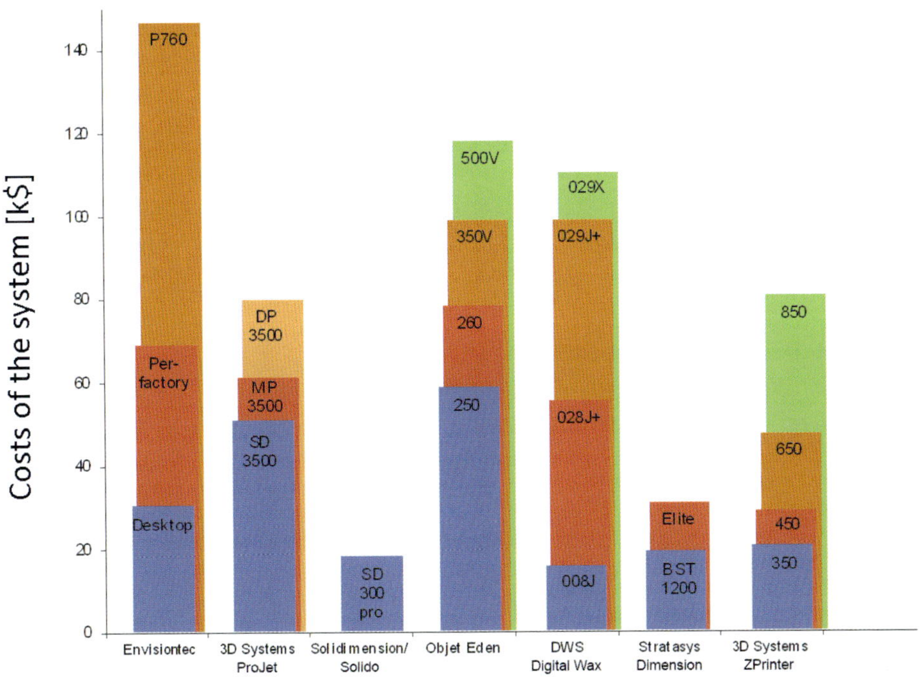

Figure 8.2 System costs, AM machines II

8.2.2.2.2 Expenses for Wear and Maintenance

Innovation cycles of additive manufacturing machines are still very short. This also brings with it a rapid succession of mostly software-related improvements (updates), or more interesting, of hardware-related upgrades. They are to be regarded as a continuous investment in the machine and must be calculated accordingly. Maintenance contracts make these costs predictable, but not necessarily cheaper. Annual costs that are about 20 % of the purchase price are quite common, so that over depreciation periods of five years, the purchase price must be reinvested. But they limit potential financial risks, especially if they include spare parts or spare parts discounts. But from the economic view, this mixes the approaches for the repair and maintenance of AM machines.

A price comparison between maintenance contracts is practically impossible because of the many different kinds of contracts. Lasers, scanners, updates, user group events, hotline use, plus more, and also sometimes consumption contingents are included, or depending on the manufacturer and type of contract, not included or discounted in different stages. Therefore, a comparison between conventional manufacturing and additive manufacturing, especially of cutting methods, always balances out in favor of the cutting machines.

8.2.2.2.3 Accommodation Expenses

A personal printer can be set up in the office, and in terms of space and noise, it can be considered as a kind of copier. Professional printers and production printers or fabricators need separate premises that, depending on the system (such as for stereolithography), must be climate controlled expensively. Anyone who has visited the production location of a service provider knows that separate facilities should be available for the cleaning and finishing. For dust and solvents, special features such as filters, hoods, and chemical storage cabinets must be provided. When operating several different machines, one must consider whether any necessary hoods must be placed separately over the roof or may be combined.

8.2.2.2.4 Personnel Expenses

Staff with different qualifications are required for the operation of an additive system. Primarily required are CAD specialists, machine operators, and modelers. Only in exceptional cases can these functions be successfully combined in one person.

For most disciplines of additive manufacturing, there is no established professional job description. Personnel development therefore must take into account a long training period and must be aligned to the manufacturer-offered, internal or self-organized external training. Any possible fluctuations in staff must be included. For these reasons, it is hardly possible to transfer personnel costs through temporary work from fixed costs into variable costs.

8.2.3 Characteristics of Additive Manufacturing and Its Impacts on Economy

8.2.3.1 Construction Time

Because the user is only interested in how quickly he can get the part in his hands, the total construction time plays a role in determining the optimum process. Therefore, apart from the net construction time of the machine, procedure-related adverse times should be taken into account. This includes preparation and setup times, such as heating and inerting times during laser sintering, downstream phases in the context of *postprocessing*, such as cooling during laser sintering or post-cross-linking in stereolithography, and system-related and therefore necessary *finishing*, such as the painting of LLM paper models. Infiltration processes are also time-consuming. This especially applies to direct metal processes, in which the phases for the expulsion of the binder and sintering of the component must be considered.

8.2.3.2 Lot Sizes and Use of Construction Space

It is uneconomic for current discontinuously discharging machines to not use the whole construction space. The consideration of the impact of not completely filled construction spaces today is more interesting from a theoretical perspective [Zäh, 06]. The practitioner will try to use a completely exploited machine or to get the additional costs from the purchaser. If in the future there will be such continuously outputting machines as the Voxeljet VXC800, the component-related consideration, as described by Zäh, is of greater importance.

As a solution, some machine concepts enable adjustment of the construction space to the current manufacturing project. There are hardware solutions, such as removable or separable construction spaces (Invision Pro, 3D Systems, M3 linear, Concept Laser) or software solutions that limit the construction process to the necessary subareas of the building platforms by limiting the route of the wiper (Smart Sweep, 3D Systems), or in the *y* direction, by printing only the necessary number of lanes (polymer printing, Objet). However, not all costs decrease, especially not proportional to the size the construction space, if the construction space in an otherwise identical machine is smaller. The costs for the inerting of the sintering machine, for example, are not proportional. Besides that, interchangeable construction spaces also require additional investments.

In contrast to the *x-y* coordinates, the *z* coordinate of the used construction space, in almost all AM processes, only depends on the height of the component. The construction process is terminated when the highest component is generated. But in many processes where components very different in height are processed at the same time, building volumes are built that contain no component. In some processes, substructures such as supports or final cover layers (such as during sintering) are required. These are mostly independent of the component height and therefore have an even greater impact the shorter the component is. These factors regarding its construction time and the material used, should be included as an expense in the calculation.

The construction time of AM processes largely depends on the construction height (z coordinate) and is less, the lower the construction height can be selected at the same volume. Therefore, the orientation of the parts in the construction space is of great importance.

However, the free placement of the component in the construction space is no longer possible with the growing importance of additive manufacturing because the placement is determined by the design.

8.2.3.3 Utilization

It is in the nature of additive manufacturing technology that the components often must be available quickly and with very short handling times. This conflicts with the principle of completely used construction spaces. It depends on the skill of the system operator whether he holds up free machine capacity for such cases or works with overbooking, as is done in aviation. Larger AM production areas include several machines that are started up with a time delay for this reason.

8.2.3.4 Material Consumption

Different AM processes need very different quantities of material for the production of the same geometry. All layer laminate processes, but also polymer printing, basically create a component-circumscribing square. It has the length (x axis) and height of the component and a process-typical width (y axis). In LLM processes, this is the width of the foil; in polymer printing it is the width of the print field.

If the maximum y dimension of the component is less than this width, the remaining part is waste. But even if this can be avoided, the process brings with it a lot of waste. A solid cylinder of a diameter D, for example, must be embedded in a cube with a base of at least $2D$. In the construction process of LLM or polymer printing, this is why 21.5 % more material is needed, without consideration of an additional frame, if necessary, than in the construction process of sintering or laser stereolithography. If, instead of a solid cylinder, a hollow cylinder with bottom is built, such as a trash can, the volume of waste inside the component is 50 times[2] than which is required for the actual component in the LLM process or in polymer printing. In polymer printing, the costs for the construction and the support material are approximately the same. However, a price for LLM that is less than 10 % of the

[2] Example: Outer diameter 300 mm, wall thickness 3 mm. Ring surface of the part: 69244.065 mm^2, surface of the inner diameter: 1405.935 mm^2. Ratio of ring surface to inner surface = 49.25. The height cancels itself out.

price of laser stereolithography materials is not sufficient in this case to compensate for the consumption disadvantage. For components with a shell character, LLM processes and polymer printing thus are disadvantageous. For the production of solid components, they are accordingly advantageous.

All extrusion processes, but also laser or lamp (DLP) stereolithography, require supports. The volume of the supports is usually much smaller than that of the circumscribing space, but the support material costs just about as much as the building material. In addition, the supports must be removed, which requires manual labor, and if necessary, there are additional investments in cleaning stations and the necessary working equipment.

Building material in the sintering process, which is located near the component but will not be processed into the component, is thermally damaged. The effect decreases with increasing distance from the component. But as the entire construction space is heated, the border is difficult to draw. In practice, therefore, all of the powder is discharged or is added with new powder in a ratio that varies individually between 50:50 and 75:25. For economic reasons, sintering processes should therefore be packed full of components as tightly as possible.

8.2.3.5 Process Safety

The process safety and the calibration sensitivity are important factors take priority over fast manufacturing. Also, the cheapest process is too expensive if it must be repeated. Therefore, it is important to know how reproducible the process is in practice and how often and how expensively it needs to be calibrated. In this set of problems, the qualification of the employees is also crucial. When operating the machine, especially in the finishing work, an entire construction process can easily be wasted. Herein, integrated process monitoring gains more and more importance.

8.2.3.6 Construction Speed

The vehement discussions about the speed of AM machines are relativized in view of the discussion of the idle times above.

From the description of the individual machines in Chapter 3, it follows that the total construction speed is determined by the z growth. Values have resulted mostly between 10 and 25 mm/h and beyond. Although this is more than a factor of two, the gross process times are not that different in light of the preparation and review times.

In Fig. 8.3, such a process comparison between laser stereolithography (SL) and selective laser sintering (LS) is represented respectively for a technical, relatively small, and a typical medical, relatively large, data set. The technical SL component was made of epoxy resin and the medical one made of acrylate. The sintered parts were made of nylon (PA) and polycarbonate.

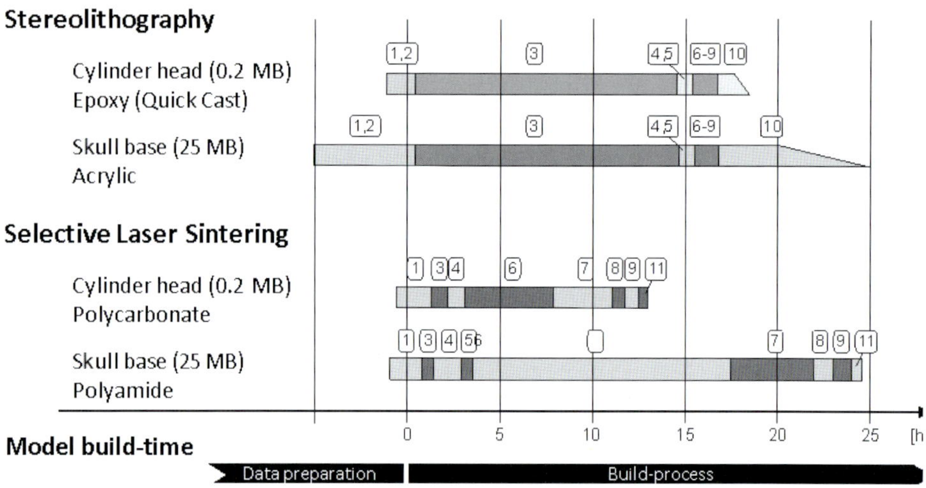

Figure 8.3 Analysis of the model construction time for different methods and model sizes

Table 8.2 Working Steps in Figure 8.3

Steps	Stereolithography	Selective Laser Sintering
1.	Preheat laser	Fill powder
2.	Fill resin	Insert wax base
3.	Generate part	Level the powder
4.	Drip resin	Inert the construction space
5.	Remove part from the machine	Heat process chamber to working temperature
6.	Clean with TPM	Generate part
7.	Remove supports	Bring the process chamber to room temperature
8.	Finish part	Remove part from the machine
9.	Clean with IPA	Remove adhering powder
10.	Post-cross-link part	Remove supports
11.		Finish part

These examples are generalized as they are from some time ago, so the construction times and the cooling times are expected to be somewhat shorter today. But they clearly show that the total construction times of laser stereolithography and laser sintering are not so far apart, in view of the possible dispersion. They also show that the differences within a process, in view of the different materials (nylon, polycarbonate), can be more pronounced than those between different systems. The scheme of Fig. 8.3 in addition is well suited as a basis for one's own analysis.

The example also shows clearly that all comparative analyses are inadequate that either trivialize or completely remove the idle times.

8.2.3.7 Technical Progress and Model Refinement

As a part of operating costs, it was mentioned that most manufacturers conduct a continuous model refinement that provides already installed machines with updates and upgrades and provides a current working basis for the user.

For the machine operator, in this context it is important whether manufacturers make technical progress compatible with one's existing investments, or whether a completely new machine reaches the market from time to time, which enforces an investment in the light of not yet written-off assets, but also involves installation, training, infrastructure changes, and stocks of spare parts.

Of course, in view of the previously rapid developments, such breaks cannot be avoided. Because most manufacturers have already been on the market for significantly more than 10 years, potential operators of AM systems can and should get an idea about this aspect of the company policy.

Although the constant renaming of machines and especially of materials, as well as inconsistent naming on machines from the same family, has no direct economic impact (especially often not the intended positive one!), it does contribute to uncertainty in an industry that is already not very concise. The increasingly frequent company acquisitions also affect this aspect negatively.

8.2.3.8 Service

One aspect when considering profitability is the customer service situation. In rapid prototyping, *rapid* takes on a special meaning. Machines that are only rapid in theory do not help the user shorten the time to market for his products. Unexpected downtimes that involve troubleshooting are especially devastating in this area.

In Germany, customer service has been a point of contention in the recent past, but has eased for most producers today. Most local and international producers have established a viable customer service system and are currently extending it. Travel times in Germany are often too long and the engineer known as a specialist for that specific problem is frequently not available.

A specific characteristic feature of rapid prototyping processes is that several sources of errors may overlap to become a total error of IT, electrical, mechanical, and materials problems, so that only specific and time- and effort-consuming tests may isolate and eliminate the actual cause of the fault once and for all. Systematic troubleshooting–not a day-to-day routine but sometimes unavoidable–is made even more complicated because for the German user, the suppliers of lasers and other components and the necessary know-how are "over the ocean"; this applies to American as well as other producers. It is advisable to decide whether the most important replacements, including expensive elements such as lasers and scanners, should be kept in stock. In the end this may prove to be the best alternative for the overall economy of the process.

The same is true for keeping sufficient amounts of construction materials in stock. The materials should be quantified and reordered at fixed times so that extraordinary, inexplicable construction faults occurring after a change of material may be systematically traced by changing the material back to the one previously used. Also, supply interruptions should be included in the analysis because the industry as a whole is still too small to give a buyer power.

8.3 Make or Buy?

The question of whether additive manufacturing should be applied at all (strategic aspect), as well as the application of the optimal process (operative aspect), depends largely on whether the models are made on the premises or whether they are bought from an external producer. The application of additive manufacturing processes on the premises has far-reaching consequences concerning personnel and infrastructure apart from the question concerning the rate of utilization—an important aspect with all technologies. For these reasons, the pros and cons of making or buying are discussed in the following paragraphs.

The following points speak for in house use:

- The components are immediately available.
- If the machine is used to high capacity, the whole profit can be realized.
- Competitive advantages can be realized by gaining specific know-how of additive manufacturing.

The following points speak for the purchase from suppliers:

- There will be fewer processes.
- Therefore the optimal technology can be applied.
- External production is carried out regardless of capacity utilization and for previously established prices.
- Order peaks are easier to work off.
- Competition between external producers ensures that the latest technology is always available.
- No qualified personnel must be available and trained.
- Extremely short innovation cycles force new investment before the machine is fully depreciated.

An absolute secrecy guarantee is also essential to require from the external producer and then the processes can be assumed to be as safe as if they had occurred on the premises.

With so-called brokers, that is, agents for additive manufacturing services, who do not have their own installation, secrecy presents more of a problem. Even if brokers work with as much secrecy as the external producers with their own installation, their livelihood depends on taking advantage of temporal and structural price gaps. They can do this if they spread their inquiries appropriately, thereby losing direct control over confidentiality; therefore, the risk of "leaks" increases.

The selection of a service provider should be done as with any other supplier selection and evaluation. A solid basis for successful purchasing is the detailed knowledge of available processes and the requirements derived therefrom of the part. Only those with the relevant expertise can judge the competence of a supplier.

The size of a supplier, expressed by the number of employees, is certainly not a general selection criterion. The corresponding compositions of Zäh [Zäh, 06], which assume that a service provider with more than 20 employees is not competitive as a result of increasing organizational problems, are in practice unproven.

Basically, the data formats and data paths as well as the shipment of finished parts should be clarified with the service provider and, if possible, a test run should be done before the time that critical contracts are awarded to that provider.

9 Future Rapid Prototyping Processes

9.1 Microcomponents

Following the main applications of additive manufacturing technology, this book has mainly focused on macroscopic or mezzo-scale components with characteristic dimensions of 10ths of mm to a few 100ths of mm. The production of microcomponents with characteristic dimensions of some 10 to 100 μm has been actively researched for some time now by Laser Center Hannover (LZH), ETH Lausanne (EPFL), TU Munich (IFG-TUM), and Wien and also by leading universities in the US and Japan.

Industrial applications are rather rare and affect mainly plastic components apart from microstereolithography (Section 3.1.6, "Microstereolithography, microTEC"). The manufacturing of microscale plastic components is mostly based on laser stereolithography. These are compiled by Varadan and his colleagues very well and extensively in his book, "Microstereolithography" [Varadan, 01].

9.1.1 Microcomponents Made of Metal and Ceramic

Processes for producing microcomponents from metals are not yet commercialized. At the Fraunhofer Institute for Laser Technology Aachen, at Mittweida, and at the Research Centre Karlsruhe (FZK), extensive work is ongoing.

The research efforts in Mittweida are focused on microsintering. In cooperation with Micromag AG in Chemnitz, they realized an industrially applicable process that will be sold by EOS. The process can be used for metals and ceramics. In the German publications with a focus on ceramic components [Exner, 06] are listed also mostly English-language sources for metal components. Figure 9.1 shows examples of metal and ceramic components.

The subsequent research efforts from the Institute for Micro Process Engineering (IMVT) at the Research Center Karlsruhe show the state of development in microapplications built on commercial platforms, especially for the machines for selective laser melting.

Figure 9.1 Microsintering: metal microlattice (left), microceramic component (right) (Source: Laser Center Mittelsachsen)

9.1.2 Microcomponents Made of Metal and Ceramics by Laser Melting

Thus far, selective laser melting has been used as additive manufacturing mainly for the production of metallic components (MCP-HEK/F&S, Section 3.2.5, "Laser Melting, Realizer GmbH"). Commercially available metal powders can be used as the starting material. The powder is locally completely melted by laser energy.

The selective laser melting process is generally also suitable for producing microstructures. The most important prerequisite is a structural resolution of less than 100 µm. The process must be able to build wall thickness or melt powder less than 100 µm that is both reproducible and gas-tight.

The structural resolution and the component density are mainly influenced by the laser spot diameter, the energy input, and the exposure strategy.

9.1.2.1 Melting Process in Selective Laser Melting

The melting process in selective laser melting cannot yet be described by a physical model. The very extensive knowledge in laser welding [Beyer, 95] [Rosenberger, 95] [VDI-TZ] is only partly transferable to the melting process in selective laser melting. With the SLM technology, a component is built up of single powder layers. A layer of powder consists of spherical powder particles having a diameter less than 50 microns. These powder grains among themselves have only point contacts, so heat conduction is much lower than compared with a solid structure of the same material. Whether such a powder layer can be completely melted by pure heat conduction with a layer thickness of 30–50 µm is very unlikely to happen with a laser beam power density in the range of 1 MW/cm^2. The top of the powder layer will vaporize while the bottom has not reached the melting temperature. To produce dense microstructures, a homogeneous melt "welding bead" is required, which means a localized melting point must be moved depending on the process parameters (laser power, scanning speed) in a stable powder layer (Fig. 9.2). A portion of the laser energy is absorbed by the powder and a portion by the melt. The top of the melt is heated to evaporation temperature by the laser beam, so high temperature gradients arise in the weld pool. The result is a convective melt pool, which leads to a thorough mixing of the melt. This effect is also observed in laser welding (heat conduction welding) and is well documented in the literature [Beyer, 95], [Rosenberger, 75].

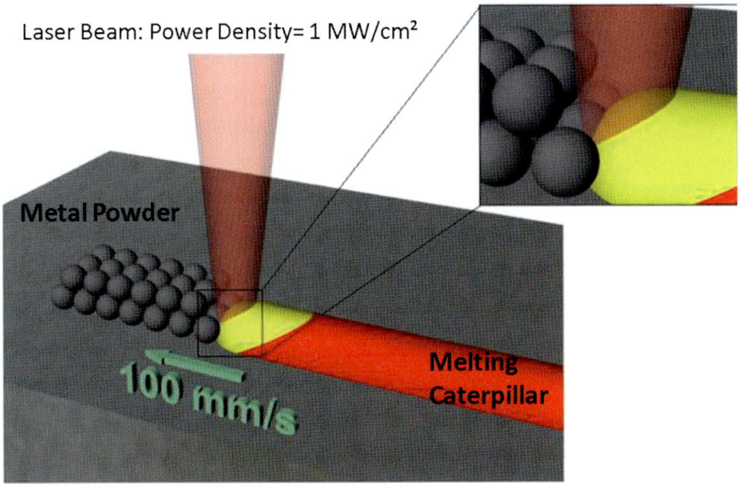

Figure 9.2 Principle of selective laser melting to produce homogeneous melt welding bead

9.1.2.2 Microstructures of Metal Powder

In the manufacture of microstructures with the SLM technique, it is necessary to build up thin walls of superimposed melting welding bead. During the building process, the quality of the melting bead can be directly affected by the laser power, scanning speed, and the path vector. These process parameters are calculated and regulated by the control software depending on the exposure strategy. Depending on the geometry of the component, the exposure strategy defines with which laser power, vector length, vector sequence, and scanning speed each melting caterpillar is melted. The quality of the melting caterpillar is greatly influenced by the exposure strategy.

At IMVT-KFK, tests were carried out for producing homogeneous melting caterpillars in order to expand the knowledge base for the production of microstructure components. A Realizer SLM 100 system (Fig. 3.43, Section 3.2.5, "Laser Melting,

Realizer GmbH") with a ytterbium fiber laser (output 30 W, wavelength 1070 nm) was used, which can be focused to a spot diameter of 12 microns and positioned to <5 microns. The total surface area corresponds to 50 × 50 mm.

The experimental results confirm the model concept that a stable molten bath is essential for a homogeneous melt caterpillar. Too-high energy excess causes a reinforced material vaporization, and the melt becomes unstable. Side-melting outbreaks and splashes are the result. A clear and demonstrable dependence of the melting caterpillars' quality on the process parameters has not been derived so far from the experiments. The test results clearly show, however, that a homogeneous structure quality of the molten metal powder can be reached.

Figure 9.3 shows a building platform with test structures, which were melted layer by layer for stainless steel powder 1.4404 (X2 CrNiMo 17 12 2). The thickness of a powder layer was 50 microns. The powder had a particle size distribution in the range of 10 microns to 30 microns. The achieved surface and melting caterpillar quality is exemplified in a microstructure component. The selected microstructure component is a cube with an edge length of 5 mm. Inside the cube, thin walls were built up vertically to the build platform. Figures 9.4 and 9.5 show the melting caterpillars of the last layer (height 5 mm) at selected locations. The total building platform was cleaned of the remaining powder in an ultrasonic bath. There was no postprocessing work.

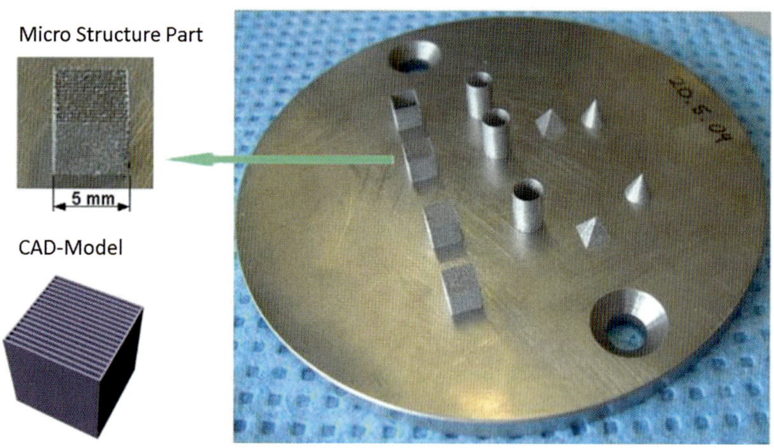

Figure 9.3 Building platform with test structures made of steel powder 1.4404

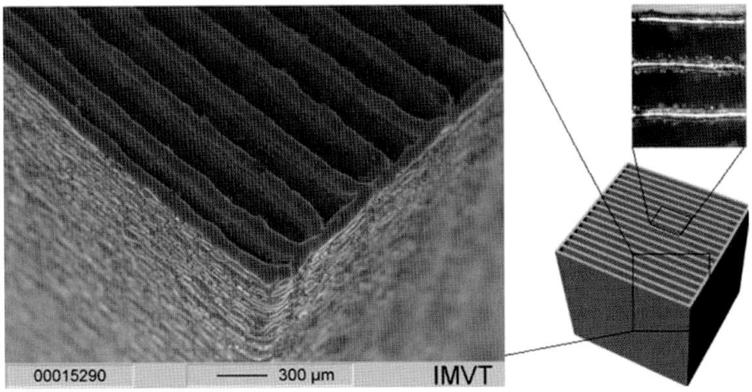

Figure 9.4 Microstructure component of stainless steel powder; melting caterpillars in the top corner area, scale: 300 µm

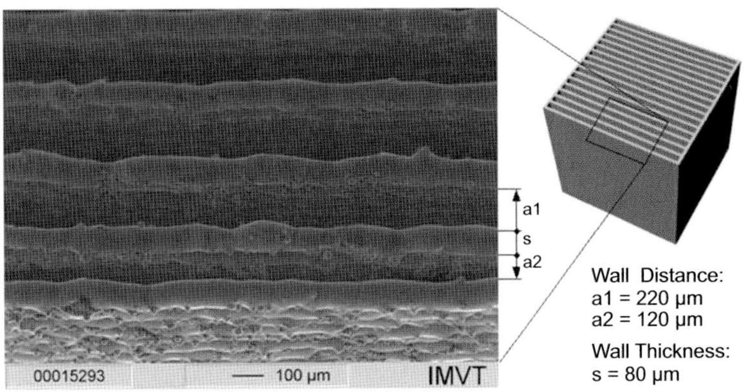

Figure 9.5 Microstructure component of stainless steel powder; melting caterpillar at the top, scale: 100 µm

Figure 9.6 shows a section of a vertical wall. Significantly, the melting of each caterpillar with 50 micron layer thickness can be identified. Powder grains adhering to the wall are only partially melted. In total, it has been shown that homogeneous melting caterpillars can be made with a width of 80 microns at a melt bead height of 50 microns with the SLM technique. This technique can be used to manufacture microstructured components or subcomponents.

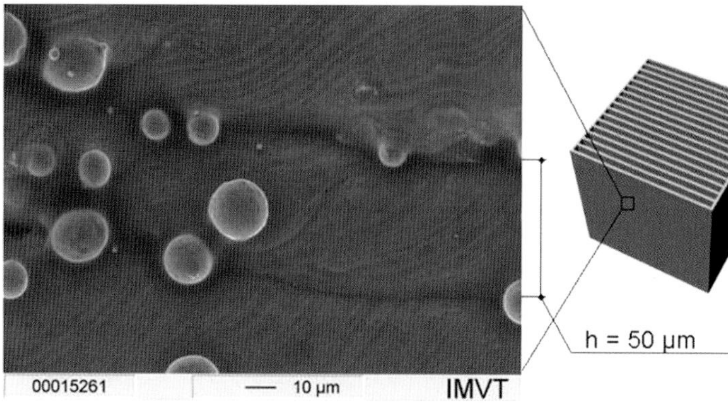

Figure 9.6 Microstructure component of stainless steel powder; melting caterpillar on the side wall, scale: 10 microns

9.1.2.3 Microstructures of Ceramic Powder

Ceramic materials are characterized by their high strength and hardness, temperature stability, and wear and corrosion resistance. In order to use these specific material properties of ceramics sensibly, a material-oriented design and an effective component manufacturing process are required [VDK].

The goal was to investigate whether the SLM technique is also suitable as a manufacturing process for the production of ceramic components.

For the experiments the SLM 100 system was used. The aim of the experiment was to prove that it is possible in principle to melt homogeneous melting caterpillars of ceramic powder. Alumina powder (Al_2O_3) was selected, a ceramic material that forms a stable melt. The powder grain size was smaller than 30 µm, and the Al_2O_3 amount was greater than 96%. As a powder layer thickness, 30 microns was selected. The laser beam power density was varied in the range of 0.6 to 1.4 MW/cm^2, and the scanning speed was 10 to 100 mm/s.

The first experiments confirmed that the laser wavelength λ = 1070 nm in an Al_2O_3 ceramic powder layer is absorbed less than in the ceramic melt. This effect meant that the melt rapidly overheats and is unstable due to the development of evaporation. Therefore the exposure strategy for Al_2O_3 must be defined so that only a small volume is melted and moved stably with reduced energy across the powder layer. Basically, the experimental results confirm that one can also melt homogeneous melting caterpillars made of Al_2O_3 powder. Figures 9.7 to 9.9 show a solidified melting caterpillar with a height of about 150 microns and a mean width of 200 microns. Noticeable is the "very smooth" surface of the solidified melt (Fig. 9.9).

In the tests that have been carried out to date, the base plate and the powder layer were not preheated. As a result, tension cracks were expected under this experimental constraint (Fig. 9.10). The cracking that depends on the exposure strategy and the preheat temperature has not yet been investigated.

With these studies, the knowledge base for additive manufacturing processing of ceramic powders could be expanded and the underlying process model confirmed. The production of microstructured components of ceramic powder with the SLM technique is therefore very realistic from today's perspective.

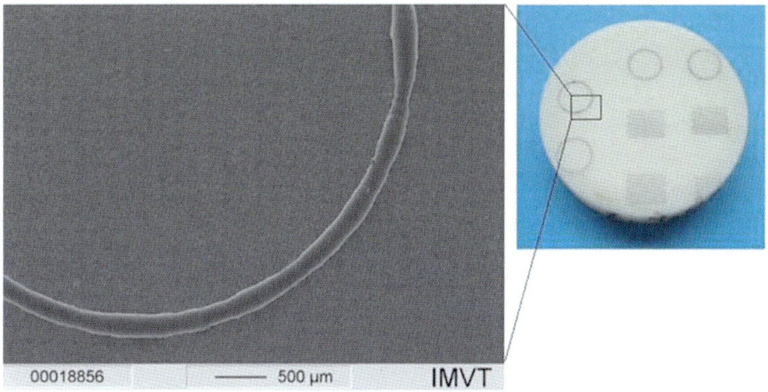

Figure 9.7 Solidified aluminum oxide melt, scale 500 microns

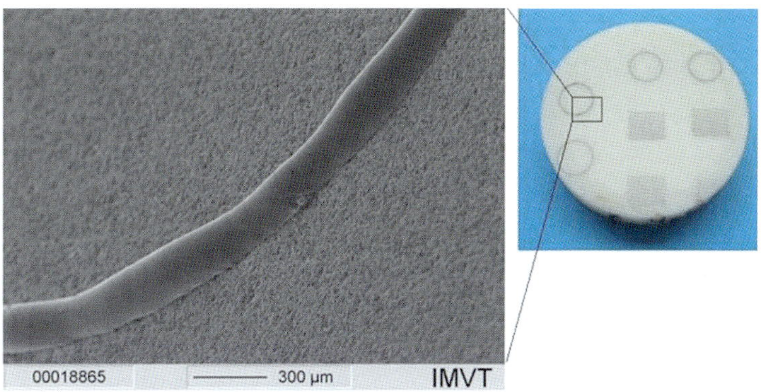

Figure 9.8 Solidified aluminum oxide melt, scale 300 microns

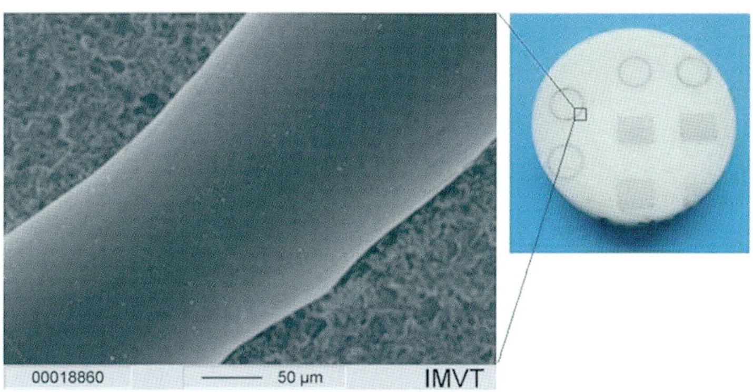

Figure 9.9 Solidified aluminum oxide melt, scale 50 microns

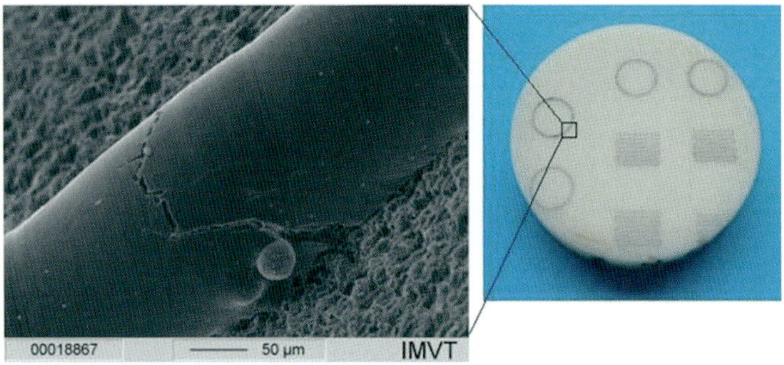

Figure 9.10 Solidified aluminum oxide melt, stress cracking, scale 50 microns

9.2 Contour Crafting

Contour Crafting (CC) is a layered fabrication technology developed by Dr. Behrokh Khoshnevis at the University of Southern California. CC uses robotic technology to create very precise computer control to fabricate very large objects, for example buildings. The extrusion of the material must have perfect precision without any waste.

Computer control in CC enables the creation of smooth and accurate, planar and free-form surfaces using the superior surface-forming capability of troweling. This process is a combination of two processes: an extrusion process to form object surfaces and a filling process (pouring or injection) to build the object core. For the extrusion process, we have an extrusion nozzle (as shown in Fig. 9.11), which consists of a top and a side trowel. During extrusion, the traversing of these trowels helps to create smooth outer and top surfaces on the layer. The side trowel is used to create nonorthogonal surfaces. This extrusion process is used for creating only the outside edges (rims) of each layer on the object. On completion of the outside edges of the given layer, the enclosed area can be filled by the filler material, such as concrete (if required).

Using CC technology makes it possible to include many different functional designs in the wall construction. Examples are cut-outs for electrical wirings, plumbing, or integrated connections for air conditioning.

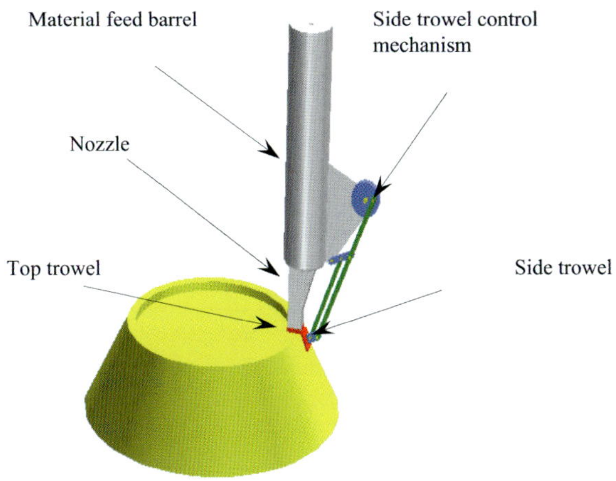

Figure 9.11
Principle of Contour Crafting process

Various reinforcements were investigated regarding their feasibility in order to strengthen the large housing structures that are built through CC. An experiment with coil reinforcement can be seen in Fig. 9.12. CC has a high extrusion pressure compared to other layered free-form fabrication techniques, which results in thorough adhesion of the extrudant around the coils without causing any internal discontinuities.

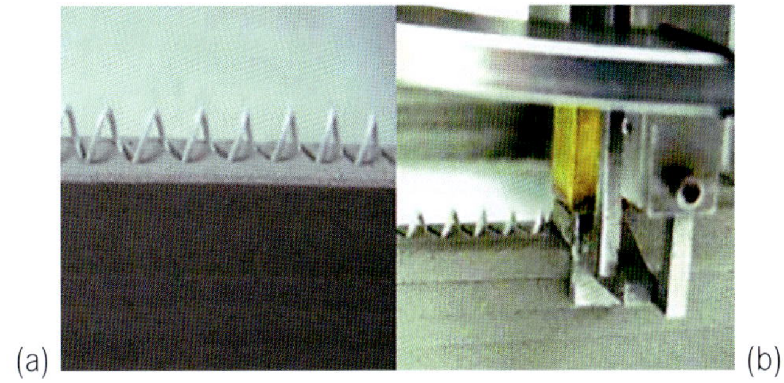

Figure 9.12 Reinforcement process of CC: (a) metal coil placed on a top layer, (b) a fresh layer of extrudant covers the coil

Based on this technology, a large version of this machine has been constructed that is capable of printing human-size walls. A Contour Crafting machine is shown in Fig. 9.13.

Figure 9.13 Contour Crafting machine

Different wall structures that fill the internal volume with material either by injection or pouring have been constructed and tested using CC technology. A wall structure printed using CC technology is shown in Fig. 9.14.

Figure 9.14
Wall structure printed using CC technology

■ 9.3 D-Shape Process

Enrico Dini is the inventor who hopes to start producing and selling 3D printers under the name D-Shape, which was set up in 2007 as Monolite UK Ltd.

The building process used by the company, D-Shape, is similar to the "3D printing" process as the system operates by injecting a binder material on a sand layer. CAD-CAM software provides the contours to the structure that holds the printer head for scanning during the building process.

D-Shape's printers use a special inorganic binder and ordinary sand as the only raw materials. This nonepoxy binder is built using two parts: sand is first mixed with a solid catalyst, and then the mixture is exposed to a liquid binding agent. The surplus sand that does not get printed acts to support the structure and can be reused for the next print.

The machine consists of a robust plane of 6 × 6 meters that is lifted along the four-column frame. The columns can be made longer, up to 9 to 12 meters. The corners of the machine's plane are provided with electropneumatic climbing equipment, which is controlled by an encoder with 0.1 mm resolution. On the lifting plan, a bridge is mounted along which the printing head runs (which is suspended perpendicular to the bridge). The print head consists of 300 servo-driven nozzles, which eject a "structural ink." The width of the print tracks is thereby dependent on the printing speed, the nozzle diameter, and the ejection pressure.

The structure is very light and transportable despite its large size. The total stroke of the printing head (6 meters) could be extended considerably regarding the bending resistance of the bridge on which it is hung. The speed of the printing head can be varied up to 500 mm/sec.

Machine details:

- Overall machine dimensions: 7.5 × 7.5 m, height: 3, 6, 9, 12, or 18 meters
- Maximum area of printing (including the shell): 6 × 6 m
- Number of nozzles: 300 at 20 mm interaxis
- Nozzle intervention time: 10–15 ms
- Command and control: by PC-PLC Siemens operating via Profibus communication protocol
- Weight (not including the feeding equipment): 1300 kg; weight including the feeding equipment: 5000 kg
- Power supply: 380 V, 220 V, 50 Hz power consumption (not including the feeding equipment): 2 kW peak power consumption; (including the feeding equipment): max 40 kW peak min/max
- Sand granulometry: 0.1–60 mm (0.3 mm during first test)
- Productivity: 15 cm/day (effective output during tests) to 30 cm/day (theoretical) with two people needed to run the machine: 2 Pixel dimension: 5 mm
- Minimum layer thickness: 5 mm ± 0.5 mm
- Maximum layer thickness 60 mm
- Theoretical output printing resolution: 25 dpi
- Effective output printing resolution during test: 4–6 dpi

Figure 9.15
D-Shape machine

As reported by Monolite, the final product is a hard and marble-like material applicable in different fields from park benches to one- or two-floor buildings. So far, the biggest structure that has been built using this technology is shown in Fig. 9.16.

Figure 9.16 Two meter tall monolithic sandstone structure printed using approximately 200 10-mm-thick layers of sandstone rejects, aggregated by a new revolutionary inorganic binder

9.4 Selective Inhibition of Sintering (SIS)

Selective inhibition of sintering (SIS) is a layered fabrication process that was invented at the University of Southern California. SIS is based essentially on inhibiting selected powder particles from sintering. This process is developed for polymer and metal applications.

9.4.1 The SIS-Polymer Process

The SIS fabrication process consists of four main steps. The process steps are also shown in Fig. 9.17.

Figure 9.17 Stages of the SIS process (left); extraction of the fabricated part (right)

1. Laying the thin powder layer

   A thin powder layer is applied on the previous layer using a roller.

2. Deposition of sintering inhibitor

   A standard multinozzle inkjet printer (for raster printing) or a single printing nozzle with a fine orifice (for vector printing) is used to deposit the sintering-inhibitor liquid on selected areas (contours, hatch patterns, or separation surfaces) of each layer.

3. Minimizing the radiation frame

   To conserve the powder material, in this step a reflective plate is applied that exposes only the desired portion of each layer to the radiation.

4. Sintering by thermal radiation

   The exposed area is treated with heat that is radiated using an electrically heated nichrome filament or a gas burner, which can sinter the selected areas of the powder layer all at once. These steps are repeated for each layer. After sintering of all layers, the final part may be extracted.

The Alpha Machine

The designing and construction of an Alpha version of the SIS machine was done after the bench-top tests. The machine includes the following elements:

- A source powder container with a step-motor-driven piston to push the powder upward with the selected magnitude.
- A build tank that receives the powder and has a stepper-motor-driven piston that lowers progressively at the layer thickness increments.
- A roller that takes the powder from the container and spreads it over the build substrate.
- An *x-y* print head that moves the print nozzle over the preset layer contours.
- A drop-on-demand nozzle for the inhibitor liquid injection.
- A heat-radiating element that can be positioned over the build substrate for the sintering process. This component can be skipped if bulk sintering method is applied.

The prototype machine while operating during sintering is shown in Fig. 9.18.

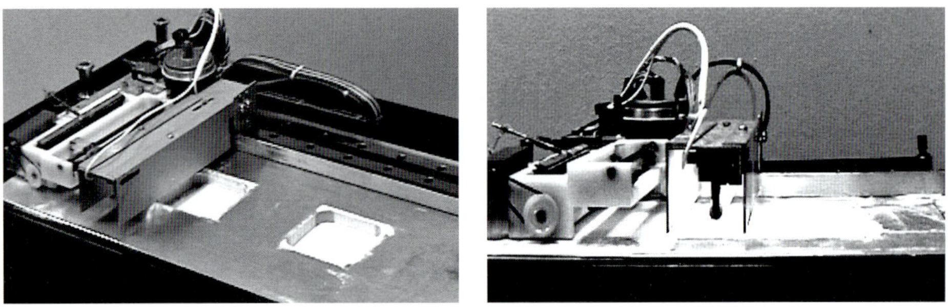

Figure 9.18 The Alpha machine and its operation while sintering using a halogen lamp

For initial trials with the Alpha machine, polystyrene powder was used. The parts shown in Fig. 9.19 have been fabricated out of polystyrene by the SIS Alpha machine.

Figure 9.19 Sample parts made by an SIS Alpha machine

Instead of layer-wise sintering (as used in the SIS-Polymer process), the bulk sintering approach is used for SIS-Metal, because it requires an oxygen-free environment for sintering, and also the cycle time for the fabrication is much longer than for SIS-Polymer.

9.4.2 The SIS-Metal Process

The SIS-Metal process regarding the microscopic mechanical inhibition is shown in Fig. 9.20.

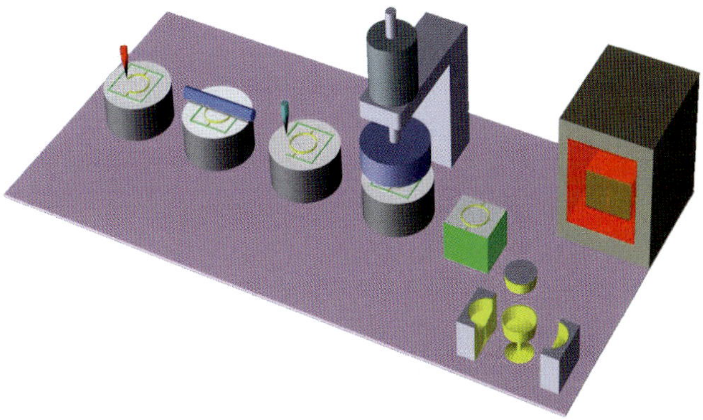

Figure 9.20 The SIS-Metal process based on microscopic mechanical inhibition

The SIS-Metal process based on the bulk sintering method has five main steps, which are as follows:

1. Printing sintering inhibitor: A deposition nozzle, including a fine orifice as well as an inkjet print head, is used to deliver a sintering-inhibitor material to the selected profiles of the powder layer.
2. Laying a thin powder layer: A thin layer of metal powder is spread over the build substrate using a blade or a roller.
3. Creation of a boundary to contain the part: The deposition of a consolidating liquid on the powder bed at the outer edges of the part is done at this step. The profile of this deposition may be a simple shape, such as a square or circle.
4. Heating and compression: The evaporation of water and other liquid additives (such as alcohol, which may be needed to break the surface tension) is conducted using a heater in the printed inhibition sections.
5. Bulk sintering in an oven: After all of the layers have been completed, a metal powder block is extracted from the build container and is then heat treated using conventional sintering oven.

A prototype machine has been developed, as shown in Fig. 9.21. A single print nozzle with an orifice size of 0.005 in. (0.127 mm) is used for injection purposes. Different inhibitors at different viscosity levels can be applied because of the large orifice size of the nozzle. Droplet printing can be conducted via an electromagnetic solenoid valve in the printing nozzle. The volume of the deposited droplets from the nozzle is 300 nanoliters. The droplets are dispensed with a frequency of 1500 Hz. The nozzle moves on the *x* and *y* axes with a speed of 31 mm/second. A heater is included to heat up every layer after the inhibitor printing process. The heater can heat up every layer up to 170 °C in order to evaporate the liquid in the inhibitor solution.

Figure 9.21
The SIS prototype machine for the metal application

The part separation process after sintering using the SIS method is shown in Fig. 9.22.

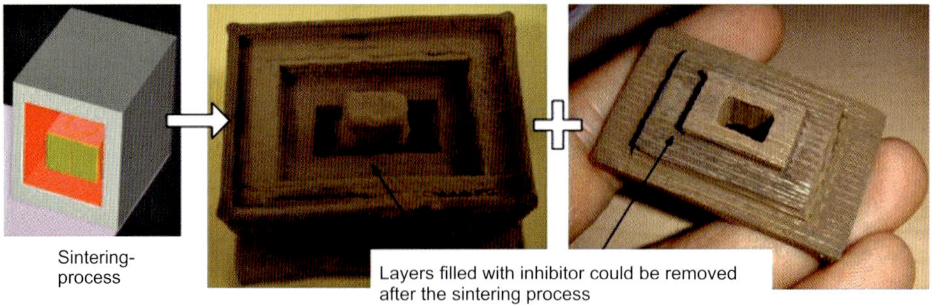

Figure 9.22 A test bronze part can be separated from other portions after sintering

■ 9.5 Free Molding

Free molding is a new technology that allows for the production of a wide range of products in different volumes, with different surfaces granting a unique character to each and every product. It uses green (recycled) plastic as a build material to create products ranging from garden furniture and home décor to light fixtures, fashion accessories, and other decorative objects. The cost of the production process is substantially lower than the well-known traditional production lines. The use of expensive heavy steel molds is substituted by lightweight cheap molds from various materials. The technology was invented by the designer Yaron Elyasi and is protected by a registered patent. The products are characterized by a random surface aesthetic. Elyasi was granted international recognition by the art and design community, and his products are being sold in design boutiques in Israel and around the world. Up until recently, the products were manufactured in a semi-manual technique by moving the mold in different directions; meanwhile production resulted in different surface designs from product to product. To move this technology toward mass production, a robotic system (Fig. 9.23) was developed that does the same process in an automated computerized way. A computer is given a specific design while a robot holds the mold and moves it under the nozzle in a predefined way, creating the surface design. This technique enables the manufacturing of identical products, but it also opens new options for custom design by the customer itself. The drawing, signature, logo, or the like is sent to the computer, which translates it into a surface texture, and it will be built automatically.

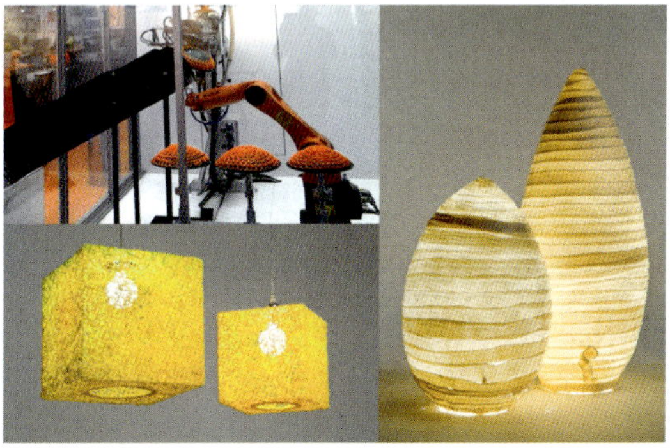

Figure 9.23 Free molding with KUKA robot and printed objects

9.6 Freeformer

One of the leading manufacturers of injection-molding machines, Arburg (Loßburg) has developed a new machine for processing plastics called the Freeformer. The Freeformer uses a newly patented manufacturing technique (Arburg Plastic Freeforming or AKF) to create 3D objects from CAD data without the need for molds. What sets this technology apart from others is the use of a clocked nozzle that opens and closes up to 100 times a second, letting only droplets of plastic out to build up the desired component layer by layer without a support structure. The process is shown in Figure 9.24. During construction, the plastic discharge unit remains in one place as the part is moved by a three- to five-axis component carrier. This allows the Freeformer to create complex geometries while incorporating colors and different types of plastics at the same time.

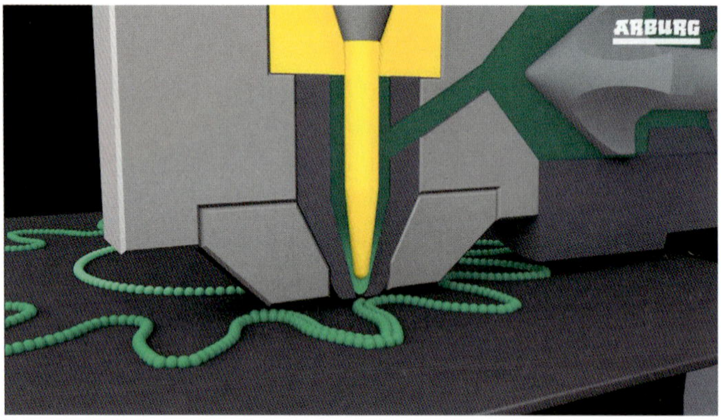

Figure 9.24 Schematic view of Freeformer process

10 Appendix

Critical success factors and competition strategies

Siegwart and Sieger [Siegwart, 91] presented a model that allowed a structured evaluation of new technologies. For this they established so-called critical success factors. These are points that can be condensed into individual factors and therefore are used as a measure for evaluating the success of a process in the context of a corporate strategy.

Although the quantification of success factors is currently controversial, they still are an effective instrument to discuss and evaluate products and product strategies.

Requirements for new products are characterized by the following critical success factors:

- Shortening time of development
- Reduction of costs
- Increasing the flexibility (product and manufacturing flexibility)
- Improving the quality

In general, this list is not obligatory, but currently it suggests a wide consensus.

Single critical success factors are in the mathematical sense not independent of each other, but rather they represent factors that, by weighting and connecting in the sense of a strategy, lead to useful statements. In a strategy, the self-concept of a company in an interplay with its competitors is expressed. Today (it might be different tomorrow) successful companies (market leaders and "hidden champions") pursue the following strategies:

1. Technology leadership (pioneer leadership)
2. Cost leadership
3. Differentiation
4. Concentration
5. Overhauling strategy

The critical success factors of time and flexibility are in the foreground of the pioneer strategy, while quality and costs are in the background. In pursuing cost leadership, the rest of the critical success factors step behind the costs. The strategy of differentiation follows for a unique product, for example with an unmistakable design. Compared to quality and flexibility, costs and time play a subordinated role. Concentration is not an independent strategy, but rather is an application of cost leadership and differentiation to a chosen market segment. Outpacing is the change between the strategies of cost leadership and differentiation, based on the situation of the competition. The "provocative outpacing" strategy first involves procuring a share of the market with inexpensive products, followed by differentiation with high-quality products. "Preventive outpacing" goes directly after cost leadership, to fend off low prices from competitors from the beginning.

A strategy known as market skimming is similar to outpacing. A product, preferably upscale, unrivaled, and introduced to the market before the competition, which is temporary and intense, is put on the market with the necessary advertising expense. When the competition begins on the market, the business is over or reduced to a low and low-priced afterbusiness. This applies especially to fancy goods, for example the mp3 player iPod (Apple). In this case, all competitive factors play a similar role, but have different weighting related to the stage.

These examinations show that most of the actually used strategies are a mixture of single projections, which are followed over the period with different weightings.

The individualization of mass products can be seen as a combination of the pioneer and differentiation strategies.

The implementation of the theory explained in the following makes it clear that time especially plays an important role.

Economic model by Siegwart and Singer

The approach of Siegwart and Singer allows for a quantitative assessment of the economic impact in the use of new technologies, taking into account the risk associated with new technologies. Steger and Conrad [Steger, 95] applied this approach to rapid prototyping. Conventional five-axis milling and stereolithography were compared, for example.

Even from an operational perspective, the authors determine the benefits of stereolithography in terms of production time and costs. The quality of the component, in particular on the surfaces, from milling is better.

The strategic assessment of the technology covers all critical success factors in the calculation. The assessment is no longer based on the costs, as in the operational view.

Procedure

An economic efficiency ratio is defined, indicating how well a strategy is supported by the application of rapid prototyping.

The economic efficiency ratio is the quotient of the growth of capacity (capacity value) and the total costs (cost value). The costs include all of the classic cost types and the cost savings that are due to the use of additive manufacturing processes. The capacity includes the critical success factors (which may be defined differently in other basic approaches) cost, quality, and flexibility. The improvement or deterioration (in percent) is calculated in each case compared with the conventional reference case (which equals 100 %).

As part of the strategic review, Steger and Conrad compare the economic importance of the use of rapid prototyping for four different competitive strategies:

1. Technology leadership (pioneer strategy)
2. Cost leadership
3. a) Differentiation or b) Concentration
4. Overtaking strategy (outpacing)

Rapid prototyping technology has an influence on the critical success factors of

1. shorter development time,
2. reduction of costs,
3. improved quality, and
4. increased flexibility

entered into the calculation. For this purpose, the effect of an additive manufacturing process is estimated on the critical success factors: for each strategy, the critical success factors of time, quality, and flexibility are connected to a capacity value and transferred to the factor cost in a cost value.

The capacity and cost factors consist of a value for the performance ratio, which indicates how much the property in question could be improved by the use of an additive manufacturing process (for example, speed increased by 80% leads to the fulfillment level 180) and a weighting factor (depending on the importance of the criterion for the strategy, between 1 and 3).

The average of the capacity values of time, quality, and flexibility is multiplied by the reciprocal value of the cost value, the efficiency index WI, which is a measure of the potential or prediction of a new technology.

An evaluation of the individual capacity values, with a probability of arriving at the predictions, leads to an economy measure WII and additionally encompasses the risk that is associated with the introduction of the new technology.

The analogous application of the process to the basic technology gives reference values for the assessment of the new technology.

Determine the numerical values

The values for the variation of the critical success factors

1. shorter development time,
2. reduced costs,
3. improved quality, and
4. increased flexibility

were estimated as follows:

No. 1: Shortening the development time

Quickly available prototypes have the potential to reduce the development time by up to 80%.

The shortening of the production time of prototypes often plays only a minor role. This advantage can be used, depending on the strategy, for

- earlier market entry,
- quality improvement, or
- cost reduction.

The total reduction of development time that is finally realized is therefore dependent on the type of strategy.

No. 2: Reducing cost

The reduction of costs is achieved in two ways, first, the effect of earlier cost-cutting, and second, through cheaper production of models. The literature states that about 50% is production costs and that the savings from an earlier assessment of the cost of product development indicate about 25% of the savings in the production model. Through following necessary processes when using the additive manufacturing process, with the conventional process the total numbers may still change.

While the cost advantage in the production always occurs, the degree of realization of the further potential, in turn, depends on the strategy.

Reducing costs, therefore, makes up at least 50% but not more than 75%.

No. 3: Improving quality

Through the use of an additive manufacturing process, a higher product maturity is attained earlier.

Methods of QFD (quality function deployment) and FMEA (failure mode and effects analysis) are thereby effectively supported, and an expensive troubleshooting later on is avoided.

The magnitude of the effect depends, as already mentioned, on a cost/benefit strategy that determines how much is spent by the time that is saved.

No. 4: Increasing flexibility

The rapid availability of prototypes influences the flexibility outward and inward. Changing market requirements can be quickly and effectively responded to by serial identical prototypes.

Shorter reaction times within the company improve communication and put the organization in a position to respond faster and better to changes of all kinds (provisions computing process, process production, environmental conditions). They are the basis for continuous learning in an operation and an important factor in increasing the rate of innovation of the company.

A quantification of these effects is very difficult.

In order to obtain meaningful answers, a competitive strategy must be defined, such as cost leadership, and depending on each critical success factor, a performance level shall be determined. In view of the fact that the maximum savings are 75% of the cost for the application of rapid prototyping, a performance level of 175% results (100% corresponds to conventional milling). Because the cost leadership strategy assumes that the time gained is fully reinvested in cost reduction, the success factors of development and quality remain unchanged (100%), which, however, are difficult to measure. Increased flexibility is assumed to be 20% (= 120%).

The performance levels thus determined are weighted by factors of 1 to 3 (1 = low importance, 2 = average meaning, 3 = very important) and hence are the importance of the individual factors of success for each competition strategy.

In the case of cost leadership, this results in a 3 for the cost, a 2 for the time, and a 1 each for quality and for flexibility.

The average of the values of capacity is determined from the product of the weight and degree of completion, and the cost value is the reciprocal of the product of the weight and the degree of satisfaction regarding the success factor costs.

The value of the cost-effectiveness ratio WI results from the ratio of capacity value and cost value. Although it contains estimates for the potential of the technology, it does not consider the risk. Therefore, for each success factor is defined a probability (<1), multiplied by the value for each success factor and the mean re-formed for the performance. By dividing by the cost value, the profitability index WII follows in terms of risk.

The performance level and the probability are specified for conventional technology with 100% or 1. By weighting and averaging, the following efficiency characteristics WI and WII result:

	WI Conventionally	WII Conventionally
Cost leadership	3.99	3.99
Differentiation strategy	2.67	2.67
Pioneer strategy	2.33	2.33
Outpacing strategy	9.00	9.00

The calculation of the cost-effectiveness ratios for the rapid prototyping technology is based on the following values:

The weighting depends on the competitive strategy.

	Costs	Time	Quality	Flexibility
Cost leadership	3	2	1	1
Differentiation strategy	1	2	3	3
Pioneer strategy	1	3	1	3
Outpacing strategy	3	3	3	3
The probability is assumed to be:	1	1	0.7	0.6

This results in the following values related to the conventional results for the profitability figures I and II:

	WI / WI Conventionally	WII / WII Conventionally
Cost leadership	1.84	1.75
Differentiation strategy	0.86	0.70
Pioneer strategy	2.14	2.02
Outpacing strategy	1.91	1.76

The pioneer, the cost leadership, and the outpacing strategies rely mainly on the factors of time and costs: such strategies that are supported by the RP process are almost twice as effective as conventional methods.

In the differentiation strategy, the advantages of the additive process, time and costs, play only a subsequent role. However, the factors of quality and flexibility, which are not positively influenced by additive processes, are very important. Consequently, the strategy of differentiation of the additive manufacturing process is worse than that supported by conventional.

Technical data and information

The following are technical data and addresses collected on the topics of

- CAD systems and software for additive manufacturing,
- additive manufacturing machines (prototypers and fabricators), and
- materials for additive manufacturing processes and casting resins.

The information was derived from the manufacturers, but occasionally also from independent research institutes and from the literature. The description is limited to selected machines and representative materials for which sufficient data are available. Therefore, they do not make any mention of the machine in the text and nor mention all of the material. Many technical details were deliberately included in the description of the machines in the text because they can be commented upon there. Also, some machines are missing on further representation in table form. If technical information is missing in the table in the appendix, reference should be made to the corresponding chapters. All prices are included in Figs. 8.1 and 8.2, "System Costs."

The information has been compiled to the best of the author's knowledge and belief. Because of the high speed of development, before using this information, it should be checked on the manufacturers' website or question the manufacturer. Therefore, wherever Internet addresses were available, they are also indicated. The data from older machines and material can be looked up in previous editions of this book.

Table A2.1, "Manufacturers of Additive Manufacturing Machines," also shows manufacturers that are not mentioned in Chapter 3. More information should be obtained directly from the manufacturer. Above all, the current partners are to be reviewed because the distribution networks are continuously expanded. Casting resins are collected in Table A3.12, "Casting Resins." Because the range here is confusingly large, one elastic, one hard, one impact resistant, one high-temperature resistant, and one clear quality resin were selected as examples. This choice does not mean that similar qualities would not be delivered by other manufacturers. In particular, no rating is associated with the selection.

CAD Systems and Software for Additive Manufacturing

It may be assumed that each of these systems has a STL-capable interface.

Additive Manufacturing Machine (Prototypers and Fabricators)

Materials for Additive Manufacturing Processes and Casting Resins

Table A1.1 3D CAD Systems

Manufacturer	Product	Website
Autodesk Ltd.	Autodesk Inventor 2013 AutoCAD Mechanical	www.autodesk.com
Bentley Systems	MicroStation	www.bentley.com
CAD Schroer	Meduas4	www.cad-schroer.com
Cimatron	CimatronE	www.cimatron.com
DATA BECKER GmbH Co. KG	BeckerCAD 8 PRO	Closed March 2014
DataSolid GmbH	CADdy++	www.datasolid.com
Delcam UK	PowerShape ArtCAM	www.delcam-na.com
Encee CAD / CAM Systeme GmbH	VX Designer VX Mold&Die	www.encee.de
Evolution Computing Inc.	FastCAD7.5	www.fastcad.com
FEGS Ltd.	CADFIX	www.fegs.co.uk
GK-Planungssoftware GmbHI	TurboCAD	www.turbocad.com
IronCAD LLC.	IronCAD	www.ironcad.com
LGPL	FreeCAD	www.freecadweb.org
MASTERCAM	MASTERCAM	www.mastercam.com
MSCsoftware	MSC/PATRAN	www.mscsoftware.com
MCS Ltd.	ANVIL EXPRESS	www.mcsaz.com
Mecadat GmbH	VISI20 CAD/CAM	www.mecadat.com
NT CADCAM Ltd.	SolidWorks	www.ntcadcam.co.uk
Bihler of America, Inc.	bihlerbNX	www.bihler.com
Pathtrace Engineering Systems Ltd.	EdgeCAM	www.edgecam.com
PTC Inc.	PTC Cero Elements/Direct CADDS 5 Pro/ENGINEER	www.ptc.com
Siemens	NX I-DEAS Solid Edge	www.siemens.com
Tebis	Tebis CAD	www.tebis.com
Tinkercad Inc.	Tinkercad	www.tinkercad.com
Tnodiana	FEMGV	www.tnodiana.com
Trimple Navigation Ltd.	SketchUp	www.sketchup.com
Toyota Caelum Inc.	Caelum	www.caelum.co.jp
VariCAD	VariCAD	www.varicad.com
3DS	CATIA ICEM Surf	www.3ds.com

Table A1.2 Additive Manufacturing Software

Manufacturer	DeskArtes Oy	Marcam Engineering	Materialise N.V.	Netfabb GmbH	Solid Concepts Inc.
Website	deskartes.fi	marcam.com	materialise.com	Netfabb.com	solidview.com
Program	3DataExpert	AutoFab	Magics	Netfabb Professional	Solid View/ Pro RP
RP Software	yes	yes	yes	yes	yes
CAD Software, Module	yes	no	no	no	no
Visualization software	yes	yes	yes	yes	yes
Interface formats					
▪ **Import**	STL, VRML, ZPR, OBJ, DXF, PLY, 3DS, Collada, FBX, 3DE	STL, IGES, VDA-FS, DXF (3D FACE), VRML (1&2), PLY	Google SketchUp 8.0, AMF, DXF, OBJ, VRML, ZPR, PLY, 3DS, 3DM (Optional: FBX, STEP, ACIS, VDA, IGES)	STL, X3D, WRL, GTS, CLI, SLI, SLC, SSL, OBJ, CLS, G-Code	DXF, DWG, SFX, SVD, STL, WRL, (Optional: IPT, IAM, SAT, SAB, SESSION, MODEL, EXP, CATPART, CATPRODUCT, CGR, IGES, IGS, VDA, PRT, ASM, STP, STEP, X_B, X_T, IDI, MCA, PAR, ASM, PSM, DFT, SLDPRT, SLDDRW, PRT
▪ **Export**	STL, VRML, ZPR, DXF, PLY, 3DE	STL, CLI, SLI, CLF, F&S, ILT, AFF, CLS, (Optional: MTT, SLM)	n. a.	SLI, CLI, CLS, SLC, SSL, USF, ABF, SLM, SVG, DXF, STL, PNG, BMP, PS, ASC	STL, PLY, SVD
Possibility to repair STL files	yes	yes	yes	yes	yes
Possibility to modify STL files	yes	yes	yes	yes	yes
Measuring objects	yes	yes	yes	yes	yes
Possibility to create slice files	yes	yes	yes	yes	yes
Support generation	yes	yes	yes	yes	yes
Tooling module	yes	no	yes	yes	yes
Hardware request	PC, Workstation	PC, Workstation	PC, Workstation	PC, Workstation	PC, Workstation
Operating system	Win NT	Windows	Windows	Windows, MAC, Linux	Windows

Table A2.1 Manufacturers of Additive Manufacturing Machines

Manufacturer	City	Country	Website
3D Systems, GmbH	64291 Darmstadt	Germany	www.3dsystems.com
3D Systems, Inc.	Rock Hill, SC 29730	USA	www.3dsystems.com
Arcam AB (publ.)	431 37 Mölndal	Sweden	www.arcam.com
Boxford Limited	Halifax, HX3 5AF	Great Britian	www.boxford.co.uk
Charlyrobot SA BP 22	74350 Cruseilles	France	www.charlyrobot.com
CMET Inc.	Kanagawa, 222-0033	Japan	www.cmet.co.jp
CONCEPT Laser GmbH	96215 Lichtenfels	Germany	www.concept-laser.de
CRP-Technology	41100 Modena	Italy	www.crptechnology.com
Cubic Technologies, Inc.	Carson, CA 907463608	USA	www.cubictechnologies.com
Cubital	The company is no longer active		
D-MEC Ltd.	Tokyo 1040045	Japan	www.d-mec.co.jp
Envisiontec GmbH	45968 Gladbeck	Germany	www.envisiontec.de
EOS GmbH	82152 Krailling	Germany	www.eos.info
F & S Stereolithography-technik GmbH	33100 Paderborn	Germany	www.fockeleandblacke.de
Kinergy Precision Engineering Co., Ltd.	Nantong	VR China	www.kinergy-nt.com
KIRA Corp. Co., Ltd.		Japan	www.kiracorp.co.jp
Kira Europe GmbH	40699 Erkrath	Germany	www.kira-europe.com
MCP HEK Tooling GmbH	23560 Lübeck	Germany	www.mcp-group.de
MEIKO Co., Ltd.		Japan	www.meiko-inc.co.jp
Microfabrica Inc.	Van Nuys, CA 91406	USA	www.Microfabrica.com
microTEC, Ges. f. Mikrotechnologie mbH	47057 Duisburg	Germany	www.microtec-d.com
Mk Technology GmbH	53501 Grafschaft	Germany	www.mk-technology.com
NEXT FACTORY S.R.L.	36015 Schio (Vi)	Italy	www.nextfactory.com
Optomec Inc.	Albuquerque, NM 87109	USA	www.optomec.com
Phenix Systems	63100 Clermont Ferrand	France	www.phenix-systems.com
POM Advanced Product Development Center	Auburn Hills, Michigan 48326	USA	www.pomgroup.com
ProMetal	Irwin, PA 15642	USA	www.prometal-rt.com
ProMetal RCT GmbH	86167 Augsburg	Germany	www.prometal-rct.com
Röders GmbH (Röders TEC)	29614 Soltau	Germany	www.roeders.de
Shanghai Union Technology Co., Ltd.	Shanghai	VR China	www.union-tek.com
Solidica	Ann Arbor, MI 481088942	USA	www.solidica.com

Table A2.1 *(continued)* Manufacturers of Additive Manufacturing Machines

Manufacturer	City	Country	Website
Solidimension Ltd.		USA, Italy, China	www.solido3d.com
Solidscape	Merrimack, NH 030544115	USA	www.solid-scape.com
Soligen	Northridge, CA 91324	USA	www.soligen.com
Speedpart	431 37 Mölndal	Sweden	www.speedpart.se
Stratasys Inc.	Eden Prairie, MN 553442020	USA	www.stratasys.com
Stratasys GmbH	60314 Frankfurt/Main	Germany	www.stratasys.com
Stratoconception/CIRTES	Saint-Dié-des-Vosges 88100	France	www.stratoconception.com
Trump Precision Machinery Co., Ltd.	Shanghai, Zhongshan	VR China	www.trumpsystem.com
TRUMPF Laser- and Systemtechnik GmbH	71254 Ditzingen	Germany	www.trumpf.com
Tschopp Technical Enginnering	Ramlinsburg	Switzerland	www.tschopptech.ch
Voxeljet	86167 Augsburg	Germany	www.voxeljet.de
Weihbrecht	74549 Wolpertsausen	Germany	www.weihbrecht.de
Zimmermann, F. GmbH	73770 Denkendorf	Germany	www.f-zimmermann.com
Z Corporation	Burlington, MA 01803	USA	www.zcorp.com

Table A2.2 Stereolithography Machines I

RP Process	Unit	SL	SL	SL	SL	SL*
Machine Designation/ Type		SLA 250	SLA 5000	SLA 7000	Viper-SLA	Viper Pro
Manufacturer		3D Systems				
Distribution in Germany		www.3dsystems.com				
Dimensions, Connection values						
Width/depth/height	m/m/m	1.24/1.64/0.69	1.88/1.19/2.02	1.88/1.22/2.03	1.34/0.86/1.78	ca. 2/1.5/2
Weight	Kg	362	1318	1196	463	
Electrical connection	V/A	230/8	200-240/15	200-140/20	220-240/16	200-140/20
Power consumption	kW	1.2	2.2	3	1.51	3
Working gas	l/min	–	–	–	–	–
Cooling water	l/min	–	–	–	–	–
Extraction	m^3/h	–	–	–	–	–
Work room temperature	°C	20-26	20-26	20-26	20-26	20-26
Relative moisture	%	< 50	< 50	< 50	< 50	< 50
Process						
Fumes, Exhaust		Closed system with active carbon filter				
Disposal, Waste		Cured polymer				
Hazardous Waste		Uncured polymer, solvent				
Contouring Element						
Kind		Laser	Laser	Laser	Laser	Laser
Type		HeCd	$Nd:YVO_4$	$Nd:YVO_4$	$Nd:YVO_4$	$Nd:YVO_4$
x-y Contour generation		Scanner	Scanner	Scanner	Scanner	Scanner
x-y Contour accuracy	mm	Depending on the material	Depending on the material	Depending on the material	Depending on the material	Depending on the material
Position repeatability	mm	–	0.013	0.001	0.0076	0.0076
z Contour accuracy	mm	0.0025	0.00177	0.00125	0.0025	0.0025
z Contour repeatable accuracy	mm	Depending on the material	Depending on the material	Depending on the material	Depending on the material	Depending on the material
Coating cycle	s	Depending on geometry	Depending on geometry	Depending on geometry	Depending on geometry	Depending on geometry
Coating arm speed	mm/s	Typ. 25-50	Typ. 25-50	Typ. 25-50	Typ. 25-50	Typ. 25-50
Power fluctuation	%	± 5	± 5	± 5	± 5	± 5
Model characteristics						
Installation space max. (Width/depth/height)	mm	254/254/254	508/508/584	508/508/600	254/254/254	min: 650/350/300 max: 1500/750/500
Model dimensions max.	mm	254/254/254	508/508/584	508/508/600	254/254/254	
Layer thickness	mm	0.075-0.15	0.05-0.15	0.025-0.15	0.025-0.15	0.05-0.15
Spot (laser diameter)	mm	ca. 0.25	ca. 0.25	ca. 0.25 plus 0.8	ca. 0.075 plus .25	0.13
Accuracy absolute	mm	0.1	0.1	0.1	0.1	0.1
Repeatability	mm	–	–	–	–	–
Additional support		yes	yes	yes	yes	yes
EDP interfaces						
Interfaces format		STL, SLC	STL, SLC	STL, SLC	STL, SLC	STL, SLC
EDP system		PC	PC	PC	PC	PC
Operating system		MS-DOS	Windows XP	Windows XP	Windows XP	Windows XP
Software		Buildstation 4.0	Buildstation 5.5	Buildstation 5.5	Buildstation 5.5	Buildstation 5.5

* The technical data, especially the building chambers, change with the equipment of the machine.

Table A2.3 Stereolithography Machines II

RP Process	Unit	Stereolithography	Stereolithography	SL
Machine designation/Type		InVision 3-D Printer	InVision HR 3-D Printer	FS-Realizer STL
Manufacturer		3D Systems		F&S Stereolitho-graphytechnik GmbH
Distribution in Germany		www.3dsystems.com		
Dimensions, Connection values				
Width/depth/height	m/m/m	0.96/1.42/1.67	0.96/1.42/1.67	1.6/0.8/2.3
Weight	kg	254	254	500
Electrical connection	V/A	100–127/15; 200–240/10	100–127/15; 200–240/10	230/16; 400/32
Power consumption	kW			1.5
Working gas	l/min			–
Cooling water	l/min			–
Extraction	m^3/h			yes
Work room temperature	°C		18–28	< 33
Relative moisture	%			45
Process				
Fumes, Exhaust				low
Disposal, Waste				
Hazardous waste				
Contouring Element				
Kind		Multijet Modeling	Multijet Modeling	Laser
Type		with UV curing	with UV curing	Nd-YAG
x-y Contour generation				Scanner
x-y Contour accuracy	dpi	328/328	656/656	± 0.02
Spot repeatability	dpi			0.002
z Contour accuracy	dpi	606	800	± 0.1
z Contour repeatable accuracy	mm			± 0.1
Coating cycle	s			< 10
Coating arm speed	mm/s			0–1000
Power fluctuation	%			± 2
Model characteristics				
Installation space max. (width/depth/height)	mm		127/178/50	400/400/300
Model dimensions max.	mm	298/185/203	64.5 cm^2 (*x*, *y*); 5.0 cm (*z*)	400/400/300
Layer thickness	mm			> 0.1
Gauge	mm			0.1
Accuracy absolute	mm			± 0.1
Repeatability	mm			± 0.1
Additional support				yes
EDP Interface				
Interface formats		Ethernet, .stl		F&S
EDP System				PC
Operating system		Windows XP, Professional, 2000, NT	Windows XP, Professional, 2000, NT	Win NT
Software		InVision print client software	InVision print client software	F&S

Table A2.4 Stereolithography Machines III

RP Process	Unit	SL	SL	SL	SL	SL*
Machine designation/Type		DigitalWax 008J	DigitalWax 009J	DigitalWax 028J	DigitalWax 028J+	DigitalWax 029J
Manufacturer		DWS Additive Manufacturing				
Distribution		www.dwssystems.com				
Dimensions, Connection values						
Width/depth/height	m/m/m	0.38/0.52/0.56	0.38/0.52/0.56	0.38/0.52/0.73	0.38/0.52/0.73	0.61/0.66/1.4
Weight	kg	43	43	56	56	150
Electrical connection	V/A	230/115	230/115	230/115	230/115	230/115
Power consumption	kW	0.2	0.2	0.4	0.4	0.5
Working gas	l/min	–	–	–	–	–
Cooling water	l/min	–	–	–	–	–
Extraction	m^3/h	–	–	–	–	–
Work room temperature	°C	22–25	22–25	22–25	22–25	22–25
Relative moisture	%	60	60	60	60	60
Process						
Fumes. Exhaust		–				
Disposal, Waste		–				
Hazardous waste		–				
Contouring Element						
Kind		Laser	Laser	Laser	Laser	Laser
Type		Solid State BluEdge® BE-100	Solid State BluEdge® BE-	Solid State BluEdge® BE-1500A/ BE-1500AHR	Solid State BluEdge® BE-1500A/ BE-1500AHR	Solid State BluEdge® BE-1700/ BE-1700HR
x-y Contour generation		Plotter	DLP® Texas Instruments Inc.	Galvanometer	Galvanometer	Galvanometer
x-y Contour accuracy	mm	Depending on the material	Depending on the material	Depending on the material	Depending on the material	Depending on the material
Position repeatability	mm	–	–	–	–	–
z Contour accuracy	mm	–	–	–	–	–
z Contour repeatable accuracy	mm	Depending on the material	Depending on the material	Depending on the material	Depending on the material	Depending on the material
Coating cycle	s	Depending on geometry	Depending on geometry	Depending on geometry	Depending on geometry	Depending on geometry
Coating arm speed	mm/s	Typ. 0-40	–	Typ. 0-2200	Typ. 0-2200	Typ. 2600
Power fluctuation	%	–	–	–	–	–
Model characteristics						
Installation space max. (width/depth/height)	mm	65/65/90	50/37/90	65/65/90	90/90/90	110/110/100
Model dimensions max.	mm	65/65/90	50/37/90	65/65/90	90/90/90	110/110/110
Layer thickness	mm	0.01-0.10	0.01-0.10	0.01-0.10	0.01-0.10	0.01-0.10
Spot (laser diameter)	mm	–	–	–	–	–
Accuracy absolute	mm	–	–	–	–	–
Repeatability	mm	–	–	–	–	–
Additional support		yes	yes	yes	yes	yes
EDP Interface						
Interface formats		STL, SLC	STL, SLC	STL, SLC	STL, SLC	STL, SLC
EDP system		PC	PC	PC	PC	PC
Operating system		32-bit Windows XP Pro, Vista, 7	Windows XP	Windows 7	Windows 7	Windows 7
Software		DigitalWax 008J Controller	DigitalWax 009J Controller	DigitalWax 028J Controller	DigitalWax 028J Controller	DigitalWax 029J Controller

Table A2.5 Stereolithography Machines IV

RP Process	Unit	SL	SL	SL	SL	SL*
Machine designation/ Type		DigitalWax 029J+	DigitalWax 030J	DigitalWax dfab	DigitalWax 008D	DigitalWax 020D
Manufacturer		DWS Additive Manufacturing				
Distribution		www.dwssystems.com				
Dimensions, Connection values						
Width/depth/height	m/m/m	0.61/0.66/1.4	1.15/1.45/2.1	-	0.38/0.52/0.56	0.38/0.52/0.81
Weight	kg	150	-	-	45	58
Electrical connection	V/A	230/115	230/115	-	230/115	230/115
Power consumption	kW	0.5	-	-	0.2	0.4
Working gas	l/min	-	-	-	-	-
Cooling water	l/min	-	-	-	-	-
Extraction	m^3/h	-	-	-	-	-
Work room temperature	°C	22-25	22-25	22-25	22-25	22-25
Relative moisture	%	60	60	60	60	60
Process						
Fumes, Exhaust		-				
Disposal, Waste		-				
Hazardous waste		-				
Contouring Element						
Kind		Laser	Laser	Laser	Laser	Laser
Type		Solid State BluEdge® BE-1800A/ BE-1800AHR	Solid State Blu Edge®	-	Solid State Blu Edge® BE-1000	Solid State Blu Edge® BE-1500C
x-y Contour generation		Galvanometer	-	-	Plotter	Galvanometer
x-y Contour accuracy	mm	Depending on the material	Depending on the material	Depending on the material	Depending on the material	Depending on the material
Position repeatability	mm	-	-	-	-	-
z Contour accuracy	mm	-	-	-	-	-
z Contour repeatable accuracy	mm	Depending on the material	Depending on the material	Depending on the material	Depending on the material	Depending on the material
Coating cycle	s	Depending on geometry	Depending on geometry	Depending on geometry	Depending on geometry	Depending on geometry
Coating arm speed	mm/s	Typ. 5000	-	-	Typ. 0-40	Typ. 0-2000
Power fluctuation	%	-	-	-	-	-
Model characteristics						
Installation space max. (width/depth/height)	mm	150/150/100	300/300/300	-	65/65/90	130/130/90
Model dimensions max.	mm	150/150/100	300/300/300	-	65/65/90	130/130/90
Layer thickness	mm	0.01-0.10	-	-	0.01-0.10	0.01-0.10
Spot (laser diameter)	mm	-	-	-	-	-
Accuracy absolute	mm	-	-	-	-	-
Repeatability	mm	-	-	-	-	-
Additional support		yes	yes	yes	yes	yes
EDP Interface						
Interface formats		STL, SLC	STL, SLC	STL, SLC	STL, SLC	STL, SLC
EDP system		PC	PC	PC	PC	PC
Operating system		Windows 7	-	-	32-bit Windows XP Pro, Vista, 7	Windows XP Professional
Software		DigitalWax 029J Controller	DigitalWax 030J Controller	-	DigitalWax 008D Controller	DigitalWax 008D Controller

Table A2.6 Stereolithography Machines V

RP Process	Unit	SL	SL	SL	SL	SL*
Machine designation/ Type		DigitalWax 028D	DigitalWax 029D	DigitalWax 030D	DigitalWax 029X	DigitalWax 030X
Manufacturer		DWS Additive Manufacturing				
Distribution		www.dwssystems.com				
Dimensions, Connection values						
Width/depth/height	m/m/m	0.38/0.52/0.73	0.61/0.66/1.4	1.1/0.7/2.0	0.61/0.66/1.4	1.15/1.45/2.1
Weight	kg	56	150	-	150	-
Electrical connection	V/A	230/115	230/115	230/115	230/115	230/115
Power consumption	kW	0.4	0.5	-	0.5	-
Working gas	l/min	-	-	-	-	-
Cooling water	l/min	-	-	-	-	-
Extraction	m^3/h	-	-	-	-	-
Work room temperature	°C	22-25	22-25	22-25	22-25	22-25
Relative moisture	%	60	60	60	60	60
Process						
Fumes, Exhaust		-				
Disposal, Waste		-				
Hazardous waste		-				
Contouring Element						
Kind		Laser	Laser	Laser	Laser	Laser
Type		Solid State BluEdge® BE-1500	Solid State Blu Edge® BE-1800B	Solid State Blu Edge®	Solid State Blu Edge® BE- 1800C	Solid State Blu Edge®
x-y Contour generation		Galvanometer	Galvanometer	-	Galvanometer	-
x-y Contour accuracy	mm	Depending on the material	Depending on the material	Depending on the material	Depending on the material	Depending on the material
Position repeatability	mm	-	-	-	-	-
z Contour accuracy	mm	-	-	-	-	-
z Contour repeatable accuracy	mm	Depending on the material	Depending on the material	Depending on the material	Depending on the material	Depending on the material
Coating cycle	s	Depending on geometry	Depending on geometry	Depending on geometry	Depending on geometry	Depending on geometry
Coating arm speed	mm/s	Typ. 0-2000	Typ. 5000	-	Typ. 5000	-
Power fluctuation	%	-	-	-	-	-
Model characteristics						
Installation space max. (width/depth/height)	mm	90/90/90	150/150/150	250/250/250	150/150/150	300/300/300
Model dimensions max.	mm	90/90/90	150/150/150	250/250/250	150/150/150	300/300/300
Layer thickness	mm	0.01-0.10	0.01-0.10	0.01-0.10	0.01-0.10	0.01-0.10
Spot (laser diameter)	mm	-	-	-	-	-
Accuracy absolute	mm	-	-	-	-	-
Repeatability	mm	-	-	-	-	-
Additional support		yes	yes	yes	yes	yes
EDP Interface						
Interface formats		STL, SLC	STL, SLC	STL, SLC	STL, SLC	STL, SLC
EDP system		PC	PC	PC	PC	PC
Operating system		Windows XP Professional	Windows XP Professional	Windows	Windows 7	Windows 7
Software		DigitalWax 028D Controller	DigitalWax 029D Controller	DigitalWax 030D Controller	DigitalWax 029X Controller	DigitalWax 030X Controller

Table A2.7 Stereolithography Machines VI

RP Process	Unit	SLA	SLA	SLA	SLA	SLA
Machine designation/ Type		Perfactory Micro	Perfactory Micro DDP	Perfactory Micro EDU	Perfactory Aureus	PixCera
Manufacturer		EnvisionTEC GmbH				
Distribution		Envisiontec.com				
Dimensions, Connection values						
Width/depth/height	m/m/m	0.23/0.18/0.58	0.23/0.23/0.52	0.28/0.25/0.61	0.45/0.78/0.45	0.45/0.78/0.45
Weight	kg	13	13	11	35	35
Electrical connection	V/A	100-240/2-3	100-240/2-3	100-240/2-3	100-240/1-2	100-240/1-2
Capacity	KW	n. a.	n. a.	n. a.	n. a.	n. a.
Work room temperature	°C	n. a.	n. a.	n. a.	n. a.	n. a.
Air moisture	%	n. a.	n. a.	n. a.	n. a.	n. a.
Model characteristics						
Installation space max. (width/depth/height)	mm	40/30/100	40/30/100	86/64.5/100	60/45/100	60/45/100
Model dimensions max.	mm	40/30/100	40/30/100	86/64.5/100	60/45/100	60/45/100
Layer thickness	mm	0.025-0.035	0.05/0.1	0.025-0.1	0.025-0.035	0.025-0.035
Accuracy absolute	mm	n. a.	n. a.	n. a.	n. a.	n. a.
EDP Interface						
Interface formats		STL	STL	STL	STL	STL
EDP system		n. a.	n. a.	n. a.	PC	PC
Operating system		n. a.	n. a.	n. a.	n. a.	n. a.
Miscellaneous						
Building materials		PIC 100, EC500, HTM140	Press-E-Cast	LS 600, HTM140V2	R5/R11, PIC 100/100G, WIC 100G, EC500, RCP30/130, HTM140	E-Dent, E-PKin-dial, Press-E-Cast
ERM pixel resolution	mm	0.031	-	-	-	-
Resolution in *x* and *y*	mm	-	0.075	-	-	-
Native pixel size	mm	-	-	0.084	0.043	0.043
Linear *z*-axis resolution	mm	0.001	-	0.001	-	-

Table A2.8 Stereolithography Machines VII

RP Process	Unit	SLA	SLA
Machine designation/ Type		Perfactory Mini DDSP	Perfactory DDSP
Manufacturer		EnvisionTEC GmbH	
Distribution		Envisiontec.com	
Dimensions, Connection values			
Width/depth/height	m/m/m	0.73/0.48/1.35	0.73/0.48/1.35
Weight	kg	79	79
Electrical connection	V/A	100-240/2.4-5.4	100-240/2.4-5.4
Capacity	KW	n. a.	n. a.
Work room temperature	°C	n. a.	n. a.
Model characteristics			
Installation space max. (width/depth/height)	mm	60/45/100	100/75/100
Model dimensions max.	mm	60/45/100	100/75/100
Layer thickness	mm	0.25-0.1	0.25-0.15
Accuracy absolute	mm	n. a.	n. a.
EDP Interface			
Interface formats		STL	STL
EDP system		n. a.	n. a.
Operating system		n. a.	n. a.
Miscellaneous			
Materials		E-Shell: 200 Series, 300 Series, 500 Series, 600, 3000 Series	E-Shell: 200 Series, 300 Series, 500 Series, 600, 3000 Series
Native pixel size, *xy*	mm	0.043	0.071

Table A2.9 Stereolithography Machines VIII

RP Process	Unit	SLA	SLA	SLA	SLA	SLA
Machine designation/ Type		DDP4	DDP4 M	DDP4 XL	Perfactory 4 Mini	Perfactory 4 Mini XL
Manufacturer		EnvisionTEC GmbH				
Distribution		Envisiontec.com				
Dimensions, Connection values						
Width/depth/height	m/m/m	0.73/0.48/1.35	0.73/0.48/1.35	0.73/0.48/1.35	0.73/0.48/1.35	0.73/0.48/1.35
Weight	kg	85	85	85	85	85
Electrical connection	V/A	100-240/ 2.4-5.4	100-240/ 2.4-5.4	100-240/ 2.4-5.4	100-240/ 2.4-5.4	100-240/ 2.4-5.4
Capacity	KW	n. a.	n. a.	n. a.	n. a.	n. a.
Work room temperature	°C	n. a.	n. a.	n. a.	n. a.	n. a.
Model characteristics						
Installation space max. (width/depth/height)	mm	115/72/160	160/100/160	192/120/160	84/52.5/230	115/72/160
Model Dimensions max.	mm	115/72/160	160/100/160	192/120/160	84/52.5/230	115/72/160
Layer thickness	mm	0.025-0.15	0.025-0.15	0.025-0.15	0.015-0.150	0.015-0.150
Accuracy absolute	mm	n. a.	n. a.	n. a.	n. a.	n. a.
EDP Interface						
Interface formats		STL	STL	STL	n. a.	n. a.
EDP system		PC	PC	PC	n. a.	n. a.
Operating system		n. a.	n. a.	n. a.	n. a.	n. a.
Miscellaneous						
Building materials		Clear Guide. E-Denstone, E-Dent, E-PKin-dial, Press-E-Cast	Clear Guide, E-Denstone, E-Dent, E-PKin-dial, Press-E-Cast	Clear Guide, E-Denstone, E-Dent, E-PKin-dial, Press-E-Cast	ABStuff,ABflex, EC500, E-Shell 200, E-Shell 300, HTM Series, LS600, Photosilver, PIC 100, R5/R5 Gray/R11, RCP30/RC31	
Native pixel size, *xy*	mm	0.060	0.083	0.1	0.019-0.044	0.060
ERM voxel size, *xy*	mm	0.030	0.042	0.050	0.010-0.022	0.030

Table A2.10 Stereolithography Machines IX

RP Process	Unit	SLA	SLA
Machine designation/ Type		Perfactory DDP3	Perfactory 3 SXGA W/ERM Mini Multi Lens
Manufacturer		EnvisionTEC GmbH	
Distribution		Envisiontec.com	
Dimensions, Connection values			
Width/depth/height	m/m/m	0.73/0.48/1.35	0.73/0.48/1.35
Weight	kg	70	70
Electrical connection	V/A	100-240/2.4-5.4	100-240/2.4-5.4
Capacity	KW	n. a.	n. a.
Work room temperature	°C	n. a.	n. a.
Model characteristics			
Installation space max. (width/depth/height)	mm	90/68/230	90/67.5/230
Model dimensions max.	mm	90/68/230	90/67.5/230
Layer thickness	mm	0.050-0.15	0.015-0.15
Accuracy absolute	mm	n. a.	n. a.
EDP Interface			
Interface formats		STL	STL
EDP system		n. a.	PC
Operating system		n. a.	n. a.
Miscellaneous			
Materials		EC-1000, WIC300,WIC402,QID200,E-Dent, RC31, LS600, HTM140 iv	ABStuff,ABflex, EC500, E-Shell 200, HTM Series, LS600, Photosilver, PIC 100, R5/R5 Gray/R11, RCP30/RC31
Native pixel size, *xy*	mm	0.050	0.032-0.060
ERM voxel size, *xy*	mm	0.025	0.016-0.042

Table A2.11 Stereolithography Machines X

RP Process	Unit	SLA	SLA	SLA	SLA	SLA
Machine designation/ Type		Perfactory 4 DSP	Perfactory 4 DSP XL	Perfactory 4 Standard	Perfactory 4 Standard XL	Perfactory 3 Digital Shell Printer
Manufacturer		EnvisionTEC GmbH				
Distribution		Envisiontec.com				
Dimensions, Connection values						
Width/depth/height	m/m/m	0.73/0.48/1.35	0.73/0.48/1.35	0.73/0.48/1.35	0.73/0.48/1.35	0.73/0.48/1.35
Weight	kg	85	85	85	85	70
Electrical connection	V/A	100-240/2.4-5.4	100-240/2.4-5.4	100-240/2.4-5.4	100-240/2.4-5.4	100-240/2.4-5.4
Capacity	KW	n. a.	n. a.	n. a.	n. a.	n. a.
Work room temperature	°C	n.a	n.a	n.a	n.a	n.a
Model characteristics						
Installation space max. (width/depth/height)	mm	160/100/160	192/120/160	160/100/160	192/120/160	140/105/230
Model dimensions max.	mm	160/100/160	192/120/160	160/100/160	192/120/160	140/105/230
Layer thickness	mm	0.025-0.150	0.015-0.15	0.025-0.150	0.025-0.150	0.050-0.150
Accuracy absolute	mm	n. a.	n. a.	n. a.	n. a.	n. a.
EDP Interface						
Interface formats		n. a.	n. a.	n. a.	n. a.	n. a.
EDP system		PC	PC	n. a.	n. a.	PC
Operating system		n. a.	n. a.	n. a.	n. a.	n. a.
Miscellaneous						
Materials		E-Shell: 200 Series. 300 Series, 500 Series, 600, 3000 Series	E-Shell: 200 Series, 300 Series, 500 Series, 600, 3000 Series	ABStuff,ABflex, EC500, E-Shell 200, E-Shell 300, HTM Series, LS600, Photosilver, PIC 100, R5/R5 Gray/R11, RCP30/RC31	ABStuff,ABflex, EC500, E-Shell 200, E-Shell 300, HTM Series, LS600, Photosilver, PIC 100, R5/R5 Gray/R11, RCP30/RC31	E-Shell: 200 Series, 300 Series, 500 Series, 600, 3000 Series
Native pixel size, *xy*	mm	0.083	0.100	0.083	0.100	0.050
ERM voxel size, *xy*	mm	0.042	0.050	0.042	0.050	0.025

Table A2.12 Stereolithography Machines XI

RP Process	Unit	SLA	SLA
Machine designation/ Type		Perfactory Xtreme	Perfactory Xede
Manufacturer		EnvisionTEC GmbH	
Distribution		Envisiontec.com	
Dimensions, Connection values			
Width/depth/height	m/m/m	1.8/1.8/2.28	1.8/1.8/2.28
Weight	kg	710	710
Electrical connection	V/A	100–240/2.4–5.4	100–240/2.4–5.4
Power consumption	kW	n. a.	n. a.
Work room temperature	°C	n. a.	n. a.
Relative moisture	%	n. a.	n. a.
Model characteristics			
Installation space max. (width/depth/height)	mm	368/229/356	444.5/355.6/500
Model dimensions max.	mm	368/229/356	444.5/355.6/500
Layer thickness	mm	0.025-0.15	0.025-0.15
Accuracy absolute	mm	n. a.	n. a.
EDP Interface			
Interface formats		STL	STL
EDP System		PC	PC
Operating system		n. a.	n. a.
Miscellaneous			
Materials		ABS, polypropylene, glass-filled nylon parts, aluminum oxide, zirconium oxide, silicon oxide, paraffin wax	ABS, polypropylene, glass-filled nylon parts, aluminum oxide, zirconium oxide, silicon oxide, paraffin wax
Native pixel size, *xy*	mm	0.143	0.103-0.130
ERM voxel size, *xy*	mm	0.072	0.052-0.065

Table A2.13 Stereolithography Machines XII

RP Process	Unit	SLA	SLA	SLA	SLA	SLA
Machine designation/ Type		3Dent with 3SP Technology	ULTRA 3SP 3D Dental Printer	ULTRA 3SP	ULTRA 3SP High Definition	ULTRA2
Manufacturer		EnvisionTEC GmbH				
Distribution in Germany		Envisiontec.com				
Dimensions, Connection values						
Width/depth/height	m/m/m	0.74/0.76/1.17	0.74/0.76/1.17	0.74/0.76/1.17	0.74/0.76/1.17	0.74/0.76/1.17
Weight	kg	90	89.8	89.8	89.8	90
Electrical connection	V/A	100-240/2-3	100-127; 200-240/8; 4	100-127; 200-240/8; 4	100-127; 200-240/8; 4	100-127; 200-240/8; 4
Power consumption	kW	n. a.	n. a.	n. a.	n. a.	n. a.
Work room temperature	°C	n. a.	n. a.	n. a.	n. a.	n. a.
Model characteristics						
Installation space max. (width/depth/height)	mm	280/180/76	279/184/76	279/184/203	279/184/203	267/165/203
Model dimensions max.	mm	280/180/76	279/184/76	279/184/203	279/184/203	267/165/203
Layer thickness	mm	0.025-0.100	0.025-0.100	0.025-0.100	0.025-0.100	0.015-0.100
Accuracy absolute	mm	n. a.	n. a.	n. a.	n. a.	n. a.
EDP Interface						
Interface formats		STL	STL	STL	STL	STL
EDP system		n. a.	n. a.	n. a.	n. a.	n. a.
Operating system		n. a.	n. a.	n. a.	n. a.	n. a.
Miscellaneous						
Materials		E-Denstone	E-Denstone	E-Denstone	E-Denstone	HTM140IV
Resolution in *xy*	mm	0.050	0.100	0.100	0.050	-
Native voxel resolution in *xy*	mm	-	-	-	-	0.045-0.090

Table A2.14 Stereolithography Machines XIII

RP Process	Unit	SLA	SLA
Machine designation/ Type		ULTRA2	ULTRA
Manufacturer		EnvisionTEC GmbH	
Distribution		Envisiontec.com	
Dimensions, Connection values			
Width/depth/height	m/m/m	0.74/0.76/1.17	0.74/0.76/1.17
Weight	kg	90	90
Electrical connection	V/A	100–127; 200–240/8; 4	100–127; 200–240/8; 4
Power consumption	kW	n. a.	n. a.
Work room temperature	°C	n. a.	n. a.
Model characteristics			
Installation space max. (width/depth/height)	mm	267/165/203	267/165/203
Model dimensions max.	mm	267/165/203	267/165/203
Layer thickness	mm	0.015–0.100	0.025–0.150
Accuracy absolute	mm	n. a.	n. a.
EDP Interface			
Interface formats		STL	STL
EDP system		n. a.	n. a.
Operating system		n. a.	n. a.
Miscellaneous			
Materials		HTM140IV	ABS-Stuff, AB-flex, SI-500, LS600, EC-500
Voxel resolution in *xy*	mm	0.021	0.139

Table A2.15 Stereolithography Machines XIV

RP Process	Unit	File Transfer Imaging (FTI)	File Transfer Imaging (FTI)	Stereo-lithography	Stereo-lithography	Stereo-lithography
Machine designation/ Type		ProJet 1000	ProJet 1500	ProJet SD 7000	ProJet HD 7000	ProJet MP 7000
Manufacturer		3D Systems				
Distribution		www.3dsystems.com				
Dimensions, Connection values						
Width/depth/height	m/m/m	0.56/0.91/0.72	0.56/0.91/0.72	0.98/0.85/1.83	0.98/0.85/1.83	0.98/0.85/1.83
Weight	kg	55.3	55.3	272	272	272
Electrical connection	V/A	100-120/3A; 220-240/1.5 A	100-120/3 A; 220-240/1.5 A	100-240/ n.a.	100-240/ n.a.	100-240/ n.a.
Capacity	KW	n.a.	n.a.	0.75	0.75	0.75
Work room temperature	°C	n.a.	n.a.	18-28	18-28	18-28
Model characteristics						
Installation space max. (width/depth/height)	mm	171/203/178	171/228/203	380/380/250	380/380/250	380/380/250
Model dimensions max.	mm	171/203/178	171/228/203	380/380/250	380/380/250	380/380/250
Layer thickness	mm	0.102	0.102-0.152	0.1-0.125	0.05-0.125	0.05-0.125
Accuracy absolute	mm	n.a.	n.a.	0.025-0.05	0.025-0.05	0.025-0.05
EDP Interface						
Interface formats		STL, CTL	STL, CTL	STL, SLC	STL, SLC	STL, SLC
EDP System		PC	PC	PC	PC	PC
Operating system		Windows XP to Windows 7	Windows XP to Windows 7	Windows XP to Windows 7	Windows XP to Windows 7	Windows XP tp Windows 7
Miscellaneous						
Pressure speed in *z*	mm/h	12.7	12.7-20.32	n.a.	n.a.	n.a.
Materials		Cream	Cream, Red, Gray, Blue, Black, Zoom	VisiJet SL: Flex, Tough, Clear, Black, HiTemp	VisiJet SL: Flex, Tough, Clear, Black, HiTemp, Jewel	VisiJet SL: Flex, Tough, Clear, Black, HiTemp, Jewel, e-Stone

Table A2.16 Polymer Printing Machines I

RP Process	Unit	Polymer Printing/Jetting		
Machine designation/ Type		EDEN260	EDEN350	EDEN500
Manufacturer		Stratasys Ltd. 2 Holtzman St. Science Park Rehovot, 76124, Israel		
Distribution		http://www.stratasys.com/		
Dimensions, Connection values				
Width/depth/height	m/m/m	0.87/0.735/1.2	1.32/0.99/1.20	1.3/0.99/1.2
Weight	kg	280	410	410
Electrical connection	V/A	230/50; 115/60	230/50; 115/60	230/50; 115/60
Power consumption	kW	1.5	1.5	3
Working gas	l/min	–	–	–
Cooling water	l/min	–	–	–
Extraction	m^3/h	–	–	–
Work room temperature	°C	18–25	18–25	18–25
Relative moisture	%	30–70	30–70	30–70
Process		–	–	–
Fumes, Exhaust		–	–	–
Disposal, Waste		Cured models	Cured models	Cured models
Hazardous waste		Residues from cleaning system	Residues from cleaning system	Residues from cleaning system
Contouring Element				
Kind		InkJet	InkJet	InkJet
Type		–	–	–
x-y Contour generation		–	–	–
x-y Contour accuracy	mm	0.042/0.084	600 dpi	0.042/0.042
Spot repeatability	mm	–	–	–
z Contour accuracy	mm	0.016	0.016/0.030	0.016/0.030
z Contour repeatable accuracy	mm	–	–	–
Coating cycle	s	–	–	–
Coating arm speed	mm/s	–	–	–
Power fluctuation	%	–	–	–
Model characteristics				
Installation space max. (width/depth/height)	mm	258/252/205	350/350/200	500/400/200
Model dimensions max.	mm	256/250/203	340/340/200	490/390/200
Layer thickness	mm	0.016	0.016/0.30	0.016/0.030
Gauge	mm	–	–	–
Accuracy absolute	mm	< 0.1	<0.1	< 0.1
Repeatability	mm	–	–	–
Additional support		Water-soluble support material	Water-soluble support material	Water-soluble support material
EDP Interface				
Interface formats		STL, LAN-TCP/IP	STL, LAN-TCP/IP	STL, LAN-TCP/IP
EDP system		Pentium IV	PC, Workstation	Pentium IV
Operating system		Windows 2000, Win XP	Windows 2000, Win XP	Windows 2000, Win XP
Software		Objet Studio	Objet Studio	Objet Studio

Table A2.17 Polymer Printing Machines II

RP Process	Unit	Polymer Printing/Jetting			
Machine designation/ Type		Connex260	Connex350	Connex500	Objet1000
Manufacturer		Stratasys Ltd. 2 Holtzman St. Science Park Rehovot, 76124, Israel			
Distribution		http://www.stratasys.com/			
Dimensions, Connection values					
Width/depth/height	m/m/m	0.87/0.735/1.2	1.4/1.12/1.13	1.42/1.12/1.13	2.8/1.8/1.8
Weight	kg	264	500	500	1950
Electrical connection	V/A	230/50; 115/60	230/50; 115/60	230/50; 115/60	-
Power consumption	kW	1.5	1.5	1.5	-
Working gas	l/min	-	-	-	-
Cooling water	l/min	-	-	-	-
Extraction	m^3/h	-	-	-	-
Work room temperature	°C	18-25	18-22	18-22	18-22
Relative moisture	%	30-70	30-70	30-70	30-70
Process		-	-	-	-
Fumes, Exhaust		-	-	-	-
Disposal, Waste		Cured models	Cured models	Cured models	Cured models
Hazardous waste		Residues from cleaning system	Residues from cleaning system	Residues from cleaning system	Residues from cleaning system
Contouring Element					
Kind		InkJet	InkJet	InkJet	InkJet
Type		-	-	-	-
x-y Contour generation		-	-	-	-
x-y Contour accuracy	mm	600	600	600	600
Spot repeatability	mm	-	-	-	-
z Contour accuracy	mm	0.016/0.030	0.016/0.030	0.016	0.016/0.030
z Contour repeatable accuracy	mm	-	-	-	-
Coating cycle	s	-	-	-	-
Coating arm speed	mm/s	-	-	-	-
Power fluctuation	%	-	-	-	-
Model characteristics					
Installation space max. (width/depth/height)	mm	260/260/200	350/350/200	500/400/200	1000/800/500
Model dimensions max.	mm	255/252/200	342/342/200	490/390/200	1000/800/500
Layer thickness	mm	0.016/0.030	0.016/0.030	0.016	0.016/0.030
Gauge	mm	-	-	-	-
Accuracy absolute	mm	<0.1	<0.1	<0.1	<0.1
Repeatability	mm	-	-	-	-
Additional support		Water-soluble support material	Water-soluble support material	Water-soluble support material	Water-soluble support material
EDP Interface					
Interface formats		STL, LAN-TCP/IP	STL, LAN-TCP/IP	STL, LAN-TCP/IP	STL, LAN-TCP/IP
EDP system		PC, Workstation	PC, Workstation	PC, Workstation	PC, Workstation
Operating system		Windows XP, Windows Vista, Windows 7	Windows XP, Windows Vista	Windows XP, Windows Vista	Windows 7 x64
Software		Objet Studio	Objet Studio	Objet Studio	Objet Studio

Table A2.18 Polymer Printing Machines III

RP Process	Unit	Polymer Printing/Jetting	
Machine designation/ Type		Objet24	Objet30 Pro
Manufacturer		Stratasys Ltd. 2 Holtzman St. Science Park Rehovot, 76124, Israel	
Distribution		http://www.stratasys.com/	
Dimensions, Connection values			
Width/depth/height	m/m/m	0.825/0.62/0.59	0.825/0.62/0.59
Weight	kg	93	93
Electrical connection	V/A	230/50; 115/60	230/50; 115/60
Power consumption	kW		1.5
Working gas	l/min	-	-
Cooling water	l/min	-	-
Extraction	m^3/h	-	-
Work room temperature	°C	18–25	18–25
Relative moisture	%	30–70	30–70
Process		-	-
Fumes, Exhaust		-	-
Disposal, Waste		Cured models	Cured models
Hazardous waste		Residues from cleaning system	Residues from cleaning system
Contouring Element			
Kind		InkJet	InkJet
Type		-	-
x-y Contour generation		-	-
x-y Contour accuracy	mm	600	600
Spot repeatability	mm	-	-
z Contour accuracy	mm	0.028	0.016/0.028
z Contour repeatable accuracy	mm	-	-
Coating cycle	s	-	-
Coating arm speed	mm/s	-	-
Power fluctuation	%	-	-
Model characteristics			
Installation space max. (width/depth/height)	mm	240/200/150	300/200/150
Model dimensions max.	mm	234/192/148.6	294/192/148.6
Layer thickness	mm	0.028	0.016/0.028
Gauge	mm	-	-
Accuracy absolute	mm	< 0.1	< 0.1
Repeatability	mm	-	-
Additional support		Water-soluble support material	Water-soluble support material
EDP Interface			
Interface formats		STL, LAN-TCP/IP	STL, LAN-TCP/IP
EDP system		PC, Workstation	PC, Workstation
Operating system		Windows XP, Windows 7	Windows XP, Windows 7
Software		Objet Studio	Objet Studio

Table A2.19 Polymer Printing Machines IV

RP Process	Unit	MJM	MJM	MJM	MJM	MJM
Machine designation/ Type		ProJet 3510 SD	ProJet 3510 HD	ProJet 3510 HDPlus	ProJet 3500 HDMax	ProJet 5000
Manufacturer		3D Systems				
Distribution		www.3dsystems.com				
Dimensions, Connection values						
Width/depth/height	m/m/m	0.75/1.2/1.51	0.75/1.2/1.51	0.75/1.2/1.51	0.75/1.2/1.51	1.53/0.91/1.45
Weight	kg	323	323	323	323	538
Electrical connection	V/A	100-127/15A; 200-240/10A	100-127/15A; 200-240/10A	100-127/15A; 200-240/10A	100-127/15A; 200-240/10A	115-240/k.a
Capacity	KW	n. a.	n. a.	n. a.	n. a.	1.2
Work room temperature	°C	18-28	18-28	18-28	18-28	18-28
Model characteristics						
Installation space max. (width/depth/height)	mm	298/185/203	298/185/203	298/185/203	298/185/203	550/393/300
Model dimensions max.	mm	298/185/203	298/185/203	298/185/203	298/185/203	550/393/300
Layer thickness	mm	0.032	0.029-0.032	0.016-0.032	0.016-0.032	0.029-0.064
Accuracy absolute	mm	0.025-0.05	0.025-0.05	0.025-0.05	0.025-0.05	0.025-0.05
Max. resolution	dpi *x*/*y*/*z*	375/375/790	750/750/890	750/750/1600	750/750/1600	750/750/890
EDP Interface						
Interface formats		STL, SLC	STL, SLC	STL, SLC	STL, SLC	STL, SLC
EDP system		PC	PC	PC	PC	PC
Operating system		Windows XP, Vista, 7	Windows XP, Vista, 7	Windows XP, Vista, 7	Windows XP, Vista, 7	Windows XP, Vista, 7
Miscellaneous						
Building materials/Supports		VisiJet: X, Crystal, Proplast, Navy, Techplast/ VisiJet S300	VisiJet: X, Crystal, Proplast, Navy, Techplast, Procast/VisiJet S300	VisiJet: X, Crystal, Proplast, Navy, Techplast, Procast/VisiJet S300	VisiJet: X, Crystal, Proplast, Navy, Techplast, Procast/VisiJet S300	

Table A2.20 Polymer Printing Machines V

RP Process	Unit	MJM	MJM	MJM	MJM
Machine designation/ Type		ProJet 3510 CP	ProJet 3510 CPX	ProJet 3510 CPCPlus	ProJet 3500 CPXMax
Area of application		RealWax process for precision casting models			
Manufacturer		3D Systems			
Distribution		www.3dsystems.com			
Dimensions, Connection values					
Width/depth/height	m/m/m	0.83/1.43/1.74	0.83/1.43/1.74	0.83/1.43/1.74	0.83/1.43/1.74
Weight	kg	323	323	323	323
Electrical connection	V/A	100-127/15A; 200-240/10A	100-127/15A; 200-240/10A	100-127/15A; 200-240/10A	100-127/15A; 200-240/10A
Capacity	KW	n. a.	n. a.	n. a.	n. a.
Work room temperature	°C	18–28	18–28	18–28	18–28
Model characteristics					
Installation space max. (width/depth/height)	mm	298/185/203	298/185/203	298/185/203	298/185/203
Model dimensions max.	mm	298/185/203	298/185/203	298/185/203	298/185/203
Layer thickness	mm	0.033	0.016-0.033	0.016-0.033	0.016-0.033
Accuracy absolute	mm	0.025-0.05	0.025-0.05	0.025-0.05	0.025-0.05
Max. resolution	dpi	375 × 375 × 775	694 × 750 × 1600	694 × 750 × 1600	694 × 750 × 1600
EDP Interface					
Interface formats		STL, SLC	STL, SLC	STL, SLC	STL, SLC
EDP system		PC	PC	PC	PC
Operating system		Windows XP, Vista, 7	Windows XP, Vista, 7	Windows XP, Vista, 7	Windows XP, Vista, 7
Miscellaneous					
Building materials/ Supports		VisiJet Prowax/ VisiJet S400	VisiJet Hi-Cast/ VisiJet S400	VisiJet Hi-Cast/ VisiJet S400	VisiJet Hi-Cast/ VisiJet S400

Table A2.21 Polymer Printing Machines VI

RP Process	Unit	MJM	MJM
Machine designation/ Type		ProJet 3510 DP	ProJet 3510 MP
Area of application		Medical and dental Applications	
Manufacturer		3D Systems	
Distribution		www.3dsystems.com	
Dimensions, Connection values			
Width/depth/height	m/m/m	0.75/1.2/1.51	0.75/1.2/1.51
Weight	kg	323	323
Electrical connection	V/A	100-127/15A; 200-240/10A	100-127/15A; 200-240/10A
Capacity	KW	n. a.	n. a.
Work room temperature	°C	18-28	18-28
Model characteristics			
Installation space max. (width/depth/height)	mm	298/185/203	298/185/203
Model dimensions max.	mm	298/185/203	298/185/203
Layer thickness	mm	0.029-0.032	0.032
Accuracy absolute	mm	0.025-0.05	0.025-0.05
Max. resolution (*x*/*y*/*z*)	dpi	750/750/890	375/375/790
EDP Interface			
Interface formats		STL, SLC	STL, SLC
EDP system		PC	PC
Operating system		Windows XP, Vista, 7	Windows XP, Vista, 7
Miscellaneous			
Building materials/ Supports		VisiJet Dentcast/ VisiJet S300	VisiJet PearlStone, Stoneplast/ VisiJet S300

Table A2.22 Sintering Machines I

RP Process	Unit	SLM	SLM
Machine designation/ Type		SPro125	SPro250
Manufacturer		3D Systems	
Distribution		www.3dsystems.com	
Dimensions, Connection values			
Width/depth/height	m/m/m	1.35/0.8/1.9	1.7/0.8/2.03
Weight	kg	900	1100
Electrical connection	V/A	208/30 3-phase	208/30 3-phase
Power consumption	kW	n. a.	n. a.
Work room temperature	°C	n. a.	n. a.
Relative moisture	%	n. a.	n. a.
Contouring Element			
Kind		Laser	Laser
Laser capacity	W	100; 200	200; 400
Type		n. a.	n. a.
Max. scan speed	mm/s	1000	1000
Build speed	cm^3/h	5–20	5–20
Model characteristics			
Installation space max. (width/depth/height)	mm	125/125/125	250/250/320
Model dimensions max.	mm	125/125/125	250/250/320
Layer thickness	mm	0.02–0.1	0.02–0.1
Spot (laser diameter)	mm	0.035	0.070

Table A2.23 Sintering Machines II

RP Process	Unit	SLS	SLS	SLS	SLS	SLS
Machine designation/ Type		sPro 60 SD	sPro 60 HD Base	sPro 60 HD-HS	sPro 140 Base	sPro 140 HS
Manufacturer		3D Systems				
Distribution		www.3dsystems.com				
Dimensions, Connection values						
Width/depth/height	m/m/m	n. a.	n. a.	n. a.	n. a.	n. a.
Weight	kg	n. a.	n. a.	n. a.	n. a.	n. a.
Electrical connection	V/A	240/n. a. 3-phase	240/n. a. 3-phase	240/n. a. 3-phase	208/n. a. 3-phase	208/n. a. 3-phase
Power consumption	kW	12.5	12.5	12.5	17	17
Working gas	l/min	54	60	108	180	300
Contouring Element						
Kind		Laser	Laser	Laser	Laser	Laser
Type		CO_2	CO_2	CO_2	CO_2	CO_2
Imaging system		High Torque Scanning Motors (analog)	ProScan™ CX (digital)	ProScan™ DX Dual Mode	ProScan™ DX Digital	ProScan™GX Dual Mode
Scanning speed	m/s	5	6	6; 12	10	15
Laser capacity	W	30	30	70	70	200
Model characteristics						
Installation space max. (width/depth/height)	mm	381/330/457	381/330/457	381/330/457	550/550/460	550/550/460
Model dimensions max.	mm	381/330/457	381/330/457	381/330/457	550/550/460	550/550/460
Layer thickness	mm	0.08–0.15	0.08–0.15	0.08–0.15	0.08–0.15	0.08–0.15

Table A2.24 Sintering Machines III

RP Process	Unit	SLS	SLS
Machine designation/ Type		sPro 230 Base	sPro 230 HS
Manufacturer		3D Systems	
Distribution		www.3dsystems.com	
Dimensions, Connection values			
Width/depth/height	m/m/m	n.a.	n.a.
Weight	kg	n.a.	n.a.
Electrical connection	V/A	208/n.a. 3-phase	208/n.a. 3-phase
Power consumption	kW	17	17
Working gas	l/min	180	300
Cooling water	l/min	–	–
Extraction	m^3/h	–	–
Work room temperature	°C	n.a.	n.a.
Moisture	%	n.a.	n.a.
Contouring Element			
Kind		Laser	Laser
Type		CO_2	CO_2
Imaging system		ProScan™ DX Digital	ProScan™ GX Dual Mode High Speed Digital
Scanning speed	m/s	10	15
Laser capacity	W	70	200
Model characteristics			
Installation space max. (width/depth/height)	mm	550/550/750	550/550/750
Model dimensions max.	mm	550/550/750	550/550/750
Layer thickness	mm	0.08–0.15	0.08–0.15

Table A2.25 Sintering Machines IV

RP Process	Unit	Metal Laser Sintering (DMLS)	Plastic Sintering	Plastic Sintering	Sand Sintering (Direct Cast)
Machine designation/ Type		EOSINT M 280	EOSINT P 395	EOSINT P 760	EOSINT S 750
Manufacturer		EOS GmbH Electro Optical Systems Robert-Stirling-Ring 1 82152 Krailling / München Germany			
Distribution		www.eos-gmbh.de			
Dimensions, Connection values					
Width/depth/height	m/m/m	2.2/1.07/2.29	1.84/1.175/2.1	2.25/1.55/2.10	1.42/1.40/2.15
Weight	kg	1250	ca. 1060	ca. 2300	ca. 1050
Electrical connection	V/A	400/32	400/32	400/32	400/32
Power consumption	kW	Max 8.5 Typical 3.2	Max 10 Typical 2.4	Max. 12 Typical 3.1	Max. 12 Typical 6
Working gas	l/min	Compressed air 20 m^3/h, argon 100 l/min	Compressed air min. 6 m^3/h	Compressed air min. 20 m^3/h	Compressed air min. 15 m^3/h
Cooling water	l/min	Depending on the option, not necessary	0	0	0
Extraction	m^3/h				

RP Process	Unit	Metal Laser Sintering (DMLS)	Plastic Sintering	Plastic Sintering	Sand Sintering (Direct Cast)
Work room temperature	°C	15–30	19–30	15–30	19–29
Relative moisture	%	80 (at 15–20 °C)	max. 50	max. 60	20–50
Process					
Fumes, Exhaust		low	low	low	low
Disposal, Waste		Remote support structures	Powder remains		Dried sand
Hazardous waste		Metal powder remains	–	–	Phenolic resin
Contouring Element					
Kind		Laser	Laser	Laser	Laser
Type		Yb fiber laser; 200 W or 400 W	CO_2; 50 W	CO_2; 2 × 50 W	CO_2; 2 × 100 W
x-y Contour generation		Galvanometer scanner	F-Theta Lens	F-Theta Lens	Galvoscanner
x-y Contour accuracy	mm	–	0.01	0.01	0.01
Spot repeatability	mm	0.005	± 0.51	± 0.51	± 0.51
z Contour accuracy	mm	–	0.1	0.1	0.1
z Contour repeatable accuracy	mm	0.005	–	–	–
Coating cycle	s	Typ. 12 s	variable	variable	variable
Coating arm speed	mm/s	Variable, max. 7000	Variable, max. 6000	Variable, max. 2 × 6000	120
Power fluctuation	%	10–100			
Model characteristics					
Installation space max. (width/depth/height)	mm	250/250/325	340/340/625	700/380/580	718/381/381
Model dimensions max.	mm	250/250/325	350/350/625	700/380/580	718/381/381
Layer thickness	mm	0.02–0.1	0.06–0.18	0.06–0.18	0.2
Gauge	mm	min. ca. 0.3			
Accuracy absolute	mm	Typ. 0.05	0.1–0.2% (Typ.)	0.1–0.2% (Typ.)	± 0.3
Repeatability	mm	–			
Additional support		for overhang > ca. 60° (depending on material)	none	none	none
EDP Interface					
Interface formats		STL, converter to all current format	STL, CLI	STL, converter to all current format	STL, CLI
EDP system		STL, CLI			
Operating system		Workstation or PC	Current Windows system	Current Windows system	
Software		EOS RP Tools		EOS RP Tools	

Table A2.26 Sintering and Melting Machines V

RP Process	Unit	SLM	SLM	SLM	Laser-cusing	Laser-cusing	Laser-cusing	Laser-cusing
Machine designation/ Type		Realizer SLM250	Realizer SLM100	Realizer SLM50	Mlab cusing	M1 cusing	M2 cusing	Xline 1000R
Manufacturer		Realizer GmbH			Concept Laser GmbH			
Distribution		www.realizer.com			Directly by Concept Laser			
Dimensions, Connection values								
Width/depth/height	m/m/m	1.8/ 1/ 2.2	0.9/ 0.8/ 2.4	0.8/ 0.7/ 0.5	0.705/ 1.833/ 0.955	1.990/ 1.790/ 1.775	2.44/ 1.63/ 1.992	4.415/ 3.9-4.5/ 3.070
Weight	kg	800	500	80	500 kg	1700 kg	1500	8000
Electrical connection	V/A	400/16	400/16	230/16	230/16	400/16	400/32	400/63
Power consumption	kW	2.5	1.5	1.0	1.5 kW	7.4 kW		13
Working gas	l/min	0.7	0.6	0.5	< 4; Ar or N_2	1.5–2.0 m^3/h; N_2	0.6–0.8 m^3/h	17–34; N_2
Cooling water	l/min	internal		none		Internal closed circulation		
Extraction	m^3/h	yes		no	intern	intern	intern	
Work room temperature	°C	RT			15–35 °C	15–35 °C	RT	
Relative moisture	%	45			n. a.	n. a.		
Process								
Fumes, Exhaust		none		none		Inside activated carbon filter		
Disposal, Waste		–				Waste		
Hazardous waste		–				no		
Contouring Element								
Kind		Laser	Laser	Laser	Laser	Laser	Laser	Laser
Type		Nd-YAG; 200 or 400 W	Fiber laser; 200 W	Fiber laser, 100 W	NdYAG, diode-pump, 100 W	NdYAG, diode-pump 200 W	Fiber laser, 200 W or 400 W	Fiber laser
x-y Contour generation		Scanner			Scanner head and linear axes	Scanner head	Galvo-scanner	
x-y Contour accuracy	mm	± 0.03				±50 micrometer		-
Spot repeatability	mm	0.002				n. a.		-
z Contour accuracy	mm	Layer thickness ± 0.05				±50 micrometer		-
z Contour repeatable accuracy	mm	Layer thickness ± 0.05				n. a.		-
Coating cycle	s	–				3–5		
Coating arm speed	mm/s	0–1000				n. a.		
Power fluctuation	%	± 2				Self-controlled system		

RP Process	Unit	SLM	SLM	SLM	Laser-cusing	Laser-cusing	Laser-cusing	Laser-cusing
Model characteristics								
Installation space max. (width/depth/height)	mm	250/ 250/ 300		Diameter: 70, Height: 40	$x = y =$ 50 × 50 70 × 70, 90 × 90; $z = 80$	150/ 220/ 200	250/ 250/ 280	630/ 400/ 500
Model dimensions max.	mm	250/ 250/ 300		Diameter: 70, Height: 40	as above	120/ 120/ 200	250/ 250/ 280	630/ 400/ 500
Layer thickness	mm	0.02-0.1	0.02-0.1	0.02-0.1	0.02-0.05	20-100 micro-meter	0.02-0.05	0.03-0.2
Gauge	mm	0.15-0.3				ca. 250 micro-meter		
Accuracy absolute	mm	± 0.1				see above		
Repeatability	mm	± 0.1			.	n. a.		
Additional support		none				dependent on the geometry		
EDP Interface								
Interface formats		F&S	F&S	F&S	STL, AMF	STL, AMF	STL, AMF	STL, AMF
EDP system		PC	PC	PC	PC	PC	PC	PC
Operating system		Windows	Windows	Windows	Windows	Windows	Windows	Windows
Software		Realizer	Realizer	Realizer	Materialise Magics	Materialise Magics	Materialise Magics	

Table A2.27 Sintering and Melting Machines VI

RP Process	Unit	Laser melting	Laser melting	Beam melting	Beam melting	Beam melting	Beam melting
Machine designation/ Type		AM125	AM250	SLM® 125 HL	SLM® 250 HL	SLM® 500 HL	SLM® 280 HL
Manufacturer		Renishaw GmbH Karl-Benz Straße 12 72124 Pliezhausen Deutschland (Germany) www.renishaw.com		SLM Solutions GmbH Roggenhorster Straße 9c 23556 Lübeck (Germany) www.slm-solutions.com			
Distribution		Directly by Renishaw		Directly by SLM Solutions			
Dimensions, Connection values							
Width/depth/height	m/m/m	1.35/0.8/ 1.9	1.7/0.8/ 2.025	1.35/0.8/ 1.9(2.4)	1.650/1.0/ 1.9(2.4)	3.0/1.5/ 2.0(2.5)	1.8/1.0/ 1.9(2.4)
Weight	kg	900	1100	700	850	2000	1000
Electrical connection	V/A	230/16	230/16	400/32	400/32	400/32	400/32
Power consumption	kW			4	5	8	6
Working gas	l/min			0.5 (Flushing: 1000 l); Ar or N_2	1.5 (Flushing: 1500 l); Ar or N_2	3.0 (Flushing: 2000 l); Ar or N_2	2.5 (Flushing: 1700 l); Ar or N_2
Cooling water	l/min						
Extraction	m^3/h						
Work room temperature	°C						
Relative moisture	%						
Process							
Fumes, Exhaust							
Disposal, Waste							
Hazardous waste							
Contouring Element							
Kind		Laser	Laser	Laser	Laser	Laser	Laser
Type		100 W or 200 W	200 W or 400 W	YLR fiber laser; 100 W or 200 W	YLR fiber laser; 200 W or 400 W	2 × YLR fiber laser; 1 × 400 W, 1 × 1000 W	YLR fiber laser; 1 × 400 W, 1 × 1000 W
x-y Contour generation							
x-y Contour accuracy	mm						
Spot repeatability	mm						
z Contour accuracy	mm						
z Contour repeatable accuracy	mm						
Coating cycle	s						
Coating arm speed	mm/s						
Power fluctuation	%						
Model characteristics							
Installation space max. (width/depth/height)	mm	120/120/ 125	245/245/ 300 (*z* = 360 mm on demand)	125/125/ 75 (125)	248/248/ 250 (350)	500/280/ 325	280/280/ 350
Model dimensions max.	mm	120/120/ 125	245/245/ 300 (*z* = 360 mm on demand)	125/125/ 75 (125)	248/248/ 250 (350)	500/280/ 325	280/280/ 350
Layer thickness	mm	0.02–0.1	0.02–0.1	0.02–0.075	0.02–0.075	0.02–0.2	0.02–0.15
Gauge	mm			min. 0.140	min. 0.150	min. 0.180	min. 0.2
Accuracy absolute	mm						

RP Process	Unit	Laser melting	Laser melting	Beam melting	Beam melting	Beam melting	Beam melting
Repeatability	mm						
Additional support							
EDP Interface							
Interface formats		STL, AMF	STL, AMF				
		PC	PC	PC	PC	PC	PC
		Windows	Windows	Windows	Windows	Windows	Windows
		Materialise, Marcam Autofab	Materialise, Marcam Autofab	AutoFabMC	AutoFabMC	AutoFabMC	AutoFabMC

Table A2.28 Sintering Machines VII

RP Process	Unit	Electron beam melting (EBM)	Electron beam melting (EBM)	SLM	SLM	SLM
Machine designation/ Type		**Q10**	**A2**	**PXS**	**PXM**	**PXL**
Manufacturer		Arcam AB Krokslätts Fabriker 27A SE 431 37 Mölndal Sweden		Phenix Systems Parc Européen d'Entreprises Rue Richard Wagner 63200 Riom Frankreich, France		
Distribution		Directly by ARCAM				
Dimensions, Connection values						
Width/depth/height	m/m/m	1.85/0.9/2.2	1.85/0.9/2.2	1.2/0.77/1.95	1.2/1.5/1.95	2.4/2.2/2.4
Weight	kg	1420	1420	1000	1500	5000
Electrical connection	V/A	3 × 400/32	3 × 400/32	230	400	400
Power consumption	kW	7	7	3 kVA	10 kVA	15 kVA
Working gas	l/min	1 l/h; He 50–75 l cooldown		6–8 bar compressed air	6–8 bar compressed air	6–8 bar compressed air
Cooling water	l/min					
Extraction	m^3/h					
Work room temperature	°C					
Relative moisture	%					
Process						
Fumes, Exhaust						
Disposal, Waste						
Hazardous waste						
Contouring Element						
Kind		Electron beam	Electron beam	Laser	Laser	Laser
Type		Single crystalline CeB_6		Fiber laser; 50 W; 1070 nm	Fiber laser; 300 W; 1070 nm	Fiber laser; 500 W; 1070 nm
x-y Contour generation						
x-y Contour accuracy	mm		0.13–0.2			
Spot repeatability	mm					
z Contour accuracy	mm					
z Contour repeatable accuracy	mm					
Coating cycle	s					
Coating arm speed	mm/s					
Power fluctuation	%					

Table A2.28 *(continued)* Sintering Machines VII

RP Process	Unit	Electron beam melting (EBM)	Electron beam melting (EBM)	SLM	SLM	SLM
Model characteristics						
Installation space max. (width/depth/height)	mm	200/200/180	250/250/400 or 350 × 350 × 250	100/100/80	140/140/100	100/100/80
Model dimensions max.	mm	200/200/180	200/200/350 or Ø300/200	100/100/80		
Layer thickness	mm					
Min. spot diameter	mm	0.1				
Accuracy absolute	mm					
Repeatability	mm			$x = 0.02$ $y = 0.02$ $z = 0.02$	$x = 0.02$ $y = 0.02$ $z = 0.02$	
Additional support						
EDP Interface						
Interface formats		STL	STL	STL, IGES, STEP	STL, IGES, STEP	STL, IGES, STEP
EDP system		PC	PC	PC	PC	PC
Operating system				Windows	Windows	Windows
Software				Phenix Processing, Phenix Manufacturing, Phenix Dental	Phenix Processing, Phenix Manufacturing, Phenix Dental	Phenix Processing, Phenix Manufacturing, Phenix Dental

Table A2.29 Extrusion Machines I

RP Process	Unit	Extrusion	Extrusion	Extrusion	Extrusion
Machine designation/ Type		Mojo 3D Printer	uPrint SE / Plus	Dimension 1200es	Dimension Elite
Manufacturer		Stratasys Inc., USA			
Distribution		www.stratasys.com			
Dimensions, Connection values					
Width/depth/height	m/m/m	0.63/0.45/0.53	0.635/0.66/0.787	0.838/0.737/1.143	0.686/0.914/1.041
Weight	kg	27	76	148	136
Electrical connection	V/A	120/6; 230/2.5	120/15; 230/7	120/15; 230/7	120/15; 230/7
Power consumption	kW	n. a.	n. a.	n. a.	n. a.
Working gas	l/min	–	–	–	–
Cooling water	l/min	–	–	–	–
Extraction	m^3/h	–	–	–	–
Work room temperature	°C	n. a.	n. a.	n. a.	n. a.
Relative moisture	%	n. a.	n. a.	n. a.	n. a.
Process					
Fumes, Exhaust		–	–	–	–
Disposal, Waste		yes	yes	yes	yes
Hazardous waste		–	–	–	–
Contouring Element					
Kind		Extruder	Extruder	Extruder	Extruder
Type					
x-y Contour generation		Belt drive *xy*	Belt drive *xy*	Belt drive *xy*	Belt drive *xy*
x-y Contour accuracy	mm	n. a.	n. a.	n. a.	n. a.
Spot repeatability	mm	–	–	–	–
z Contour accuracy	mm	n. a.	n. a.	n. a.	n. a.
z Contour repeatable accuracy	mm	n. a.	n. a.	n. a.	n. a.
Coating cycle	s	0	0	0	0
Coating arm speed	mm/s	–	–	–	–
Power fluctuation	%	0	0	0	0
Model characteristics					
Installation space max. (width/depth/height)	mm	127/127/127	203/152/152	254/254/305	203/203/305
Model dimensions max.	mm	127/127/127	203/152/152	254/254/305	203/203/305
Layer thickness	mm	0.178	0.254	0.33; 0.254	0.254; 0.178
Gauge	mm	n. a.	n. a.	n. a.	n. a.
Accuracy absolute	mm	± n. a.	± n. a.	± n. a.	± n. a.
Repeatability	mm	n. a.	n. a.	n. a.	n. a.
Additional support		yes	yes	yes	yes
EDP Interface					
Interface formats		STL	STL	STL	STL
EDP system		PC, Workstation	PC, Workstation	PC, Workstation	PC, Workstation
Operating system		Windows XP, 7	Windows XP, 7	Windows XP, 7	Windows XP, 7
Software		Print Wizard, Control Panel	CatalystEX	CatalystEX	CatalystEX

Table A2.30 Extrusion Machines II

RP Process	Unit	Extrusion	Printing Extrusion	Extrusion	Extrusion
Machine designation/ Type		Fortus 250mc	Fortus 360mc	Fortus 400mc	Fortus 900mc
Manufacturer		Stratasys Inc., USA			
Distribution		www.stratasys.com			
Dimensions, Connection values					
Width/depth/height	m/m/m	0.838/0.737/1.143	1.281/0.896/1.962	1.281/0.896/1.962	2.772/1.683/2.027
Weight	kg	186	786	786	3287
Electrical connection	V/A	120/15; 230/7	230/16/3-phase	230/16/3-phase	230/40
Power consumption	kW	n. a.	n. a.	n. a.	n. a.
Working gas	l/min	–	–	–	–
Cooling water	l/min	–	–	–	–
Extraction	m^3/h	–	–	–	–
Work room temperature	°C	30	n. a.	n. a.	n. a.
Relative moisture	%	30–70	n. a.	n. a.	n. a.
Process					
Fumes, Exhaust		–	–	–	–
Disposal, Waste		yes	yes	yes	yes
Hazardous waste		–	–	–	–
Contouring Element					
Kind		Extruder	Extruder	Extruder	Extruder
Type					
x-y Contour generation		Belt drive *xy*	Belt drive *xy*	Belt drive *xy*	Belt drive *xy*
x-y Contour accuracy	mm	n. a.	n. a.	n. a.	n. a.
Spot repeatability	mm	–	–	–	–
z Contour accuracy	mm	n. a.	n. a.	n. a.	n. a.
z Contour repeatable accuracy	mm	n. a.	n. a.	n. a.	n. a.
Coating cycle	s	0	0	0	0
Coating arm speed	mm/s	–	–	–	–
Power fluctuation	%	0	0	0	0
Model characteristics					
Installation space max. (width/depth/height)	mm	254/254/305	406/355/406	406/355/406	914/610/914
Model dimensions max.	mm	254/254/305	406/355/406	406/355/406	914/610/914
Layer thickness	mm	0.33; 0.254; 0.178	0.33; 0.254; 0.178; 0.127	0.33; 0.254; 0.178; 0.127	0.33; 0.254; 0.178
Gauge	mm	n. a.	n. a.	n. a.	n. a.
Accuracy absolute	mm	± 0.241	± 0.127	± 0.127	± 0.09
Repeatability	mm	n. a.	n. a.	n. a.	n. a.
Additional support		yes	yes	yes	yes
EDP Interface					
Interface formats		STL	STL	STL	STL
EDP System		PC, Workstation	PC, Workstation	PC, Workstation	PC, Workstation
Operating system		n. a.	n. a.	n. a.	n. a.
Software		Vorhanden Insight	Vorhanden Insight	Vorhanden Insight	Vorhanden Insight

Table A2.31 Commercial 3D Printer: "Fabber" I

RP Process	Unit			
Machine designation/ Type		The Replicator™	The Replicator™ 2	The Replicator™ 2X
Manufacturer		Makerbot Objet® Industries LLC		
Distribution		http://www.makerbot.com/		
Dimensions, Connection values				
Width/depth/height	m/m/m	320/467/381	490/420/380	490/420/531
Weight	kg	14.5	11.5	12.6
Kit		-	-	-
Electrical connection	V	100-240	100-240	100-240
Power consumption	kW	n. a.	n. a.	n. a.
Working gas	l/min	-	-	-
Cooling water	l/min	-	-	-
Extraction	m^3/h	-	-	-
Work room temperature	°C	-	n. a.	n. a.
Build platform temperature max.	°C	120	-	180
Relative moisture	%	-	-	n. a.
Process				
Fumes, Exhaust		-	-	-
Disposal, Waste		yes	yes	yes
Hazardous waste		-	-	-
Contouring Element				
Kind		Extruder	Extruder	Extruder
Type				
x-y Contour generation		Belt drive *xy*	Belt drive *xy*	Belt drive *xy*
x-y Contour accuracy	mm	0.011	0.011	0.011
Spot repeatability	mm	-	-	-
z Contour accuracy	mm	0.0025	0.0025	0.0025
z Contour repeatable accuracy	mm	n. a.	n. a.	n. a.
Coating cycle	s	-	-	-
Pressure speed	mm/s	40	n. a.	n. a.
Power fluctuation	%	n. a.	n. a.	0
Nozzle diameter	mm	0.4	0.4	0.4
Model characteristics				
Installation space max. (width/depth/height)	mm	225/145/150	285/153/155	250/160/150
Model Dimensions max.	mm	-	-	-
Layer thickness	mm	0.2-0.3	0.1-0.34	0.1-0.34
Minimum wall thickness	mm	n. a.	n. a.	n. a.
Accuracy absolute	mm	± n. a.	± n. a.	± n. a.
Repeatability	mm	n. a.	n. a.	n. a.
Additional support/ Extruder		no	no	no
Material form		Filament	Filament	Filament
Filament diameter	mm	1.75	1.75	1.75

Table A2.31 *(continued)* Commercial 3D Printer: "Fabber" I

RP Process	Unit			
EDP Interface				
Interface formats		STL, GCode	STL, Obj., thing	STL, Obj., thing
EDP system		PC, Workstation	PC, Workstation	PC, Workstation
Connectivity		USB, SD Card	USB, SD card (max. 2 GB)	USB, SD card (included)
Operating system		Linux, OSX, Windows	Windows (XP/7), Mac OS X (10.6/10.7/10.8), Linux (Ubuntu 12.04+)	Windows (XP/7), Mac OS X (10.6+), Linux (Ubuntu 11, 10+) 7
Software		ReplicatorG™	MakerBot Makerware™ Bandle 1.0	MakerBot Makerware™ 2.0

Table A2.32 Commercial 3D Printer: "Fabber" II

RP Process	Unit			
Machine designation/ Type		Fabbster	Cube	Delta Tower
Manufacturer		Sinteringmask GmbH	3D Systems	3DreamFactory
Distribution		http://www.fabbster.de	www.3dsystems.com	www.3dreamfactory.com
Dimensions, Connection values				
Width/depth/height	m/m/m	590/470/540	260/260/340	d55/d55/124
Weight	kg	n. a.	4.3	22
Electrical connection	V	110–240	n. a.	85–264
Power consumption	kW	0.35	n. a.	0.3
Working gas	l/min	–		–
Cooling water	l/min	–	–	–
Extraction	m^3/h	–	–	–
Work room temperature	°C	–	n. a.	–
Build platform temperature max.	°C	n. a.	n. a.	n. a.
Relative moisture	%	–	–	–
Fumes, Exhaust		–	–	–
Disposal, Waste		yes	yes	yes
Hazardous waste		–	–	–
Contouring Element				
Kind		Extruder	Extruder	Extruder
Type				
x-y Contour generation		Belt drive *xy*	Belt drive *xy*	Delta Kinematik
x-y Contour accuracy	mm	n. a.	n. a.	n. a.
Spot repeatability	mm	–	–	–
z Contour accuracy	mm	n. a.	n. a.	n. a.
z Contour repeatable accuracy	mm	n. a.	n. a.	n. a.
Coating cycle	s	–	–	–
Pressure speed	mm/s	600	n. a.	n. a.
Power fluctuation	%	n. a.	n. a.	n. a.
Nozzle diameter	mm	0.4	n. a.	0.4/0.5
Model characteristics				
Installation space max. (width/depth/height)	mm	225/225/210	n. a.	200/200/580, 350/350/450
Model Dimensions max.	mm	–	140/140/140	–

RP Process	Unit			
Layer thickness	mm	0.044-0.176	0.2	0.1
Minimum wall thickness	mm	0.6	n. a.	n. a.
Accuracy absolute	mm	± n. a.	± n. a.	± n. a.
Repeatability	mm	n. a.	n. a.	n. a.
Additional support / Extruder		no	yes	no
Material form		Sticks	ABS-, PLA-CKindridges	Filament
Filament diameter	mm	-	-	1.75, 3
EDP Interface				
Interface formats		STL, GCode	n. a.	GCode
EDP system		PC, Workstation	PC, Workstation	PC, Workstation
Connectivity		USB	USB, Wireless (WiFi)	USB
Operating system		Windows (XP/7/8), Mac OS X	Windows (XP with SP 3/7), Mac OS X 10.8	Windows, Mac, Linux
Software		Netfabb Engine	Cube Software	Open Source (Marlin, Cura/Slic3r, Repetier Host)

Table A2.33 Commercial 3D Printer: "Fabber" III

RP Process	Unit	FDM	FDM	FDM
Machine designation/ Type		Rapman 3.1	Cube	CubeX
Manufacturer		3D Systems		
Distribution		www.3dsystems.com		
Dimensions, Connection values				
Width/depth/height	m/m/m	0.65/0.57/0.82	0.26/0.26/0.34	0.52/0.52/0.6
Weight	kg	17	4.3	36-38
Electrical connection	V/A	12/5	n. a./n. a.	110-240/n. a.
Capacity	KW	0.06	n. a.	n. a.
Work room temperature	°C	n. a.	n. a.	n. a.
Model characteristics				
Installation space max. (width/depth/height)	mm	270/205/210	140/140/140	275/265/240
Model dimensions max.	mm	270/205/210	140/140/140	275/265/240
Layer thickness	mm	0.125	0.2	0.1
Accuracy absolute	mm	+/- 0.2	n. a.	+/- 0.2
EDP Interface				
Interface formats		STL	CUBE	STL
EDP system		PC	PC	PC
Operating system		n. a.	Windows XP, Vista, 7 and Mac OS 10.8	Windows XP, Vista, 7
Miscellaneous				
Materials		n. a.	PLA/ABS or Tough recyclable or compostable plastic	PLA/ABS/Dissolvable natural PLA

Table A2.34 Layer Laminate Machines I

RP Process	Unit	LOM	LOM	PLT	PLT
Machine designation/ Type		LOM-1015P	LOM-2030H (TÜV/GS)	PLT-20 KATANA	PLT-A4
Manufacturer		Helisys Inc., USA		KIRA Europe GmbH	
Distribution		http://cubictechnologies.com/		http://www.kiracorp.co.jp	
Dimensions, Connection values					
Width/depth/height	m/m/m	1.1/1.0/1.47	2.08/1.47/1.42	320	800/920/1563
Weight	kg	750	1250	100/15	450
Electrical connection	V/A	230/16	230/32		100 V
Power consumption	kW	1.5–2.5	3–5		
Working gas	l/min	–	–		–
Cooling water	l/min	no	yes (internal)		–
Extraction	m^3/h	yes	yes	10–30	–
Work room temperature	°C	RT	RT	ca. 60	10–30
Relative moisture	%	40–50 (preferably)			35–75
Process					
Fumes, Exhaust		yes	yes		–
Disposal, Waste		yes	yes		yes
Hazardous waste		–	–		–
Contouring Element					
Kind		Laser	Laser	Cutting blade	Cutting blade
Type		CO_2	CO_2		
x-y Contour generation		Plotter	Plotter		Plotter
x-y Contour accuracy	mm	0.025	0.025		± 0.025
Spot repeatability	mm	–	–		
z Contour accuracy	mm	Layer thickness 0.05–0.2		0.1	
z Contour repeatable accuracy	mm	Layer thickness	Layer thickness		0.1
Coating cycle	s	–	–		
Coating arm speed	mm/s	130–200	130–200		
Power fluctuation	%	± 2	± 2		
Model characteristics					
Installation space max. (width/depth/height)	mm	254/380/355	559/812/508	280/180/150	297/210/200
Model dimensions max.	mm	250/376/355	555/808/508		280/190/200
Layer thickness	mm	0.05–0.2	0.05–0.2		0.08
Gauge	mm	arbitrary	arbitrary		
Accuracy absolute	mm	± 0.15	± 0.15		± 0.2
Repeatability	mm	± 0.05	± 0.05		± 0.025
Additional support		no	no		
EDP Interface					
Interface formats		STL	STL	STL	STL
EDP system		PC	PC		PC
Operating system		Win NT	Win NT		Win NT/95
Software		LOMSlice	LOMSlice		RP-CAD

Table A2.35 Layer Laminate Machines II

RP Process	Unit	RPS	RPS	RPS
Machine designation/ Type		ZIPPY I	ZIPPY II	ZIPPY III
Manufacturer		KINERGY-NT, Nantong		
Distribution		-		
Dimensions, Connection values				
Width/depth/height	m/m/m	1730/1000/1600	2570/1860/2000	2100/1500/1800
Weight	kg	800	1.5 K	2.5 K
Electrical connection	V/A	220/20	380/30	380/25
Power consumption	kW	4.4	11.4	9.5
Working gas	l/min			
Cooling water	l/min	7.50	7.50	7.5
Extraction	m^3/h	9	12	12
Work room temperature	°C	20-28	20-28	20-28
Relative moisture	%	60	60	60
Process				
Fumes, Exhaust				
Disposal, Waste				
Hazardous waste				
Contouring Element				
Kind		Laser	Laser	Laser
Type		CO_2	CO_2	CO_2
x-y Contour generation		CKindesian Robot	CKindesian Robot	CKindesian Robot
x-y Contour accuracy	mm	0.01	0.01	0.01
Spot repeatability	mm	0.01	0.01	0.01
z Contour accuracy	mm	0.01	0.01	0.01
z Contour repeatable accuracy	mm	0.01	0.01	0.01
Coating cycle	s			
Coating arm speed	mm/s			
Power fluctuation	%			
Model characteristics				
Installation space max. (width/depth/height)	mm			
Model dimensions max.	mm	400/300/350	1180/730/550	
Layer thickness	mm	0.12	0.125	0.122
Gauge	mm	350	880	600
Accuracy absolute	mm	0.15	0.3	0.25
Repeatability	mm	0.15	0.3	0.25
Additional support				
EDP Interface				
Interface formats		STL	STL	STL
EDP System		PC	PC	PC
Operating system		Win 98	Win 98	Win 98
Software		RPP-S016	RPP-S026	RPP-S036

Table A2.36 Layer Laminate Machines III

RP Process	Unit	Sheet Lamination	
Machine designation/ Type		LD 3D Printer	Graphtec XD 700
Manufacturer		3D Systems	
Distribution		www.3dsystems.com	
Dimensions, Connection values			
Width/depth/height	m/m/m	0.465/0.77/0.42	0.465/0.77/0.42
Weight	kg	36	45
Electrical connection	V/A		200-240
Power consumption	kW	0.66	0.62
Working gas	l/min		
Cooling water	l/min		
Extraction	m^3/h		
Work room temperature	°C	35	18–30
Relative moisture	%		
Process			
Fumes, Exhaust			
Disposal, Waste			
Hazardous waste			
Contouring Element			
Kind			
Type			
x-y Contour generation		± 0.25	
x-y Contour accuracy	mm		
Spot repeatability	mm		
z Contour accuracy	mm		± 0.15
z Contour repeatable accuracy	mm		
Coating cycle	s		
Coating arm speed	mm/s		
Power fluctuation	%		
Model characteristics			
Installation space max. (width/depth/height)	mm		
Model dimensions max.	mm	160/210/135	160/210/135
Layer thickness	mm	0.15	0.168
Gauge	mm		
Accuracy absolute	mm		
Repeatability	mm		
Additional support			
EDP Interface			
Interface formats		STL, 3DS	STL, 3DS
EDP system		PC	PC
Operating system		Windows 2000, XP	Windows 2000, XP
Software			SDview

Table A2.37 3D Printer I

RP Process	Unit	3DP	3DP	3DP	3DP	3DP
Machine designation/ Type		ProJet 160	ProJet 260C	ProJet 360	ProJet 460Plus	ProJet 660Pro
Manufacturer		3D Systems				
Distribution in Germany		www.3dsystems.com				
Dimensions, Connection values						
Width/depth/height	m/m/m	0.74/0.79/1.4	0.74/0.79/1.4	1.22/0.79/1.4	1.22/0.79/1.4	1.88/0.74/1.45
Weight	kg	165	165	179	193	340
Electrical connection	V/A	90–100/7.5; 110–120/5.5; 208–240/4	90–100/7.5; 110–120/5.5; 208–240/4	90–100/7.5; 110–120/5.5; 208–240/4	90–100/7.5; 110–120/5.5; 208–240/4	100–240/ 7.5–15
Capacity	KW	n. a.	n. a.	n. a.	n. a.	n. a.
Work room temperature	°C	13–24	13–24	13–24	13–24	13–24
Air moisture	%	20–55	20–55	20–55	20–55	20–55
Model characteristics						
Installation space max. (width/depth/height)	mm	236/185/127	236/185/127	203/254/203	203/254/203	254/381/203
Model dimensions max.	mm	236/185/127	236/185/127	203/254/203	203/254/203	254/381/203
Layer thickness	mm	0.1	0.1	0.1	0.1	0.1
Accuracy absolute	mm	n. a.	n. a.	n. a.	n. a.	n. a.
EDP Interface						
Interface formats		STL, VRML, PLY, 3DS, FBX, ZPR	STL, VRML, PLY, 3DS, FBX, ZPR	STL, VRML, PLY, 3DS, FBX, ZPR	STL, VRML, PLY, 3DS, FBX, ZPR	STL, VRML, PLY, 3DS, FBX, ZPR
EDP system		PC	PC	PC	PC	PC
Operating system		Windows Vista, 7	Windows Vista, 7	Windows Vista, 7	Windows Vista, 7	Windows Vista, 7
Miscellaneous						
Building material		VisiJet PXL	VisiJet PXL	VisiJet PXL	VisiJet PXL	VisiJet PXL
Colors		Monochrome	64 Colors	Monochrome	> 2.8 Million colors	> 6 Million colors
Quantity of nozzles		304	604	304	604	1520
Printer heads		1	2	1	2	5
Vertical build speed	mm/h	20	20	20	23	28
Minimum feature size	mm	0.4	0.4	0.15	0.15	0.1

Table A2.38 3D Printer II

RP Process	Unit	3DP
Machine designation/ Type		ProJet 860Pro
Manufacturer		3D Systems
Distribution in Germany		www.3dsystems.com
Dimensions, Connection values		
Width/depth/height	m/m/m	1.19/1.16/1.62
Weight	kg	363
Electrical connection	V/A	100-240/7.5-15
Capacity	KW	n.a.
Work room temperature	°C	13-24
Air moisture	%	20-55
Model characteristics		
Installation space max. (width/depth/height)	mm	508/381/229
Model dimensions max.	mm	508/381/229
Layer thickness	mm	0.1
Accuracy absolute	mm	n.a.
EDP Interface		
Interface formats		STL, VRML, PLY, 3DS, FBX, ZPR
EDP system		PC
Operating system		Windows Vista, 7
Miscellaneous		
Building materials		VisiJet PXL
Colors		>6 Million
Quantity of nozzles		1520
Printer heads		5
Vertical build speed	mm/h	5-15
Minimum feature size	mm	0.1

Table A2.39 3D Printer III

RP Process	Unit	3D Printing						
Machine designation/ Type		Z™150/ 250 3D Printer	Z™310 3D Printer	Z™350 3D Printer	Z™450 3D Printer	Z™650 3D Printer	Z™650 3D Printer	Z 850 System
Manufacturer		Z Corporation www.zcorp.com						
Distribution in Germany								
Dimensions, Connection values								
Width/depth/height	m/m/m	0.74/ 0.79/1.4	0.74/ 0.81/1.09	1.22/ 0.79/1.4	1.2/0.79/ 1.4	1.07/ 0.79/1.27	1.88/ 0.74/1.45	1.19/ 1.16/1.62
Weight	kg	165	179	193	179	204	340	363
Electrical connection	V/A	230/16	230/16	230/16	230/16	230/16	100/15; 240/7.5	100/15; 240/7.5
Power consumption	kW		<1					
Working gas	l/min	-	-	-	-	-	-	-
Cooling water	l/min	-	-	-	-	-	-	-
Extraction	m^3/h	-	-	-	-	-	-	-
Work room temperature	°C	18-25	18-25	18-25	18-25	18-25	18-25	18-25
Relative moisture	%	30-80	30-80	30-80	30-80	30-80	30-80	30-80

RP Process	Unit	3D Printing						
Process								
Fumes, Exhaust		–	–	–	–	–	–	–
Disposal, Waste		yes	yes	yes	yes	yes	yes	yes
Hazardous waste		–	–	–	–	–	–	–
Contouring Element								
Kind		1 or 2 Inkjet print heads	1 Inkjet print head	2 Inkjet print heads	1 Inkjet print head	4 Inkjet print heads	5 Inkjet print heads	5 Inkjet print heads
Type		HP DeskJet	HP DeskJet	HP DeskJet	HP DeskJet	HP DeskJet	HP DeskJet	HP DeskJet
x-y Contour generation		Printer/ Plotter	Printer/ Plotter	Printer/ Plotter	Printer/ Plotter	Printer/ Plotter	Printer/ Plotter	Printer/ Plotter
x-y Contour accuracy	dpi	450 × 300 dpi	450 × 300 dpi	450 × 300 dpi	450 × 300 dpi	300 × 450 dpi	600 × 540 dpi	600 × 540 dpi
Spot repeatability	dpi							
z Contour accuracy	mm							
z Contour repeatable accuracy	mm							
Coating cycle	s	3	3	3	3	3		
Coating arm speed	mm/s							
Power fluctuation	%	–	–	–	–	–	–	–
Model characteristics								
Installation space max. (width/depth/height)	mm	236/185/ 127	203/254/ 203	203/254/ 203	203/254/ 203	203/254/ 203	254/381/ 203	508/381/ 229
Model dimensions max.	mm	236/185/ 127 (monochrome+ color)	203/254/ 203 (monochrome)	203/254/ 03 (monochrome + color)	203/254/ 203 (monochrome)	203/254/ 203 (monochrome + color)	254/381/ 203 (monochrome + color)	508/381/ 229 (monochrome + color)
Layer thickness	mm	0.1	0.09–0.1	0.1	0.09–0.1	0.09–0.1	0.09–0.1	0.09–0.1
Gauge	mm		10	10	10	10		
Accuracy absolute	mm		0.20%	0.20%		0.20%		
Repeatability	mm							
Additional support		–	–	–	–	–	–	–
EDP Interface								
Interface formats		STL/ VRML/ PLY/3DS/ FBX/ZPR	STL/ VRML/ PLY/3DS/ FBX/ZPR	STL/ VRML/ PLY/3DS/ FBX/ZPR	STL/ VRML/ PLY/3DS/ FBX/ZPR	STL/ VRML/ PLY/3DS/ FBX/ZPR	STL/ VRML/ PLY/3DS/ FBX/ZPR	STL/ VRML/ PLY/3DS/ FBX/ZPR
EDP system		PC	PC	PC	PC	PC	PC	PC
Operating system		Windows 7/Vista/ XP	Windows 7/Vista/ XP	Windows 7/Vista/ XP	Windows 7/Vista/ XP	Windows 7/Vista/ XP	Windows 7/Vista/ XP	Windows 7/Vista/ XP
Software		Zprint	Zprint	Zprint	Zprint	Zprint	Zprint	Zprint

Table A2.40 3D Printer IV

RP Process	Unit	3D Printing				
Machine designation/ Type		S-Max	S-Print	M-Lab	M-Print	ThermoJet
Manufacturer		ExOne				3D Systems
Distribution in Germany		www.exone.com				www.3dsys-tems.com
Dimensions, Connection values						
Width/depth/height	m/m/m	7/3.6/2.9	2.5/2.6/2.15	0.97/0.71/1.07	2.3/2.6/2.12	1.37/1.12/0.76
Weight	kg	6500	2500	-	2500	375
Electrical connection	V/A	400/-	400/32	120/1	400/-	100/12.5; 115/10; 230/6.3
Power consumption	kW	Max. 5	5	-	12	-
Working gas	l/min	-	-	-	-	-
Cooling water	l/min	-	-	-	-	-
Extraction	m^3/h	-	-	-	-	-
Work room temperature	°C	-	-	-	-	-
Relative moisture	%	-	-	-	-	-
Process						
Fumes, Exhaust		-	-	-	-	no
Disposal, Waste		-	-	-	-	yes
Hazardous waste		-	-	-	-	no
Contouring Element						
Kind		Inkjet print head	Inkjet print head	Inkjet print head	Inkjet print head	Multijet modeling
Type		-	-	-	-	
x-y Contour generation		Printer/Plotter	Printer/Plotter -	Printer/Plotter -	Printer/Plotter -	Printer
x-y Contour accuracy	dpi	-	-	-	-	300/400
Spot repeatability	dpi	-	-	-	-	-
z Contour accuracy	mm	-	-	-	-	600
z Contour repeatable accuracy	mm	-	-	-	-	-
Coating cycle	s	-	-	-	-	-
Coating arm speed	mm/s	-	28	-	-	ca. 1
Power fluctuation	%		-		-	-
Model characteristics						
Installation space max. (width/depth/height)	mm	1000/1800/ 700	380/750/400	40/60/35	390/760/400	250/190/200
Model dimensions max.	mm	1000/1800/ 700	380/750/400	40/60/35	390/760/400	250/190/200
Layer thickness	mm	0.280.5	-	0.05	-	0.04
Gauge	mm	-	-	-	-	-
Accuracy absolute	mm	0.07-0.98	-	-	-	-
Repeatability	mm	-	-	-	-	-
Additional support		-	-	-	-	Automatic configuration
EDP Interface						
Interface formats		STL	STL, CLI	STL	STL, CLI	TCP/IP, Ethernet
EDP system		Workstation	Workstation	Workstation	Workstation	PC/.stl
Operating system		Windows 7	Windows 7	Windows 7	Windows 7	Windows 98/2000/XP
Software						ThermoJet print client software

Table A2.41 3D Printer V

RP Process	Unit	3D Printer				
Machine designation/Type		VX200	VX500	VX1000	VXC800	VX4000
Manufacturer		Voxeljet Systems				
Distribution in Germany		www.voxeljet.de				
Dimensions, Connection values						
Width/depth/height	m/m/m	0.9/1.7/1.5	1.8/1.8/1.7	2.8/2.4/2	2.8/4/2.2	3.8/19.5/7
Weight	kg	450	1200	3500	2500	–
Electrical connection	V/A	–	–	–	–	–
Power consumption	kW	–	–	–	–	–
Working gas	l/min	–	–	–	–	–
Cooling water	l/min	–	–	–	–	–
Extraction	m^3/h	–	–	–	–	–
Work room temperature	°C	22 +/-2	22 +/-2	22 +/-2	22 +/-2	22 +/-2
Relative moisture	%	55-75	55-75	55-75	55-75	55-75
Process						
Fumes, Exhaust		–	–	–	–	–
Disposal, Waste		–	–	–	–	–
Hazardous waste		–	–	–	–	–
Contouring Element						
Kind		Piezo Printhead System	Printhead	Printhead	Printhead	Printhead
Type		–	–	–	–	–
x-y Contour generation		Printer/Plotter	Printer/Plotter	Printer/Plotter	Printer/Plotter	Printer/Plotter
x-y Contour accuracy	dpi	300	600	600	600	600
Spot repeatability	dpi	–	–	–	–	–
z Contour accuracy	mm	–	–	–	–	–
z Contour repeatable accuracy	mm	–	–	–	–	–
Coating cycle	s	–	–	–	–	–
Coating arm speed	mm/s	–	–	–	–	–
Power fluctuation	%	–	–	–	–	–
Model characteristics						
Installation space max. (width/depth/height)	mm	200/300/150	400/500/300	600/1060/500	500/850/1500/2000	2000/4000/1000
Model dimensions max.	mm	200/300/150	400/500/300	600/1060/500	500/850/1500/2000	2000/4000/1000
Layer thickness	mm	0.15	0.08-0.15	0.1-0.3	0.3	0.12-0.3
Gauge	mm	21	112	112/450	112	1210
Accuracy absolute	mm	–	–	–	–	–
Repeatability	mm	–	–	–	–	–
Additional support		–	–	–	–	–
EDP Interface						
Interface formats		STL	STL	STL	STL	STL
EDP system		Workstation	Workstation	Workstation	Workstation	Workstation
Operating system		Current Windows	Current Windows	Current Windows	Current Windows	Current Windows
Software						

Materials for Additive Manufacturing Processes and Casting Resins

The materials described in the following represent only a selection of the available materials. Because in this field there are frequent new and further developments, at this point, one must refer to the manufacturer.

Table A3.1 Stereolithography Materials I

Properties	Unit	3D Systems Accura Bluestone	3D Systems AccuGen	3D Systems Acura 50	Vantico RhenShape SL7580	Vantico RhenShape SL7565	DSM Somos ProtoTool 20L	DSM Somos NanoForm 15120	DSM Somos WaterShed 11120	DSM Somos Somos 9120
Type		Photopolymer	Photopolymer	Photopolymer	Photopolymer	Photopolymer	Photopolymer UV aftercure	Photopolymer UV + thermal aftercure	Photopolymer	Photopolymer
Color		Opaque blue	Translucent amber	Amber-white/ gray	White	Transparent amber	Gray	Gray	Transparent green	Translucent amber
E-Modulus/Tensile modulus/ Young's modulus	MPa	7600–11700	2280	2480–2690	2300	2000–2500	10100–11200	5900	2650–2880	1227–1462
Breaking elongation	%	1.4–2.4	15	5.3–15	11	>20	1.2–1.3	1.2	11–20	15–25
Flexural modulus	MPa	8300–9800	1720	2210–2340	2400	1900–2100	9240–9600	4450	2040–2370	1310–1455
Flexural strength	MPa	124–154	63	72–77	83	60–66	118–123	129	63–74	41–48
Charpy impact strength	kJ/m^2									
Charpy notch impact strength	kJ/m^2	13–17	25	16.5–28.1	34	27–38				48–53
Tensile strength	MPa	66–68	50	48–50	53	46–54	72–79	53	47–54	30–32
Hardness at 23 °C	Shore A	92	82	86	84	84				80–82
Residual ash	%									
Melting point	°C									
Heat resistance										
▪ at 0.45 MPa	°C						269	257–259	45.9–54.5	52–61
▪ at 1.84 Mpa	°C	66	n.a.	43–46	52		115	83–94	49–49.70	n.a.
Glass transition temperature	°C	65–66	n.a.	49–53	63	48–52	49	80	39–56	n.a.

Table A3.2 Stereolithography Materials II

Properties	Unit	Envisiontec RHarz	Objet FullCure M-720	Objet FullCure M-830	Objet FullCure M-840	Objet TangoBlack	Objet TangoGray
Type		Photopolymer	Photopolymer	Photopolymer	Photopolymer	Photopolymer	Photopolymer
Color			Transparent	White	Blue-gray	Black	Gray
E-Modulus/Tensile modulus/ Young's modulus	MPa	1980	2870	2495	2740	–	–
Breaking elongation	%		15–25%	15–25%	15–25%	47.7 %	47%
Flexual moduleus	MPa	2450	1718	2137	1983	–	–
Flexural strength	MPa	75	75.8	74.6	83.6	–	–
Charpy impact strength	kJ/m^2		–	–	–	–	–
Charpy notch impact strength	kJ/m^2		39.6	37.5	42.5	–	–
Tensile strength	MPa		60.3	49.8	55.1	2.0	4.36
Hardness at 23 °C	Shore A	85	83	83	83	61	75
Residual ash	%		< 0.01	< 0.4	< 0.3	–	–
Melting point	°C		–	–	–	–	–
Heat resistance							
▪ at 0.45 MPa	°C		48.4	47.6	48.8	n. a.	n. a.
▪ at 1.84 MPa	°C	64	44.4	43.6	44.8	n. a.	n. a.
Glass transition temperature	°C	135	48.7	58	48.7	n. a.	n. a.

Table A3.3 Stereolithography Materials III

Property	Condition	Unit	Accura ClearVue	Accura PEAK Plastic	Accura Saphire
Appearance			Transparent	Amber colors	Sapphire blue
Density, liquid	at 25 °C	g/cm^3	1.1	1.32	1.1
Density, solid	at 25 °C	g/cm^3	1.17	1.36	1.18
Viscosity	at 30 °C	cps	235-260	605	160–200
Penetration depth		mils	6.1	5.6	2.9
Critical energy		mJ/cm^2	9.5	11.5	8.23
Tested build styles					Viper Hi-Resolution Exact™, Viper High-Resolution Thin Layer™
Tensile strength	ASTM D 638	MPa	46–53	57–78	20–24
Modulus of elasticity	ASTM D 638	MPa	2270–2640	4220–4790	910–1110
Breaking elongation	ASTM D 638	%	3–15	1.3–2.5	9–16
Flexural strength	ASTM D 790	MPa	72–84	77–126	28–38
Flexural modulus	ASTM D 790	MPa	1980–2310	4180–4790	1080–1420
Impact strength	ASTM D 256	J/m	40–58	21.3–27.3	29–40
Heat resistance	ASTM D 648				
	at 66 PSI	°C	51	78	38
	with heat treatment			153	
	at 264 PSI	°C	50	59	33
	with heat treatment			124	
Hardness, Shore D			80	86	72
Coefficient of thermal expansion	ASTM E 831-93				
		µm/m °C	TMA (T<Tg, 25–50 °C): 122	TMA (T<Tg, 0–50 °C): 48	TMA (T<Tg, 25–50 °C): 135
		µm/m °C	TMA (T<Tg, 50–100 °C): 155	TMA (T<Tg, 50–120 °C): 92	TMA (T<Tg, 75–140 °C): 165
Glass transition temperature	DMA, E″	°C	62	104, with heat treatment at 120 °C: 110	58
Water absorption	ASTM D 570-98	%	0.3		

Table A3.4 Stereolithography Materials IV

Property	Condition	Unit	Accura Xtreme White 200	Accura Xtreme	Accura 60
Appearance			White	Gray	Transparent
Density, liquid	at 25 °C	g/cm^3	1.12	1.13	1.13
Density, solid	at 25 °C	g/cm^3	1.18	1.19	1.21
Viscosity	at 30 °C	cps	650–750	250–300	150–180
Penetration depth		mils	4.6	4.1	6.3
Critical energy		mJ/cm^2	8.3	11.7	7.6
Tested build Styles				EXACT™	EXACT™, FAST™, QuickCast™
Tensile strength	ASTM D 638	MPa	45–50	38–44	58–68
Modulus of elasticity	ASTM D 638	MPa	2300–2630	1790–1980	2690–3100
Breaking elongation	ASTM D 638	%	7–20	14–22	5–13
Flexural strength	ASTM D 790	MPa	75–79	52–71	87–101
Flexural modulus	ASTM D 790	MPa	2350–2550	1520–2070	2700–3000
Impact strength	ASTM D 256	J/m	55–66	35–52	15–25
Heat resistance	ASTM D 648				
	at 66 PSI	°C	47	62	53–55
	at 264 PSI	°C	42	54	48–50
Hardness, Shore D			78–80		86
Coefficient of thermal expansion	ASTM E 831-93				
		µm/m °C	TMA (T<Tg, 30–50 °C): 95		TMA (T<Tg, 0–20 °C): 71
		µm/m °C	TMA (T<Tg, 70–140 °C): 180		TMA (T<Tg, 75–140 °C): 153
Glass transition temperature	DMA, E"	°C	52		58
Water absorption	ASTM D 570-98	%	0.38		

Table A3.5 Stereolithography Materials V

Property	Condition	Unit	Accura CeraMAX	Accura 48HTR	Accura e-Stone
Appearance			White	Transparent amber	Green
Density, liquid	at 25 °C	g/cm³	1.59	1.17	1.13
Density, solid	at 25 °C	g/cm³	1.62	1.23	1.19
Viscosity	at 30 °C	cps	1500–2000	200–250	200–300
Penetration depth		mils	5.7	5.5	4.2
Critical energy		mJ/cm²	7.2	7.4	10.5
Tested build styles				Viper™-EXACT™, EXACT™-HR, Viper™ Pro-EXACT™	EXACT™
Tensile strength	ASTM D 638	MPa	78–87	64–67	37–39
Modulus of elasticity	ASTM D 638	MPa	9460–9680	2800–3980	1500–1750
Breaking elongation	ASTM D 638	%	1–1.5	4–7	10–23
Flexural strength	ASTM D 790	MPa	137–145	105–118	54–59
Flexural modulus	ASTM D 790	MPa	8270–8370	2760–3400	1350–1750
Notched impact strength	ASTM D 256	J/m	14.5–17.9	22–29	18–25
Heat resistance	ASTM D 648				
	at 66 PSI	°C	148	65	58–63
	with heat treatment		220	130	
	at 264 PSI	°C	95	57	51–55
	with heat treatment		97	110	
Hardness, Shore D			89	86	80
Coefficient of thermal expansion	ASTM E 831-93				
		µm/m °C	TMA ($T<T_g$, 25–57 °C): 31.1	TMA ($T<T_g$, <50 °C): 115	
		µm/m °C	TMA ($T<T_g$, 70–200 °C): 87.4	TMA ($T>T_g$, >120 °C): 165	
Glass transition temperature	DMA, E"	°C	108–110, with heat treatment at 120 °C: 112–114	62, with heat treatment at 160 °C: 132	60

Table A3.6 Stereolithography Materials VI

Property	Condition	Unit	Accura CastPro	Accura 55	Accura 25
Appearance			Transparent	White	White
Density, liquid	at 25 °C	g/cm^3	1.08	1.13	1.14
Density, solid	at 25 °C	g/cm^3	1.17	1.20	1.19
Viscosity	at 30 °C	cps	240–260	155–185	250
Penetration depth		mils	6.2	5.2	4.2
Critical energy		mJ/cm^2	8.7	7.4	10.5
Tested build styles				EXACT™, FAST™; EXACT-HR	FAST™, EXACT™, EXACT-HR
Tensile strength	ASTM D 638	MPa	52–53	63–68	38
Modulus of elasticity	ASTM D 638	MPa	2490–2620	3200–3380	1590–1660
Breaking elongation	ASTM D 638	%	4.1–8.3	5–8	13–20
Flexural strength	ASTM D 790	MPa	82–84	88–110	55–58
Flexural modulus	ASTM D 790	MPa	2310–2340	2690–3240	1380–1660
Notched impact strength	ASTM D 256	J/m	43–49.5	12–22	19–24
Heat resistance	ASTM D 648				
	at 66 PSI	°C	51	55–58	58–63
	at 264 PSI	°C	50	51–53	51–55
Hardness, Shore D			85	85	80
Coefficient of thermal expansion	ASTM E 831-93				
		µm/m °C	102.9	TMA ($T>T_g$, >120 °C): 61	TMA ($T>T_g$, >120 °C): 107
		µm/m °C	160.2	TMA ($T>T_g$, >120 °C): 163	TMA ($T>T_g$, >120 °C): 151
Glass transition temperature	DMA, E"	°C	70	56	60
Water absorption	ASTM D 570-98	%	0.33		

Table A3.7 Stereolithography Materials VII

Property	Condition	Unit	Accura Bluestone	Accura Amethyst
Appearance			Blue	Purple
Density, liquid	at 25 °C	g/cm^3	1.70	1.1
Density, solid	at 25 °C	g/cm^3	1.78	1.23
Viscosity	at 30 °C	cps	1200–1800	350
Penetration depth		mils	4.1	3.7
Critical energy		mJ/cm^2	6.9	14.4
Tested build styles			EXACT™	FAST™, EXACT™, EXACT-HR
Tensile strength	ASTM D 638	MPa	66–68	22–38
Modulus of elasticity	ASTM D 638	MPa	7600–11700	3514–3996
Breaking elongation	ASTM D 638	%	1.4–2.4	0.56–1.04
Flexural strength	ASTM D 790	MPa	124–154	87–125
Flexural modulus	ASTM D 790	MPa	8300–9800	3652–3721
Notched impact strength	ASTM D 256	J/m	13–17	9–12
Heat resistance	ASTM D 648			
	at 66 PSI	°C	65–66	77
	with heat treatment		267–284	
	at 264 PSI	°C	65	62
	with heat treatment			
Hardness, Shore D			92	87
Coefficient of thermal expansion	ASTM E 831-93			
		µm/m °C	TMA ($T>T_g$, >120 °C): 33–44	TMA ($T>T_g$, >120 °C): 57
		µm/m °C	TMA ($T>T_g$, >120 °C): 81–98	TMA ($T>T_g$, >120 °C): 133
Glass transition temperature	DMA, E″	°C	71–83	103

Table A3.8 Stereolithography Materials VIII

Property	Condition	Unit	RenShape SL7800		RenShape SL7810	
Appearance					White	
Density, liquid	at 25 °C	g/cm³	1.12		1.13	
Density, solid	at 25 °C	g/cm³	1.15		1.16	
Viscosity	at 30 °C	cps	205		210	
Penetration depth		mils			5.6	
	on SLA 7000/5000		5.74			
	auf SLA 500/3500/Vipser Si_2 System		5.67			
Critical energy		mJ/cm²			9.9	
	auf SLA 7000/5000		9.98			
	auf SLA 500/3500/Vipser Si_2 System		9.51			
			90 minute + UV postcure	90 minute UV + 2 h at 80 °C thermal treatment	90 minute UV postcure	90 minute UV + 2 h at 80 °C thermal treatment
Tensile strength	ASTM D 638	MPa	41–47	44	35–51	39–51
Modulus of elasticity	ASTM D 638	MPa	2075–2400	2180–2250	1793–2400	2000–2400
Breaking elongation	ASTM D 638	%	10–18	10–15	10–20	10–14
Flexural strength	ASTM D 790	MPa	69–74	72–74	58–69	62–69
Flexural modulus	ASTM D 790	MPa	2280–2650	2380–2640	1896–2400	2096–2400
Notched impact strength	ASTM D 256	J/m	37–58	37–48	44–49	44–50
Heat resistance	ASTM D 648					
	at 66 PSI	°C	62	62	51	50
Hardness, Shore D			87	87	86	87
Coefficient of thermal expansion	ASTM E 831-93					
		µm/m °C	100	109	96	99
		µm/m °C				
Glass transition temperature	DMA, E″	°C	57	59	62	62
Water absorption	ASTM D 570-98	%				

Table A3.9 Stereolithography Materials IX

Property	Condition	Unit	RenShape SL7820		RenShape SL7840	
Appearance			White		Clear	
Density, liquid	at 25 °C	g/cm³	1.13		1.13	
Density, solid	at 25 °C	g/cm³	1.16		1.16	
Viscosity	at 30 °C	cps	210		320	
Penetration depth		mils	4.5		5	
Critical energy		mJ/cm²	10		15	
			90 minute + UV post-cure	90 minute UV + 2 h at 80 °C thermal treatment	90 minute + UV postcure	90 minute UV + 2 h at 80 °C thermal treatment
Tensile strength	ASTM D 638	MPa	36–51	39–51	36–45	37–45
Modulus of elasticity	ASTM D 638	MPa	1900–2400	2000–2500	1700–2200	1600–2200
Breaking elongation	ASTM D 638	%	8–18	9–14	11–17	11–18
Flexural strength	ASTM D 790	MPa	59–80	62–80	65–80	62–80
Flexural modulus	ASTM D 790	MPa	2000–2400	2100–2500	1600–2200	1600–2200
Impact strength	ASTM D 256	J/m	42–48	30–49	37–60	35–60
Heat resistance	ASTM D 648					
	at 66 PSI	°C	51	50	48	46
Hardness, Shore D			86	87	86	87
Coefficient of thermal expansion	ASTM E 831-93	µm/m °C	93	93	100	110
Glass transition temperature	DMA, E″	°C	62	62	58	59

Table A3.10 Stereolithography Materials X

Property	Condition	Unit	RenShape SL7870	
Appearance			Clear	
Density, liquid	at 25 °C	g/cm^3	1.13	
Density, solid	at 25 °C	g/cm^3		
Viscosity	at 30 °C	cps	180	
Penetration depth		mils	7.2	
Critical energy		mJ/cm^2	10.6	
			90 minute + UV postcure	90 minute UV + 2 h at 80 °C thermal treatment
Tensile strength	ASTM D 638	MPa	38–42	39–42
Modulus of elasticity	ASTM D 638	MPa	1930–2020	1860–2030
Breaking elongation	ASTM D 638	%	10–22	10–23
Flexural strength	ASTM D 790	MPa	65–71	65–74
Flexural modulus	ASTM D 790	MPa	1980–2310	2010–2350
Impact strength	ASTM D 256	J/m	45–61	43–60
Heat resistance	ASTM D 648			
	at 66 PSI	°C	48	49
	at 264 PSI	°C		
Hardness, Shore D			86	87
Glass transition temperature	DMA, E″	°C	56	57
Water absorption	ASTM D 570-98	%		

Table A3.11 Sintering Materials I

Properties	Unit	3D Systems DuraForm PA	3D Systems DuraForm GF	3D Systems DuraForm Flex*	3D Systems CastForm PS	3D Systems LaserForm A6
Type		Polyamide	Polyamide, glass-filled	Elastomer	Wax infiltrated	Bronze infiltrated
Manufacturer/Distributor/Info		3D Systems				
Color		White	Gray	Natural white	White	Gray
Average particle size	µ	58	48	85	62	34
Particle size	95% < µ	92	92	138	106	58
Particle size area	µ	25-92	10-92	21-138	25-106	20-58
Moisture absorption	%	0.41	0.3			
Specific weight	g/cm^3	1.01-1.02	1.49	0.44		7.8
Density (powder)	g/cm^3	0.59	0.84	0.58		8.15
Density (sintered material)	g/cm^3	0.97	1.4	0.91		7.5
E-Modulus	MPa	1600	4068	7.4		138×10^3
Breaking elongation	%	14	1.4	110		10.0-20.0
Flexural modulus	MPa	1387	3106	5.9		
Flexural strength	MPa		37			
Charpy impact strength	kJ/m^2	336	123			
Charpy notch impact strength	kJ/m^2	32	41			
Tensile strength	MPa	44	27	1.8		610
Hardness	Shore A	73	77	60		HRC 22
Residual ash	%				0.03	
Melting point	°C	184	185	192		
Glass transition temperature	°C	-			89	
Dimensional stability temperature	°C	180	175			

* All values for DuraForm Flex in infiltrated state.

Table A3.12 Sintering Materials II

Properties	Unit	EOS PrimePKind	EOS PA 2200	EOS PA 3200GF	EOS Alumide	EOS PrimeCast 100
Type		PA 12	PA 12	Glass-filled PA 12	Aluminum-filled PA 12	Polystyrene
Color		White	White	White	Gray-silver	White
Average particle size	µm	56	60	60	60	80
Particle size	95% < µ					
Particle size area	µm					
Moisture absorption at 23 °C	%					
Specific weight	g/cm^3					
Bulk density (powder)	g/cm^3	0.44	0.44	0.61	0.64	0.61
Density (laser-sintered material)	g/cm^3	0.92	0.93	1.26	1.36	0.78
E-Modulus/Tensile modulus/ Young's modulus	MPa	1700	1700	3200	3800	1600
Breaking elongation	%	15	20	6	3.5	0.4
Flexual moduleus	MPa	1240	1240	2100	3000	
Flexural strength	MPa				74 ± 2	
Charpy impact strength	kJ/m^2	53	53	35	29	
Charpy notch impact strength	kJ/m^2	4.8	4.8	5.4	4.6	
Tensile strength	MPa	47	45	48	46	1.2–5.5
Hardness	Shore D	75	75	80	76	
Residual ash	%					0.002
Melting point	°C	172–180	172–180	172–180	172–180	
Dimensional stability						
according to Vicat Last A/ 50 K/Std.	°C	181	181	179		
according to Vicat Last B/ 50 K/Std.	°C	163	163	166	169	
Glass transition temperature	°C					105 ± 1

All data should be regarded as guide values.

Table A3.13 Sintering Materials III

Properties	Unit	EOS DirectMetal 20	EOS DirectSteel 20	EOS DirectSteel H20	EOS CoCr*	EOS Stainless steel*
Type		Bronze-based metal powder	Steel-based metal powder	Steel-based metal powder	CoCr alloy metal powder	Stainless steel powder
Color		Brown	Gray	Gray	Gray	Gray
Average particle size	µm	< 20	< 20	< 20	< 20	< 20
Particle size	95% < µ					
Particle size area	µm					
Moisture absorption at 23 °C	%					
Specific weight	g/cm^3					
Density (powder)	g/cm^3					
Density (sintered material)	g/cm^3	up to 7.6	up to 7.6	up to 7.8		
E-Modulus/Tensile modulus/ Young's modulus	MPa	80	130	180		
Breaking elongation	%				5–12	30
Flexural modulus	MPa					
Flexural strength	N/mm^2	700	1000	2000		
Charpy impact strength	kJ/m^2					
Charpy notch impact strength	kJ/m^2					
Tensile strength	MPa	up to 400	up to 600	up to 1100	1300	1000
Hardness		110 HB, 115 HV	220 HB, 225 HV	350–420 HV, 35–42 HRC	30–52 HRC	240 HB
Residual ash	%					
Melting point	°C					
Heat resistance						
▪ at 0.45 MPa	°C					
▪ at 1.84 MPa	°C					
Glass transition temperature	°C					

All data should be regarded as guide values.
* Preliminary values
** Hardness testing by Vickers (HV), Rockwell B (HRB), and Rockwell C (HRC).

Table A3.14 Sintering Materials IV

Properties	Unit	EOS Ceramics 5.2	EOS QuKindz 4.2	EOS QuKindz 5.7
Type		Aluminum silicate sand	Quartz sand	Quartz sand
Color		Brown	Brown	Brown
Average particle size	µm	130 ± 20	140 ± 20	160 ± 10
Particle size	95% < µ			
Particle size area	µm			
Moisture absorption at 23 °C	%			
Specific weight	g/cm^3			
Bulk density (powder) (by VDGP27)	g/cm^3	1.64 ± 0.05	1.4 ± 0.04	1.35 ± 0.03
Density (sintered material)	g/cm^3			
E-Modulus/Tensile modulus/ Young's modulus	MPa			
Breaking elongation	%			
Flexual moduleus	MPa			
Hot flexural strength (according to VDGP74)	N/cm^2	800 ± 80	600 ± 60	550 ± 55
Cold flexural strength (nach VDGP74)	N/cm^2	2000 ± 200	1600 ± 160	1500 ± 150
Charpy impact strength	kJ/m^2			
Charpy notch impact strength	kJ/m^2			
Tensile strength	MPa			
Hardness at 23 °C	Shore A			
Residual ash	%			
Melting point (sand base material)	°C	1825	1700	1700
Heat resistance				
▪ at 0.45 MPa	°C			
▪ at 1.84 MPa	°C			
Glass transition temperature	°C			
Loss on ignition (according to VDG-P74)	%	5.2 ± 0.2	4.2 ± 0.2	5.7 ± 0.2

Table A3.15 Sintering Materials V

Properties	Unit	Windform XT	Windform Pro B Polyamide	ARCAM H13	ARCAM Ti6A114V
Type		Carbon composite		CrMoVa Leg.	Titan Leg.
Manufacturer/Vertreiber/Infos		CRP-Technology www.windform.it		ARCAM Sweden www.arcam.com	
Color		Black	Gray	Gray	Gray
Average particle size	µ				
Particle size	95% < µ				
Particle size area	µ			15–40	20–50
Moisture absorption	%				
Specific weight	g/cm^3			7.8	4
Density (powder)	g/cm^3				
Density (sintered material)	g/cm^3				
E-Modulus	MPa	7320	3612	210,000	128,000
Breaking elongation	%	2.6	3.81		10
Flexural modulus	MPa	1387	3106		
Flexural strength	MPa	131.5	96.1		
Charpy impact strength	kJ/m^2				
Charpy notch impact strength	kJ/m^2	4.73	2.69		
Tensile strength	MPa	77.85	47.05	1300	930
Hardness	Shore A	92		HRC 48–50	HRC 30–35
Residual ash	%				
Melting point	°C	179.33	180		
Glass transition temperature	°C	–			
Dimensional stability temperature	°C				

* All values for DuraForm Flex in infiltrated state.

Table A3.16 FDM Materials

Properties	Unit	Polycarbonat	ABS	ABSplus	ABSi	PPSU
Manufacturer/Contact/Web		www.stratasys.com				
E-Modulus	MPa	2000	1627	2265	1915	2068
Breaking elongation	%	3	6	4	3.1	3
Flexural modulus	MPa	2137	1834	2198	1820	2206
Flexural strength	MPa	97	41	52	61	110
Notch impact strength	kJ/m^2	53.4	106	96	101.0	58.73
Tensile strength	MPa	52	22	36	37	55
Heat resistance						
▪ at 0.45 MPa	°C	127	96.00		95	189
▪ at 1.84 MPa	°C					
Glass transition temperature	°C	161	104	n. a.	116	230
Density	g/cm^3	1.2	1.05	1.04	1.08	1.28

Special Features:
The materials are available in colors: white, black, red, blue, green, yellow, and custom-made in any RAL color.
ABSi: meets all FDA USP CLASS VI requirements, gamma-sterilizable.

Table A3.17 Model Maker Materials

Properties	Unit	Protobuild	Protosupport
Type		Green building materials	Red supports
Manufacturer		Solidscape	
Density	g/cm^3	1.25	0.93
Melting temperature	°C	92–106	50–72
Flame temperature	°C	> 175	> 175
Shore hardness	Shore A	45	33
Coefficient of thermal expansion			
▪ at 50 °C	ppm/°C	82.7	351.3
▪ at 55 °C	ppm/°C		403.2
▪ at 60 °C	ppm/°C	87.6	
▪ at 70 °C	ppm/°C	91.4	

Table A3.18 LOM Materials

Properties	Unit	LPH 042		LXP 050		LGF 045	
Type		Paper		Polyester		Glass fiber	
Manufacturer		Helisys		Helisys		Helisys	
Model characteristics		Fiber	Transverse	Fiber	Transverse	Fiber	Transverse
Density	g/cm^3	1.449		1.0–1.3		1.3	
E-Modulus	MPa	2524		3435		–	
Tensile strength	MPa	26	1.4	85		> 124.1	4.8
Compressive strength	MPa	15.1	115.3	17	52	–	–
Compression modulus	MPa	2192.9	406.9	2460	1601	–	–
Max. deformation under pressure	%	1.01	40.4	3.58	2.52	–	–
Flexural strength	MPa	2.8–4.8		4.3–9.7		–	
Glass transition temperature	°C	30		–		53–127	
Coefficient of expansion	ppm/K	3.7	185.4	17.2	229	*x*: 3.9 *y*: 15.5	*z*: 111.1
Thermal conductivity	W/mK	0.117		–		–	
Ash content	%	2.7–3.1		–		–	

Table A3.19 3D Printer Materials

Properties	Unit	zp130*
Manufacturer		Z Corporation
Color		White
Average particle size	µm	50.00
Particle size	95% < µ	
Particle size area	µm	
Moisture absorption at 23 °C	%	
Specific weight	g/cm^3	
Density (Powder)	g/cm^3	1.60
Density (sintered material)	g/cm^3	
E-Modulus/Tensile modulus/ Young's modulus	MPa	
Breaking elongation	%	
Flexural modulus	MPa	
Flexural strength	MPa	
Charpy impact strength	kJ/m^2	
Charpy notch impact strength	kJ/m^2	
Tensile strength	MPa	
Hardness at 23 °C	Shore A	
Residual ash	%	
Melting point	°C	

* All data without infiltration.
The component receives its individual properties only through infiltration.

Table A3.20 Casting Resins

Property	Unit	PX 774	2170	Biresin G 55	MG 8051	Megithan 2MD787
Manufacturer		Axson	HEK	SIKA	ebalta	BIOTOOL
Pronounced property		Elastic	Hard	Impact / flexural strength	High-temperature resistant	Clear (UV resistant)
Aspect/Color		Black	Yellowish	Opaque	Black	Transparent
Mixing ratio A:B	g:g	44:100	100:150	80:100	100:70	100:300
Pot life (g; °C)	min	2-3 (100; 25)	4 (100; 25)	4 (300; 20)	5-7 (200; 20)	15-20 (200; 20)
Viscosity (25 °C)						
Component A	Pa · s	0.025-0.045	0.63	1.45 (23 °C)	0.7 ± 0.10	1.5
Component B	Pa · s	0.8-1.2	0.23	0.3 (23 °C)	0.7 ± 0.10	0.8-2.2
Spec. Weight (25°)						
Component A	kg/dm^3	1.18-1.22	1.11	1.06 (23 °C)	1.14 ± 0.2	-
Component B	kg/dm^3	1.03-1.07	1.22	1.22 (23 °C)	1.16 ± 0.2	-
Hardness	Shore A	75 (23 °C)				
	Shore D		82 (23 °C)	81-84 (23 °C)	80 ± 3 (23 Grad)	85-87 (RT)
Flexural strength	MPA		120	110-120	100 ± 5	
Flexural modulus	MPA		2690	2800	1930 ± 100	
Tensile strength	MPA	7	78.5	70-80	67 ± 5	90
Tension/E-Modulus	MPA		3530			
Impact strength	kJ/m^2			> 100	63 ± 10	
Breaking elongation	%	300	6.5	6-8	10.5 ± 2	
Stretching	%		5.5			
Shrink	%		0.1	0.25-0.35	ca. 0.3	0.1
Thermal conductivity	W/m K		224			
Heat resistance	°C	80	85	80	120 ± 3	90-100

Due to the wide product range, every manufacturer and therefore all products cannot be listed and characterized.
The casting materials listed here are only examples representing what different pronounced properties can be crucial for the right choice. There are certainly other materials that are similar or have properties even better than expected. For specific information, the manufacturer should be contacted directly.

Glossary

In this list are frequently used terms and their abbreviations that are often used in the literature but not always clearly explained there.

Abbreviation	Definition	Explanation
3DP	Three-Dimensional Printing	Rapid prototyping process. Layer formation by injecting liquid into powder. Trademark of Massachusetts Institute of Technology
ACES	Accurate Clear Epoxy Solids	Stereolithography building style (3D Systems)
AF	Additive Fabrication	Production by adding volume elements
AF	Anatomic Facsimile	Medical anatomical (operations) models
AFM	Anatomic Facsimile Models	Medical anatomical (operations) models
AIM	ACES Injection Molding	(Injection molding) tool of stereolithography material (see ACES) (3D Systems)
AM	Additive Manufacturing	Production by adding volume elements
AM	Agile Manufacturing	Proprietary designation from 3D Systems for process for direct production. The point is to emphasize the effectiveness of the process
BASS	Break Away Support System	Support structure for the FDM process, which can be removed by manually (Stratasys)
BIS	Beam Interference Solidification	Rapid prototyping process. Polymerization point by two intersecting laser beams in the polymerization
BPM	Ballistic Particle Manufacturing	Rapid prototyping process. Layer formation by stacking shooting of volume elements
	Buy-to-fly ratio	Ratio of the volume of material to be paid for to the volume used; a measure of the waste in a manufacturing process
CAD	Computer-Aided Design	Computer-aided design, preferably in the sense of geometric construction
CAE	Computer-Aided Engineering	Computer-aided design in terms of an interpretation defined using geometric and mathematical basic elements
CAL	Computer-Aided Logistics	Computer-aided material supply
CAM	Computer-Aided Manufacturing	Computer-aided manufacturing
CAMOD	Computer-Aided Modeling Devices	Software and hardware elements for the production of models
CAP	Computer-Aided Production	Computer-aided production
CAQ	Computer-Aided Quality Assurance	Computer-aided quality assurance

Abbreviation	Definition	Explanation
CAS	Chemical Abstracts Service	US-American nomenclature and institution for the identification of chemical substances
CAS	Computer-Aided Styling	Computer-aided shaping (styling within the meaning of the German conception of design)
CAT	Computer-Aided Testing	Computer-aided measurement and testing
CAx	Computer-Aided ... Computer-Assisted ...	Abbreviation for any kind of computer-aided process
	AutoFabs	Automated fabricator(s). Equivalent to fabricator
CD	Concurrent Design	Parallelized design and construction classifications
CEM	Contract Electronics Manufacturing	Contract manufacturing for electronic components and products
CIM	Computer-Integrated Manufacturing	Production based on a closed chain CAD-CAM. Equivalent to ICAM
CMB	Controlled Metal Buildup	Rapid prototyping process. Laser generation and subsequent contour milling of metal (FhGIPT)
CMC	Computer-Mediated Communication	Computer-mediated communication
CP	Centrum für Prototypenbau GmbH	Rapid prototyping service providers located in Erkelenz/Dusseldorf (Germany)
CPDM	CIMATRON Product Data Management	PDM System Co. CIMATRON
CS ...	Computer-Supported ...	Computer-aided process
CSCW	Computer-Supported Cooperative Work	Computer-aided communication system
CSG	Constructive Solid Geometry	Description of a complex body by combining simple body about Boolean operations
CT	Computed Tomography	X-ray film process, preferably in medicine
DCM	Direct Composite Manufacturing	Direct fabrication of components made of composite material. Preferably used by OptoForm for the M3D Process
DMLS	Direct Metal Laser Sintering	Sintering process company EOS for direct sintering of metal
DMU	Digital Mock-Up	Digitale prototype, mostly in the sense of animatable in virtual reality
D_p	Cure Depth	For stereolithography, the optical penetration depth for resin
DSPC	Direct Shell Production Casting	Hardware and software solution from the company Soligen for direct sintering of ceramic molds
DTM	Desktop Manufacturing	Process for fast production of three-dimensional physical models "on the desk"
DTM	DTM Corp., Austin, TX, USA	Manufacturer of laser sintering machines; now 3D Systems
DXF	Drawing Exchange Format	Data format for storing CAD drawings
E_C	Critical Energy	For stereolithography; the threshold energy at which a chemical reaction occurs

Abbreviation	Definition	Explanation
EDM	Electronic Data Management; Engineering Data Management	Program systems for managing large, multiple beneficiaries of simultaneously processed files
EDM	Electrical Discharge Machining	Electrical discharge machining
EDM	Electronic Document Management	Electronic archiving of documents
EOS	Electrical Optical Systems GmbH, Planegg/Munich, Germany	Manufacturer of sintering machinery
ERM	Enterprise Resource Management	System and software programs for the management and planning of logistics supply chains
ERP	Enterprise Resource Planning	System and software programs for the management and planning of logistics supply chains
	Fabber	Short form of fabricator
	Fabrikator (Fabricator)	Additive manufacturing machine for the direct production of components with product character
FDM	Fused Deposition Modeling	FLM process for the company Stratasys
FFF	Fast Freeform Fabrication	Process for the rapid production of three-dimensional physical models
FLM	Fused Layer Modeling	Rapid prototyping process. Film formation by extrusion of molten thermoplastics and solidification due to heat conduction
FM	Facsimile Models	Scale mapping of existing geometry
FRP	Foam Reaction Prototyping	Process for layered application of self-reacting foams. Preferably proprietary designation of Herback (Hennings 2004)
	Grower	*For growing;* designation for additive manufacturing machinery, mainly used by Solidscape for the T66
HIS	Holographic Interference Solidification	Rapid prototyping process. Polymerization by projecting holographic images onto photosensitive materials
HPGL	Hewlett Packard Graphic Language	Plotter interface
HSC	High-Speed Cutting	High-speed milling
HSPC	High-Speed Precision Cutting	HSC of the company Kern Microtechnik
ICAM	Integrated Computer-Aided Manufacturing	Production based on a closed-chain CAD-CAM; equivalent to CIM
IGES	Initial Graphics Exchange Specification	File format for the exchange of neutral geometry data between CAD systems
KMG	Coordinate Measuring Machine	3D Coordinate measuring machine
LCVD	Laser Chemical Vapor Deposition	Rapid prototyping process. Laser-assisted deposition of material in the gas phase

Abbreviation	Definition	Explanation
LENS	Laser-Engineered Net Shaping	Rapid prototyping process. Laser-generated metal (Company OPTOMEC)
LLM	Layer Laminate Manufacturing	Rapid prototyping process. Contouring by laser, knife, or cutter. Layer formation by bonding
LM	Laminate Manufacturing	Collective term for rapid prototyping processes after layer laminate processes
LMC	Layer Milling Center	Layer milling machine of the company Zimmermann
LMP	Layer Milling Process	Layer milling process of the company Zimmermann/ Pauser
LMPM	Low Melting Point Metal	Low-melting metal alloy
LMS	Laser Model System	Stereolithography process of the company Fockele & Blacke
LMT	Layer Manufacturing Technologies & Techniques	General term for layer-oriented operating manufacturing processes
LOM	Laminated Object Manufacturing	LLM process of the company Helisys
LS	Laser Sintering	Rapid prototyping process. Layer formation by local melting and subsequent solidification of powder-like materials
LSM	Laser Surface Melting	Laser-sintering process of the company Fockele & Blacke for the production of metal prototypes
MEMS	Micro Electromechanical Systems	Microelectromechanical systems
MIM	Metal Injection Molding	Injection-molding process based on plastifiable metal-plastic mixtures
MIM/ MAM/ MDM/	Material Increase Manufacturing/ Material Addition Manufacturing/ Material Deposition Manufacturing	Production by the juxtaposition and successive joining of volume elements
MJM	MultiJet Modeling	FLM process of the company 3D Systems
MJS	Multiphase Jet Solidification	FLM process of the company ITP
	Modden (Modding)	Fitting of accessories, ornamentation, and so on, especially in their function of unchanged products. Most often externally, for example on the housing of computers (case modding). Equivalent to tune, style, pimp
MRT	Magnetic Resonance Tomography	Medical imaging process, preferably for the examination of soft tissue
NMR	Nuclear Magnetic Resonance	Process used in medical imaging
OEM	Original Equipment Manufacturer	Supplier for products that the original manufacturer sells under its own name
	Pimp	Especially externally spectacular treatment of a product that is otherwise unchanged, especially in its function. See "Pimp My Ride" TV show (off air)

Abbreviation	Definition	Explanation
PDM	Product Data Management	EDP systems for product data management
PET	Positron Emission Tomography	Imaging process, preferably for medicine
PPS	Production Planning System	Production planning and control system
	Prototyper	Additive manufacturing machine for the direct production of prototypes, patterns, and dummies
RMM	Rapid Mock-up Machine	Additive manufacturing machine for the production of prototypes, dummies (mock-ups), and models. Preferably from the Kira Corp. for the *Katana* term coined
RIM	Reaction Injection Molding	Injection-molding process on the basis of plasticizable chemically reactive metal-plastic mixtures
RM	Rapid Modeling	Process for the rapid production of models
RP	Rapid Prototyping	Technology that deals with processes and methods for production of layered models directly from 3D CAD data
RP	Reinforced Plastics	Preferably fiber-reinforced plastics (as opposed to nonreinforced plastics)
RP&M	Rapid Prototyping & Manufacturing	Process for the rapid production of prototypes and their immediate production
RPD	Rapid Product Development	Rapid product development
RPro	Rapid Production	Rapid manufacturing or production
RPT	Rapid Prototyping Techniques/Technologies	Processes and methods for the implementation of rapid prototyping
RT	Rapid Tooling	Process for the manufacturing of tools by rapid prototyping methods
SAHP	Selective Adhesive and Hot Press Process	LLM process of the company KIRA
SDU	Shell Design Unit	Hardware and software combination for the calculation of negatives for ceramic molds (Soligen)
SE	Simultaneous Engineering	Methodic approach for parallel work of several people or teams on a (development-) task
SET	Standard d'Echange et de Transfer	File format for neutral data exchange of geometry data between CAD systems
SFF	Solid Freeform Fabrication	Designation of (additive manufacturing) process for production of physical volume models (solids)
SFM	Solid Freeform Manufacturing	Designation of (additive manufacturing) process for production of physical volume models
SFP	Solid Foil Polymerization	Additive manufacturing process. Layer formation contoured by gluing films by polymerization
SGC	Solid Ground Curing	Stereolithography process of the company Cubital
SL	Stereolithography	Additive manufacturing process. Layer formation by local solidification of photosensitive resins (photopolymerization)

Abbreviation	Definition	Explanation
SLA	Stereolithography Apparatus	Stereolithography system of the company 3D Systems
SLPR	Selective Laser Powder Remelting	Additive manufacturing process. Film formation by melting and then solidifying of one-component metal powder (FhGILT)
SLS	Selective Laser Sintering	Additive manufacturing process. Layer formation by local melting and subsequent solidification of powder-like material
SOM	Stratified Object Manufacturing	Software system for undercut-free and optimal treatment by milling a complex body (ERATZ)
SOUP	Solid Object Ultraviolet Laser Plotter	Additive manufacturing process and the same stereolithography system of the company CMET
SPECT	Single Photon Emission Computed Tomography	Imaging process, preferably in medicine
SPF	Super Plastic Forming	"Inflation" of sandwich structures
STAR-Weave	Staggered Alternated Retracted Hatch	Stereolithography building style (3D Systems)
STEP	Standard of Exchange of Product Model Data	File format for the neutral exchange of complete product data between CAx systems
STL	Stereolithography Language	Interface format for the exchange of geometry data between CAD systems and additive manufacturing machines. Originally was "standard transformation language." Initially developed for simple shading of 3D CAD structures. Additive manufacturing process, especially stereolithography, to engross, to pocket
TCT	Time-Compressing Technologies	Collective designation for all processes capable of shortening the product development time
THESA	Thermoelastic Stress Analysis	Process for the experimental verification of component stress by measuring thermal effects
TI	Taylored Implants	Individual "tailored" implants
TP	Thermal Polymerization	Polymerization by heat
UV	Ultraviolet	Range of wavelengths from 38 nm (sensitivity limit of the eye in the ultraviolet) to 19 nm. In the range 60–10 nm, overlaid with X-rays
VDAFS	Association of Automobile Manufacturer, surface interface	CAD interface dedicated to the transmission of freeform surfaces
VDAIS	Association of Automobile Manufacturer, IGES Interface	CAD interface, which represents a subset of the elements defined in the IGES
VR	Virtual Reality	Realistic simulation of components, assemblies or whole products on the computer, usually combined with real-time animation. Input and output through data gloves, 3D projection, or the like

11 Bibliography

Inside this book, the references are in the format: [name of the main author, last two digits of the year of publication]. This bibliography is accordingly alphabetical by the name of the main author and then optionally sorted by year of publication.

Abouliatim, Y., Chartier, T., Chaput, C., Delage, C.: Fabrication of Ceramics by Stereolithography, Rapid. Techol. Anwendertagung (Conference), Session 5, Lecture 2, May 22–23, 2007, Erfurt trade fair, Erfurt, 2007

Ader, Ch.: Untersuchungen zum Lasersintern, Dissertation, RWTH Aachen, 2006

Altmann, O.: Kunststoffteile mit Simultaneous Engineering kostenbewusst entwickeln, Kunststoffe 84 (1994) 12, pp. 1728–1736

Ambos, E., Hornig, T. et al.: Erfahrungen zum Einsatz von Stereolithographiemodellen in der Gießereitechnik, Gießerei-Praxis No. 13/14, 1995

Argawalw, M. K., Bourell, D. L., Wu, B. et al.: An Evaluation on the Mechanical Behaviour of Bronze-Ni Composites Produced by Selective Laser Sintering Contribution to the Solid Freeform Fabrication Symposium, Austin, TX, 1991

ASTM F2792-09e1: Standard Terminology for Additive Manufacturing Technologies, 2010

ASTM F2915-12: Standard Specification for Additive Manufacturing File Format (AMF) Version 1.1, 2011, ANSI, American National Standards Insitute, West 43 Street, New York, NY, 10036

Aubin, R. F.: A World Wide Assessment of Rapid Prototyping Technologies, Contribution to the IMS International Conference on Rapid Product Development, Stuttgart, 1994

Automobil-Produktion: Wahnsinn mit Methode. Zögerlicher Kampf gegen Variantenvielfalt, Automobil-Produktion, Jan. 2005, pp. 38 ff., MI Publisher, 2005

Back, A.: Reverse Engineering als Bindeglied zwischen der virtuellen und realen Produktentwicklung, RTejournal, Issue 2, 2005, *Online Article: www.rtejournal.de/archiv/ausgabe2/98/*

Bauerle, D.: Laser Processing and Chemistry, 3rd ed., Springer Publisher, Berlin, 2000

Beitz, W., Lam, A., Ratfisch, U. et al.: Eine Systemumgebung zur Unterstützung von Simultaneous Engineering, VDI-report 1148, VDI-Publisher, Düsseldorf, 1994, pp. 439–455

Bergmann, F.: Neue Arbeit, neue Kultur. ISBN 392419596X, Arbor Publisher, Freiamt, 2004

Bernard, A., Taillander, G.: Le prototypage rapide, HERMES ed., Paris, 1998

Beyer, E., Wissenbach, K.: Oberflächenbehandlung mit Laserstrahlung, Series: Laser in Technik und Forschung, Springer Publisher, Berlin, 1998

Beyer, E.: Schweißen mit Laser, Grundlagen, Springer Publisher, Berlin, 1995

Bode, M.: Abstimmbare Einfrequenz-Strahlquellen hoher Stabilität im infraroten, sichtbaren und ultravioletten Spektralbereich, Laser und Optoelektronik 30, 5, 1998, pp. 41–47

Boehler, P., Levy, G. N., Martinoni, R., Schleiss, P.: A New Reinforced PA12 Powder for High Temperature and High Stiffness SLS Applications. Univ. of St. Gallen, FHSG, 2007, *www.fhsg.ch/rpd*

Breitinger, F.: Zeitorientierte Strategien zur Herstellung von Prototypen-Spritzgußwerkzeugen, Thesis, Technical Univ. of Munich, 1995

Breuninger, J.; Becker, R.; Wolf, A.; Rommel, S.; Verl, A.: Generative Fertigung mit Kunststoffen, Springer Publisher, Berlin, 2013

BRITE/EuRAM, N. N.: "Rapid Prototyping Techniques," Layer Interface, Version 1.3, 1994

Bullinger, H. J.: What Makes Product Development Rapid? Proceedings of the International Conference on RPD Rapid Product Development, Stuttgarter trade fair and Convention mbH, Stuttgart, 1995

Burns, M.: Automated Fabrication – Improving Productivity in Manufacturing, PTR Prentice Hall, Englewood Cliffs, New Jersey, 1993

Carter, W. T. Jr., Jones, M. G.: Direct Laser Sintering of Metals, Contribution to the Solid Freeform Fabrication Symposium, Austin, Texas, 1991

Christen, H. R.: Chemie, Diesterweg Publisher, 1984

Chua, C. K., Leong, K. F., Lim, C. S.: Rapid Prototyping, Principles and Applications World Scientific, 2nd ed., New Jersey, 2003

CP-GmbH Erkelenz, Verbundprojekt MAC4U–Mass Customization für individualisierte Produkterweiterungen, Teilvorhaben: Additive Manufacturing für Konsumenten, BMBF, 2012–2014

Daas, M.: Entwicklungsbegleitendes Rapid Prototyping. Proceedings of the Konferenz: Von der Virtual Reality zum Rapid Prototyping, Nov. 10–11, 1998, Düsseldorf Euroforum, Düsseldorf, 1998

Dahmen, R.: Charakteristiken und Bewertung der (laserunterstützten) Rapid Prototyping Verfahren, insbes. der LLM-Verfahren, Thesis, Fachhochschule (Univ. Of Applied Sciences), Aachen, 1998

Degner, W., Lutze, H., Smejkal, E.: Spanende Formgebung. Carl Hanser Publisher, Munich, 2002

Dickens, Ph.: Investment Casting of Rapid Prototype Models, EARP Letter 7/1995, c/o Bent Mieritz, DTI, Aarhus, Denmark

Dillon, K., Terchek, R.: Dimensionally Controlled Cobalt-Containing Precision Molded Metal Article. U. S. Patent No. 4 431 449, on Feb. 14, 1984, Patent issued to 3M Corp.

DIN 8580: Fertigungsverfahren. Begriffe, Einteilung, DEUTSCHE NORM, Sept. 2003, DIN Deutsches Institut für Normung e. V., Berlin, Ref. No. DIN 8580:200309, Beuth Publisher GmbH, 10772 Berlin

Domack, M. S., Baughman, J. M.: Development of Nickel-Titanium Graded Composition Components, Rapid Prototyping J., Vol. 11, No. 1/2005, pp. 41–51, Emerald Group Publishing Limited, ISSN 1355254, 2005
Full text: *www.emeraldinsight.com/Insight/ViewContentServlet?Filename=Published/EmeraldFullTextArticle/Articles/1560110106.html*

Donahue, R. J., Turner, R. S.: Proceedings of the Second International Conference on Rapid Prototyping, Univ. of Daytona, Daytona, FL, 1991, p. 221

DRFAB13: *http://3druck.com/drucker-und-produkte/weitere-delta-rostock-3d-drucker-projekte-509964/*

Drummer, D., Drexler, M., Rietzel, D.: Basic considerations for the modeling of transient thermal processes during the selective laser sintering of thermoplastics, , RTeJournal 8, 2011*EOS:* Pressemitteilung zur Euromold '98, EOS, Munich, 1998, *www.eos.info*

Exner, H., Regenfuß, P., Ebert, R., Hartwig, L., Streek, A., Klötzer, S., Horn, M.: Lasermikrosintern von keramischen Materialien, In: RTejournal – Forum für Rapid Technologie 3, Issue 3, May 2006, ISSN 16140923, URN urn:nbn:de:0009-2-3470, URL: *www.rtejournal.de/aktuell/archiv/ausgabe3/347/*

Fassbender, M.: Ermittlung von prozeß- und bauteilspezifischen Parametern für die Prototypenfertigung von Turbinenteilen auf einer Stereolithographieanlage, Diplomarbeit, Rheinisch-Westfälische Technische Hochschule Aachen, 1994

Fockele, M., Schwarze, D.: Neue Wege im Rapid Prototyping, Contribution to the 27th ISATA, Aachen, 1994

Föllinger, M.: Konzepte und Wege zum Prototyping am Beispiel von Pkw-Handschaltgetrieben, Proceedings of the Konferenz: Rapid Prototyping in der Anwendung, June 4–5, 1997 in Bad Homburg, Euroforum, Düsseldorf, 1997

FORWERKZEUG – Teilprojekt: Optimierung des Indirekten Metall Lasersinterns zur schnellen Herstellung hochgenauer Formeinsätze für den Werkzeugbau, Zwischenpräsentation, Technical Univ. Munich, 2006

FORM + WERKZEUG: Urmodell sofort in Serie bauen, Rapid Manufacturing mit Course4, Form + Werkzeug, Jan. 2007, pp. 58–60, Carl Hanser Publisher, Munich, 2007

Freieck, U.: Metall-LOM_Technologie – Revolution im Werkzeugbau, Rapid. Tech. Anwendertagung (Conference), Session 3, Lecture 1, May 22–23, 2007, Erfurt trade fair, Erfurt, 2007

Gartzen, J., Gebhardt, A. et al.: Optimisation using THESA – Thermo Elastic Tension Analysis is now being used for Laser-Sintered Components. Prototyping Technology International, 1998. UK & International Press, Dorking, Surrey, UK, 1998

Gebhardt, A.: Durch dick und dünn, Laser in der Materialbearbeitung. WT – Produktion und Management 85, 1995, 6, pp. 314–317

Gebhardt, A., Mathar, W.: Quality Parts on a small Scale. Prototyping Technology International, UK & International Press, Dorking, Surrey, UK 1997, p. 280 ff.

Gebhardt, A.: Vision Rapid Prototyping. Generative Verfahren zur Herstellung von Keramikbauteilen, In: *Kriegsmann, J. (Ed.).* Technische Keramische Werkstoffe 96. Suppl. material, Section 3.4.2.3, pp. 1–21, Deutscher Wirtschaftsdienst, Ellerau, 2007

Gershenfeld, N.: When Things Start to Think, Henry Holt and Co., Inc., ISBN: 080505880X, New York, NY, 1999

Grimm, T.: Stereolithography, Selective Laser Sintering and PolyJet: Evaluating and Applying the Right Technology, *www.time-compression.com/articles/html/stereolithography.html*

Haferkamp, H., Bach, F. W., Gerken, J.: 3D-Laserstrahl-Pulver-Auftragsschweißen metallischer Komponenten, Schweißen und Schneiden, 47, 1995, p. 3

Heissler, E., Fischer, F. S., Bolouri, S., Lehmann, T., Mathar, W., Gebhardt A., Lanksch, W., Bier, J.: Custom-made cast titanium implants produced with CAD/CAM for the reconstruction of cranium defects. The Berlin Craniofacial Centre, Clinic for Maxillofacial and Plastic Surgery, Virchow-Hospital, Medical Faculty of the Humboldt-University Berlin, Germany, Int. J. Oral. Maxillofac. Surg., 27, 1998, pp. 334–338

Heller, R.: Experimentelle Bestimmung der Harzkennwerte für das Harz HS 660, Personal communication, 1995

Hennings, D., Wirth, I., Stührmann, S.: ecoMold, Rapid Tooling für den Spritzguss- und Al-Druckgusswerkzeugbau, RTejournal 1, 2004, 28, Online: *www.rtejournal.de*

Hoffmann, J.: Vom Rapid Prototyping bis zur Serienproduktion, Maschinenmarkt 101, 1995, 16, pp. 30–34

IAO: Kundenorientiertes Qualitätsmanagement im FuE-Bereich, Deutschlandweite Unternehmensstudie, FhG-IAO, Stuttgart, 1997

Irle, M.: Die Alternative, Wirtschaftsmagazin “brand eins,” Vol. 6, Issue 07, Sept. 2004, pp. 74–78, *www.brandeins.de*, Hamburg, 2004

Jacobs, P. F.: Rapid Prototyping & Manufacturing, Society of Manufacturing Engineers, Dearborn, MI, 1992

Jacobs, P. F., Partaman, J., Bedal, B.: Oberflächenqualität, The Edge, Vol. 3, No. 3, p. 6, 1994

Jacobs, P. F.: Recent Advances in Rapid Tooling from Stereolithography, 3D-Systems, Darmstadt, 1997

Jahn, H.: Erzeugnisqualität, die logische Folge von Arbeitsqualität, VDI-Z 130, 1988, 4, pp. 4–12

JTEC: Panel Report on Rapid Prototyping in Europe and Japan, Vol. 1, Analytic Chapters, March 1997, Loyola College in Maryland, Maryland, 1997

Karjalainen, J., Tuomi, J. et al.: Quantifying the savings in product development, Contribution to the 7èmes assises du prototypage rapide, Nov. 19–20, Paris, 1998

Klare, M., Altmann, R.: Rapid Manufacturing in der Hörgeräteindustrie, RTejournal, Feb. 2005, (urn:nbn:de:0009-2-1049), *www.rtejournal.de*, Aachen, 2005

Klare, M.: Rapid Manufacturing in der Medizintechnik. – Neue Verfahren und Materialkonzepte, Rapid.Tech Anwendertagung (Conference), Session 5, Lecture 3, May 22–23, 2007, Erfurt trade fair, Erfurt, 2007

Klosterman, Ch., Osborne, G., Lightman, H., Akos, St.: Development of a Curved Layer LOM Process for Monolithic Ceramics and Ceramic Matrix Composites Rapid Prototype Development Laboratory, Univ. of Dayton, OH, Journal of Rapid Prototyping, Vol. 5, No. 2, 1999, pp. 61–71

Kochan, D. (Ed.): Solid Freeform Manufacturing – Advanced Rapid Manufacturing. Manufacturing Research and Technology, Vol. 19, Elsevier Science Publishers, 1993

König, W., Celi, J., Celiker, T. et al.: Rapid Metal Prototyping – Verfahren zur Direktherstellung metallischer Bauteile und Prototypwerkzeuge, VDI-Z 136, No. 7/8, 1994, pp. 57–60

Kramer, B.: Kalibrierung der spannungsoptischen Konstanten für die Fotopolymere XB 50811, XB 51341, XB 5143, und SOMOS 3100 des Stereolithografie-Verfahrens, Thesis, Univ. of Kassel, Kassel, 1992

Krause, F.-L., et al.: A Three-Stage Conceptual Design Process Using Virtual Environments, WSCG 2004, Feb. 2–6, 2004, (wscg.zcu.cz/wscg2004/Papers_2004_Poster/ K17.pdf), Plzen, Czech Republic, 2004

Kreicher, D. M., Miller, W. D.: Laser Engineered Net Shaping (LENS): Beyond Rapid Prototyping to direct fabrication, Contribution to the 7èmes assises du prototypage rapide, Nov. 19–20, Paris, 1998

Kruth, J. P.: Material Incress Manufacturing by Rapid Prototyping Techniques, CRIP Annuals, Vol. 40/2, 1991

Kunze, H.-D.: Competitive Advantages by Near-Net-Shape Manufacturing, DGM information society publisher, Frankfurt, 1997

Lehmann, O., Stuke, M.: Three-dimensional laser direct writing of electrically conducting and isolating microstructures. Mater. Lett. 21, Oct. 1994, pp. 131–136

Lenz, J.: Laser-Sintern. Die Schlüsseltechnologie des e-Manufacturing. RTejournal 1/2004 (urn:nbn:de:0009-2-141), *www.rtejournal.de*, Aachen, 2004

Lenz, J., Shellabear, M.: e-Manufacturing mit Lasersintern – bis zur Serienfertigung und darüber hinaus, Lecture 3, VDI-Seminar Rapid Manufacturing – Innovation für die Produktion, Aachen, IPT, March 1–2, 2005. Seminar band, VDI Wissensforum Düsseldorf, 2005

Lucke, St.: Rapid Prototyping – Einsatz der Stereolithographie-Technik zur Herstellung von Spritzgießwerkzeugen für Prototypteile, Thesis, Univ. of Stuttgart, 1995

Maihack, St.: Schnittstellentools für neutrale CAD-Schnittstellen, CAD-CAM-Report 13, 1994, p. 9

Mareco: Mareco b. v., NL, Press release from July 15, 1998, Venlo, 1998, *www.mareco.nl/mareco_pages/page_press.html*

Mariques-Frayre, J. A., Bourell, D. L.: Selective Laser Sintering of Cu-Pb/Sn Solder Powders, Article from the Solid Freeform Fabrication Symposium, Austin, TX, 1991

Michaeli, W.: Einführung in die Kunststoffverarbeitung, 5th ed., Carl Hanser Publisher, Munich, 2006

Ming, L. W., Gibson, I.: Specification of VRML in Color Rapid Prototyping, Int. J. of CAD/CAM, Vol. 1, No. 1, pp. 1–9, 2002

Nachtrodt, M.: Rapid-Prototyping mit SOM, Contribution to the Conference: Die Wirtschaft im Dialog mit der Forschung, Nov. 25–27, 1998, Forum Niederberg, Velbert, 1998

NCG: RP Systemanbieter weltweit auf einer Übersicht, RPD-Report NCG, Vol. 5, No. 8, 1998, NCG e. V., Ulm, 1998

Nöken, S.: Technologie des Selektiven Lasersinterns von Thermoplasten, In: 3rd. ed., e aus der Produktionstechnik, Band 8/97, Shaker Publisher, Aachen, 1997

Poprawe, R.: Lasertechnik für die Fertigung, Springer Publisher, Berlin, 2004

Rasche, T.: Rapid Manufacturing im Feinguss. Konstruieren + Giessen 30, 2005, No. 1, pp. 15–17, Zentrale für Gussverwendung, Düsseldorf, 2005

Reinertsen, D.G.: Whodunit? The Search for the New-Product Killers, Electronic Business, July 1983, pp. 62–66

Reinhart, G., Führer, J., Breitinge, F.: Rapid Tooling für den Magnesiumdruckguß, *www.apppex.com/de/knowhow/aktikel4/artikel4.htm*

Rosenberger, D.: Technische Anwendungen des Lasers, Springer Publisher, 1975

Sachs, E.M., Haggerty, J.: Three-Dimensional Printing Techniques, U.S.M.J. Cima and P. Williams, Patent No. 5204055, April 20, 1993, Full text in: *www.freepatentsonline.com/5204055.html* and *www.freepatentsonline.com/5340656.html*

Sändig, S., Leutbecher, T., Wiesner, P.: Laminated Tool Manufacturing by Laser Cutting and Diffusion Bonding. 20th International Congress on Applications of Laser & Electro-Optics, ICALEO 2001, Jacksonville, FL, Oct. 15–18, 2001, p. 10, Proc. LIA Vol. and CD-ROM ISBN: 0912035730

Schoormans, A.: Die dreidimensionale Reproduktion meines Skeletts. "body incoming," ISBN 392479054x, Skulpturenmuseum Glaskasten Marl, Marl, 2001

Schröder, O.E.: Mehr Sicherheit im Betrieb durch Objektgestaltung und Verhaltenssteuerung – zur Psychologie der Arbeitssicherheit im betrieblichen Alltag, Vorlesungsskript, 4th ed., Hochschule der Post, Dillenburg, 1993

Schulz, H.: Hochgeschwindigkeitsbearbeitung, Carl Hanser Publisher, Munich, 1996

Serbin, J., Wilkening, Ch., Bretsch, Ch. et al.: STEREOS and EOSINT, The New Developments and State of the Art, Produ Aachen, 1995

Steinchen, W., Kupfer, G., Kramer, B.: Photoelastic Investigations by means of Stereolithography, Beitrag zur 27th ISATA, Aachen, 1994

Susebach, K., Schneck, R.: Untersuchung von Fotopolymeren bezüglich der Kriechgeschwindigkeit und räumliche, spannungsoptische Unterschung von Motorpleueln aus Fotopolymeren und Araldit B. Diplomarbeit, Universität Gh Kassel, 1993

Varadan, V., Xiaoning, J., Vasundra, V.: Microstereolithography and other Fabrication Techniques for 3D MEMS. ISBN0 471 52185 X, John Wiley & Sons, LTD, Chicester, 2001

VDI-TZ: Handbuchreihe, Band 6, Fügen mit Festkörperlasern. VD k I-Technologiezentrum Physikalische Technologien, VDI, Düsseldorf, 2006, *www.vditz.de/home/kompetenzen/forschungsfoerderung/laser_und_optikforschung/publikationen/handbuchreihe/index.html*

VDI 3404: VDI-Richtlinie 3404: Generative Fertigungsverfahren-Rapid Technologien (Rapid Prototyping); Grundlagen, Begriffe, Qualitätskenngrößen, Liefervereinbarungen; (Eng.: Additive fabrication), Rapid technologies (rapid prototyping), Fundamentals, terms and definitions, quality, parameters, supply agreements), VDI Gesellschaft Produktionstechnik (ADB), Postfach 101139, D-40002 Düsseldorf, 2009, *http://www.vdi.de/uploads/tx_vdirili/pdf/1392434.pdf*

VDI 2221: VDI-Richtlinie 2221: Methodik zum Entwickeln und Konstruieren technischer Systeme und Produkte, VDI-Publisher, Düsseldorf, 1993

VDK: Verband der Keramischen Industrie e.V.: Brevier Technische Keramik, *www.keramverband.de/*

VRML97: Die Normung der Virtual Reality Modeling Language besteht aus zwei Teilen, Part 1 (ISO/IEC 147721) definiert die Basis-Funktionalitäten die Kodierung für VRML; Part 2 (ISO/IEC FDIS 147722) definiert die Basis-Funktionalitäten und alle Verknüpfungen mit dem VRML External Authoring Interface, *www.web3d.org/x3d/specifications/vrml/*

Walz, P.: Digital Imaging Terminologie, CompArt98 Computer Art Faszination, 1998, pp. 283 ff., *www.dot-online.de*

Weihnacht, W.: Cyclone – die neue Technik zur superschnellen Schalenherstellung beim Feing uss. urn:nbn:de:0009-2-1088, Rtejournal, 2005, *www.rtejournal.de/archiv/ausgabe2/index_html/108/*

Wildung, D.: Nefertitit's Mother-in-Law – Rapid Prototyping and Ancient Egypt, Proceedings of the International Conference on RPD Rapid Product Development, Stuttgarter trade fair and Convention mbH, Stuttgart, 1995

Willinger, H. M.: Kostenrechnung Rapid-Prototyping Verfahren, Diplomarbeit, Technical Univ. of Munich, 1995

Wirth, J.: Rapid Modeling, Carl Hanser Publisher, Munich, 2002

Witt, G.: Taschenbuch der Fertigungstechnik, Fachbuch Publisher Leipzig, 2006

Wohlers, T.: Worldwide Developments and Trends in Rapid Prototyping and Tooling, Proceedings of the ICRPM '98, International Conference on Rapid Prototyping, Manufacturing and Rapid Tooling, July 21–23, 1998, in Beijing, Shaanxi Science and Technology Press, Xian, 1998

Wohlers, T.: État de l'art mondial et perspectives en prototypage rapide, 7ème Assises Européennes du Prototypage Rapide, Nov. 19–20, 1998, École Centrale de Paris, Paris, 1998

Wohlers, T.: Rapid Prototyping & Tooling, State of the Industry: 1999, Worldwide Progress Report, Wohlers Associates Inc., Fort Collins, CO, 1999

Zäh, M. F.: Wirtschaftliche Fertigung mit Rapid-Technologien, Carl Hanser Publisher, Munich, 2006

Suggested Reading

Alscher, G.: Das Verhalten teilkristalliner Thermoplaste beim Lasersintern, Dissertation, Univ. of Essen, 2000

Fockele, M., Schwarze, D.: Rapid Prototyping auf dem Vormarsch – Neue Methoden der Datenaufbereitung, Der Konstrukteur, April 1993, pp. 64–66

Gebhardt, A. et al.: Produktionstechnische Aspekte bei der Herstellung stereolithographischer Modelle für die Medizin, Proceedings of the 1st International Conference on Stereolithography in Medicine, Zürich, 1993

Guidot, R.: Industrial Design, Techniques and Materials, Éditions Flammarion, Paris, 2006

Hochgeladen, R., Scherer, J.: Neue Wege im Modellbau? Form + Werkzeug, May 1994, pp. 20–21

Hoyer, J., Uhl, Chr., Beyer, Chr.: Virtual & Rapid Prototyping, Bestandsaufnahme aktueller Produkte und deren Einsatz. Fortschritt, 3rd. ed., VDI No. 656, Düsseldorf, 2006

Jacobs, P. F.: Rapid Prototyping & Manufacturing – Fundamentals of Stereolithography. Society of Manufacturing Engineers, Dearborn, 1992

Kochan, D.: Solid Freeform Manufacturing – Advanced Rapid Prototyping. Elsevier Science Publishers B. V., Amsterdam, 1993

McDonald, J. A., Ryall, C. J., Wimpenny, D. I.: Rapid Prototyping Casebook. Professional Engineering Publishing Ltd, Burry St., Edmunds, 2001

Roller, et. al.: Fachkunde Modellbau, 4th ed., 2003, Europa Lehrmittel, Haan-Gruiten, 2003

Spencer, J. D.: Surface Finishing Techniques for Rapid Prototyping, Rapid Prototyping Conference, Dearborn, 1993

Steger, W.: Neue Verfahren der Prototypherstellung und ihre Integration in der Unternehmensstrategie, Spektrum der Wissenschaft, April 1995, pp. 90–95

Wood, L.: Rapid Automated Prototyping: An Introduction, Industrial Press Inc., New York, 1993

Wyatt, M., Dishart, K.: Choosing a Solvent to Clean Stereolithography Parts, Prototyping International 98, pp. 107–109, UK & International Press, Dorking, UK, 1998

Index

Q

R

S

T